XXVI International Conference on Integrable Systems and Quantum Symmetries (ISQS-26)

Journal of Physics: Conference Series Volume 1416

Prague, Czech Republic
8 - 12 July 2019

ISBN: 978-1-7138-1321-7
ISSN: 1742-6588

Printed from e-media with permission by:

Curran Associates, Inc.
57 Morehouse Lane
Red Hook, NY 12571

Some format issues inherent in the e-media version may also appear in this print version.

Institute of Physics
Dirac House, Temple Back
Bristol BS1 6BE UK

Phone: 44 1 17 929 7481
Fax: 44 1 17 920 0979

techtracking@iop.org

Additional copies of this publication are available from:

Curran Associates, Inc.
57 Morehouse Lane
Red Hook, NY 12571 USA
Phone: 845-758-0400
Fax: 845-758-2633
Email: curran@proceedings.com
Web: www.proceedings.com

TABLE OF CONTENTS

Journal of Physics: Conference Series 1416 (2019) 011001 doi:10.1088/1742-6596/1416/1/011001

XXVI International Conference on Integrable Systems and Quantum Symmetries (ISQS-26)

Č. Burdík[1], **O. Navrátil**[2], **S. Pošta**[1]

[1] Department of Mathematics, Faculty of Nuclear Sciences and Physical Engineering, Czech Technical University in Prague, Trojanova 13, 12000 Prague, Czech Republic

[2] Department of Applied Mathematics, Faculty of Transportation Sciences, Czech Technical University in Prague, Na Florenci 25, 11000 Prague, Czech Republic

E-mail: cestmir.burdik@fjfi.cvut.cz

Abstract.

The XXVI International Conference on Integrable Systems and Quantum Symmetries (ISQS-26), organized by the Department of Mathematics, Faculty of Nuclear Sciences and Physical Engineering, Czech Technical University Prague and the Bogoliubov Laboratory of Theoretical Physics of the Joint Institute for Nuclear Research, belongs to the successful series of conferences held at the Czech Technical University which began in 1992 and is devoted to problems of mathematical physics related to the theory of integrable systems, quantum groups and quantum symmetries. During the last 5 years, each of the conferences gathered around 110 scientists from all over the world. 45 papers of plenary lectures and contributions presented at ISQS-26 are published in the present issue of Journal of Physics: Conference Series.

Journal of Physics: Conference Series **1416** (2019) 011002 doi:10.1088/1742-6596/1416/1/011002

Peer review statement

All papers published in this volume of *Journal of Physics: Conference Series* have been peer reviewed through processes administered by the proceedings Editors. Reviews were conducted by expert referees to the professional and scientific standards expected of a proceedings journal published by IOP Publishing.

Journal of Physics: Conference Series **1416** (2019) 012001 doi:10.1088/1742-6596/1416/1/012001

Bound state solutions of the Klein-Fock-Gordon equation with the sum of Manning-Rosen potential and Yukawa potential within SUSYQM

A. I. Ahmadov[1,2], M. Demirci[3] and S. M. Aslanova[1]

[1] Department of Theoretical Physics, Baku State University, Z. Khalilov st. 23, AZ-1148, Baku, Azerbaijan

[2] Institute for Physical Problems, Baku State University, Z. Khalilov st. 23, AZ-1148, Baku, Azerbaijan

[3] Department of Physics, Karadeniz Technical University, TR61080 Trabzon, Turkey

E-mail: ahmadovazar@yahoo.com, mehmetdemirci@ktu.edu.tr, sariyya.aslanova@mail.ru

Abstract. In this study, the bound state solutions of the Klein-Fock-Gordon equation are examined for the sum of Manning-Rosen and Yukawa potential by using a recent improved scheme to deal with the centrifugal term. For any $l \neq 0$, the energy eigenvalues and corresponding radial wave functions are determined under the condition of equal scalar and vector potentials. In order to obtain bound state solutions, we use two different methods called supersymmetric quantum mechanics (SUSYQM) and Nikiforov-Uvarov (NU) methods. The identical expressions for the energy eigenvalues are obtained, and the expression of radial wave functions transformations to each other is revealed via both methods. For arbitrary l states, the energy levels and the corresponding normalized eigenfunctions are given in terms of the Jacobi polynomials. A closed form of the normalized wave function is also obtained. It is seen that the energy eigenvalues and eigenfunctions are sensitive to n_r radial and l orbital quantum numbers.

1. Introduction

Since the early years of quantum mechanics (QM), the exactly solvable problems for physical potentials have received much attention in many branches of physics. In the QM, it is very important to obtain an analytical solution of wave equation since the wave function contains all the necessary information for the exact definition of a quantum system. Furthermore, in the relativistic QM's applications to particle and nuclear physics, the behavior of particles at high energies can be predicted from the relativistic wave equations. The Klein-Fock-Gordon (KFG) equation [1, 2, 3, 4] is a fundamental relativistic wave equation that is well known to describe the motion of spin-0 particles, for example, pions, Higgs boson and so forth. Thereby, the analytical solutions of the KFG equation with interaction potentials are very important for relativistic QM.

Many techniques have been developed to obtain the solutions of relativistic and non-relativistic wave equations under physical potentials. Some of these are as follows: Nikiforov-Uvarov (NU) method [5], supersymmetry QM (SUSYQM) [6, 7], shifted 1/N expansion approach [8, 9], the path integral method [10], asymptotic iteration method [11], factorization [12] and Laplace transform approach [13]. With help of these techniques, many works have been carried

Journal of Physics: Conference Series **1416** (2019) 012001 doi:10.1088/1742-6596/1416/1/012001

out to obtain exact or approximate solutions with some combined potentials such as: Manning-Rosen plus ring-shaped potential [14], Hulthen plus a Ring-Shaped like potential [15] and Hulthen plus Yukawa potentials [16], and references therein. In the present study, we attempt to study bound state solutions of KFG equation with a potential formed by summing two different potentials which are called as the Manning-Rosen and Yukawa potentials. The Manning-Rosen potential can be used to describe the continuum and bound-states in an interaction system, is given [17]

$$V_{MR}(r) = \frac{\hbar^2}{2Mb^2}\left[\frac{\alpha(\alpha-1)e^{-2r/b}}{(1-e^{-r/b})^2} - \frac{Ae^{-r/b}}{(1-e^{-r/b})}\right]. \tag{1}$$

Our other potential is the Yukawa potential which was proposed by Yukawa [18] as an effective non-relativistic potential describing the strong interactions of nucleons. It is given by:

$$V(r) = -\frac{V_0 e^{-\delta r}}{r}. \tag{2}$$

The linear combination of these two potentials can be used to examine the interactions of deformed-pair of the nucleus and spin-orbit coupling for a particle in the potential field. Another fascinating perspective of this potential can be utilized in the description of vibrations in the side of the hadronic system. Inspired by all motivations and works, we consider a sum of Manning-Rosen and Yukawa potentials:

$$V(r) = \frac{\hbar^2}{2Mb^2}\left[\frac{\alpha(\alpha-1)e^{-2r/b}}{(1-e^{-r/b})^2} - \frac{Ae^{-r/b}}{(1-e^{-r/b})}\right] - \frac{V_0 e^{-\delta r}}{r}. \tag{3}$$

In this study, our aim is to examine such a large quantum system. Therefore, we can investigate the considered quantum system at a long distance. To achieve this goal, we use both NU method and SUSYQM [19, 20], taking account of a recent improved scheme to deal with the centrifugal term. Thereby, the radial wave functions and energy eigenvalues are found for any orbital angular momentum l. However, the same problem have been investigated within ordinary quantum mechanics in Refs. [21, 22] as well, but our results disagree with their results.

The remainder of the paper is organized as follows: In Sec. 2, the KFG equation for the Manning-Rosen potential plus the Yukawa potential is presented. In Sec. 3, the exact analytic bound state energies and the corresponding radial solution of KFG equation are derived in the context of NU method (Sec. 3.1) and SUSYQM method (Sec. 3.2), respectively. Finally, the results are summarized and discussed in Sec. 4.

2. The KFG Equation with the Manning-Rosen potential plus the Yukawa potential

Two different type potentials could be presented into KFG equation since it includes two objects; the scalar rest mass and the four-vector linear momentum operator. These are a scalar potential (S)(which introduce via scalar coupling) and a vector potential (V) (which introduce via minimal coupling). In the spherical coordinates system, the KFG equation with scalar potential and vector potential is formulated by

$$[-\nabla^2 + (M + S(r))^2]\psi(r,\theta,\phi) = [E - V(r)]^2\psi(r,\theta,\phi), \tag{4}$$

where M is the rest mass of a scalar particle and E denotes the relativistic energy of the system.

Substituting this into Eq.(4), the radial part of KFG equation is obtained as

$$\chi''(r) + [(E^2 - M^2) - 2(M\cdot S(r) + E\cdot V(r)) + (V^2(r) - S^2(r)) - \frac{l(l+1)}{r^2}]\chi(r) = 0. \tag{5}$$

Journal of Physics: Conference Series **1416** (2019) 012001 doi:10.1088/1742-6596/1416/1/012001

In present work, the vector potential is assumed to be equal to scalar potential, namely, $V(r) = S(r)$. Hence, equation (5) becomes

$$\chi''(r) + [(E^2 - M^2) - 2(M+E)\cdot V(r) - \frac{l(l+1)}{r^2}]\chi(r) = 0. \tag{6}$$

The solution of KFG equation (6) with the potential (3) can be exactly obtained by using suitable approximation scheme to overcome the centrifugal term. In order to effectively use the Yukawa potential in this system, the following improved approximation scheme must be used for its centrifugal term. For $\delta r << 1$, the approximation scheme [23, 24, 25] is given by,

$$\frac{1}{r} \approx 2\delta\left[\frac{e^{-\delta r}}{(1-e^{-2\delta r})}\right] \text{ and } \frac{1}{r^2} \approx 4\delta^2\frac{e^{-2\delta r}}{(1-e^{-2\delta r})^2}, \tag{7}$$

In the Manning-Rosen potential we assume $1/b = 2\delta$ and hence one gets:

$$V_{MR}(r) = \frac{\hbar^2}{2Mb^2}\left[\frac{\alpha(\alpha-1)e^{-2r/b}}{(1-e^{-r/b})^2} - \frac{Ae^{-r/b}}{(1-e^{-r/b})}\right] = \frac{V_{01}e^{-4\delta r}}{(1-e^{-2\delta r})^2} - \frac{V_{02}e^{-2\delta r}}{1-e^{-2\delta r}}. \tag{8}$$

where $V_{01} = \frac{2\delta^2\alpha(\alpha-1)}{M}$ and $V_{02} = \frac{2\delta^2 A}{M}$.

By using Eq.(7), then we can write Yukawa potential in the following form:

$$V_Y(r) = -\frac{2\delta Be^{-2\delta r}}{1-e^{-2\delta r}} = -\frac{V_{03}e^{-2\delta r}}{1-e^{-2\delta r}} \tag{9}$$

where $V_{03} = 2\delta B$.

As a result, sum of Manning-Rosen potential and Yukawa potential is written as

$$\begin{aligned} V(r) &= V_{MR}(r) + V_Y(r) \\ &= \frac{V_{01}e^{-4\delta r}}{(1-e^{-2\delta r})^2} - \frac{(V_{02}+V_{03})e^{-2\delta r}}{1-e^{-2\delta r}} = \frac{V_{01}e^{-4\delta r}}{(1-e^{-2\delta r})^2} - \frac{V_{023}e^{-2\delta r}}{1-e^{-2\delta r}} \end{aligned} \tag{10}$$

where $V_{023} = (V_{02} + V_{03})$. Substituting the potential (10) into Eq.(6), we have

$$\chi''(r) + \left[(E^2-M^2) - 2(M+E)\left(\frac{V_{01}e^{-4\delta r}}{(1-e^{-2\delta r})^2} - \frac{V_{023}e^{-2\delta r}}{1-e^{-2\delta r}}\right) - \frac{4l(l+1)\delta^2 e^{-2\delta r}}{(1-e^{-2\delta r})^2}\right]\chi(r) = 0. \tag{11}$$

Thereby, the effective potential is defined as:

$$V_{\text{eff}}(r) = 2\left((M+E)\frac{V_{01}e^{-4\delta r}}{(1-e^{-2\delta r})^2} - \frac{V_{023}e^{-2\delta r}}{1-e^{-2\delta r}}\right) + \frac{4l(l+1)\delta^2 e^{-2\delta r}}{(1-e^{-2\delta r})^2}. \tag{12}$$

3. Bound State Solutions of the Radial Klein-Fock-Gordon Equation

3.1. Implementation of Nikiforov-Uvarov Method

To apply the NU method, we should transform Eq.(11) to the following form:

$$\chi''(s) + \frac{\tilde{\tau}}{\sigma}\chi'(s) + \frac{\tilde{\sigma}}{\sigma^2}\chi(s) = 0. \tag{13}$$

Equation (11) can be further simplified by using a new variable $s = e^{-2\delta r}$, and $r \in [0, \infty)$ and $s \in [0, 1]$. For bound states, it can be used the following notations:

$$\varepsilon = \frac{\sqrt{M^2 - E^2}}{2\delta} > 0\,, \quad \alpha = \frac{\sqrt{2(E+M)V_{01}}}{2\delta} > 0\,, \quad \beta = \frac{\sqrt{2(E+M)V_{023}}}{2\delta} > 0, \tag{14}$$

where E and ε should be, respectively, smaller and bigger than M and zero. The boundary conditions are $\chi(0) = 0$ and $\chi(\infty) \to 0$. Using Eq. (7) and Eq. (14), the Eq. (11) can be rewritten with the transformation $s = e^{-2\delta r}$ as folows:

$$\chi''(s) + \frac{1-s}{s(1-s)}\chi'(s) + \left[\frac{1}{s(1-s)}\right]^2 \left[-\varepsilon^2(1-s)^2 - \alpha^2 s^2 + \beta^2 s(1-s) - l(l+1)s\right]\chi(s) = 0. \quad (15)$$

As we expected, this equation has a suitable form to apply the NU method. By comparing Eq.(15) with Eq.(13) we get:

$$\tilde{\tau}(s) = 1-s, \quad \sigma(s) = s(1-s), \quad \tilde{\sigma}(s) = -\varepsilon^2(1-s)^2 - \alpha^2 s^2 + \beta^2 s(1-s) - l(l+1)s. \quad (16)$$

In order to obtain the exact solution to Eq.(13), the wave function $\chi(s)$ is set as follows:

$$\chi(s) = \phi(s)y(s), \quad (17)$$

for the appropriate function $\phi(s)$. Then, substituting Eq.(17) into Eq.(13) reduces Eq.(13) into the well known hypergeometric-type equation,

$$\sigma(s)y''(s) + \tau(s)y'(s) + \bar{\lambda}y(s) = 0. \quad (18)$$

The appropriate function $\phi(s)$ should satisfy the condition as follows

$$\frac{\phi'(s)}{\phi(s)} = \frac{\pi(s)}{\sigma(s)}, \quad (19)$$

where $\pi(s)$, which can be at most first-order polynomial, is given by

$$\pi(s) = \frac{\sigma' - \tilde{\tau}}{2} \pm \sqrt{\left(\frac{\sigma' - \tilde{\tau}}{2}\right)^2 - \tilde{\sigma} + k\sigma}. \quad (20)$$

Consequently, the equation, where $y(s)$ is one of its solutions, takes the form of hypergeometric-type, if the polynomial $\bar{\sigma}(s) = \tilde{\sigma}(s) + \pi^2(s) + \pi(s)[\tilde{\tau}(s) - \sigma'(s)] + \pi'(s)\sigma(s)$ can be divisible by $\sigma(s)$, i.e., $\bar{\sigma} = \bar{\lambda}\sigma(s)$. The constant $\bar{\lambda}$ and polynomial $\tau(s)$ in Eq.(13) are given by

$$\bar{\lambda} = k + \pi'(s), \quad \tau(s) = \tilde{\tau}(s) + 2\pi(s), \quad (21)$$

respectively. For our problem, $\pi(s)$ is obtained as

$$\pi(s) = \frac{-s}{2} \pm \sqrt{s^2[a-k] - s[b-k] + c} \quad (22)$$

where $a = \frac{1}{4} + \varepsilon^2 + \alpha^2 + \beta^2$, $b = 2\varepsilon^2 + \beta^2 - l(l+1)$ and $c = \varepsilon^2$. The parameter k could be determined by using the condition that the discriminant of the expression Eq.(22) under the square root is equal to zero. Therefore, we get:

$$k_{1,2} = (b-2c) \pm 2\sqrt{c^2 + c(a-b)}\ . \quad (23)$$

We can obtain four different expressions of $\pi(s)$ from Eq.(22) for the above four different values of k.

$$\pi(s) = \frac{-s}{2} \pm \begin{cases} (\sqrt{c} - \sqrt{c+a-b})s - \sqrt{c} & for\ k = (b-2c) + 2\sqrt{c^2 + c(a-b)}, \\ (\sqrt{c} + \sqrt{c+a-b})s - \sqrt{c} & for\ k = (b-2c) - 2\sqrt{c^2 + c(a-b)}. \end{cases} \quad (24)$$

Journal of Physics: Conference Series **1416** (2019) 012001 doi:10.1088/1742-6596/1416/1/012001

However, we choose only one of the above expressions so that the function $\tau(s)$ has the negative derivative. Another expressions have no physical meaning.

Given a nonnegative integer n_r, the hypergeometric-type equation has an unique polynomial solution of degree n_r, if

$$\bar{\lambda} = \bar{\lambda}_{n_r} = -n_r\tau' - \frac{n_r(n_r-1)}{2}\sigma'', (n_r = 0, 1, 2...), \tag{25}$$

and $\bar{\lambda}_m \neq \bar{\lambda}_n$ for $m = 0, 1, 2, ..., n_r - 1$, it follows that,

$$\begin{aligned}\bar{\lambda}_{n_r} &= b - 2c - 2\sqrt{c^2 + c(a-b)} - \left[\frac{1}{2} + \sqrt{c} + \sqrt{c+a-b}\right] \\ &= 2n_r\left[1 + \left(\sqrt{c} + \sqrt{c+a-b}\right)\right] + n_r(n_r-1).\end{aligned} \tag{26}$$

We can solve Eq.(26) explicitly for c by using the relation $c = \varepsilon^2$ which brings

$$\varepsilon^2 = \left[\frac{\beta^2 - l(l+1) - 1/2 - n_r(n_r+1) - (2n_r+1)\sqrt{\frac{1}{4} + \alpha^2 + l(l+1)}}{2n_r + 1 + 2\sqrt{\frac{1}{4} + \alpha^2 + l(l+1)}}\right]^2. \tag{27}$$

After inserting ε^2 into Eq.(14) for energy levels we get:

$$M^2 - E^2_{n_r,l} = \left[\frac{\beta^2 - l(l+1) - 1/2 - n_r(n_r+1) - (2n_r+1)\sqrt{\frac{1}{4} + \alpha^2 + l(l+1)}}{n_r + \frac{1}{2} + \sqrt{\frac{1}{4} + \alpha^2 + l(l+1)}} \times \delta\right]^2. \tag{28}$$

The energy levels $E_{n_r,l}$ can be obtained from the energy equation Eq.(28), which is rather complicated than the square equation.

Let us now begin to obtain the radial eigenfunctions by using the NU-method. After substituting $\pi(s)$ and $\sigma(s)$ into Eq.(19), the function $\phi(s)$ is easily obtained as a solution of the first order differential equation:

$$\phi(s) = s^{\varepsilon}(1-s)^K, \tag{29}$$

where $K = 1/2 + \sqrt{\frac{1}{4} + \alpha^2 + l(l+1)}$. Beyond that, the other part of the wave function, $y_n(s)$, is the hypergeometric-type function and its polynomial solutions are given by Rodrigues relation:

$$y_n(s) = \frac{C_n}{\rho(s)}\frac{d^n}{ds^n}\left[\sigma^n(s)\rho(s)\right], \tag{30}$$

where C_n is a normalization constant and $\rho(s)$ denotes the weight function. Note that $\rho(s)$ is a solution of the Pearson differential equation. The Pearson differential equation and $\rho(s)$ for this problem are given by

$$(\sigma\rho)' = \tau\rho, \tag{31}$$

$$\rho(s) = (1-s)^{2K-1}s^{2\varepsilon}. \tag{32}$$

By substituting the Eq.(32) into Eq.(30), one get

$$y_{n_r}(s) = C_{n_r}(1-s)^{1-2K}s^{-2\varepsilon}\frac{d^{n_r}}{ds^{n_r}}\left[s^{2\varepsilon+n_r}(1-s)^{2K-1+n_r}\right]. \tag{33}$$

Then by using the definition of the Jacobi polynomials [26] and substituting Eq.(33) into the Eq.(17) then for $\chi_{n_r}(s)$ obtain:

$$\chi_{n_r}(s) = C_{n_r} s^{\varepsilon}(1-s)^K P_{n_r}^{(2\varepsilon, 2K-1)}(s). \tag{34}$$

Also $\chi_{n_r}(s)$ written in terms of the hypergeometric function as follows [26]

$$\chi_{n_r}(s) = C_{n_r} s^{\varepsilon}(1-s)^K \frac{\Gamma(n_r + 2\sqrt{c}+1)}{n_r!\Gamma(2\varepsilon+1)} \underset{21}{F}\left(-n_r, 2\varepsilon + 2K + n_r, 1 + 2\varepsilon; s\right). \tag{35}$$

The constant C_{n_r} can be calculated from the following normalization condition

$$\int_0^{\infty} |R(r)|^2 r^2 dr = \int_0^{\infty} |\chi(r)|^2 dr = \frac{1}{2\delta}\int_0^1 \frac{1}{s}|\chi(s)|^2 ds = 1, \tag{36}$$

After simple calculations, the normalization constant is obtained by

$$C_{n_r} = \sqrt{\frac{2\delta n_r!(n_r + K + \varepsilon)\Gamma(2\varepsilon+1)\Gamma(n + 2\varepsilon + K - 1)}{(n_r + K)\Gamma(2\varepsilon)\Gamma(n_r + 2K)}}. \tag{37}$$

3.2. Implementation of SUSY Quantum Mechanics

According to SUSYQM, the eigenfunction of ground state $\chi_0(r)$ is given by

$$\chi_0(r) = N exp\left(-\int W(r)dr\right), \tag{38}$$

where $W(r)$ is called as superpotential function and N is a normalization constant. The relationships between the superpotential $W(r)$ and the supersymmetric partner potentials $V_-(r)$, $V_+(r)$ are defined as [6, 7]:

$$V_-(r) = W^2(r) - W'(r)\ , \quad V_+(r) = W^2(r) + W'(r). \tag{39}$$

The particular solution of the Riccati equation Eq.(39) is given by

$$W(r) = \left(F - \frac{Ge^{-2\delta r}}{1 - e^{-2\delta r}}\right), \tag{40}$$

where G and F are unknown constants. By using $V_-(r) = V_{\text{eff}}(r) - (E^2 - M^2)$, the expressions (12) and (40) are inserted into the Eq.(39). Then, for G and F, we get the following relations from comparison of compatible quantities in the left and right sides of the equation:

$$F^2 = 4\delta^2\varepsilon^2, \tag{41}$$

$$2FG + 2\delta G = 4\delta^2\beta^2 - 4\delta^2 l(l+1), \quad G^2 - 2\delta G = 4\delta^2\alpha^2 + 4\delta^2 l(l+1). \tag{42}$$

If taking into account of extremity conditions for wave functions, it reaches $G > 0$ and $F < 0$. Solving Eq.(42) gives

$$G = \frac{2\delta \pm \sqrt{4\delta^2 + 16\delta^2(\alpha^2 + l(l+1))}}{2} = \delta \pm 2\delta\sqrt{\frac{1}{4} + l(l+1) + \alpha^2}. \tag{43}$$

Journal of Physics: Conference Series **1416** (2019) 012001 doi:10.1088/1742-6596/1416/1/012001

When consider $G > 0$, from Eq.(42) we find

$$2FG + G^2 = 4\delta^2(\alpha^2 + \beta^2) \Rightarrow F = -\frac{G}{2} + \frac{2\delta^2(\alpha^2 + \beta^2)}{G}. \tag{44}$$

Next, from Eq. (41) and Eq. (44), we find

$$\varepsilon^2 = \frac{1}{4\delta^2}\left[-\frac{\delta + 2\delta\sqrt{\frac{1}{4} + \alpha^2 + l(l+1)}}{2} - \frac{2\delta^2(\alpha^2 + \beta^2)}{\delta + 2\delta\sqrt{\frac{1}{4} + \alpha^2 + l(l+1)}}\right]^2. \tag{45}$$

After inserting Eq. (45) into the Eq. (14) for energy eigenvalue, we obtain

$$M^2 - E^2 = \left[-\frac{\delta + 2\delta\sqrt{\frac{1}{4} + \alpha^2 + l(l+1)}}{2} - \frac{2\delta(\alpha^2 + \beta^2)}{1 + 2\sqrt{\frac{1}{4} + \alpha^2 + l(l+1)}}\right]^2. \tag{46}$$

When $r \to \infty$, the superpotential $W(r) \to F$. Inserting the Eq.(40) into Eq.(39), the supersymmetric partner potentials $V_-(r)$ and $V_+(r)$ can be found in the following forms.

$$V_{\mp}(r) = W^2(r) \mp W'(r) = \left[F^2 - \frac{(2FG \pm 2\delta G)e^{-2\delta r}}{1 - e^{-2\delta r}} + \frac{(G^2 \mp 2\delta G)e^{-4\delta r}}{(1 - e^{-2\delta r})^2}\right]. \tag{47}$$

By using the superpotential $W(r)$ in Eq.(40), the radial eigenfunction $\chi_0(r)$ is obtained as

$$\chi_0(r) = N\exp\left[-\int W(r)dr\right] = Ne^{Fr}\exp\left[\frac{G}{\delta}\int\frac{d(1 - e^{-2\delta r})}{1 - e^{-2\delta r}}\right] = Ne^{Fr}(1 - e^{-2\delta r})^{\frac{G}{2\delta}}, \tag{48}$$

where $r \to 0$; $\chi_0(r) \to 0, G > 0$ and $r \to \infty$; $\chi_0(r) \to 0, F < 0$. The potentials $V_-(r)$ and $V_+(r)$ differ from each other with additive constants. They have the same functional form called the invariant potentials [19, 20]. The invariant forms of potentials $V_-(r)$ and $V_+(r)$ in Eq.(39) are given by

$$R(G_1) = V_+(G, r) - V_-(G_1, r) = \left[-\frac{G}{2} + \frac{2\delta^2(\alpha^2 + \beta^2)}{G}\right]^2 + \left[\frac{G + 2\delta}{2} + \frac{2\delta^2(\alpha^2 + \beta^2)}{G + 2\delta}\right]^2, \tag{49}$$

$$\begin{aligned} R(G_i) &= V_+[G + (i-1) \times 2\delta, r] - V_-[G + i \times 2\delta, r] \\ &= \left(-\frac{G + (i-1) \times 2\delta}{2} + \frac{2(\alpha^2 + \beta^2)\delta^2}{G + (i-1) \times 2\delta}\right)^2 - \left(\frac{G + i \times 2\delta}{2} - \frac{2(\alpha^2 + \beta^2)\delta^2}{G + i \times 2\delta}\right)^2, \end{aligned} \tag{50}$$

where the reminder $R(G_i)$ is independent of r. If this procedure is continued by using $G_{n_r} = G_{n_r - 1} + 2\delta = G + 2n_r\delta$, then the whole discrete spectrum of Hamiltonian $H_-(G)$ can be written as:

$$E_{n_r}^2 = E_0^2 + \sum_{i=1}^{n_r} R(G_i) = M^2 - \left(-\frac{G + 2n_r\delta}{2} + \frac{2\delta^2(\alpha^2 + \beta^2)}{G + 2n_r\delta}\right)^2. \tag{51}$$

Substitute Eq.(43) into Eq.(51) then finally for energy eigenvalues as in ordinary quantum mechanics, we found the following form:

$$M^2 - E_{n_r,l}^2 = \left[\frac{\beta^2 - l(l+1) - 1/2 - n_r(n_r+1) - (2n_r+1)\sqrt{\frac{1}{4} + \alpha^2 + l(l+1)}}{n_r + \frac{1}{2} + \sqrt{\frac{1}{4} + \alpha^2 + l(l+1)}} \times \delta\right]^2. \tag{52}$$

As seen from Eq.(52), it is exactly identical with Eq.(28).

4. Results and Discussion

In this study, we have used alternative two method to solve the KFG equation for the Manning-Rosen plus Yukawa potential by using the developed scheme to overcome the centrifugal part. The energy eigenvalues and corresponding eigenfunctions of a mentioned quantum system have been analytically obtained for any l and n_r quantum numbers by using both NU and SUSYQM methods. The same expressions for the energy eigenvalues are obtained, and the expression of radial wave functions transformations to each other is revealed via both methods. It is seen that the energy eigenvalues and eigenfunctions are sensitive to the quantum numbers n_r and l.

A few important points should be emphasized as follows: When $l = 0$, the approximation centrifugal term is $l(l+1)\delta^2 \frac{e^{-2\delta r}}{(1-e^{-2\delta r})^2} = 0$. If we let $l = 0$ in Eq.(28) and Eq.(35), they describe the unnormalized radial wave functions and the exact energy spectrum formula for the bound states of s-wave KFG equation. It is obvious from Eq.(28) that the bound states are more stable for Manning-Rosen plus Yukawa potential compared to each of these potentials (Manning-Rosen and Yukawa potentials). Two important cases are seen at the combined potential: In the first case which $V_{03} = 0$ the potentials turn to central Manning-Rosen potential. In the second case, if $V_{01} = V_{02} = 0$ then the potentials turn to central Yukawa potential.

References

[1] Klein O 1926 *Z. Phys.* **37** 895-906
[2] Gordon W 1926 *Z. Phys.* **40** 117-33
[3] Fock V 1926 *Z. Phys.* **38** 242-50
[4] Fock V 1926 *Z. Phys.* **39** 226-32
[5] Nikiforov A F and Uvarov V B 1988 *Special Functions of Mathematical Physics* (Birkhäuser, Boston: Basel)
[6] Cooper F, Khare A and Sukhatme U 2001 *Supersymmetry in Quantum Mechnics* (World Scientific)
[7] Cooper F, Khare A and Sukhatme U 1995 *Phys. Rep.* **251** 267-385
[8] Tang A Z and Chan F T 1987 *Phys. Rev.* A **35** 911
[9] Roy B and Roychoudhury R 1987 *J. Phys. A: Math. Gen.* **20** 3051
[10] Cai J M, Cai P Y and Inomata A 1986 *Phys. Rev.* A **34** 4621
[11] Ciftci H, Hall R L and Saad N 2003 *J. Phys. A: Math. Gen.* **36** 11807
[12] Dong S H 2007 *Factorization Method in Quantum Mechanics* (Dordrecht: Springer)
[13] Schiff J L 1999 *The Laplace Transform: Theory and Applications* (New York: Springer).
[14] Ahmadov A I, Aydin C and Uzun O 2014 *Int. J. Mod. Phys.* A**29**, 1450002
[15] Ahmadov A I, Nagiyev Sh M, Qocayeva M V, Uzun K and Tarverdiyeva V A 2018 *Int. J. Mod. Phys. A* **33**, 1850203
[16] Ahmadov A I, Aslanova S M, Orujova M Sh Badalov S V and Dong S H 2019 *Phys. Lett. A* **383**, 3010-7
[17] Manning M F and Rosen N 1933 *Phys. Rev.* **44** 953
[18] Yukawa H 1935 *Proc. Phys. Math. Soc. Jap.* **17** 48-57
[19] Gendenshtein L E 1983 *JETP Lett.* **38** 356-9
[20] Gendenshtein L E and Krive I V 1985 *Sov. Phys. Usp.* **28** 645-66
[21] Ita B I, Louis H, Amos P I, Magu T O and Nzeata-ibe N A 2017 *Phys. Sci. Int. J.* **15** 1-6
[22] Louis H, Ita B I, Amos P I, Akakuru O U, Orosun M M, Nzeata-Ibe N A and Philip M 2018 *Int.J. Chem. and Phys. Sci.* **7** 33
[23] Qiang W C and Dong S H 2007 *Phys. Lett.* A **363** 169-76
[24] Wei G F and Shi-Hai Dong S H 2008 *Phys. Lett.* A **373** 49-53
[25] Qiang W C and Dong S H 2009 *Phys. Scr.* **79** 045004
[26] Abramowitz M and Stegun I A 1965 *Handbook of Mathematical Functions with Formulas, Graphs and Mathematical Tables* (New York: National Bureau of Standards Applied Mathematics series 55)

Journal of Physics: Conference Series **1416** (2019) 012002 doi:10.1088/1742-6596/1416/1/012002

Covariant stochastic products of quantum states

Paolo Aniello

Dipartimento di Fisica "Ettore Pancini", Università di Napoli "Federico II", and INFN – Sezione di Napoli, C. Universitario di Monte S. Angelo, via Cintia, I-80126 Napoli, Italy

E-mail: `paolo.aniello@na.infn.it`

Abstract. A notion of *stochastic product* of quantum states — a binary operation on the set of density operators preserving the convex structure — is discussed. We describe, in particular, a class of group-covariant, associative stochastic products: the *twirled products*. Each binary operation in this class can be constructed by means of a square integrable projective representation of a locally compact group, a probability measure on this group and a fiducial density operator in the Hilbert space of the representation. By suitably extending this operation from the convex set of density operators to the full Banach space of trace class operators, one obtains a Banach algebra, which is commutative in the case where the relevant group is abelian.

1. Introduction

The properties of operator algebras and of some related mathematical structures have shown to play a central role in modern theoretical physics. They turn out to be fundamental — in particular — in quantum mechanics, quantum field theory, quantum statistical mechanics and non-commutative geometry [1–5]. In this context, quantum *states* can be defined as *normalized positive functionals* on the C*-algebra $\mathcal{B}(\mathcal{H})$ of (bounded) *observables* [1–3] — the Banach space of bounded operators in a separable complex Hilbert space $\mathcal{H}$, endowed with the ordinary product (composition) and with the involution $A \mapsto A^*$ (the adjoining map). In several applications — including quantum information, quantum control and quantum measurement theory [6, 7] — it is convenient to deal with a distinct class of states, the *normal* states [2], that can be realized as normalized, positive trace class operators (the density operators). They form a convex set $\mathcal{S}(\mathcal{H}) \subset \mathcal{B}_1(\mathcal{H})$, with $\mathcal{B}_1(\mathcal{H})$ denoting the complex Banach space of *trace class* operators in $\mathcal{H}$.

Endowing the *selfadjoint part* $\mathcal{B}(\mathcal{H})_{\mathbb{R}}$ of the C*-algebra $\mathcal{B}(\mathcal{H})$ with the pair of (non-associative) products $A \circ B := (AB + BA)/2$ (Jordan) and $A \diamond B := (AB - BA)/2\mathrm{i}$ (Lie), one obtains the *Jordan-Lie Banach algebra* [1] $(\mathcal{B}(\mathcal{H})_{\mathbb{R}}, (\cdot)\circ(\cdot), (\cdot)\diamond(\cdot))$ of 'true' observables (and, conversely, every Jordan-Lie Banach algebra gives rise, upon suitable complexification, to a C*-algebra [1]); but *states are only indirectly involved in this algebraic structure* (as functionals).

In fact, let us observe that, for any $\rho, \sigma \in \mathcal{S}(\mathcal{H})$,

- the selfadjoint trace class operator $\rho \circ \sigma$ is a density operator if and only if $\rho = \sigma \equiv P$, where P is a *pure* state (i.e., a rank-one projection) [8];
- the selfadjoint trace class operator $\rho \diamond \sigma$ cannot be a density operator because $\mathrm{tr}(\rho \diamond \sigma) = 0$.

Based on these considerations, a natural question is the following [8,9]: One may ask whether it is possible to endow the Banach space $\mathcal{B}_1(\mathcal{H})$ with some *binary operation* such that

(i) one obtains an *algebra* structure and, moreover, *the product of two states is a state too.*

In order to achieve a noteworthy class of examples, such as to justify our interest in this issue, in addition to the previous condition (i) it is reasonable to require that

(ii) the aforementioned algebra be *associative*;

(iii) the algebra product be *continuous* wrt some suitable topology;

(iv) this product be — in general (i.e., apart from special cases or trivial examples) — a *genuinely binary* operation, namely, it should effectively depend on *both* its arguments.

Taking into account the remarkable parallelism of classical and quantum descriptions of physical systems [1], another natural question next arises: Does an analogous product exist in the *classical setting*?

The answer is in the affirmative. Consider, indeed, the *convolution* [10] $\mu \circledcirc \nu$ of two *complex Radon measures* μ, ν on a *locally compact group* G. The Banach space $\mathcal{M}(G)$ formed by all such measures, endowed with this binary operation, is promoted to a *Banach algebra*. Moreover, if μ, ν are *classical states* — i.e., probability measures — then $\mu \circledcirc \nu$ is a classical state too. Clearly, $G = \mathbb{R}^n \times \mathbb{R}^n$ — the group of translations on phase space — is the standard case.

Convolution is a group-theoretical notion. Thus, regarding the classical-quantum parallelism as a good guiding principle, and considering the fundamental role of *symmetry transformations* in quantum physics [11–13], we will also stipulate that

(v) the product of quantum states we are going to discuss is group-theoretical and enjoys some nice *covariance* property wrt a *symmetry action* of a certain abstract group G.

A binary operation on $\mathcal{B}_1(\mathcal{H})$ satisfying (i) will be called a *stochastic product*, and the Banach space $\mathcal{B}_1(\mathcal{H})$, endowed with a stochastic product satisfying (ii), will be called a *stochastic algebra*. It turns out that for such an algebra condition (iii) is automatically verified wrt the natural topology induced by the norm ((iii) being actually a consequence of condition (i) alone). We will argue that, by a suitable group-theoretical construction, one obtains a class of associative stochastic products — the *twirled products* [8,9] — satisfying requirements (iv) and (v), as well. These group-theoretical stochastic products give rise to the class of *twirled stochastic algebras*.

2. Stochastic maps and stochastic products

A map $\Phi\colon \mathcal{S}(\mathcal{H}) \to \mathcal{S}(\mathcal{H})$ is called *stochastic* if it preserves the convex structure of the space of quantum states $\mathcal{S}(\mathcal{H})$; namely, if it is *convex-linear*. Such a map can be extended to a unique (bounded) *trace-preserving, positive linear map* $\Phi_{\mathrm{ext}}\colon \mathcal{B}_1(\mathcal{H}) \to \mathcal{B}_1(\mathcal{H})$ (equivalently, a linear map Φ_{ext} in $\mathcal{B}_1(\mathcal{H})$ mapping $\mathcal{S}(\mathcal{H})$ into itself) [7,9]. Analogously, we define a *stochastic product* on $\mathcal{S}(\mathcal{H})$ as a map $(\cdot) \odot (\cdot)\colon \mathcal{S}(\mathcal{H}) \times \mathcal{S}(\mathcal{H}) \to \mathcal{S}(\mathcal{H})$ that is *convex-linear in both its arguments*; i.e., for all $\rho, \sigma, \tau, \upsilon \in \mathcal{S}(\mathcal{H})$ and all $\alpha, \epsilon \in [0,1]$,

$$(\alpha\rho + (1-\alpha)\sigma) \odot (\epsilon\tau + (1-\epsilon)\upsilon) = \alpha\epsilon\, \rho \odot \tau + \alpha(1-\epsilon)\, \rho \odot \upsilon + (1-\alpha)\epsilon\, \sigma \odot \tau + (1-\alpha)(1-\epsilon)\, \sigma \odot \upsilon. \tag{1}$$

We will also say that a binary operation $(\cdot) \boxdot (\cdot)$ on $\mathcal{B}_1(\mathcal{H})$ is *state-preserving* if it is such that $\mathcal{S}(\mathcal{H}) \boxdot \mathcal{S}(\mathcal{H}) \subset \mathcal{S}(\mathcal{H})$. A stochastic product is automatically continuous wrt a natural topology and admits a bilinear extension that is analogous to the linear extension of a stochastic map [9]:

Proposition 1. *Every stochastic product is continuous wrt the topologies on* $\mathcal{S}(\mathcal{H})$ *and on* $\mathcal{S}(\mathcal{H}) \times \mathcal{S}(\mathcal{H})$ *induced, respectively, by the metrics*

$$\mathrm{d}_1(\rho, \sigma) := \|\rho - \sigma\|_1 \quad \text{and} \quad \mathrm{d}_{1,1}((\rho,\tau),(\sigma,\upsilon)) := \max\{\|\rho-\sigma\|_1, \|\tau-\upsilon\|_1\}, \tag{2}$$

where $\|\cdot\|_1$ *is the norm of* $\mathcal{B}_1(\mathcal{H})$. *For each stochastic product* $(\cdot) \odot (\cdot)\colon \mathcal{S}(\mathcal{H}) \times \mathcal{S}(\mathcal{H}) \to \mathcal{S}(\mathcal{H})$, *moreover, there is a unique stochastic map (or product)* $(\cdot) \boxdot (\cdot)\colon \mathcal{B}_1(\mathcal{H}) \times \mathcal{B}_1(\mathcal{H}) \to \mathcal{B}_1(\mathcal{H})$ — *namely, a state-preserving bilinear map on* $\mathcal{B}_1(\mathcal{H})$ — *such that* $\rho \odot \sigma = \rho \boxdot \sigma$, *for all* $\rho, \sigma \in \mathcal{S}(\mathcal{H})$.

The Banach space $\mathcal{B}_1(\mathcal{H})$, endowed with a binary operation $(\cdot)\boxdot(\cdot)\colon \mathcal{B}_1(\mathcal{H})\times\mathcal{B}_1(\mathcal{H}) \rightarrow \mathcal{B}_1(\mathcal{H})$ that is both *stochastic* (bilinear, state-preserving) and *associative*, is called a *stochastic algebra.*

By Proposition 1, every stochastic product $(\cdot)\odot(\cdot)$ on $\mathcal{S}(\mathcal{H})$ is the restriction of a unique stochastic product $(\cdot)\boxdot(\cdot)$ on $\mathcal{B}_1(\mathcal{H})$, which is *associative* if and only if $(\cdot)\odot(\cdot)$ is associative. Thus, one may equivalently define a stochastic algebra as a Banach space of trace class operators $\mathcal{B}_1(\mathcal{H})$ together with an *associative* stochastic product defined on the convex subset $\mathcal{S}(\mathcal{H})$.

Let us now consider the Banach space $\mathscr{BL}_1(\mathcal{H})$ of all bounded bilinear maps on $\mathcal{B}_1(\mathcal{H})$, endowed with the norm $\|(\cdot)\boxdot(\cdot)\|_{(1)} := \sup\{\|A\boxdot B\|_1 \colon \|A\|_1, \|B\|_1 \leq 1\}$. Denoting by $\mathcal{B}_1(\mathcal{H})_{\mathbb{R}}$ the real Banach space of all *selfadjoint* trace class operators, the following result holds [9]:

Proposition 2. *Every stochastic map* $(\cdot)\boxdot(\cdot)\colon \mathcal{B}_1(\mathcal{H})\times\mathcal{B}_1(\mathcal{H})\rightarrow\mathcal{B}_1(\mathcal{H})$ *is bounded and its norm satisfies* $\|(\cdot)\boxdot(\cdot)\|_{(1)} \leq 2$*; whereas, its restriction* $(\cdot)\boxminus(\cdot)$ *to a bilinear map on* $\mathcal{B}_1(\mathcal{H})_{\mathbb{R}}$ *is such that* $\|(\cdot)\boxminus(\cdot)\|_{(1)} = 1$*. Thus, if the stochastic map is associative, then* $(\mathcal{B}_1(\mathcal{H})_{\mathbb{R}}, (\cdot)\boxminus(\cdot))$ *is a Banach algebra, because* $\|A\boxminus B\|_1 \leq \|A\|_1\,\|B\|_1$, $\forall A, B \in \mathcal{B}_1(\mathcal{H})_{\mathbb{R}}$.

Remark 1. The restriction of a stochastic map $(\cdot)\boxdot(\cdot)$ on $\mathcal{B}_1(\mathcal{H})$ to a bilinear map on $\mathcal{B}_1(\mathcal{H})_{\mathbb{R}}$ is well defined, because $(\cdot)\boxdot(\cdot)$, being bilinear and state-preserving, is also *adjoint-preserving*; i.e., $\mathcal{B}_1(\mathcal{H})_{\mathbb{R}}\boxdot\mathcal{B}_1(\mathcal{H})_{\mathbb{R}}\subset\mathcal{B}_1(\mathcal{H})_{\mathbb{R}}$. Moreover, the inequality $\|(\cdot)\boxdot(\cdot)\|_{(1)}\leq 2$ may *not* be saturated. E.g., $(\mathcal{B}_1(\mathcal{H}), (\cdot)\boxdot(\cdot))$ may be a Banach algebra; see the second assertion of Theorem 1 below.

3. A technical detour: square integrable representations

Let U be an irreducible — in general, projective — representation of a locally compact, second countable topological group (in short, a l.c.s.c. group) G, acting in a separable complex Hilbert space $\mathcal{H}$. We will denote by μ_G (a normalization of) the left Haar measure on G and by $\langle\cdot,\cdot\rangle$ the scalar product in $\mathcal{H}$. Given two vectors $\psi,\phi\in\mathcal{H}$, consider the bounded Borel *coefficient* function $c_{\psi\phi}\colon G\ni g\mapsto\langle\mathsf{U}(g)\,\psi,\phi\rangle\in\mathbb{C}$. The functions of this kind allow us to define the set

$$\mathscr{A}(\mathsf{U}) := \{\psi\in\mathcal{H}\mid \exists\phi\in\mathcal{H}\colon\ \phi\neq 0,\ c_{\psi\phi}\in \mathrm{L}^2(G,\mu_G;\mathbb{C})\} \tag{3}$$

of *admissible vectors* for U, and U is called *square integrable* [8,14–16] if $\mathscr{A}(\mathsf{U})\neq\{0\}$.

If G is unimodular — i.e., the Haar measure μ_G on G is both left and right invariant — and U is square integrable, then $\mathscr{A}(\mathsf{U}) = \mathrm{L}^2(G,\mu_G;\mathbb{C})$ (*all* vectors are admissible); moreover, *all* coefficient functions are square integrable wrt μ_G and satisfy the *orthogonality relations*; namely,

$$\int_G \mathrm{d}\mu_G(g)\,\langle\eta,\mathsf{U}(g)\phi\rangle\langle\mathsf{U}(g)\psi,\chi\rangle = \mathsf{c}_{\mathsf{U}}\,\langle\eta,\chi\rangle\langle\psi,\phi\rangle,\quad \forall\eta,\chi,\psi,\phi\in\mathcal{H}, \tag{4}$$

where c_{U} is a strictly positive constant depending on U (and on the normalization of μ_G).

Remark 2. The non-unimodular case is slightly more complicated. In general, the orthogonality relations involve a positive selfadjoint operator D_{U} in $\mathcal{H}$ (the *Duflo-Moore operator* [8,14–16]), whose domain coincides with $\mathscr{A}(\mathsf{U})$, and that is bounded if and only if G is unimodular. In such case, D_{U} is just a positive multiple of the identity — $D_{\mathsf{U}} = \mathsf{d}_{\mathsf{U}}I$, $\mathsf{d}_{\mathsf{U}}\equiv\mathsf{c}_{\mathsf{U}}^{1/2}$ — and we get (4).

Remark 3. In the following we will consider, in particular, two remarkable classes of square integrable representations of unimodular l.c.s.c. groups (see the examples in sect. 4):

- *Every* irreducible unitary representation of a *compact* group G is (finite-dimensional and) square integrable, because the associated Haar measure μ_G is *finite*. If μ_G is normalized as a probability measure, then, according to the Peter-Weyl theorem [10], $\mathsf{c}_{\mathsf{U}} = \dim(\mathcal{H})^{-1}$.
- The irreducible projective representations of the group $\mathbb{R}^n\times\mathbb{R}^n$ of *phase-space translations* with *symplectic multiplier* $\gamma_{\mathsf{h}}(q,p;\tilde{q},\tilde{p}) = \exp(\mathrm{i}(q\cdot\tilde{p}-p\cdot\tilde{q})/2\mathsf{h})$, where h is any nonzero real number, are square integrable too and form a unitary equivalence class (two representations in the unitary equivalence classes associated, respectively, with h and $-\mathsf{h}$ are *anti*-unitarily equivalent) [16,17]. For $\hbar\equiv\mathsf{h}>0$, this parameter can be interpreted as Planck's constant, and an irreducible representation of $\mathbb{R}^n\times\mathbb{R}^n$ with multiplier $\gamma_\hbar$ is called a ($\hbar$-) *Weyl system.*

4. Covariant stochastic products from square integrable representations

We now construct the *twirled products* [9]. Let G be a *unimodular* l.c.s.c. group, admitting square integrable representations; let us denote by $\mathscr{B}(G)$ the Borel σ-algebra of G and by $\mathcal{PM}(G)$ the set of Borel probability measures on G. Then, the main ingredients of our construction are:

(i) We choose an (irreducible) *square integrable projective representation* U of G in a (separable) complex Hilbert space $\mathcal{H}$. U gives rise to an isometric representation of G in $\mathcal{B}_1(\mathcal{H})$: $\mathsf{U}\vee\mathsf{U}(g)A := \mathsf{U}(g)A\,\mathsf{U}(g)^*$ (a *symmetry action* [11–13] of G on the space where the quantum states live). Notice that, although U is in general projective — $\mathsf{U}(gh) = \gamma(g,h)\,\mathsf{U}(g)\mathsf{U}(h)$ — where $\gamma(g,h) \in \mathbb{T}$ is a group multiplier, $\mathsf{U}\vee\mathsf{U}$ behaves like an ordinary group representation: $\mathsf{U}\vee\mathsf{U}(gh) = \mathsf{U}\vee\mathsf{U}(g)\;\mathsf{U}\vee\mathsf{U}(h)$. We will also use the shorthand notation $\rho_g \equiv \mathsf{U}\vee\mathsf{U}(g)\rho$.

(ii) We choose a *fiducial density operator* $\upsilon \in \mathcal{S}(\mathcal{H})$ and a *measure* $\varpi \in \mathcal{PM}(G)$. We denote by ϖ^g (ϖ_g) the left (right) g-translate of ϖ: $\varpi^g(\mathscr{E}) := \varpi(g^{-1}\mathscr{E})$ ($\varpi_g(\mathscr{E}) := \varpi(\mathscr{E}g)$), $\mathscr{E} \in \mathscr{B}(G)$.

Proposition 3. *For every* $\mu \in \mathcal{PM}(G)$*, the linear map*

$$\mu[\mathsf{U}]\colon \mathcal{B}_1(\mathcal{H}) \ni A \mapsto \int_G \mathrm{d}\mu(g)\,(\mathsf{U}\vee\mathsf{U}(g)A) \in \mathcal{B}_1(\mathcal{H}) \tag{5}$$

— *where a Bochner integral on* G *is understood — is stochastic; namely:* $\mu[\mathsf{U}]\,\mathcal{S}(\mathcal{H}) \subset \mathcal{S}(\mathcal{H})$.

Proposition 4. *For a suitable normalization of the Haar measure* μ_G *on* G*, and for every* $\rho \in \mathcal{S}(\mathcal{H})$*, the map*

$$\nu_{\rho,\upsilon}\colon \mathscr{B}(G) \ni \mathscr{E} \mapsto \int_{\mathscr{E}} \mathrm{d}\mu_G(g)\;\mathrm{tr}\big(\rho\,(\mathsf{U}\vee\mathsf{U}(g)\upsilon)\big) \tag{6}$$

is a Borel probability measure on G.

Remark 4. Proposition 4 relies on the *square-integrability* of the representation U [9], whereas Proposition 3 does not [9, 18]. The stochastic map $\mu[\mathsf{U}]$, see (5), is called *twirling operator* [18].

Taking into account these results, we can define a *binary operation* on $\mathcal{S}(\mathcal{H})$ by setting [9]

$$\rho \underset{\varpi}{\overset{\upsilon}{\odot}} \sigma := \big((\nu_{\rho,\upsilon}\circledcirc\varpi)[\mathsf{U}]\big)\,\sigma = \int_G \mathrm{d}(\nu_{\rho,\upsilon}\circledcirc\varpi)(g)\,(\mathsf{U}\vee\mathsf{U}(g)\sigma), \quad \forall \rho,\sigma \in \mathcal{S}(\mathcal{H}). \tag{7}$$

Here, we have considered the fact that $\nu_{\rho,\upsilon}$ is a probability measure associated with the density operators ρ and υ (Proposition 4), then we have formed the convolution $\mu \equiv \nu_{\rho,\upsilon}\circledcirc\varpi$ of this measure with the previously fixed probability measure ϖ and, finally, we have constructed the stochastic map $(\nu_{\rho,\upsilon}\circledcirc\varpi)[\mathsf{U}]$ according to Proposition 3. More explicitly, we have:

$$\rho \underset{\varpi}{\overset{\upsilon}{\odot}} \sigma = \int_G \mathrm{d}\mu_G(g) \int_G \mathrm{d}\varpi(h)\;\mathrm{tr}\big(\rho\,(\mathsf{U}\vee\mathsf{U}(g)\upsilon)\big)\,(\mathsf{U}\vee\mathsf{U}(gh)\sigma). \tag{8}$$

We stress that, in this formula, *Bochner integrals* of $\mathcal{B}_1(\mathcal{H})$-valued functions are understood and the Haar measure μ_G is supposed to be normalized in such a way that $\mathsf{c}_{\mathsf{U}} = 1$ [9]; recall Remark 4 and relation (4). In particular, for $\varpi = \delta \equiv \delta_e$ (the Dirac measure at the identity e of G), we get:

$$\rho \overset{\upsilon}{\odot} \sigma \equiv \rho \underset{\delta}{\overset{\upsilon}{\odot}} \sigma = \int_G \mathrm{d}\mu_G(g)\;\mathrm{tr}\big(\rho\,(\mathsf{U}\vee\mathsf{U}(g)\upsilon)\big)\,(\mathsf{U}\vee\mathsf{U}(g)\sigma). \tag{9}$$

Remark 5. Notice that we do not claim that the members of family of stochastic products

$$\{(\cdot) \underset{\varpi}{\overset{\upsilon}{\odot}} (\cdot) \colon \upsilon \in \mathcal{S}(\mathcal{H}),\ \varpi \in \mathcal{PM}(G)\} \tag{10}$$

are all distinct one another. On the contrary, using the fact that the iterated integrals in (8) can be interchanged (by Fubini's theorem for Bochner integrals), next performing the change of variables $g \mapsto gh^{-1}$ and, finally, inverting the integration order again and exploiting the fact that the Bochner integral wrt the measure ϖ commutes with both the trace functional and the bounded operator $\mathsf{U}\vee\mathsf{U}(g)$ in $\mathcal{B}_1(\mathcal{H})$, one finds out that

$$\rho \overset{\upsilon}{\underset{\varpi}{\odot}} \sigma = \rho \overset{\check{\varpi}[\mathsf{U}]\upsilon}{\odot} \sigma, \tag{11}$$

where $\check{\varpi} \in \mathcal{PM}(G)$ is defined by $\check{\varpi}(\mathscr{E}) := \varpi(\mathscr{E}^{-1})$, $\mathscr{E} \in \mathscr{B}(G)$, and

$$\check{\varpi}[\mathsf{U}]\upsilon := \int_G \mathrm{d}\check{\varpi}(g)\, (\mathsf{U}\vee\mathsf{U}(g)\upsilon) = \int_G \mathrm{d}\varpi(g)\, (\mathsf{U}\vee\mathsf{U}(g^{-1})\upsilon) \in \mathcal{S}(\mathcal{H}). \tag{12}$$

Moreover, certain choices of $\upsilon \in \mathcal{S}(\mathcal{H})$ or $\varpi \in \mathcal{PM}(G)$ may 'trivialize' the associated products; see Example 1 below.

We now formulate our main result [9].

Theorem 1. *For every $\upsilon \in \mathcal{S}(\mathcal{H})$ and $\varpi \in \mathcal{PM}(G)$, the binary operation*

$$(\cdot) \overset{\upsilon}{\underset{\varpi}{\odot}} (\cdot)\colon \mathcal{S}(\mathcal{H}) \times \mathcal{S}(\mathcal{H}) \to \mathcal{S}(\mathcal{H}) \quad (\text{'twirled product'}) \tag{13}$$

is an associative stochastic product that is left-covariant wrt the representation U, namely,

$$\rho_g \overset{\upsilon}{\underset{\varpi}{\odot}} \sigma = \left(\rho \overset{\upsilon}{\underset{\varpi}{\odot}} \sigma\right)_g, \quad \forall g \in G,\ \forall \rho, \sigma \in \mathcal{S}(\mathcal{H}). \tag{14}$$

Extending the twirled product to a stochastic product on the space $\mathcal{B}_1(\mathcal{H})$ of trace class operators, one gets a Banach algebra that, in the case where the l.c.s.c. group G is abelian, is commutative.

Beside covariance, there are two properties of *families of* twirled products: *invariance* and *equivariance*. Let X be a G-space wrt to some group action $(\cdot)[\cdot]\colon G \times X \ni (g,\xi) \mapsto g[\xi] \in X$. Suppose that the points of X label a family of stochastic products:

$$\{(\cdot) \overset{\xi}{\odot} (\cdot)\colon \mathcal{S}(\mathcal{H}) \times \mathcal{S}(\mathcal{H}) \to \mathcal{S}(\mathcal{H})\}_{\xi \in X}. \tag{15}$$

We say that this family of products is *invariant* wrt the action $(\cdot)[\cdot]$ if

$$\rho \overset{\xi}{\odot} \sigma = \rho \overset{g[\xi]}{\odot} \sigma, \quad \forall g \in G,\ \forall \xi \in X,\ \forall \rho, \sigma \in \mathcal{S}(\mathcal{H}). \tag{16}$$

We say that $\{(\cdot) \overset{\xi}{\odot} (\cdot)\}_{\xi \in X}$ is *right inner equivariant* wrt the pair $((\cdot)[\cdot], \mathsf{U})$ if

$$\rho \overset{\xi}{\odot} (\mathsf{U}\vee\mathsf{U}(g^{-1})\sigma) = \rho \overset{g[\xi]}{\odot} \sigma, \quad \forall g \in G,\ \forall \xi \in X,\ \forall \rho, \sigma \in \mathcal{S}(\mathcal{H}). \tag{17}$$

Then, we have the following [9]:

Proposition 5. *The family of twirled products*

$$\{(\cdot) \overset{\upsilon}{\underset{\varpi}{\odot}} (\cdot)\colon \upsilon \in \mathcal{S}(\mathcal{H}),\ \varpi \in \mathcal{PM}(G)\} \tag{18}$$

is invariant wrt the group action

$$g[(\upsilon, \varpi)] := (\upsilon_g \equiv \mathsf{U}\vee\mathsf{U}(g)\upsilon, \varpi^g); \tag{19}$$

namely,

$$\rho \underset{\varpi}{\overset{\upsilon}{\odot}} \sigma = \rho \underset{\varpi^g}{\overset{\upsilon_g}{\odot}} \sigma, \quad \forall g \in G,\ \forall \rho, \upsilon, \sigma \in \mathcal{S}(\mathcal{H}),\ \forall \varpi \in \mathcal{PM}(G). \tag{20}$$

It is right inner equivariant wrt the pair $((\cdot)[\cdot], \mathsf{U})$*, where this time* $(\cdot)[\cdot]$ *is the group action*

$$g[(\upsilon, \varpi)] := (\upsilon, \varpi_g); \tag{21}$$

namely,

$$\rho \underset{\varpi}{\overset{\upsilon}{\odot}} \sigma_{g^{-1}} = \rho \underset{\varpi_g}{\overset{\upsilon}{\odot}} \sigma, \quad \forall g \in G,\ \forall \rho, \upsilon, \sigma \in \mathcal{S}(\mathcal{H}),\ \forall \varpi \in \mathcal{PM}(G). \tag{22}$$

Remark 6. The invariance relation (20) can be regarded as a consequence of (11) and of the fact that $\check{\varpi}[\mathsf{U}]\upsilon = \check{\varpi}^g[\mathsf{U}]\upsilon_g$. Moreover, in the case where G is *abelian*, one can derive further symmetry relations:

$$\rho_g \underset{\varpi}{\overset{\upsilon}{\odot}} \sigma = \left(\rho \underset{\varpi}{\overset{\upsilon}{\odot}} \sigma\right)_g = \rho \underset{\varpi}{\overset{\upsilon}{\odot}} \sigma_g = \rho \underset{\varpi_{g^{-1}}}{\overset{\upsilon}{\odot}} \sigma = \rho \underset{\varpi^g}{\overset{\upsilon}{\odot}} \sigma = \rho \underset{\varpi}{\overset{\upsilon_{g^{-1}}}{\odot}} \sigma, \tag{23}$$

$$\rho \underset{\varpi}{\overset{\upsilon}{\odot}} \sigma = \rho \underset{\varpi^g}{\overset{\upsilon_g}{\odot}} \sigma = \rho \underset{\varpi_{g^{-1}}}{\overset{\upsilon_g}{\odot}} \sigma = \rho \underset{\varpi}{\overset{\upsilon_g}{\odot}} \sigma_g = \left(\rho \underset{\varpi}{\overset{\upsilon_g}{\odot}} \sigma\right)_g = \rho_g \underset{\varpi}{\overset{\upsilon_g}{\odot}} \sigma, \tag{24}$$

$$\rho_{g^{-1}} \overset{\upsilon}{\odot} \sigma = \left(\rho \overset{\upsilon}{\odot} \sigma\right)_{g^{-1}} = \rho \overset{\upsilon}{\odot} \sigma_{g^{-1}} = \rho \overset{\upsilon_g}{\odot} \sigma. \tag{25}$$

Example 1. Suppose that the topological group G is *compact* and U is an *irreducible unitary representation* of G; hence, $\gamma \equiv 1$ (the multiplier of U is trivial) and $\dim(\mathcal{H}) = n < \infty$. Because, in this case, the Haar measure μ_G on G is *finite*, U is square integrable. By the Peter-Weyl theorem [10], assuming that $\mu_G(G) = 1$ (i.e., $\mu_G \in \mathcal{PM}(G)$), we have: $\mathsf{c}_{\mathsf{U}} = \dim(\mathcal{H})^{-1} = n^{-1}$; therefore:

$$\rho \underset{\varpi}{\overset{\upsilon}{\odot}} \sigma = n \int_G \mathrm{d}\mu_G(g) \int_G \mathrm{d}\varpi(h)\ \mathrm{tr}\big(\rho\,(\mathsf{U} \vee \mathsf{U}(g)\upsilon)\big)\,(\mathsf{U} \vee \mathsf{U}(gh)\sigma). \tag{26}$$

For the *maximally mixed state* $\Omega := n^{-1} I$ (where I is the identity operator), we have [9]:

$$\rho \underset{\varpi}{\overset{\upsilon}{\odot}} \Omega = \Omega, \quad \Omega \underset{\varpi}{\overset{\upsilon}{\odot}} \sigma = \Omega, \quad \rho \underset{\varpi}{\overset{\Omega}{\odot}} \sigma = \Omega, \quad \forall \rho, \upsilon, \sigma \in \mathcal{S}(\mathcal{H}),\ \forall \varpi \in \mathcal{PM}(G). \tag{27}$$

The last of the previous relations shows that the choice of Ω as a fiducial state 'trivializes' the twirled product. Similarly, putting $\varpi = \mu_G$, one obtains a trivial stochastic product too:

$$\rho \underset{\mu_G}{\overset{\upsilon}{\odot}} \sigma = \Omega, \quad \forall \rho, \upsilon, \sigma \in \mathcal{S}(\mathcal{H}). \tag{28}$$

Exploiting relation (11), with $\varpi = \mu_G$, and then the fact that $\mu_G[\mathsf{U}]\upsilon = \Omega$ ($\mu_G = \check{\mu}_G$), for all $\upsilon \in \mathcal{S}(\mathcal{H})$ — see formula (128) of [9] — this statement can be derived as a consequence of the last of relations (27), with $\varpi = \delta$. Using, say, the irreducible unitary representations of the compact group SU(2), one can construct twirled products for every *finite* Hilbert space dimension.

Example 2. We now discuss an infinite-dimensional example. Let G be the *group of translations on phase space* — $G = \mathbb{R}^n \times \mathbb{R}^n$ — let us identify $\mathcal{H}$ with the Hilbert space $\mathrm{L}^2(\mathbb{R}^n)$ and, putting $\hbar \equiv \mathsf{h} = 1$, let U be the standard Weyl system [16, 17, 19, 20] defined by

$$\big(\mathsf{U}(q,p) f\big)(x) := \mathrm{e}^{-\mathrm{i}q\cdot p/2}\, \mathrm{e}^{\mathrm{i}p\cdot x} f(x-q), \quad (q,p) \in \mathbb{R}^n \times \mathbb{R}^n; \tag{29}$$

i.e., $\mathsf{U}(q,p) = \mathrm{e}^{-\mathrm{i}q\cdot p/2}\, \mathrm{e}^{\mathrm{i}p\cdot\hat{q}}\, \mathrm{e}^{-\mathrm{i}q\cdot\hat{p}}$, where $\hat{q}$, $\hat{p}$ are the ordinary (vector) position and momentum operators in $\mathrm{L}^2(\mathbb{R}^n)$. This is a strongly continuous, irreducible projective representation — with symplectic multiplier $\gamma(q,p;\tilde{q},\tilde{p}) = \exp(\mathrm{i}(q\cdot\tilde{p} - p\cdot\tilde{q})/2)$ — that, as previously mentioned, is square integrable. Setting $\mathrm{L}^2(G) = \mathrm{L}^2(\mathbb{R}^n\times\mathbb{R}^n, (2\pi)^{-n}\mathrm{d}^n q\,\mathrm{d}^n p;\mathbb{C})$, we have that $\mathsf{c}_{\mathsf{U}} = 1$; thus:

$$\tau = \rho \underset{\varpi}{\overset{\upsilon}{\odot}} \sigma = (2\pi)^{-n}\int \mathrm{d}^n q\,\mathrm{d}^n p\ \mathrm{tr}\big(\rho\,(\mathrm{e}^{\mathrm{i}p\cdot\hat{q}}\,\mathrm{e}^{-\mathrm{i}q\cdot\hat{p}}\,\upsilon\,\mathrm{e}^{\mathrm{i}q\cdot\hat{p}}\,\mathrm{e}^{-\mathrm{i}p\cdot\hat{q}})\big)$$
$$\times\int \mathrm{d}\varpi(\tilde{q},\tilde{p})\ \big(\mathrm{e}^{\mathrm{i}(p+\tilde{p})\cdot\hat{q}}\,\mathrm{e}^{-\mathrm{i}(q+\tilde{q})\cdot\hat{p}}\,\sigma\,\mathrm{e}^{\mathrm{i}(q+\tilde{q})\cdot\hat{p}}\,\mathrm{e}^{-\mathrm{i}(p+\tilde{p})\cdot\hat{q}}\big). \tag{30}$$

This product — the so-called *phase-space stochastic product* [9] — is *commutative*, since it stems from a representation of an *abelian* group. For $\varpi = \delta \equiv \delta_e$, we get the *quantum convolution* [9]:

$$\tau = \rho \overset{\upsilon}{\odot} \sigma = (2\pi)^{-n}\int \mathrm{d}^n q\,\mathrm{d}^n p\ \mathrm{tr}\big(\rho\,(\mathrm{e}^{\mathrm{i}p\cdot\hat{q}}\,\mathrm{e}^{-\mathrm{i}q\cdot\hat{p}}\,\upsilon\,\mathrm{e}^{\mathrm{i}q\cdot\hat{p}}\,\mathrm{e}^{-\mathrm{i}p\cdot\hat{q}})\big)\,\big(\mathrm{e}^{\mathrm{i}p\cdot\hat{q}}\,\mathrm{e}^{-\mathrm{i}q\cdot\hat{p}}\,\sigma\,\mathrm{e}^{\mathrm{i}q\cdot\hat{p}}\,\mathrm{e}^{-\mathrm{i}p\cdot\hat{q}}\big). \tag{31}$$

To justify this term, let us express this operation in terms of the *Wigner distributions* [8,16,19–23] $\mathcal{W}_\rho$, $\mathcal{W}_\upsilon$, $\mathcal{W}_\sigma$, $\mathcal{W}_\tau$ of the states ρ, υ, σ, τ. Setting $\widehat{\mathcal{W}_\upsilon}(x) := \mathcal{W}_\upsilon(-x)$, $x\equiv(q,p)\in\mathbb{R}^{2n}$, we find:

$$\mathcal{W}_\tau(z) = \int \mathrm{d}^{2n}x\,\Big(\int \mathrm{d}^{2n}y\ \mathcal{W}_\rho(y)\,\widehat{\mathcal{W}_\upsilon}(x-y)\Big)\,\mathcal{W}_\sigma(z-x). \tag{32}$$

Notice that we get a *double* convolution of Wigner functions, where $\mathcal{W}_\upsilon$ plays a *pivotal role* with no analogue in the classical case. The function $(q,p)\mapsto \int \mathrm{d}^n\tilde{q}\,\mathrm{d}^n\tilde{p}\ \mathcal{W}_\rho(\tilde{q},\tilde{p})\,\widehat{\mathcal{W}_\upsilon}(q-\tilde{q},p-\tilde{p})$ is a probability distribution wrt the Lebesgue measure on $\mathbb{R}^n\times\mathbb{R}^n$. E.g., putting $n=1$, and choosing the pure state $|\psi\rangle\langle\psi|$, with $\psi(q) = (2\pi)^{-1/4}\,\mathrm{e}^{-q^2/4}$, as the fiducial state υ — whose Wigner function is $\mathcal{W}_\upsilon(q,p)\equiv\mathcal{W}_\psi(q,p) = \pi^{-1}\,\mathrm{e}^{-(q^2+p^2)}$ — we find the probability distribution $\mathcal{Q}_\rho(q,p) = \pi^{-1}\int \mathrm{d}\tilde{q}\,\mathrm{d}\tilde{p}\ \mathcal{W}_\rho(\tilde{q},\tilde{p})\,\mathrm{e}^{-(q-\tilde{q})^2-(p-\tilde{p})^2}$, the so-called *Husimi-Kano function* associated with ρ [23]. Thus, in this case the quantum convolution, in terms of phase-space functions, is of the form

$$\mathcal{W}_\tau(q,p) = \int \mathrm{d}\tilde{q}\,\mathrm{d}\tilde{p}\ \mathcal{Q}_\rho(\tilde{q},\tilde{p})\,\mathcal{W}_\sigma(q-\tilde{q},p-\tilde{p}), \quad \tau = \rho \overset{|\psi\rangle\langle\psi|}{\odot} \sigma. \tag{33}$$

5. Final remarks and perspectives

Quantum measurements are described by POVMs and (quantum) instruments [7]. The mapping

$$\mathscr{B}(G)\ni\mathscr{E}\mapsto\int_{\mathscr{E}}\mathrm{d}\mu_G(g)\,(\mathsf{U}\vee\mathsf{U}(g)\upsilon) =: \mathsf{E}_\upsilon(\mathscr{E}) \quad (\mathsf{U}\text{ square integrable, }\upsilon\in\mathcal{S}(\mathcal{H})) \tag{34}$$

is a *covariant quantum observable*; i.e., a POVM *covariant* wrt the irreducible representation U: $\mathsf{E}_\upsilon(g\mathscr{E}) = \mathsf{U}\vee\mathsf{U}(g)\,\mathsf{E}_\upsilon(\mathscr{E})$, $\forall g\in G$, $\forall\mathscr{E}\in\mathscr{B}(G)$. The function $\mathsf{p}_{\rho,\upsilon}(g) = \mathrm{tr}\big(\rho\,(\mathsf{U}\vee\mathsf{U}(g)\upsilon)\big)$, that we used in the construction of a stochastic product, is the *probability density* on G of the observable E_υ wrt the state ρ; namely, the Radon-Nikodym derivative of the probability measure $\nu_{\rho,\upsilon}\colon\mathscr{E}\mapsto\int_{\mathscr{E}}\mathrm{d}\mu_G(g)\ \mathrm{tr}\big(\rho\,(\mathsf{U}\vee\mathsf{U}(g)\upsilon)\big)$ (recall Proposition 4) wrt μ_G. Moreover, the mapping

$$\mathscr{B}(G)\ni\mathscr{E}\mapsto\Big(\mathcal{B}_1(\mathcal{H})\ni A\mapsto\int_{\mathscr{E}}\mathrm{d}\mu_G(g)\ \mathrm{tr}\big(A\,(\mathsf{U}\vee\mathsf{U}(g)\upsilon)\big)\,(\mathsf{U}\vee\mathsf{U}(g)\sigma) =: \mathcal{I}^{\upsilon,\sigma}_{\mathscr{E}}A\Big) \tag{35}$$

— where $\mathcal{I}^{\upsilon,\sigma}_{\mathscr{E}}\colon\mathcal{B}_1(\mathcal{H})\to\mathcal{B}_1(\mathcal{H})$ is a quantum operation [7] (in particular, a stochastic map, for $\mathscr{E}=G$) — is a quantum instrument; specifically, a U-*covariant quantum instrument* based on G, because $\mathcal{I}^{\upsilon,\sigma}_{g\mathscr{E}}(\mathsf{U}\vee\mathsf{U}(g)A) = \mathsf{U}\vee\mathsf{U}(g)\big(\mathcal{I}^{\upsilon,\sigma}_{\mathscr{E}}A\big)$, $\forall g\in G$, $\forall\mathscr{E}\in\mathscr{B}(G)$. Observe that the twirled product generated by the triple $(\mathsf{U},\upsilon,\delta)$ can be recovered by setting $\mathscr{E}=G$; i.e., $\rho\overset{\upsilon}{\odot}\sigma = \mathcal{I}^{\upsilon,\sigma}_{G}\rho$. Also note that the *associativity* of the twirled stochastic product translates into the following

Journal of Physics: Conference Series **1416** (2019) 012002 doi:10.1088/1742-6596/1416/1/012002

relation involving the pair of stochastic maps $\mathcal{I}_G^{\upsilon,\rho}$ and $\mathcal{I}_G^{\upsilon,\sigma}$: $\mathcal{I}_G^{\upsilon,\sigma} \circ \mathcal{I}_G^{\upsilon,\rho} = \mathcal{I}_G^{\upsilon,\tau}$, where $\tau = \rho \overset{\upsilon}{\odot} \sigma$. If $\varpi \neq \delta$, the quantum observable E_υ is replaced with the *smeared observable* $\mathsf{E}_{\upsilon|\varpi}$,

$$\mathsf{E}_{\upsilon|\varpi}(\mathscr{E}) \equiv \mathsf{E}_\upsilon \circledcirc \varpi(\mathscr{E}) := \int_G \mathrm{d}\varpi(h)\, \mathsf{E}_\upsilon(\mathscr{E}h) = \int_G \mathrm{d}\varpi(h) \int_{\mathscr{E}} \mathrm{d}\mu_G(g)\, (\mathsf{U} \vee \mathsf{U}(gh^{-1})\upsilon), \tag{36}$$

and the probability density $\mathsf{p}_{\rho,\upsilon}(g)$ is replaced with $\mathsf{p}_{\rho,\upsilon|\varpi}(g) = \big(\mathsf{p}_{\rho,\upsilon} \circledcirc \varpi\big)(g) = \int \mathrm{d}\varpi(h)\, \mathsf{p}_{\rho,\upsilon}(gh^{-1})$. An analogous smearing takes place, when $\varpi \neq \delta$, for the quantum instrument $\mathscr{E} \mapsto \mathcal{I}_{\mathscr{E}}^{\upsilon,\sigma}$. It would be worth studying further aspects of stochastic products (in particular, of twirled products) [9]:

- There is a natural notion of *complete* positivity for a stochastic product.
- The classification of covariant stochastic products is an interesting issue.
- Twirled products generated by representations of non-unimodular groups can be considered.
- It is natural to wonder whether the twirled products are entropy-nondecreasing [24, 25].

References

[1] Landsman N P 1998 *Mathematical Topics Between Classical and Quantum Mechanics* (New York: Springer)

[2] Moretti V 2017 *Spectral Theory and Quantum Mechanics* 2nd edn (Cham: Springer)

[3] Haag R 1996 *Local Quantum Physics* 2nd edn (Berlin: Springer)

[4] Bratteli O and Robinson D W 1987 *Operator Algebras and Quantum Statistical Mechanics 1* 2nd edn (Berlin: Springer)

[5] Connes A 1994 *Noncommutative Geometry* (Boston, MA: Academic Press)

[6] Nielsen M A and Chuang I L 2000 *Quantum Computation and Quantum Information* (Cambridge: Cambridge University Press)

[7] Heinosaari T and Ziman M 2012 *The Mathematical Language of Quantum Theory* (Cambridge: Cambridge University Press)

[8] Aniello P 2019 Discovering the manifold facets of a square integrable representation: from coherent states to open systems *J. Phys.: Conf. Ser.* **1194** 012006

[9] Aniello P 2019 A class of stochastic products on the convex set of quantum states *J. Phys. A: Math. Theor.* **52** 305302

[10] Folland G B 1995 *A Course in Abstract Harmonic Analysis* (Boca Raton, FL: CRC Press)

[11] Bargmann V 1964 Note on Wigner's theorem on symmetry operations *J. Math. Phys.* **5** 862–8

[12] Aniello P and Chruściński D 2017 Symmetry witnesses *J. Phys. A: Math. Theor.* **50** 285302

[13] Aniello P 2018 A notion of symmetry witness related to Wigner's theorem on symmetry transformations *J. Phys.: Conf. Ser.* **965** 012004

[14] Aniello P, Cassinelli G, De Vito E and Levrero A 1998 Square-integrability of induced representations of semidirect products *Rev. Math. Phys.* **10** 301–13

[15] Aniello P 2006 Square integrable projective representations and square integrable representations modulo a relatively central subgroup *Int. J. Geom. Meth. Mod. Phys.* **3** 233–67

[16] Aniello P 2018 Square integrable representations, an invaluable tool *Coherent States and Their Applications: a Contemporary Panorama* (*Springer Proceedings in Physics* vol 205) ed J-P Antoine *et al* (Cham: Springer) chapter 2 pp 17–40

[17] Aniello P 2010 On the notion of Weyl system *J. Russ. Laser Res.* **31** 102–16

[18] Aniello P, Kossakowski A, Marmo G and Ventriglia F 2010 Brownian motion on Lie groups and open quantum systems *J. Phys. A: Math. Theor.* **43** 265301

[19] Aniello P, Man'ko V I and Marmo G 2008 Frame transforms, star products and quantum mechanics on phase space *J. Phys. A: Math. Theor.* **41** 285304

[20] Aniello P 2009 Star products: a group-theoretical point of view *J. Phys. A: Math. Theor.* **42** 475210

[21] Aniello P 2015 Playing with functions of positive type, classical and quantum *Phys. Scr.* **90** 074042

[22] Aniello P 2016 Functions of positive type on phase space, between classical and quantum, and beyond *J. Phys.: Conf. Ser.* **670** 012004

[23] Aniello P, Man'ko V, Marmo G, Solimeno S and Zaccaria F 2000 On the coherent states, displacement operators and quasidistributions associated with deformed quantum oscillators *J. Opt. B: Quantum Semiclass. Opt.* **2** 718–25

[24] Aniello P and Chruściński D 2016 Characterizing the dynamical semigroups that do not decrease a quantum entropy *J. Phys. A: Math. Theor.* **49** 345301

[25] Aniello P 2017 Quantum entropies, Schur concavity and dynamical semigroups *J. Phys.: Conf. Ser.* **804** 012003

Journal of Physics: Conference Series **1416** (2019) 012003 doi:10.1088/1742-6596/1416/1/012003

Darboux transformation and soliton solution for generalized Konno-Oono equation

G Bekova[1,3], **G Shaikhova**[1], **K Yesmakhanova**[2], **R Myrzakulov**[1]

[1] Eurasian International Center for Theoretical Physics and Department of General Theoretical Physics, Eurasian National University, Nur-Sultan, 010000, Kazakhstan
[2] Eurasian International Center for Theoretical Physics and Department of Higher Mathematics, Eurasian National University, Nur-Sultan, 010000, Kazakhstan
[3] Kh. Dosmukhamedov Atyrau state university, Atyrau, 060000, Kazakhstan

E-mail: `bekovaguldana@gmail.com`

Abstract. In this paper, the generalized Konno-Oono equation is investigated. Using the Lax pair, we obtain a new Darboux transformation for the dispersionless equation. A soliton solution of the generalized Konno-Oono equation is obtained on the basis of the Darboux transformation. Corresponding graphs are also built.

1. Introduction

Nowadays dispersionless equations are equations that describes various problems of quantum mechanics, fluid mechanics, plasma physics, propagation of shallow water waves, optical fibers, electricity, etc. Therefore, investigations of dispersionless equations are becoming more and more interesting in nonlinear sciences day by day. However, not all equations posed of these models are solvable [1,2]. As a result, many new techniques have been successfully developed by diverse groups of mathematicians and physicists, such as: the Hirotas bilinear transformation method, the Darboux transformation (DT), the sine-cosine method, the tanh-function method and so forth. [3-6]. DT is one of the most useful method to get soliton solutions of the many developed equations from a trivial seed [7-14]. In this paper, we constructed DT for the generalized Konno-Oono equation [15] and we get single soliton, double soliton and three soliton solutions are derived by assuming a trivial seed solution by the DT.

2. The generalized Konno-Oono equation

The Konno-Oono equation (KOE) [15] in form

$$q_{xt} - \rho q = 0, \qquad (1)$$

$$\rho_x + 0.5(q^2)_t = 0, \qquad (2)$$

was proposed by Konno and Oono. The KOE is also known as the coupled dispersionless equation (CDE). Later, Konno and Kakuhata proposed a generalization of this equation having the following form

$$q_{xt} - \rho q = 0, \tag{3}$$
$$r_{xt} - \rho r = 0, \tag{4}$$
$$\rho_x + 0.5(rq)_t = 0. \tag{5}$$

This set of equations we call the generalized Konno-Oono equation (GKOE) or the generalized CDE (GCDE). GKOE is integrable by IST and its Lax representation (LR) can be expressed as

$$\Psi_x = U\Psi, \tag{6}$$
$$\Psi_t = V\Psi, \tag{7}$$

where

$$U = \begin{pmatrix} \frac{i}{4\lambda} & -0.5q \\ 0.5r & -\frac{i}{4\lambda} \end{pmatrix}, \quad V = -i\lambda \begin{pmatrix} \rho & q_t \\ r_t & -\rho \end{pmatrix}. \tag{8}$$

The compatibility condition of equations (3)-(5) given us

$$U_t - V_x + [U, V] = 0,$$

by direct calculation of above equation, we can yield the generalized Konno-Oono equation.

3. Darboux transformation

In this section, firstly, we consider the transformation of linear function Ψ by

$$\Psi' = T\Psi. \tag{9}$$

New function Ψ' is supposed to satisfy

$$\Psi'_x = U'\Psi',$$
$$\Psi'_t = V'\Psi'.$$

Here

$$U' = \begin{pmatrix} \frac{i}{4\lambda} & -0.5q' \\ 0.5r' & -\frac{i}{4\lambda} \end{pmatrix}, \quad V' = -i\lambda \begin{pmatrix} \rho' & q'_t \\ r'_t & -\rho' \end{pmatrix}.$$

Then matrix T must satisfy following identities

$$T_x = U'T - TU, \tag{10}$$
$$T_t = V'T - TV. \tag{11}$$

Taking

$$T = \frac{1}{\lambda}I - M,$$

where I is an unit matrix and $S = H\Lambda H^{-1}$.

Now we come to construct a concrete transformation. Let $\lambda = \lambda_1$, the solution of Lax pairs (6) and (7) is $(h_1, h_2)^T$. It is easy to see that $(h_2, -h_1)^T$ is the solution of $\lambda = -\lambda_1$. Suppose that

$$H - \begin{pmatrix} h_1 & h_2 \\ h_2 & -h_1 \end{pmatrix} \quad \text{and} \quad \Lambda = \begin{pmatrix} \frac{1}{\lambda_1} & 0 \\ 0 & -\frac{1}{\lambda_1} \end{pmatrix}.$$

Let $\sigma = \frac{h_2}{h_1}$ and

$$S = \frac{1}{\lambda_1(1+\sigma^2)}\begin{pmatrix} 1-\sigma^2 & 2\sigma \\ 2\sigma & \sigma^2-1 \end{pmatrix}. \tag{12}$$

In simplicity, noting $\tan\frac{\theta}{2} = \sigma$, then Eq. (12) will become

$$S = \frac{1}{\lambda_1}\begin{pmatrix} \cos\theta & \sin\theta \\ \sin\theta & -\cos\theta \end{pmatrix}.$$

Then our Darboux matrix have the form

$$T = \begin{pmatrix} \frac{1}{\lambda} - \frac{1}{\lambda_1}\cos\theta & -\frac{1}{\lambda_1}\sin\theta \\ -\frac{1}{\lambda_1}\sin\theta & \frac{1}{\lambda} + \frac{1}{\lambda_1}\cos\theta \end{pmatrix}. \tag{13}$$

With the help of

$$\sigma_x = 0.5(r - q\sigma^2) - \frac{i}{2\lambda}\sigma,$$

we find θ_x for eq. (10)

$$\theta_x = \frac{r(1+\cos\theta) - q(1-\cos\theta)}{2} - \frac{i}{2\lambda}\sin\theta.$$

Then by using of the above transformation T and eq. (11), we can get

$$\begin{aligned} \sigma_t &= -i\lambda r_t + 2i\lambda\rho\sigma + i\lambda q_t\sigma^2, \\ \theta_t &= i\lambda q_t(1-\cos\theta) - i\lambda r_t(1+\cos\theta) + 2i\lambda\rho\sin\theta. \end{aligned}$$

Then the Darboux transformation on q, r and ρ can be get by eq. (10)-(11), and we obtain

$$q' = q - \frac{i}{\lambda_1}\frac{2\sigma}{(1+\sigma^2)^2}, \tag{14}$$

$$r' = r - \frac{i}{\lambda_1}\frac{2\sigma^3}{(1+\sigma^2)^2}, \tag{15}$$

$$\rho' = -\rho - \frac{2\lambda}{\lambda_1}\frac{(1-\sigma^2)(q_t\sigma^2 - r_t + 2i\lambda\rho\sigma)}{(1+\sigma^2)^2}. \tag{16}$$

or

$$q' = q - \frac{i}{2\lambda_1}\sin\theta(\cos\theta + 1), \tag{17}$$

$$r' = r + \frac{i}{2\lambda_1}\sin\theta(\cos\theta - 1), \tag{18}$$

$$\rho' = -\rho - \frac{i}{\lambda_1}(\cos\theta)_t. \tag{19}$$

Using the this DT we can found soliton solution of the generalized Konno-Oono equation.

Journal of Physics: Conference Series **1416** (2019) 012003 doi:10.1088/1742-6596/1416/1/012003

4. The soliton solutions

For a given solution of the coupled dispersionless Eqs.(3)-(5), suppose we have known the compatible basic solution old Lax pairs (6)-(7)

$$\Psi(x,t,\lambda)=\begin{pmatrix}\psi_{11}(x,t,\lambda) & \psi_{12}(x,t,\lambda)\\ \psi_{21}(x,t,\lambda) & \psi_{22}(x,t,\lambda)\end{pmatrix},$$

λ_1, μ_1 are the arbitrary constants, let

$$\sigma=\frac{\psi_{22}(x,t,\lambda_1)+\mu_1\psi_{21}(x,t,\lambda_1)}{\psi_{12}(x,t,\lambda_1)+\mu_1\psi_{11}(x,t,\lambda_1)},$$

be the proportion of the two component of a solution for Lax pairs (6)-(7).

Now we take the trivial solution $q=0$, $r=0$, $\rho=1$, the corresponding compatible basic solution od the Lax pairs (6)-(7) can be written as

$$\Psi(x,t,\lambda)=\begin{pmatrix}\exp(\frac{i}{4\lambda}x-i\lambda t) & 0\\ 0 & \exp(-\frac{i}{4\lambda}x+i\lambda t)\end{pmatrix},$$

for the constant $\lambda_1\neq 0$, and $\mu_1=\exp(2\alpha_1)>0$. We know

$$\sigma=\exp\left(-\frac{i}{2\lambda}x+2i\lambda t-2\alpha_1\right),$$

end

$$T=\frac{1}{\lambda}I-\frac{1}{\lambda_1\cosh\nu_1}\begin{pmatrix}\sinh\nu_1 & 1\\ 1 & -\sinh\nu_1\end{pmatrix},$$

with

$$\nu_1=\exp(\frac{i}{2\lambda}x-2i\lambda t+2\alpha_1).$$

By using eqs. (14)-(16), we get the one soliton solution of the generalized Konno-Oono equation.

$$q'=-\frac{i}{\lambda_1}sech^2(-\frac{i}{2\lambda}x+2i\lambda t-2\alpha_1), \tag{20}$$

$$r'=-\frac{i}{\lambda_1}sech^2(-\frac{i}{2\lambda}x+2i\lambda t-2\alpha_1)\exp(-\frac{i}{2\lambda}x+2i\lambda t-2\alpha_1), \tag{21}$$

$$\rho'=-1-\frac{2i\lambda^2}{\lambda_1}\frac{sh(\exp(-\frac{i}{2\lambda}x+2i\lambda t-2\alpha_1))}{ch^2(\exp(-\frac{i}{2\lambda}x+2i\lambda t-2\alpha_1))}. \tag{22}$$

Visualization of single soliton solutions (20)-(22) of the GKOE is shown in figure 1.

5. Conclusion

In this paper, using the Lax pair, we obtained a new Darboux transformation for the generalized Konno-Oono equation (3)-(5). The single soliton solution for the GKOE have been constructed by the DT. We provided graphical representation not only for all obtained solutions, but also for double and three soliton solutions (see figure 2, 3).

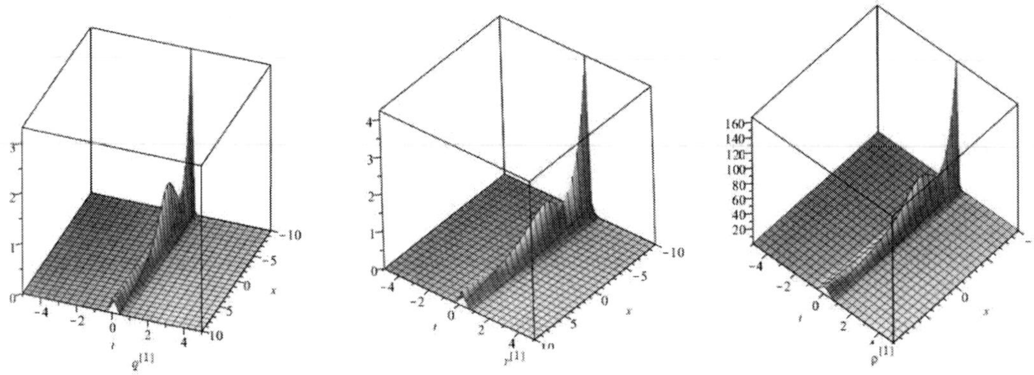

Figure 1. The propagation of one-soliton solution $q^{[1]}$, $r^{[1]}$ and $\rho^{[1]}$ when $\lambda_1 = -3i + 1$, $\alpha_1 = 1$.

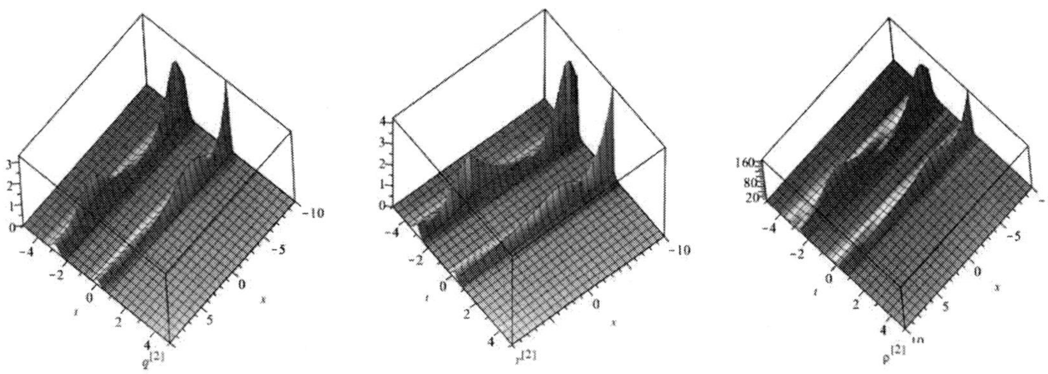

Figure 2. The propagation of two-soliton solutions $q^{[2]}$, $r^{[2]}$ and $\rho^{[2]}$ when $\lambda_2 = 2i - 2$, $\alpha_2 = 5$.

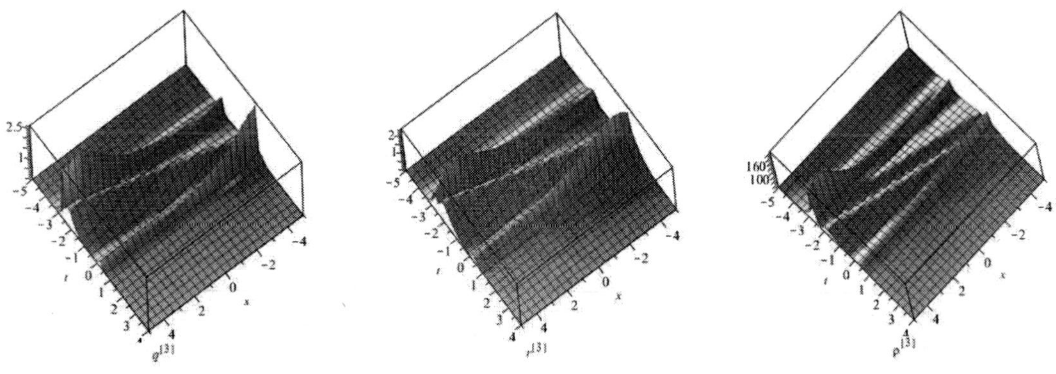

Figure 3. The propagation of two-soliton solutions $q^{[3]}$, $r^{[3]}$ and $\rho^{[3]}$ when $\lambda_3 = -i$, $\alpha_3 = -1$.

6. Acknowledgments

This work was performed as part of the financial support of the scientific and technical program (F. 0811, No 0118RK00935) of the Ministry of Education and Science of the Republic of Kazakhstan.

References

[1] Md. Abul Bashar, Gobinda Mondal, Kamruzzaman Khan, and Ahmet Bekir 2016 *New Trends in math. Scien.* **4**, 2 296-303.
[2] Aihua Chena, Xuemei Li 2007 *Phys. Lett. A* **370** 281-6.
[3] Hirota R 1973 *J. of Math. Phys.* **14** 805-10.
[4] Hirota R, Satsuma J 1981 *Phys. Lett. A* **85** 404-8.
[5] Malfliet W and Hereman W 1996 *Physica Scripta* **54** 563-8.
[6] Zheng X D, Xia T C and Zhang H Q 2002 *Appl. Math. E-Note* **2** 45-50.
[7] Zhaqilao, Zhao Yin-Long, and Li Zhi-Bin 2009 *Chin. Phys. B* **18** 5.
[8] Bekova G, Yesmakhanova K, Shaikhova G 2018 *AIP Conf. Proc.* **1997** 020039.
[9] Yesmakhanova K, Bekova G, Ybyraiymova S, Shaikhova G 2017 *J. of Phys.: Conf. Ser.* **936** 012098.
[10] Bekova G, Yesmakhanova K, Shaikhova G, Myrzakulov R *J. of Phys.: Conf. Ser.* 2017 **804** 012004.
[11] Yesmakanova K R, Shaikhova G N, Bekova G T and Myrzakulov R 2017 *J. of Phys.: Conf. Ser.* **936** 012045.
[12] Yesmakhanova K, Shaikhova G, Bekova G. 2016 *AIP Conf. Proceedings* **1759** 020147 (1-5).
[13] Nugmanova G, Sagidullayeva Zh, Myrzakulov R 2017 *J. of Phys. Conf. Series* **804** 012035.
[14] Nugmanova G, Sagidullayeva Zh 2017 *Bulletin of the Karaganda University-Mathematics* **86** 2.
[15] Bekova G, Nugmanova G, Shaikhova G, Yesmakhanova K, Myrzakulov R 2019 [arXiv:1901.01470].

Journal of Physics: Conference Series **1416** (2019) 012004 doi:10.1088/1742-6596/1416/1/012004

Cho–Maison magnetic monopole: BPS limit and lower mass bound

Petr Beneš[1] **and Filip Blaschke**[1,2]

[1] Institute of Experimental and Applied Physics,
Czech Technical University in Prague,
Husova 240/5, 110 00 Prague 1, Czech Republic
[2] Institute of Physics and Research Centre of Theoretical Physics and Astrophysics,
Faculty of Philosophy and Science, Silesian University in Opava,
Bezručovo náměstí 1150/13, 746 01 Opava, Czech Republic

E-mail: petr.benes@utef.cvut.cz

Abstract. We construct a class of effective extensions of the Standard Model that support finite-mass Cho–Maison magnetic monopole in the Bogomol'nyi–Prasad–Sommerfield (BPS) limit. We present an example of exact analytic monopole solution and derive a universal lower mass bound $M \geq 2\pi v/g \approx 2.37$ TeV of the Cho–Maison magnetic monopole.

1. Introduction

It's been a well known fact that in the Standard Model (SM) there is no magnetic monopole of the 't Hooft–Polyakov type [1, 2]. Reason is that the second homotopy group of the vacuum manifold is trivial, $\pi_2\big(SU(2)_{\rm L} \times U(1)_{\rm Y}/U(1)_{\rm em}\big) = \{1\}$. However, it was shown by Cho and Maison [3] that the desired topology that could support the solitonic monopole solution can be found elsewhere, namely in the Higgs doublet field H itself. In order to see that, it is instrumental to write H as

$$H = \frac{v}{\sqrt{2}}\rho\,\xi\,, \tag{1}$$

where ξ is a complex doublet normalized as $\xi^\dagger\xi = 1$ and $v = 246\,\mathrm{GeV}$ is the vacuum expectation value of H (the Higgs potential is $\propto (|H|^2 - v^2/2)^2$). It can be shown that, due to the $U(1)_{\rm Y}$ symmetry, the field ξ can be regarded as a $\mathbb{C}\mathrm{P}^1$ coordinate. But $\pi_2(\mathbb{C}\mathrm{P}^1) = \mathbb{Z}$, so that stable soliton solutions with the properties of a magnetic monopole can, potentially, exist in SM.

Indeed, such a solution can be found. Alas, it turns out that it has an infinite mass:

$$M = \frac{2\pi}{g'^2}\int_0^\infty \frac{\mathrm{d}r}{r^2} + \text{finite terms} = \infty\,. \tag{2}$$

Thus, the common wisdom that there is no (physical, finite-mass) magnetic monopole in SM seems justified.

Thus, one must step beyond the SM and modify it in order to get a finite-mass monopole. One way to do that is due to Cho, Kim, and Yoon (CKY) [4] and consists of modifying the hypercharge kinetic term as

$$-\frac{1}{4g'^2}B_{\mu\nu}B^{\mu\nu} \quad \longrightarrow \quad -\frac{1}{4g'^2}\epsilon\big(|H|^2\big)B_{\mu\nu}B^{\mu\nu}\,, \tag{3}$$

where ϵ is some positive function of the Higgs field squared, normalized as $\epsilon(v^2/2) = 1$ (in order to recover the SM in the vacuum). Crucially, if $\epsilon \to 0$ sufficiently fast as $|H|^2 \to 0$, the mass of the Cho–Maison monopole comes out finite, as desired. In fact, models beyond SM of this type have indeed been proposed in the literature [5, 6].

Since ϵ is otherwise virtually arbitrary, also the corresponding monopole mass is in principle arbitrary. E.g., Ellis, Mavromatos and You [7] were varying ϵ (while keeping consistency with experimental bounds on $H \to \gamma\gamma$ production) and found a whole variety of monopole masses reaching as low as to 5.5 TeV. Thus, a question of phenomenological importance arises whether one can go with monopole mass arbitrarily low or if there exists some limit.

Indeed, as showed recently, it turns out that there exists a definite lower bound for the monopole mass [8]. In order to derive it, one has to first construct a Bogomol'nyi–Prasad–Sommerfield (BPS) limit of the Cho–Maison monopole and show that such the mass bound holds also in the CKY theory.

2. Construction of the BPS limit

In order to construct a BPS limit of a given theory, it is usually sufficient to express the Hamiltonian of a static configuration as a perfect square plus a total derivative and nullify what remains (typically the potential). For the Cho–Maison monopole (with CKY regularization) this approach doesn't work, though. It turns out that besides removing the potential one has to modify non-trivially the gauge sector of the theory, too. Instead, we turn the typical procedure of constructing the BPS theory upside down and start from the desired result: We first write down the BPS equation of motion and subsequently derive the BPS theory from it.

Thus, we postulate the most general BPS equation

$$D_i H \;=\; \frac{1}{\rho}f_1(\rho)\,M_i\xi + f_2(\rho)\,(\xi^\dagger M_i\,\xi)\xi + f_3(\rho)\,G_i\xi\,, \tag{4}$$

where we defined "magnetic fields" of $SU(2)_{\mathrm{L}}$ and $U(1)_{\mathrm{Y}}$

$$M_i \;=\; \frac{1}{2}\varepsilon_{ijk}F^a_{jk}\sigma_a\,, \qquad G_i \;=\; \frac{1}{2}\varepsilon_{ijk}B_{jk}\,. \tag{5}$$

Notice presence of as yet arbitrary functions f_1, f_2, f_3 of $\rho = \sqrt{2}|H|/v$.

The equation (4) is gauge-covariant, linear in the gauge fields and first-order in derivatives. Thus, it qualifies as a candidate for a valid BPS equation, that is, equation of motion for a BPS theory. This operationally means that Hamiltonian of a static configuration of such a theory should be possible to write in the form

$$\mathcal{H} \;=\; \Big|D_iH - \frac{1}{\rho}f_1\,M_i\xi - f_2\,(\xi^\dagger M_i\,\xi)\xi - f_3\,G_i\xi\Big|^2 + \partial_i X_i \tag{6}$$

for some X_i, that is, as a sum of squares that vanish when the BPS equation is satisfied, plus a total derivative. Moreover, one also requires that the Hamiltonian be non-negative. This can indeed be done, but only if the functions f_i are no longer independent, but f_2 is related to f_1 as

$$f_2 \;=\; \frac{1}{2}f_1' - \frac{1}{\rho}f_1 \tag{7}$$

(prime denotes derivative with respect to ρ). The three vector X_i is then given as

$$X_i = \frac{v}{\sqrt{2}} f_1 \left(\xi^\dagger M_i \xi\right) + \sqrt{2} v F_3 G_i , \qquad \text{where} \quad F_3' = f_3 , \tag{8}$$

and the Hamiltonian can be written in a manifestly non-negative form

$$\mathcal{H} = \left|D_i H\right|^2 + \left|\frac{1}{\rho} f_1 \left(\mathbb{1} - \xi\xi^\dagger\right) M_i \xi + \frac{1}{2} f_1' (\xi^\dagger M_i \xi)\xi + f_3 G_i \xi\right|^2 . \tag{9}$$

Now, when we know the BPS Hamiltonian, we can write down the corresponding Lagrangian:

$$\begin{aligned} \mathcal{L}_{\mathrm{BPS}} &= |D_\mu H|^2 - \frac{v^2}{4g^2|H|^2} h^2 \left(\mathrm{Tr}\left[F_{\mu\nu}^2\right] - \frac{2}{|H|^4} \mathrm{Tr}\left[F_{\mu\nu} H H^\dagger\right]^2 \right) \\ &\quad - \frac{1}{4} \left(\frac{h'}{g|H|^2} \mathrm{Tr}\left[F_{\mu\nu} H H^\dagger\right] + \frac{f'}{g'} B_{\mu\nu} \right)^2 , \end{aligned} \tag{10}$$

where we rescaled for convenience the functions f_1 and f_3 in favor of functions f' and h as

$$f_1 = \frac{\sqrt{2}}{g} h , \qquad f_3 = \frac{1}{\sqrt{2} g'} f' . \tag{11}$$

We assume $f'(1) = h(1) = 1$ in order to have standard normalization of kinetic term in vacuum ($\rho = 1$). The BPS equation reads

$$D_i H = \frac{\sqrt{2}}{g\rho} h(\rho) \left(M_i - \xi^\dagger M_i \xi\right)\xi + \frac{1}{\sqrt{2} g} h'(\rho) \left(\xi^\dagger M_i \xi\right)\xi + \frac{1}{\sqrt{2} g'} f'(\rho)\, G_i \,\xi . \tag{12}$$

A consequence of this equation is that the $U(1)_{\mathrm{Y}}$ gauge field B_i has the form

$$B_i = 2\mathrm{i}\xi^\dagger \left(\partial_i + \frac{1}{2} \mathrm{i} A_i^a \sigma^a \right) \xi , \tag{13}$$

i.e., it is entirely expressed in terms of the $SU(2)_{\mathrm{L}}$ gauge field A_i^a and ξ.

If the BPS equation is satisfied, the Hamiltonian is given only by the surface term, i.e.,

$$\mathcal{H} = \partial_i \left[\frac{v}{g} h(\rho)\, \xi^\dagger M_i \xi + \frac{v}{g'} f(\rho)\, G_i \right] , \tag{14}$$

i.e., the Bogomol'nyi bound is saturated.

3. Exact monopole solution - An example

The BPS equation is first order, not second order as a usual equation of motion, so there is a hope that it might be exactly solvable, at least for suitable f' and h. Indeed, several exact solutions have been found [8].

Adopting the spherically symmetric Ansatz for the $SU(2)_{\mathrm{L}}$ gauge field

$$A_i^a = \left(1 - K\right) \varepsilon_{iak} \frac{x_k}{r^2} , \tag{15a}$$

where $K = K(r)$ is some function of the radial coordinate, and assuming ξ to have the form (in spherical coordinates)

$$\xi = \mathrm{i} \begin{pmatrix} \sin(\theta/2)\, \mathrm{e}^{-\mathrm{i}\varphi} \\ -\cos(\theta/2) \end{pmatrix} \tag{15b}$$

(so that ξ is a member of the lowest non-trivial homotopy class), the problem is reduced to finding the two unknown functions of the radial coordinate, K and ρ, subject to the boundary conditions

$$K(0) = 1\,, \qquad \rho(0) = 0\,, \tag{16a}$$
$$K(\infty) = 0\,, \qquad \rho(\infty) = 1\,, \tag{16b}$$

dictated by requirements of regularity of the solution in the origin and of finiteness of the total energy. It is not hard to show that with this Anstaz the equation (13) for B_i yields

$$B_i = 2i\xi^\dagger\partial_i\xi = (1-\cos\theta)\partial_i\varphi \;=\; -\varepsilon_{ij3}\partial_j\log(r+z)\,, \tag{17}$$

regardless of the form of K.

Here we present, for the sake of an illustration, only the simplest solution, corresponding to the BPS theory defined by functions

$$f'(\rho) \;=\; \rho^2 \;=\; \frac{2|H|^2}{v^2}\,, \qquad h(\rho) \;=\; 1\,.$$

The resulting ρ and K read

$$\rho(r) \;=\; \frac{1}{1+\dfrac{1}{\mu r}}\,, \tag{18a}$$
$$K(r) \;=\; \exp\left\{-\frac{g}{g'}\left[\frac{\mu r}{2}\,\frac{2+\mu r}{1+\mu r} - \log\left(1+\mu r\right)\right]\right\}, \tag{18b}$$

where we defined for convenience the scale $\mu \equiv vg' \approx 86.1$ GeV. This solution corresponds to a monopole with the magnetic charge

$$q \;=\; \frac{4\pi}{e}\,, \tag{19}$$

where $e = gg'/\sqrt{g^2+g'^2}$ is the electric charge, and mass

$$M \;=\; 4\pi v\left(\frac{1}{2g}+\frac{1}{3g'}\right) \;\approx\; 5.32\ \text{TeV}\,. \tag{20}$$

4. Monopole mass and its lower bound

Recall that energy density $\mathcal{H}$ is given by expression (14). Since it is a total derivative, the total energy (mass) of the system $M = \int_{\mathbb{R}^3}\mathrm{d}^3x\,\mathcal{H}$ doesn't depend, due to the Gauss–Ostrogradsky theorem, on details of A_i^a, B_i and H in the whole space volume, but only on their asymptotic behavior.

For our spherically symmetric Ansatz (15) this asymptotic behavior is given solely by the boundary conditions (16). Thus, it is possible, without having to solve the equations of motion at all, to arrive at a simple formula for the magnetic monopole mass:

$$M \;=\; 4\pi v\left[\frac{1}{2g}+\frac{1}{g'}\left|\int_0^1\mathrm{d}\rho\, f'(\rho)\right|\right]. \tag{21}$$

(Curiously, this expression depends only on f', not on h.) From this result it immediately follows that there is a lower bound on the mass of the monopole:

$$M \;\geq\; \frac{2\pi v}{g} \;\approx\; 2.37\ \text{TeV}\,. \tag{22}$$

Journal of Physics: Conference Series **1416** (2019) 012004 doi:10.1088/1742-6596/1416/1/012004

Thus, in a BPS theory (10) the mass of a magnetic monopole cannot be lower than $2\pi v/g$, regardless of the choice of functions f' and h. However, the BPS theory doesn't describe the real world. The question therefore arises whether this bound holds also in a more realistic non-BPS theory, that is, in particular, in the SM with CKY regularization [4], whose bosonic part reads

$$\mathcal{L}_{\rm CKY} \;=\; |D_\mu H|^2 - V\big(|H|^2\big) - \frac{1}{4g^2}\,{\rm Tr}\left[F_{\mu\nu}^2\right] - \frac{1}{4g'^2}\epsilon(\rho) B_{\mu\nu}^2\,, \tag{23}$$

where, again, $\rho = \sqrt{2}|H|/v$.

Usually (e.g., for the 't Hooft–Polyakov monopole) the only difference between a BPS and non-BPS theory is lack or presence of the scalar potential. Since it is positive, the energy density (Hamiltonian) of a BPS theory is never greater than that of a non-BPS theory. Thus, any lower mass bound holding in the former holds necessarily also in the latter.

In the present case the situation is more complicated. One might wish that

$$\mathcal{H}_{\rm BPS}[f', h] \;\leq\; \mathcal{H}_{\rm CKY}[\epsilon] \tag{24}$$

for *arbitrary* f', h and ϵ. This, however, turns out not to be the case. Nevertheless, this requirement is in fact too strong anyway. It suffices to demand that for each ϵ there exist f' and h such that the inequality (24) holds. This weaker requirement already proves to be true. E.g., for f', h defined in terms of ϵ as

$$f'(\rho) \;=\; \sqrt{\frac{3}{4}\epsilon(\rho)}\,, \qquad h(\rho) \;=\; \rho \tag{25}$$

the inequality (24) holds indeed. This proves that the magnetic monopole bound (22) holds also in a more realistic non-BPS theory.

5. Conclusions

In this note, we have presented a family of effective extensions of the SM that contain BPS magnetic monopoles as classical solutions. These solutions are the BPS extensions of Cho–Maison monopoles presented in [4] and further studied in [7]. We have also presented an explicit example of monopole solution in analytic form.

Critically, we obtained a universal lower bound on the mass of the monopole as $M \geq 2\pi v/g \approx 2.37$ TeV and argued that this bound also applies to phenomenologically relevant non-BPS models, in particular to the CKY model presented in [4] and [7].

Our work opens up the possibility of studying multi-particle configurations of Cho–Maison monopoles, which are sometimes characterized as conceptually being something between Dirac's monopole and the 't Hooft–Polyakov monopole. To construct a multi-monopole configuration of Dirac's monopoles is not particularly challenging. On the other hand, to obtain a 't Hooft–Polyakov multi-monopole configuration requires the use of highly sophisticated tools, such as the Nahm construction [9]. It would make an interesting future study to elaborate how difficult is to write down a multi-monopole configuration of Cho–Maison monopoles in the BPS limit and whether one can adopt the Nahm approach or other techniques.

Furthermore, our model can be a useful tool to explore non-topological solutions of the electroweak model, such as the spharelon [10]. Indeed, the BPS limit lends itself to the possibility that we can find unstable static solutions in the analytic form. We plan to investigate this in the future work.

As far as we know our theory is not a bosonic part of any known supersymmetric theory. It would be interesting to see if there actually exists a SUSY extension of our model and whether the BPS Cho–Maison monopole can be constructed as a 1/2, 1/4, or other SUSY state. We leave this puzzle for the future.

Journal of Physics: Conference Series **1416** (2019) 012004 doi:10.1088/1742-6596/1416/1/012004

Acknowledgments

This work was supported by the Albert Einstein Centre for Gravitation and Astrophysics financed by the Czech Science Agency Grant No. 14-37086G (F. B.) and by the program of Czech Ministry of Education, Youth and Sports INTEREXCELLENCE Grant No. LTT17018 (F. B., P. B.). One of the authors, P. B., thanks TJ Balvan Praha for support and the organizers of the ISQS-26 for the opportunity to speak at the conference.

References

[1] 't Hooft G 1974 *Nucl. Phys.* **B79** 276–84
[2] Polyakov A M 1974 *JETP Lett.* **20** 194–5 [Pisma Zh. Eksp. Teor. Fiz.20,430(1974)]
[3] Cho Y M and Maison D 1997 *Phys. Lett.* **B391** 360–5 (*Preprint* `hep-th/9601028`)
[4] Cho Y M, Kim K and Yoon J H 2015 *Eur. Phys. J.* **C75** 67 (*Preprint* `1305.1699`)
[5] Arai M, Blaschke F, Eto M and Sakai N 2017 (*Preprint* `1703.00351`)
[6] Arai M, Blaschke F, Eto M and Sakai N 2017 *PTEP* **2017** 053B01 (*Preprint* `1703.00427`)
[7] Ellis J, Mavromatos N E and You T 2016 *Phys. Lett.* **B756** 29–35 (*Preprint* `1602.01745`)
[8] Blaschke F and Beneš P 2018 *PTEP* **2018** 073B03 (*Preprint* `1711.04842`)
[9] Nahm W 1981 *12th NATO Advanced Summer Institute on Theoretical Physics: Structural Elements in Particle Physics and Statistical Mechanics Freiburg, Germany, August 31-September 11, 1981* p 301
[10] Klinkhamer F R and Manton N S 1984 *Phys. Rev.* **D30** 2212

Journal of Physics: Conference Series **1416** (2019) 012005 doi:10.1088/1742-6596/1416/1/012005

Dynamical generation of flavor vacuum and Lorentz invariance

M Blasone[1], **P Jizba**[2] **and L Smaldone**[1]

[1] Dipartimento di Fisica, Università di Salerno, Via Giovanni Paolo II, 132 84084 Fisciano, Italy & INFN Sezione di Napoli, Gruppo collegato di Salerno, Italy

[2] FNSPE, Czech Technical University in Prague, Břehová 7, 115 19 Praha 1, Czech Republic

E-mail: blasone@sa.infn.it

E-mail: p.jizba@fjfi.cvut.cz

E-mail: luca.smaldone1991@gmail.com

Abstract. In this paper we review dynamical generation of field mixing after chiral symmetry breaking. We also study the explicit form of Lorentz boosts transformations of flavor states in a two-flavor scalar model with field mixing. We find that Lorentz symmetry is spontaneously broken on flavor vacuum because of its dynamically generated condensate structure.

1. Introduction

Lorentz invariance of flavor mixing and oscillations has been considered in modern theoretical physics literature [1, 2, 3, 4]. Within the context of neutrino oscillations in quantum field theory (QFT) [5, 6, 7, 8, 9, 10, 11], deformations of the Lorentz dispersion relations were considered in Ref. [2, 3]. This result was achieved implementing the formalism of Ref. [12, 13]. Similar deformations were also found in string theory [14], which predicts deformations of the Lorentz dispersion relation of the order of some power of E/M_P, where E is the energy of a particle and M_P is the Planck mass. In that context, the scattering between open strings and $D0$-branes is described by an effective action and field mixing is dynamically generated via flavor vacuum condensates [15, 16, 17, 18, 19, 20]. This mechanism was generalized in Refs. [21, 22], with algebraic methods.

Standard Model extension (SME) [23] also derives from effective string theory scenarios. In that context, Lorentz violating terms were added to the Standard Model Lagrangian, looking at possible phenomenological consequences, e.g., for neutrino oscillations [24] and for other physical systems [25, 26, 27]. Bounds on SME can be also put in connection [28] with the generalized uncertainty principle [29, 30, 31, 32, 33].

In this paper we briefly review the dynamical mechanism of flavor vacuum condensate and field mixing generation presented in Ref. [21]. As a consequence, Lorentz symmetry is spontaneously broken. We here limit our study to the case of Lorentz boosts in a simple effective model of two flavor scalar fields with mixing [34, 35]. We explicitly show that this model presents a spontaneous symmetry breaking (SSB) of Lorentz symmetry, which is an exact symmetries of the Lagrangian. To this end, the transformation properties of flavor creation and annihilation

Journal of Physics: Conference Series **1416** (2019) 012005 doi:10.1088/1742-6596/1416/1/012005

operators are explicitly derived. It is also shown that QFT flavor oscillation formula is Lorentz invariant.

The paper is organized as it follows: In Section 2 we review the mechanism of dynamical mixing generation described in Ref. [21]. In Section 3 quantization of flavor (scalar) fields is reviewed [34, 35]. Lorentz boosts in the flavor representation are studied in Section 4. Finally, in Section 5, conclusions and perspectives are presented.

2. Dynamical field mixing generation

Following Ref.[21], let us take a Lagrangian density $\mathcal{L}$ which is invariant under the global *chiral-flavor* group $G = SU(2)_L \times SU(2)_R \times U(1)_V$. We here introduce the fermion flavor doublet

$$\psi = \begin{bmatrix} \tilde{\psi}_1 \\ \tilde{\psi}_2 \end{bmatrix} . \tag{1}$$

The action of a generic chiral-group transformation $\mathbf{g}$ on ψ is given by [36]

$$\psi' = \mathbf{g}\psi = \exp\left[i\left(\phi + \boldsymbol{\omega}\cdot\frac{\boldsymbol{\sigma}}{2} + \boldsymbol{\omega}_5\cdot\frac{\boldsymbol{\sigma}}{2}\gamma_5\right)\right]\psi . \tag{2}$$

Here $\sigma_j\,, j = 1,2,3$ are the Pauli matrices and $\phi, \boldsymbol{\omega}, \boldsymbol{\omega}_5$ are real-valued transformation parameters of G. From Noether's theorem it follows that the vector and axial currents

$$J^\mu = \overline{\psi}\gamma^\mu\psi\,, \qquad \boldsymbol{J}^\mu = \overline{\psi}\gamma^\mu\frac{\boldsymbol{\sigma}}{2}\psi\,, \qquad \boldsymbol{J}_5^\mu = \overline{\psi}\gamma^\mu\gamma_5\frac{\boldsymbol{\sigma}}{2}\psi\,, \tag{3}$$

and the ensuing charges

$$Q = \int d^3\mathbf{x}\,\psi^\dagger\psi\,, \qquad \boldsymbol{Q} = \int d^3\mathbf{x}\,\psi^\dagger\frac{\boldsymbol{\sigma}}{2}\psi\,, \qquad \boldsymbol{Q}_5 = \int d^3\mathbf{x}\,\psi^\dagger\frac{\boldsymbol{\sigma}}{2}\gamma_5\psi\,, \tag{4}$$

are conserved quantities. The Lie algebra of the chiral-flavor group G is thus recovered:

$$\begin{aligned} &[Q_i, Q_j] = i\,\varepsilon_{ijk}Q_k\,, \quad [Q_i, Q_{5,j}] = i\,\varepsilon_{ijk}Q_{5,k}\,, \quad [Q_{5,i}, Q_{5,j}] = i\,\varepsilon_{ijk}Q_k\,, \\ &[Q, Q_{5,j}] = [Q, Q_j] = 0\,. \end{aligned} \tag{5}$$

Here $i,j,k = 1,2,3$ and ε_{ijk} is the Levi-Civita pseudo-tensor.

SSB is characterized by the existence of some local operator(s) $\phi(x)$ [36] so that

$$\langle[N_i, \phi(0)]\rangle = \langle\varphi_i(0)\rangle \equiv v_i \neq 0\,, \tag{6}$$

where $\langle\ldots\rangle \equiv \langle\Omega|\ldots|\Omega\rangle$ and $|\Omega\rangle$ is the vacuum state. Here v_i are the *order parameters* and N_i represent group generators from the quotient space G/H, with H being the vacuum stability group. In our case N_i will be identified with $\boldsymbol{Q}$ and $\boldsymbol{Q}_5$ according to the SSB scheme discussed.

By analogy with quark condensation in QCD [36], we will limit our considerations to order parameters that are condensates of fermion-antifermion pairs. In order to simplify the notation, we introduce the following composite operators

$$\Phi_k = \overline{\psi}\sigma_k\psi\,, \qquad \Phi_k^5 = \overline{\psi}\sigma_k\gamma_5\psi\,, \qquad \sigma_0 \equiv \mathbb{1}\,, \tag{7}$$

with $k = 0,1,2,3$. For simplicity we now assume $\langle\boldsymbol{\Phi}^5\rangle = 0$.

Let us now consider three specific SSB schemes $G \to H$:

Journal of Physics: Conference Series **1416** (2019) 012005 doi:10.1088/1742-6596/1416/1/012005

i) SSB sequence corresponding to a single mass generation is [36, 37]

$$SU(2)_L \times SU(2)_R \times U(1)_V \longrightarrow U(2)_V \,. \tag{8}$$

The broken-phase symmetry (which corresponds to dynamically generated mass matrix $M = m_0 \mathbb{I}$) is characterized by the order parameter

$$\langle \Phi_0 \rangle \;=\; v_0 \neq 0 \,, \qquad \langle \Phi_k \rangle \;=\; 0 \,, \quad k = 1,2,3 \,. \tag{9}$$

One can easily check that this is invariant under the residual symmetry group $H = U(2)_V$ but not under the full chiral group G.

ii) As a second case we consider the SSB scheme

$$SU(2)_L \times SU(2)_R \times U(1)_V \longrightarrow U(1)_V \times U(1)^3_V \,, \tag{10}$$

which is responsible for the dynamical generation of different masses m_1, m_2. In this case the order parameters take the form

$$\langle \Phi_0 \rangle \;=\; v_0 \neq 0 \,, \qquad \langle \Phi_3 \rangle \;=\; v_3 \neq 0 \,. \tag{11}$$

iii) Finally, we consider the SSB pattern

$$SU(2)_L \times SU(2)_R \times U(1)_V \longrightarrow U(1)_V \times U(1)^3_V \longrightarrow U(1)_V \,, \tag{12}$$

which is responsible for the dynamical generation of field mixing.

We now define

$$\Phi_{k,m} \;=\; \overline{\psi}\, \sigma_k \, \psi \,, \qquad k = 1,2,3 \,, \tag{13}$$

where m indicates that ψ is now a doublet of fields $\psi = [\psi_1 \, \psi_2]^T$ in the mass basis. The SSB condition now reads

$$\langle \Phi_{1,m} \rangle \;\equiv\; v_{1,m} \neq 0 \,. \tag{14}$$

Therefore, we understand that a *necessary condition* for dynamical generation of field mixing within chiral symmetric systems, is the presence of pairs in the vacuum which mix fermions and antifermions with different masses [21]:

$$\langle \overline{\psi}_i(x) \, \psi_j(x) \rangle \neq 0 \,, \qquad i \neq j \,. \tag{15}$$

In other words, *field mixing generation requires mixing at the level of the vacuum condensate structure.* This statement is consistent with similar results obtained in the QFT treatment of neutrino oscillations. There, a *flavor vacuum* was defined, which has a non-trivial condensate structure, where neutrinos and antineutrinos with different masses are mixed [5, 6, 7, 8, 9, 10, 11]. Moreover, the present result agrees with Ref. [15, 16, 17, 18], where the condensate structure was dynamically generated by string-brane scattering. We remark that the above achievement is almost model independent (the only strong hypothesis made was the global chiral symmetry), and it is non-perturbative.

We can see the strong analogy with neutrino physics [6, 7, 8, 9, 10, 11], by looking at mean-field approximation [21]. In that case, the vacuum responsible for (14) is the aforementioned flavor vacuum

$$\begin{aligned} |0\rangle_{e,\mu} \;=\; \prod_{\mathbf{k}} \prod_{r} \Big[& (1 - \sin^2\theta \, V^2_{\mathbf{k}}) - \eta^r \sin\theta \, \cos\theta \, V_{\mathbf{k}} (\alpha^{r\dagger}_{\mathbf{k},1} \beta^{r\dagger}_{-\mathbf{k},2} + \alpha^{r\dagger}_{\mathbf{k},2} \beta^{r\dagger}_{-\mathbf{k},1}) \\ & + \eta^r \sin^2\theta \, V_{\mathbf{k}} U_{\mathbf{k}} (\alpha^{r\dagger}_{\mathbf{k},1} \beta^{r\dagger}_{-\mathbf{k},1} - \alpha^{r\dagger}_{\mathbf{k},2} \beta^{r\dagger}_{-\mathbf{k},2}) + \sin^2\theta \, V^2_{\mathbf{k}} \alpha^{r\dagger}_{\mathbf{k},1} \beta^{r\dagger}_{-\mathbf{k},2} \alpha^{r\dagger}_{\mathbf{k},2} \beta^{r\dagger}_{-\mathbf{k},1} \Big] |0\rangle_{1,2} \,, \end{aligned} \tag{16}$$

where $\alpha^{r\dagger}_{\mathbf{k},j}$, $\beta^{r\dagger}_{\mathbf{k},j}$, $\alpha^{r}_{\mathbf{k},j}$, $\beta^{r}_{\mathbf{k},j}$ are the ladder operators, and

$$U_{\mathbf{k}} \;=\; \mathcal{A}_{\mathbf{k}}\left(1+\frac{|\mathbf{k}|^2}{(\omega_{\mathbf{k},1}+m_1)\,(\omega_{\mathbf{k},2}+m_2)}\right), \quad V_{\mathbf{k}} \;=\; \mathcal{A}_{\mathbf{k}}\left(\frac{|\mathbf{k}|}{\omega_{\mathbf{k},1}+m_1}-\frac{|\mathbf{k}|}{\omega_{\mathbf{k},2}+m_2}\right), \tag{17}$$

with $\mathcal{A}_{\mathbf{k}}=\sqrt{\frac{(\omega_{\mathbf{k},1}+m_1)(\omega_{\mathbf{k},2}+m_2)}{4\omega_{\mathbf{k},1}\omega_{\mathbf{k},2}}}$.

In this case [21]:

$$v_{1,m} \;=\; 2\sin 2\theta \int \! d^3\mathbf{k}\left(\frac{m_2}{\omega_{\mathbf{k},2}}-\frac{m_1}{\omega_{\mathbf{k},1}}\right). \tag{18}$$

Let us remark that QFT approach to flavor oscillations can be equally well applied to the study of the boson mixing [35]. Therefore, just for technical simplicity, we here study the properties of flavor vacuum in this simplified framework. However, it is implicit that the origin of such condensate and the consequent Lorentz SSB have to be retraced in a dynamical generation mechanism similar to the one just described.

3. Scalar field mixing

Let us consider the effective Lagrange density

$$\mathcal{L}(x) \;=\; \partial^\mu\varphi_f^\dagger(x)\partial_\mu\varphi_f(x) \;-\; \varphi_f^\dagger(x)\,M^2\,\varphi_f(x)\,, \tag{19}$$

where

$$\varphi_f(x) \;=\; \begin{bmatrix}\varphi_A(x)\\ \varphi_B(x)\end{bmatrix}, \qquad M^2 \;=\; \begin{bmatrix} m_A^2 & m_{AB}^2\\ m_{AB}^2 & m_B^2\end{bmatrix}, \tag{20}$$

which describes the dynamics of two linearly coupled (mixed) scalar fields. Here we use the name *flavor fields*, in analogy with the case of quark and neutrino physics. The Lagrange density (20) can be diagonalized with the transformation:

$$\varphi_f(x) \;=\; U\,\varphi_m(x)\,, \qquad U=\begin{bmatrix}\cos\theta & \sin\theta\\ -\sin\theta & \cos\theta\end{bmatrix}, \tag{21}$$

where $\tan 2\theta \;=\; 2m_{AB}^2/\left(m_B^2-m_A^2\right)$. Therefore, $\mathcal{L}$ becomes

$$\mathcal{L}(x) \;=\; \partial^\mu\varphi_m^\dagger(x)\,\partial_\mu\varphi_m(x) \;-\; \varphi_m^\dagger(x)\,M_d^2\,\varphi_m(x)\,, \tag{22}$$

where

$$\varphi_m(x) \;=\; \begin{bmatrix}\varphi_1(x)\\ \varphi_2(x)\end{bmatrix}, \qquad M_d^2 \;=\; \begin{bmatrix} m_1^2 & 0\\ 0 & m_2^2\end{bmatrix}. \tag{23}$$

The Lagrange density (22) describes two free scalar fields with masses m_1 and m_2. These can be Fourier expanded as:

$$\varphi_j(x) \;=\; \int\frac{d^3\mathbf{k}}{2\omega_{\mathbf{k},j}(2\pi)^3}\left[a_{\mathbf{k},j}\,e^{-i\omega_{\mathbf{k},j}t} \;+\; b^\dagger_{-\mathbf{k},j}\,e^{i\omega_{\mathbf{k},j}t}\right]e^{i\mathbf{k}\cdot\mathbf{x}}\,, \quad j=1,2$$

where ladder operators satisfy the Lorentz invariant commutation relations:

$$\left[a_{\mathbf{k},i}\,,\,a^\dagger_{\mathbf{p},j}\right] \;=\; \left[b_{\mathbf{k},i}\,,\,b^\dagger_{\mathbf{p},j}\right] \;=\; 2\omega_{\mathbf{k},i}\,(2\pi)^3\,\delta(\mathbf{k}-\mathbf{p})\,\delta_{ij}\,. \tag{24}$$

The *mass vacuum* is defined by the relations:

$$a_{\mathbf{k},j}|0\rangle_{1,2} \;=\; b_{\mathbf{k},j}|0\rangle_{1,2} \;=\; 0\,. \tag{25}$$

We now expand flavor fields in a similar way:

$$\varphi_\sigma(x) \;=\; \int \frac{\mathrm{d}^3\mathbf{k}}{2\omega_{\mathbf{k},\sigma}(2\pi)^3}\left[a_{\mathbf{k},\sigma}(t)\,e^{-i\omega_{\mathbf{k},\sigma}t} \;+\; b^\dagger_{-\mathbf{k},\sigma}(t)\,e^{i\omega_{\mathbf{k},\sigma}t}\right]e^{i\mathbf{k}\cdot\mathbf{x}}\,,\qquad \sigma=A,B \tag{26}$$

with $\omega_{\mathbf{k},\sigma}=\sqrt{|\mathbf{k}|^2+\mu_\sigma^2}$ and μ_σ are mass parameters which have to be specified. From mixing transformation (21), it follows[1]:

$$a_{\mathbf{k},A}(t) \;=\; \int \mathrm{d}^3\mathbf{x}\,e^{i(\omega_{\mathbf{k},A}t-\mathbf{k}\cdot\mathbf{x})}\,i\,\overleftrightarrow{\partial_0}\left(\cos\theta\varphi_1(x)\;+\;\sin\theta\varphi_2(x)\right)\,, \tag{27}$$

and similar expressions work for the other operators. Explicitly:

$$\begin{bmatrix} a_{\mathbf{k},A}\\ b^\dagger_{-\mathbf{k},A}\\ a_{\mathbf{k},B}\\ b^\dagger_{-\mathbf{k},B}\end{bmatrix} = \begin{bmatrix} c_\theta\,\rho^{\mathbf{k}}_{A1} & c_\theta\,\lambda^{\mathbf{k}}_{A1} & s_\theta\,\rho^{\mathbf{k}}_{A2} & s_\theta\,\lambda^{\mathbf{k}}_{A2}\\ c_\theta\,\lambda^{\mathbf{k}}_{A1} & c_\theta\,\rho^{\mathbf{k}}_{A1} & s_\theta\,\lambda^{\mathbf{k}}_{A2} & s_\theta\,\rho^{\mathbf{k}}_{A2}\\ -s_\theta\,\rho^{\mathbf{k}}_{B1} & -s_\theta\,\lambda^{\mathbf{k}}_{B1} & c_\theta\,\rho^{\mathbf{k}}_{B2} & c_\theta\,\lambda^{\mathbf{k}}_{B2}\\ -s_\theta\,\lambda^{\mathbf{k}}_{B1} & -s_\theta\,\rho^{\mathbf{k}}_{B1} & c_\theta\,\lambda^{\mathbf{k}}_{B2} & c_\theta\,\rho^{\mathbf{k}}_{B2}\end{bmatrix}\begin{bmatrix} a_{\mathbf{k},1}\\ b^\dagger_{-\mathbf{k},1}\\ a_{\mathbf{k},2}\\ b^\dagger_{-\mathbf{k},2}\end{bmatrix}, \tag{28}$$

where $c_\theta\equiv\cos\theta$, $s_\theta\equiv\sin\theta$,

$$\rho^{\mathbf{k}}_{\sigma j} \;=\; |\rho^{\mathbf{k}}_{\sigma j}|e^{i(\omega_{\mathbf{k},\sigma}-\omega_{\mathbf{k},j})t}\,,\qquad \lambda^{\mathbf{k}}_{\sigma j} \;=\; |\lambda^{\mathbf{k}}_{\sigma j}|e^{i(\omega_{\mathbf{k},\sigma}+\omega_{\mathbf{k},j})t}\,,\quad (\sigma,j)=(A,1),(B,2), \tag{29}$$

and

$$|\rho^{\mathbf{k}}_{\sigma j}| \;=\; \frac{1}{2}\left(\frac{\omega_{\mathbf{k},\sigma}}{\omega_{\mathbf{k},j}}+1\right)\,,\qquad |\lambda^{\mathbf{k}}_{\sigma j}| \;=\; \frac{1}{2}\left(\frac{\omega_{\mathbf{k},\sigma}}{\omega_{\mathbf{k},j}}-1\right)\,. \tag{30}$$

One can verify that Eq.(28) is a canonical transformation:

$$\left[a_{\mathbf{k},\sigma}(t)\,,\,a^\dagger_{\mathbf{p},\rho}(t)\right] \;=\; \left[b_{\mathbf{k},\sigma}(t)\,,\,b^\dagger_{\mathbf{p},\rho}(t)\right] \;=\; 2\omega_{\mathbf{k},\sigma}\,(2\pi)^3\,\delta(\mathbf{k}-\mathbf{p})\,\delta_{\sigma\rho}\,. \tag{31}$$

As it was previously mentioned, mass parameters μ_σ have to be specified [6, 7, 8, 9, 10, 11, 39]: as in QFT in curved spacetime [40], unitarily inequivalent representations of canonical commutation relation are connected by improper Bogolyubov transformations. In Ref. [41], it was shown that different mass parameters μ_σ correspond to different forms of the Casimir force between two plates.

Flavor vacuum is defined, at a fixed time t, as

$$a_{\mathbf{k},\sigma}(t)\,|0(t)\rangle_{A,B} \;=\; b_{\mathbf{k},\sigma}(t)\,|0(t)\rangle_{A,B} \;=\; 0\,. \tag{32}$$

This is characterized by a boson-condensate structure, similar to the one shown in Eq. (16). In the case $\mu_A=m_1,\mu_B=m_2$ one has [35]:

$${}_{A,B}\langle 0(t)|a^\dagger_{\mathbf{k},j}\,a_{\mathbf{k},j}|0(t)\rangle_{A,B} \;=\; {}_{A,B}\langle 0(t)|b^\dagger_{-\mathbf{k},j}\,b_{-\mathbf{k},j}|0(t)\rangle_{A,B} \;=\; \sin^2\theta\,|V_{\mathbf{k}}|^2\,2\,\omega_{\mathbf{k},j}\,(2\pi)^3\,, \tag{33}$$

where $|V_{\mathbf{k}}|\;\equiv|\lambda_{\mathbf{k}}|_{\mu_A=m_1,\mu_B=m_2}$, analogous to the function encountered in Eq.(17). This structure is responsible for the Lorentz symmetry breaking discussed below.

[1] Here the time dependence of creation and annihilation operators indicates that flavor fields are interacting fields. Actually, this interacting model can be solved exactly, without perturbation expansion.

Journal of Physics: Conference Series **1416** (2019) 012005 doi:10.1088/1742-6596/1416/1/012005

Flavor states are defined as excitations of the flavor vacuum:

$$|a_{\mathbf{k},\sigma}(t)\rangle \;\equiv\; a^{\dagger}_{\mathbf{k},\sigma}(t)|0(t)\rangle_{A,B}\,. \tag{34}$$

These are eigenstates of the flavor charges

$$\begin{aligned} Q_\sigma(t) &= i\int \mathrm{d}^3\mathbf{x} :\varphi^{\dagger}_\sigma(x)\,\overleftrightarrow{\partial_0}\,\varphi_\sigma(x): \\ &= \int \frac{\mathrm{d}^3\mathbf{k}}{2\omega_{\mathbf{k},\sigma}\,(2\pi)^3}\left(a^{\dagger}_{\mathbf{k},\sigma}(t)\,a_{\mathbf{k},\sigma}(t) - b^{\dagger}_{\mathbf{k},\sigma}(t)\,b_{\mathbf{k},\sigma}(t)\right), \end{aligned} \tag{35}$$

at fixed time t

$$Q_\sigma(t)\,|a_{\mathbf{k},\sigma}(t)\rangle \;=\; |a_{\mathbf{k},\sigma}(t)\rangle\,, \qquad \sigma = A, B. \tag{36}$$

4. Flavor vacuum and Lorentz invariance

Let us now analyze the Lorentz boost transformation properties of flavor operators. The generator of a boost along the l-th axis is:

$$K_l \;=\; \int \mathrm{d}^3\mathbf{x}\left(x^0\,T^{0l} - x^l\,T^{00}\right). \tag{37}$$

This can also be written as

$$K_l \;=\; x^0\,P^l \;-\; \int \mathrm{d}^3\mathbf{x}\,x^l\,\mathcal{H}. \tag{38}$$

In our case:

$$\begin{aligned} K_l \;=\; &\Big[x^0\int \mathrm{d}^3\mathbf{x}\left(\boldsymbol{\pi}^{\dagger}_f(x)\,\partial^l\boldsymbol{\varphi}_f(x) + \partial^l\boldsymbol{\varphi}^{\dagger}_f(x)\,\boldsymbol{\pi}_f(x)\right) \\ &- \int \mathrm{d}^3\mathbf{x}\,x^l\left(\boldsymbol{\pi}^{\dagger}_f(x)\,\boldsymbol{\pi}_f(x) + \nabla\boldsymbol{\varphi}^{\dagger}_f(x)\cdot\nabla\boldsymbol{\varphi}_f(x) + \boldsymbol{\varphi}^{\dagger}_f(x)\,M^2\,\boldsymbol{\varphi}_f(x)\right)\Big], \end{aligned} \tag{39}$$

A generic boost can be thus expressed as:

$$U(L) \;=\; \exp\left(-i\sum_{l=1}^{3}\xi^l\,K_l\right), \tag{40}$$

where $L(\xi)$ indicates the Lorentz transformation:

$$x^\mu \;\to\; x'^\mu \;=\; L^\mu{}_\nu(\xi)\,x^\nu\,. \tag{41}$$

Then, we get

$$U(L)\,\varphi_\sigma(x)\,U^{-1}(L) \;=\; \varphi_\sigma\left(x'\right), \tag{42}$$

i.e. φ_σ behaves as scalar under Lorentz boosts. From Eq.(26) we get:

$$\varphi_\sigma\left(x'\right) \;=\; \int \frac{\mathrm{d}^3\mathbf{k}}{2\omega_{\mathbf{k},\sigma}(2\pi)^3}\left[a_{\mathbf{k},\sigma}\left(t'\right)e^{-ik\,x'} + b^{\dagger}_{\mathbf{k},\sigma}\left(t'\right)e^{ik\,x'}\right]. \tag{43}$$

Here and in the following we formally use the notation $L\mathbf{k}$ to indicate $L^j_\mu k^\mu$ (j=1,2,3), respectively. This equation should be actually written in the form:

$$\varphi_\sigma\left(x'\right) \;=\; \int \frac{\mathrm{d}^4k}{(2\pi)^4}\,(2\pi)\,\delta^4(k^2-\mu^2_\sigma)\theta(k_0)\left[a_{k,\sigma}\left(t'\right)e^{-ik\,x'} + b^{\dagger}_{k,\sigma}\left(t'\right)e^{ik\,x'}\right]. \tag{44}$$

Performing the change of variables $k \to k' = L^{-1}k$ [38], we get:

$$\varphi_\sigma(x') = \int \frac{\mathrm{d}^4 k}{(2\pi)^4} (2\pi)\, \delta^4(k^2 - \mu_\sigma^2)\theta(k_0) \left[a_{Lk,\sigma}(t')\, e^{-ikx} + b^\dagger_{Lk,\sigma}(t')\, e^{ikx} \right]. \quad (45)$$

By integrating once more on k_0 we find:

$$\varphi_\sigma(x') = \int \frac{\mathrm{d}^3 \mathbf{k}}{2\omega_{\mathbf{k},\sigma}(2\pi)^3} \left[a_{L\mathbf{k},\sigma}(t')\, e^{-ikx} + b^\dagger_{L\mathbf{k},\sigma}(t')\, e^{ikx} \right]. \quad (46)$$

Therefore we get

$$U(L)\, a_{\mathbf{k},\sigma}(t)\, U^{-1}(L) = a_{L\mathbf{k},\sigma}(t'), \quad (47)$$

$$U(L)\, b_{\mathbf{k},\sigma}(t)\, U^{-1}(L) = b_{L\mathbf{k},\sigma}(t'). \quad (48)$$

We see immediately that flavor vacuum is not left unchanged by a Lorentz boost, i.e. $U(L)|0(t)\rangle_{A,B} \neq |0(t)\rangle_{A,B}$. However, we show that flavor oscillations can be equivalently described in every Lorentz frame.

Let us consider a flavor wavepacket:

$$|a_\sigma(x_0)\rangle \equiv \int \frac{\mathrm{d}^3 \mathbf{k}}{2\omega_{\mathbf{k},\sigma}(2\pi)^3}\, e^{-ikx_0}\, f(\mathbf{k})\, a_{\mathbf{k},\sigma}(t_0)|0(t_0)\rangle_{A,B}. \quad (49)$$

We suppose that the momentum space distribution $f(\mathbf{k})$ were Lorentz invariant. Therefore:

$$|a'_\sigma(x_0)\rangle \equiv U(L)|a_\sigma(x_0)\rangle = |a_\sigma(x'_0)\rangle, \quad (50)$$

as one can derive from Eqs.(47),(48). Covariant oscillation formula should be written as:

$$\mathcal{J}^\mu_{\sigma\to\rho}(x - x_0) = \langle a_\sigma(x_0)|J^\mu_\rho(x)|a_\sigma(x_0)\rangle, \quad (51)$$

where $J^\mu_\rho(x)$ are the flavor currents defined as [35]

$$J^\mu_\rho(x) \equiv i\varphi^\dagger_\rho(x)\, \overleftrightarrow{\partial^\mu}\, \varphi_\rho(x). \quad (52)$$

In the primed Lorentz frame Eq.(51) reads

$$\langle a'_\sigma(x_0)|J^\mu_\rho(x')|a'_\sigma(x_0)\rangle = \langle a_\sigma(x'_0)|J^\mu_\rho(x')|a_\sigma(x'_0)\rangle = \mathcal{J}^\mu_{\sigma\to\rho}(x' - x'_0). \quad (53)$$

Therefore, flavor oscillation formula in the primed Lorentz frame is the same as in the unprimed one.

5. Conclusions

In this paper, we reviewed basic facts on dynamical generation of field mixing via flavor vacuum condensate, as described in Ref. [21]. Then, we have studied the properties of flavor creation and annihilation operators under Lorentz boost transformations, in an effective model describing a flavor scalar doublet with mixing. We have proved that the Lagrangian under consideration is Lorentz invariant, while flavor vacuum is not. This SSB can be driven by a mechanism analogous to the one described in Section 2, in the fermionic case. However, we also proved that flavor oscillation formula Eq. (51) is Lorentz invariant and the flavor oscillations can be equivalently described in every Lorentz frame.

As it was mentioned in the introduction, in Refs. [23, 24], explicit Lorentz violating terms were added to the Standard Model Lagrangian, in order to analyze, at an effective level, signal of Lorentz SSB. An interesting perspective would be to find a possible connection of these researches with the present approach.

The problem of the interplay of Poincaré transformations and flavor states was recently treated in Ref. [42], where Standard Model was generalized in an appropriate way. It is then evident that the analysis of vacuum structure in connection with dynamical mixing generation [15, 16, 17, 18, 19, 20, 21, 22] could be important to understand new physics beyond Standard Model.

Acknowledgments
P.J. was supported by the Czech Science Foundation Grant No. 17-33812L.

References
[1] Giunti C 2004 *Am. J. Phys.* **72** 699
[2] Blasone M, Magueijo J and Pires-Pacheco P 2005 *Europhys. Lett.* **70** 600
[3] Blasone M, Magueijo J and Pires-Pacheco P 2005 *Braz. J. Phys.* **35** 447
[4] Blasone M, Di Mauro M and Lambiase G 2005 *Acta Phys. Polon.* B **36** 3255
[5] Blasone M and Vitiello G 1995 *Ann. Phys.* **244** 283
[6] Fujii K, Habe C and Yabuki T 1999 *Phys. Rev.* D **59** 113003
[7] Fujii K, Habe C and Yabuki T 2001 *Phys. Rev.* D **64** 013011
[8] Hannabuss K C and Latimer D C 2000 *J. Phys.* A **33** 1369
[9] Hannabuss K C and Latimer D C 2003 *J. Phys.* A **36** L69
[10] Ji C R and Mishchenko Y 2002 *Phys. Rev.* D **65** 096015
[11] Ji C R and Mishchenko Y 2005 *Ann. Phys.* **315** 488
[12] Magueijo J and Smolin L 2002 *Phys. Rev. Lett.* **88** 190403
[13] Magueijo J and Smolin L 2003 *Phys. Rev.* D 044017
[14] Ellis J R, Mavromatos N E and Westmuckett M 2004 *Phys. Rev.* D **70** 044036
[15] Mavromatos N E and Sarkar S 2008; *New J. Phys.* **10** 073009
[16] Mavromatos N E, Sarkar S and Tarantino W 2009 *Phys. Rev.* D **80** 084046
[17] Mavromatos N E, Sarkar S and Tarantino W 2011 *Phys. Rev.* D **84** 044050
[18] Mavromatos N E, Sarkar S and Tarantino W 2013 Mod. Phys. Lett. A **28** 1350045.
[19] Blasone M, Jizba P, Lambiase G and Mavromatos N E 2014 *J. Phys. Conf. Series* **538** 012003
[20] Blasone M, Jizba P and Smaldone L 2015 *Nuovo Cim. C* **38** 201
[21] Blasone M, Jizba P, Mavromatos N E and Smaldone L 2019 *Phys. Rev.* D **100** 045027
[22] Blasone M, Jizba P, Mavromatos N E and Smaldone L 2019 *J. Phys. Conf. Series* **1194** 012014
[23] Colladay D and Kostelecky V A 1997 *Phys. Rev.* D **55** 6760; 1998 *Phys. Rev.* D **58** 116002
[24] Kostelecky V A and Mewes M 2004 *Phys. Rev.* D **69** 016005
[25] Coleman S and Glashow S L 1997 *Phys. Lett.* B **405** 249; 1999 *Phys. Rev.* D **59** 116008
[26] Katori T [MiniBooNE Collaboration] 2012 *Mod. Phys. Lett.* A **27** 1230024
[27] Antonelli V, Miramonti L and Torri M D C 2018 *Eur. Phys. J.* C **78** 667
[28] Lambiase G and Scardigli F 2018 *Phys. Rev.* D **97** 075003
[29] Kempf A, Mangano G and Mann R B 1995 *Phys. Rev.* D **52** 1108
[30] Scardigli F and Casadio R 2015 *Eur. Phys. J.* C **75** 425
[31] Scardigli F, Blasone M, Luciano G G and Casadio R 2018 *Eur. Phys. J.* C **78** 728
[32] Luciano G G and Petruzziello L 2019 *Eur. Phys. J.* C **79** 283
[33] Buoninfante L, Luciano G G and Petruzziello L 2019 *Eur. Phys. J.* C **79** 663
[34] Binger M and Ji C R 1999 *Phys. Rev.* D **60** 056005
[35] Blasone M, Capolupo A, Romei O and Vitiello G 2001 *Phys. Rev.* D **63** 125015
[36] Miransky V A 1993 *Dynamical Symmetry Breaking in Quantum Field Theories* (World Scientific)
[37] Fujimoto Y and Papastamatiou N J 1977 *Nuovo Cim.* A **4** 468
[38] Peskin M E, Schroeder D V 1995 *An introduction to quantum field theory* (Westview)
[39] Blasone M 2011 *J. Phys. Conf. Ser.* **306** 012037
[40] Birrell N D and Davies P C W 1984 *Quantum Fields in Curved Space* (Cambridge U. Press, Cambridge)
[41] Blasone M, Luciano G G, Petruzziello L and Smaldone L 2018 *Phys. Lett.* B **786** 278
[42] Lobanov A E 2019 *Ann. Phys.* **403** 82

Journal of Physics: Conference Series **1416** (2019) 012006 doi:10.1088/1742-6596/1416/1/012006

On the status of pointlike fields in integrable QFTs

Henning Bostelmann

University of York, Department of Mathematics, York YO10 5DD, United Kingdom

E-mail: henning.bostelmann@york.ac.uk

Abstract. In integrable models of quantum field theory, local fields are normally constructed by means of the bootstrap-formfactor program. However, the convergence of their n-point functions is unclear in this setting. An alternative approach uses fully convergent expressions for fields with weaker localization properties in spacelike wedges, and deduces existence of observables in bounded regions from there, but yields little information about their explicit form. We propose a new, hybrid construction: We aim to describe pointlike local quantum fields; but rather than exhibiting their n-point functions and verifying the Wightman axioms, we establish them as closed operators affiliated with a net of local von Neumann algebras that is known from the wedge-local approach. This is shown to work at least in the Ising model.

1. Introduction

Quantum field theory (QFT), describing the behaviour of subatomic particles at relativistic speeds, is usually formulated in terms of the eponymous quantum fields: A quantum field $\Phi(x)$—and we will restrict to scalar Bose fields in all what follows—is a quantum observable localized at the spacetime point x, where localization manifests itself in commutativity at spacelike distances:

$$[\Phi(x), \Phi(y)] = 0 \quad \text{if } x \text{ is spacelike separated from } y. \tag{1}$$

However, localization at a single point x is an unphysical over-idealization. By Heisenberg's uncertainty relation, one expects a quantity localized arbitrarily sharp in time and space to be arbitrarily *delocalized* in energy and momentum; in other words, these fields should have an infinite energy-momentum transfer.

From a mathematical perspective, this over-idealization is reflected in singularities of the fields. While often referred to as "operators on a Hilbert space" in the physics literature, the $\Phi(x)$ are actually only defined as quadratic forms: Even in the simplest examples, $\Phi(x)\psi$ is never a normalizable vector in the Hilbert space of the theory, regardless of the vector ψ, not even if $\psi = \Omega$ is the vacuum vector. Only matrix elements $\langle\psi_1, \Phi(x)\psi_2\rangle$ make sense as finite numbers if both $\psi_{1,2}$ are sufficiently "well-behaved" vectors, for example, with an energy cutoff. Consequently, products $\Phi(x)\Phi(y)$ of fields do not exist in general (or require some regularization procedure), and there is no notion of spectral projections or (quasi-)eigenvalues of $\Phi(x)$, all of which would be fundamental to a physical interpretation.

The usual remedy to this problem is to average the fields (or, mathematically, to treat them as operator-valued distributions): If g is a smooth, compactly supported function on spacetime, then $\Phi(g) := \int dx\, g(x)\Phi(x)$ can often be defined as an operator, at least on a dense set of vectors. In other words, smearing in spacetime improves the high-energy singularities of the fields, and allows us to define their products etc. This approach, first formalized by Wightman [1], is known to work at least in free theories as well as in the interacting $P(\Phi)_2$ and Φ_3^4 models [2]. However, can this procedure work in all QFTs with self-interaction? Even if that were the case in principle, the operator domains of the $\Phi(g)$ are often hard to access in concrete models, and their functional analytic properties remain difficult to obtain.

Here we investigate this problem in a specific, simplified model of interaction, namely integrable QFTs in $1+1$ spacetime dimensions. (We will recall their construction in Sec. 2.) Traditionally, in this situation, one tries to define the fields $\Phi(g)$ or $\Phi(x)$ by specifying their n-point functions, which are given as an infinite series. However, convergence of this series is hard to control—yet another reflection of the singular nature of pointlike fields.

We present an alternative approach: We define the quantum field theory indirectly via a net of local von Neumann algebras associated with spacetime regions. Then, we show that the $\Phi(g)$ exist as (unbounded) operators that are *affiliated* with the algebras (Sec. 3), thus bypassing the treatment of products of fields or of their n-point functions. In this way, we ameliorate (though not eliminate) the convergence problem of the series, breaking it down to a problem that is tractable at least in the simplest example, the massive Ising model.

The present paper contains a brief and mostly non-technical summary of these results; the reader is referred to [3] for details.

2. Construction of integrable models

Integrable QFTs are quantum field theories on 1+1 dimensional Minkowski space with a simplified kind of self-interaction: They describe particles under elastic scattering on the two-particle level only; scattering between an arbitrary number of particles is merely a combination of two-particle scattering processes, i.e., the model has a factorizing scattering matrix. There is no particle creation or annihilation; the particle number as well as the momenta of individual particles are conserved at all times. Hence, in order to specify such a theory, it is enough to know the particle spectrum (masses and spin of elementary particles) and the scattering matrix at the two-particle level. In the present article, we focus on one species of scalar particle of mass $\mu > 0$, with no bound states. The only input then required is the two-particle scattering function S. In this section, we recall how a corresponding quantum field theory is constructed from this data.

2.1. General setup

Our model is specified by the particle mass μ and a two-particle scattering function S, that is, a meromorphic function which is analytic and bounded in the physical strip $\mathbb{R}+i[0,\pi]$ and fulfills

$$S(\zeta)^{-1} = S(-\zeta) = \overline{S(\bar{\zeta})} = S(\zeta + i\pi). \tag{2}$$

Examples include the massive Ising model, where $S = -1$, and the sinh-Gordon model, where $S(\zeta) = \frac{\sinh\zeta - i\sin B\pi/2}{\sinh\zeta + i\sin B\pi/2}$ with some $B \in (0,1)$.

Given such function S, one defines a representation of the Zamolodchikov-Faddeev algebra, a modified CCR algebra with creators and annihilators $z^\dagger(\theta)$, $z(\theta)$, depending on rapidities θ, which fulfill

$$z^{\#}(\theta_1)z^{\#}(\theta_2) = S(\theta_1-\theta_2)\, z^{\#}(\theta_2)z^{\#}(\theta_1) \qquad (z^{\#} = z^\dagger, z), \tag{3}$$

$$z(\theta_1)z^\dagger(\theta_2) = S(\theta_2-\theta_1)\, z^\dagger(\theta_2)z(\theta_1) + \delta(\theta_1-\theta_2)\cdot \mathbf{1}. \tag{4}$$

These act on a modified Fock space $\mathcal{H}$ spanned by n-particle vectors of the form

$$\psi_n = \int d^n\theta\, f(\theta_1,\ldots,\theta_n)\, z^\dagger(\theta_1)\cdots z^\dagger(\theta_n)\Omega, \tag{5}$$

where Ω is the Fock vacuum. If the support of f is such that the rapidities are in ascending, respectively descending, order, then ψ_n can be interpreted as an outgoing, respectively incoming, particle configuration with wave function f, although this interpretation can be justified only after introducing local observables [4, Sec. 6].

Spacetime symmetries are represented on $\mathcal{H}$ as follows: translations T_x (for $x \in \mathbb{R}^2$) and boosts B_λ (for $\lambda \in \mathbb{R}$) are fixed by the relations

$$U(T_x)z^\dagger(\theta)U(T_x)^* = e^{ip(\theta)\cdot x} z^\dagger(\theta), \qquad U(B_\lambda)z^\dagger(\theta)U(B_\lambda)^* = z^\dagger(\theta+\lambda) \tag{6}$$

and $U(T_x)\Omega = U(B_\lambda)\Omega = \Omega$; here $p(\boldsymbol{\theta}) = \mu(\cosh\theta, \sinh\theta)$. Most importantly, spacetime reflections R act on ψ_n (as in (5)) by

$$U(R)\psi_n = \int d^n\theta\, \overline{f(\theta_1,\ldots,\theta_n)}\, z^\dagger(\theta_n)\cdots z^\dagger(\theta_1)\Omega. \tag{7}$$

The generator of time translations, i.e., the Hamiltonian H, is thus given by

$$H\psi_n = \int d^n\theta \Big(\sum_{j=1}^n \mu\cosh\theta_j\Big) f(\theta_1,\ldots,\theta_n)\, z^\dagger(\theta_1)\cdots z^\dagger(\theta_n)\Omega. \tag{8}$$

2.2. Form factor program

In order to construct quantum fields, the longest established approach is the *form factor program* (see [5] for a review). Here one makes an *ansatz* for the local fields as follows. One formally expands their n-point functions via an intermediate basis of sharp-rapidity states, for example for $n = 2$,

$$\langle\Omega, \Phi(x)\Phi(y)\Omega\rangle = \sum_{k=0}^{\infty}\int \frac{d\theta_1\cdots d\theta_k}{k!} \big|\underbrace{\langle z^\dagger(\theta_1)\cdots z^\dagger(\theta_k)\Omega|\Phi(0)\Omega\rangle}_{=:F_k(\boldsymbol{\theta})}\big|^2 e^{i(y-x)\cdot\sum_j p(\theta_j)}\,. \tag{9}$$

It turns out that the expansion coefficients F_k, called the *form factors* (for which we use slightly different conventions than usual), already determine all n-point functions. The expected properties of the field Φ, such as locality and covariance under U, lead to restrictions on the F_k, the *form factor equations*. Namely, the F_k have meromorphic continuations, analytic where $\operatorname{Im}\zeta_1 < \ldots < \operatorname{Im}\zeta_k < \operatorname{Im}\zeta_1 + \pi$ with at most first-order poles at the boundary of this region, and they fulfill, $1 \le j < k$,

$$F_k(\boldsymbol{\zeta}) = S(\zeta_{j+1}-\zeta_j)F_k(\zeta_1,\ldots,\zeta_{j+1},\zeta_j,\ldots,\zeta_k), \tag{10}$$

$$F_k(\boldsymbol{\zeta}) = F_k(\zeta_2,\ldots,\zeta_k,\zeta_1+2\pi i), \tag{11}$$

$$\operatorname*{res}_{\zeta_2-\zeta_1=i\pi} F_k(\boldsymbol{\zeta}) = -\frac{1}{2\pi i}\Big(1-\prod_{j=1}^k S(\zeta_1-\zeta_j)\Big)F_{k-2}(\zeta_3,\ldots,\zeta_k). \tag{12}$$

Given a specific S, explicit solutions of these equations can be found in relevant cases.

However, now returning to the infinite series (9), only very partial results are known about its convergence [6], which remains the open mathematical point in the construction.

2.3. Algebraic construction

An alternative approach, initiated by Schroer [7], uses an indirect route to describe local observables. We temporarily depart from pointlike localized objects; as already seen for smeared

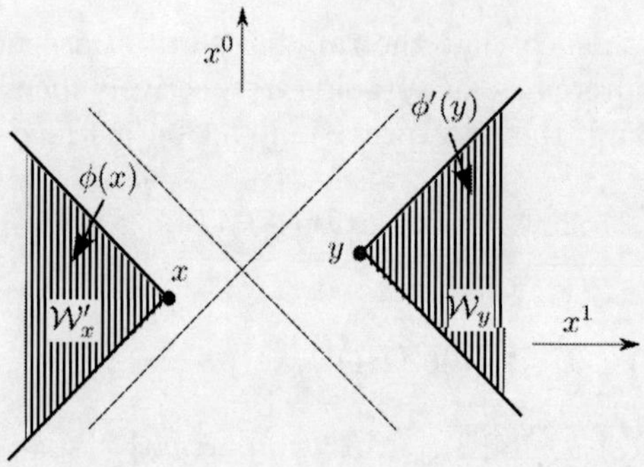

Figure 1. Fields localized in spacelike wedges

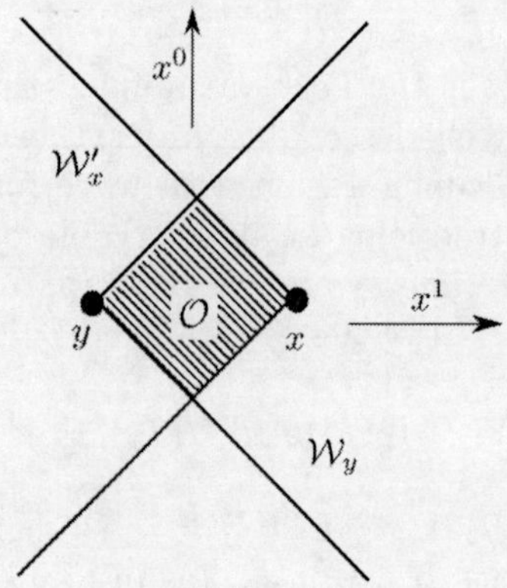

Figure 2. Intersection of two wedges

fields $\Phi(g)$, weaker localization properties lead to milder high-energy singularities. An even less strictly localized object is the following field:

$$\phi(x) := \int d\theta \left(e^{ip(\theta)\cdot x} z^\dagger(\theta) + e^{-ip(\theta)\cdot x} z(\theta) \right). \tag{13}$$

This ϕ is *not* local in the sense of Eq. (1). However, with $\phi'(x) := U(R)\phi(-x)U(R)$, one finds

$$[\phi(x), \phi'(y)] = 0 \quad \text{if } x \text{ is spacelike separated } \textit{to the left} \text{ of } y. \tag{14}$$

This can usefully be interpreted as follows: $\phi(x)$ is localized in the spacelike *wedge* $\mathcal{W}_x'$ with tip at x, opening to the left, and $\phi'(y)$ is localized in the wedge $\mathcal{W}_y$ with tip at y, opening to the right; the fields commute if the wedges are spacelike separated (see Fig. 1).

As the main advantage of these wedge-local quantities, ϕ is easy to control mathematically—it behaves almost like a free field and can be treated with same methods [8, Sec. X.7]. In particular, the smeared field $\phi(g)$ is essentially selfadjoint, hence one can pass to its bounded functions $\exp i\phi(g)$, $\operatorname{supp} g \subset \mathcal{W}_x'$. We call $\mathcal{A}(\mathcal{W}_x')$ the von Neumann algebra generated by all $\exp i\phi(g)$ with $\operatorname{supp} g \subset \mathcal{W}_x'$, and analogously we define $\mathcal{A}(\mathcal{W}_y)$. Importantly, they fulfill

$$\mathcal{A}(\mathcal{W}_x') = \mathcal{A}(\mathcal{W}_x)', \tag{15}$$

the prime on the right-hand side denoting the commutant of the von Neumann algebra.

Now for a bounded region $\mathcal{O} = \mathcal{W}_x' \cap \mathcal{W}_y$ as in Fig. 2, set

$$\mathcal{A}(\mathcal{O}) := \mathcal{A}(\mathcal{W}_x') \cap \mathcal{A}(\mathcal{W}_y). \tag{16}$$

This algebra $\mathcal{A}(\mathcal{O})$ then contains strictly (though not pointlike) local operators. One can check without much effort that these algebras are indeed local and covariant in the sense of the Haag-Kastler axioms [9]. It is not so immediate that $\mathcal{A}(\mathcal{O})$ contains any operator except multiples of the identity; but at least for certain S and for sufficiently large regions, one can establish that $\mathcal{A}(\mathcal{O})\Omega$ is indeed dense in $\mathcal{H}$ [4, 10].

This approach solves all mathematical convergence problems, but it has a quite different shortcoming: the explicit form of the local operators $A \in \mathcal{A}(\mathcal{O})$ remains unclear. Their matrix elements $\langle z^\dagger(\theta_1) \cdots z^\dagger(\theta_k)\Omega, A\Omega \rangle$ do fulfill the form factor equations [11]; but in the end, these operators are "constructed" using the axiom of choice, and no further information about their relation to pointlike fields or other generators is available.

3. A combined approach

From the two approaches to local observables described in the previous section, it seems that we face a binary choice: We can either describe pointlike fields explicitly, but lose control of convergence issues at high particle numbers; or we can fully treat all convergence issues to obtain local operators, but have no access to their explicit form.

The aim of this section is to present a construction that combines aspect of those two approaches, and leads to fully convergent but explicitly accessible observables. We are going to define these using the explicit expressions for pointlike fields known from the form factor program. However, we want to interpret them differently: Our aim is not to compute the n-point functions of the fields $\Phi(x)$, nor to investigate whether products of two fields $\Phi(g_1)\Phi(g_2)$ exist. Rather, we make use of the underlying (abstract) von Neumann algebras, and show that the $\Phi(g)$ are *affiliated* with the local algebra $\mathcal{A}(\mathcal{O})$ if $\operatorname{supp} g \subset \mathcal{O}$. Here being affiliated (the precise definition will be recalled in Sec. 3.4 below) is the closest possible way to say that $\Phi(g)$ is an element of $\mathcal{A}(\mathcal{O})$: It cannot literally be contained in the algebra, since $\Phi(g)$ is unbounded but $\mathcal{A}(\mathcal{O})$ consists of bounded operators; but when writing $A = V\int \lambda\, dP(\lambda)$ in polar decomposition, one demands that V and all $P(\lambda)$ are contained in $\mathcal{A}(\mathcal{O})$.

For obtaining this result, we will proceed as follows.

(1) Define the field $\Phi(g)$ as a quadratic form. To this end, we use the well-known solutions of the form factor equations;

(2) Show that $\Phi(g)$ is local. Here locality will be defined relative to the wedge-local field ϕ;

(3) Show that $\Phi(g)$ is closable. That is, we want to extend both $\Phi(g)$ and its adjoint to operators on a dense domain, removing one of the particle number cutoffs. It is at this point that convergence issues of an infinite series show up in the construction;

(4) Show that $\Phi(g)$ is affiliated with $\mathcal{A}(\mathcal{O})$. This will follow from locality (2) and closability (3) by an abstract argument.

In the remainder of this section, we will comment briefly on each of these four aspects. Details of the construction can be found in [3].

3.1. Definition of fields

Our starting point are the meromorphic functions F_k which are solutions of the form factor equations (10)–(12). Let us give an example for the massive Ising model (i.e., for the case $S=-1$): There, for the basic field or order parameter $\Phi(g)$, the form factors are given by

$$F_{2k+1}(\boldsymbol{\zeta}) := \frac{1}{(2\pi i)^k}\tilde{g}(p(\boldsymbol{\zeta})) \prod_{1\leq i<j\leq 2k+1} \tanh\frac{\zeta_i-\zeta_j}{2}, \qquad F_{2k}(\boldsymbol{\zeta}) = 0. \tag{17}$$

Compared with [12, 13], note the extra factor $\tilde{g}(p(\boldsymbol{\zeta}))$ resulting from averaging in configuration space. Also, it is evident from (17) that the F_{2k+1} have poles at $\zeta_j - \zeta_i = i\pi$, the so-called *kinematic poles* required by Eq. (12); they will play a role in the following.

Given such a set of form factors F, we can then define our field $\Phi(g)$ as (cf. [14, 15])

$$\Phi(g) := \sum_{m,n=0}^{\infty}\int \frac{d^m\theta\, d^n\eta}{m!n!} F_{m+n}(\boldsymbol{\theta}+i\mathbf{0}, \boldsymbol{\eta}+i\boldsymbol{\pi}-i\mathbf{0}) z^\dagger(\theta_1)\cdots z^\dagger(\theta_m) z(\eta_1)\cdots z(\eta_n). \tag{18}$$

This gives us $\Phi(g)$ as a quadratic form, with well-defined matrix elements $\langle\psi_1, \Phi(g)\psi_2\rangle$ if ψ_j are "suitable" vectors: First, they should both include a particle number cutoff, so that the sum (18) is actually finite in matrix elements; hence no convergence issues arise at this point. Second we demand that they decay sufficiently fast in momentum space; technically, we require that $\|\exp(H^\alpha)\psi_j\| < \infty$ where H is the Hamiltonian (8), and $\alpha \in (0,1)$ is fixed.

3.2. Locality

We need to establish that our fields are local in a suitable sense. This cannot be done in the usual way (1), since products of fields $\Phi(g)$, and hence their commutators, are not defined at this point. Instead, we consider locality relative to the wedge-local field ϕ; we want to establish that

$$[\phi(h_L), \Phi(g)] = 0 = [\phi'(h_R), \Phi(g)] \tag{19}$$

when h_L is supported spacelike to the left and h_R to the right of g. Since ϕ changes the particle number only by one, and hence preserves particle-number cutoffs, the commutators in (19) are well-defined in matrix elements for suitable h_L, h_R. That the commutators do actually vanish is then quite direct to verify; it is a consequence of the form factor equations, see [11, Secs. 4.3 and 5.3].

3.3. Closability

We now want to pass from a quadratic form $\Phi(g)$, where only matrix elements $\langle\psi_1, \Phi(g)\psi_2\rangle$ are defined, to an (unbounded) operator on a dense domain, that is, $\Phi(g)\psi$ should be a normalizable vector for vectors ψ from a dense set. In other words, we wish to remove the particle cutoff and energy damping from one side of the matrix element. In mathematical terms, we are aiming at a densely defined, closed operator (cf. [16, Ch. 3.5]) that extends $\Phi(g)$, with a corresponding extension for its adjoint; see [3, Def. 3.1] for the exact technical conditions.

In this step, convergence aspects of the infinite sum (18) play a role. A sufficient condition on closability (i.e., the existence of such extension) turns out to be that

$$\sum_{m=0}^{\infty} \frac{2^{m/2}}{\sqrt{m!}} \Big(\|F_{m+n}\|_{m\times n}^{(\alpha)} + \|F_{m+n}\|_{n\times m}^{(\alpha)} \Big) < \infty \tag{20}$$

for all $n \in \mathbb{N}$, where

$$\|F_{m+n}\|_{m\times n}^{(\alpha)} := \frac{1}{2}\|T\exp(-H^\alpha)\| + \frac{1}{2}\|\exp(-H^\alpha)T\| \tag{21}$$

and T is the integral operator with kernel $F_{m+n}(\boldsymbol{\theta} + i\mathbf{0}, \boldsymbol{\eta} + i\boldsymbol{\pi} - i\mathbf{0})$, $\boldsymbol{\theta} \in \mathbb{R}^m$, $\boldsymbol{\eta} \in \mathbb{R}^n$.

This summation condition can in fact be verified in the example (17) from the Ising model; to that end, the test function g needs to be chosen such that $\tilde{g}$ and its derivatives decay faster than $\exp(-\|p\|^\alpha)$ in momentum space, i.e., g needs to be of Jaffe class [17]. A main difficulty in establishing (20) are the kinematical poles of the form factors: They mean that the kernels $F_{m+n}(\boldsymbol{\theta} + i\mathbf{0}, \boldsymbol{\eta} + i\boldsymbol{\pi} - i\mathbf{0})$ are boundary values of meromorphic functions at a pole, i.e., the T are singular integral operators of non-convolution type. Estimating the norms $\|F_{m+n}\|_{m\times n}^{(\alpha)}$ therefore requires considerable technical effort; see [3, Sec. 5.2] for the techniques used.

3.4. Affiliation

Our last step is to establish a relation between the operator $\Phi(g)$ and the abstractly defined von Neumann algebras $\mathcal{A}(\mathcal{O})$, as introduced in Sec. 2.3. As mentioned, this is by way of affiliation.

We recall the definition: Let A be a closed operator on a dense domain $D(A) \subset \mathcal{H}$, and let $\mathcal{N} \subset \mathfrak{B}(\mathcal{H})$ be a von Neumann algebra. Since A is closed, it can be written in its polar decomposition: $A = V|A| = V\int \lambda\, dP(\lambda)$, where V is a partial isometry, and $|A| = \int \lambda\, dP(\lambda)$ is a positive operator written in spectral decomposition, with spectral projectors $P(\lambda)$. Now A is said to be affiliated with $\mathcal{N}$ if both V and all $P(\lambda)$ are elements of $\mathcal{N}$.

Equivalently [18, Thm. 3.16], A is affiliated with $\mathcal{N}$ if the following holds: For every operator $B \in \mathcal{N}'$ (the commutant of $\mathcal{N}$), we have

$$BD(A) \subset D(A) \qquad \text{and } AB\psi = BA\psi \ \text{ for all } \psi \in D(A) \tag{22}$$

Thus, "A is affiliated with $\mathcal{N}$", for unbounded operators A, generalizes the notion "A is an element of $\mathcal{N}$" in two equivalent ways: First, its polar data are elements of $\mathcal{N}$; second, it commutes with the commutant of $\mathcal{N}$.

The task is now to deduce that our field operators $\Phi(g)$ are actually affiliated with $\mathcal{A}(\mathcal{O})$, where $\operatorname{supp} g \subset \mathcal{O}$. This turns out to be a consequence of locality (relative to the wedge-local field) and closability. The argument is roughly as follows; details can be found in [3, Sec. 3].

Let $\mathcal{O} = \mathcal{W}_x' \cap \mathcal{W}_y$ as in Fig. 2 and $\operatorname{supp} g \subset \mathcal{O}$. We restrict our attention to the Ising model, where the wedge-local fields $\phi(h)$ are in fact bounded operators; they generate the algebras $\mathcal{A}(\mathcal{W}_y')$ if h varies over all test functions with $\operatorname{supp} h \subset \mathcal{W}_y'$. It follows from locality (19) that $[\Phi(g), \phi(h)] = 0$, at least in matrix elements. Closability of $\Phi(g)$ allows us to maintain this relation in the sense of operators on the domain of the unbounded operator $\Phi(g)$. By criterion (22), $\Phi(g)$ is then affiliated with $\mathcal{A}(\mathcal{W}_y')' = \mathcal{A}(\mathcal{W}_y)'' = \mathcal{A}(\mathcal{W}_y)$; we have used (15) here. By a similar reasoning, $\Phi(g)$ is affiliated with $\mathcal{A}(\mathcal{W}_x')$ as well. It is thus affiliated with $\mathcal{A}(\mathcal{W}_y) \cap \mathcal{A}(\mathcal{W}_x')$, which equals $\mathcal{A}(\mathcal{O})$ by Eq. (16).

Outside the Ising model, where $\phi(h)$ is not bounded, the argument is somewhat more complicated (see [3, Prop. 3.5]). One needs to work with the bounded generators $\exp i\phi(h)$, which in turn can be approximated by polynomials in the $\phi(h)$. Indeed, an extra technical assumption is needed here, since it is not clear a priori that $B = \exp i\phi(h)$ maps the domain of $A = \Phi(g)$ into itself, as required for criterion (22).

4. Conclusions and outlook

As we have summarized in this paper (and as is laid out in more detail in [3]), we have established the existence of smeared fields $\Phi(g)$ in the Ising model as closed operators affiliated with the local algebras $\mathcal{A}(\mathcal{O})$. This was done for the basic field (order parameter), but the same arguments apply to descendant fields, which are obtained by multiplying the form factors (17) with symmetric Laurent polynomials in the variables e^{ζ_j}.

This construction bypasses the convergence problems of n-point functions that are inherent in the form factor program, and replaces them with the summability criterion (20) which can in fact be verified rigorously. On the other hand, we still retain explicit control about the matrix elements of our local observables.

The existence of these fields in the Ising model is maybe not very surprising; the same model has been constructed both in a Euclidean setting [19] and in the algebraic framework [20]. Our point is rather a conceptual one: It is possible, and technically feasible, to treat the model directly in Minkowski space, giving sense to the local fields *without* referring to their n-point functions.

Whether the same construction applies for more general choices of S, for example in the sinh-Gordon model [21], remains a point for further investigation. Our present results certainly give hope that the problem could be settled also in this case, replacing the convergence problem of the n-point functions with a milder one to show affiliation with the local algebras. However, additional technical difficulties remain to be overcome. Not only is the structure of the form factors increasingly intricate, complicating the estimates, but as mentioned in Sec. 3.4, additional criteria on the domain of the field operators would need to be verified.

References

[1] Streater R F and Wightman A S 1964 *PCT, Spin and Statistics, and All That* (New York: Benjamin)
[2] Glimm J and Jaffe A 1987 *Quantum Physics – A functional integral point of view* 2nd ed (New York: Springer)
[3] Bostelmann H and Cadamuro D 2019 *Ann. H. Poincaré* (to appear) (*Preprint* `arXiv:1806.00269`)
[4] Lechner G 2008 *Commun. Math. Phys.* **277** 821–60 (*Preprint* `math-ph/0601022`)
[5] Babujian H M, Foerster A and Karowski M 2006 *SIGMA* **2** 082 (*Preprint* `hep-th/0609130`)
[6] Babujian H M and Karowski M 2004 *Int. J. Mod. Phys.* **A19S2** 34–49 (*Preprint* `hep-th/0301088`)

[7] Schroer B and Wiesbrock H W 2000 *Rev. Math. Phys.* **12** 301–26 (*Preprint* hep-th/9812251)
[8] Reed M and Simon B 1975 *Methods of Modern Mathematical Physics* vol II: Fourier Analysis, Self-Adjointness (New York: Academic Press)
[9] Haag R 1996 *Local Quantum Physics* 2nd ed (Berlin: Springer)
[10] Alazzawi S and Lechner G 2017 *Commun. Math. Phys.* **354** 913–56 (*Preprint* arXiv:1608.02359)
[11] Bostelmann H and Cadamuro D 2015 *Commun. Math. Phys.* **337** 1199–240 (*Preprint* arXiv:1402.6127)
[12] Schroer B and Truong T T 1978 *Nucl. Phys.* **B144** 80–122
[13] Berg B, Karowski M and Weisz P 1979 *Phys. Rev. D* **19** 2477–9
[14] Lashkevich M Y 1994 Sectors of mutually local fields in integrable models of quantum field theory arXiv:hep-th/9406118
[15] Bostelmann H and Cadamuro D 2013 *Journal of Physics A* **46** 095401 (*Preprint* arXiv:1208.4763)
[16] Kato T 1984 *Perturbation Theory for Linear Operators* 2nd ed (New York: Springer)
[17] Jaffe A M 1967 *Phys. Rev.* **158**(5) 1454–61
[18] Jorgensen P and Tian F 2017 *Non-commutative Analysis* (Singapore: World Scientific)
[19] Palmer J and Tracy C 1983 *Advances in Applied Mathematics* **4** 46–102
[20] Lechner G 2005 *J. Phys. A: Math. Gen.* **38** 3045 (*Preprint* math-ph/0405062)
[21] Fring A, Mussardo G and Simonetti P 1993 *Nucl. Phys.* **B393** 413–41 (*Preprint* hep-th/9211053)

Discretization of Generalized Chebyshev Polynomials of (Anti)symmetric Multivariate Sine Functions

Adam Brus[1], Jiří Hrivnák[1] and Lenka Motlochová[1]

[1] Department of Physics, Faculty of Nuclear Sciences and Physical Engineering, Czech Technical University in Prague, Břehová 7, 115 19 Prague 1, Czech Republic

E-mail: adam.brus@fjfi.cvut.cz, jiri.hrivnak@fjfi.cvut.cz, lenka.motlochova@fjfi.cvut.cz.

Abstract. The multivariate antisymmetric and symmetric trigonometric functions allow to generalize the four kinds of classical Chebyshev polynomials to multivariate settings. The four classes of the bivariate polynomials, related to the symmetrized sine functions, are studied in detail. For each of these polynomials, the weighted continuous and discrete orthogonality relations are shown. The related cubature formulas for numerical integration together with further model examples and properties of selected special cases are discussed.

1. Introduction

The four discrete Fourier transforms of antisymmetric and symmetric multivariate trigonometric functions are introduced in [1]. Multivariate extensions of all sixteen types of one-dimensional discrete trigonometric transforms [2] together with multivariate generalizations of the four kinds of Chebyshev polynomials [3] are developed in [4, 5]. Interpolation methods for both functions and polynomials are inherent in the construction of the discrete transforms. The corresponding method of numerical integration [6–9] produces 32 types of efficient cubature rules, including eight optimal Gaussian formulas.

This paper demonstrates the types of developed transforms and constructions of the polynomials on specific bivariate examples. Two-dimensional cases of the discrete trigonometric transforms of types I–IV are already derived and tested on model examples in [10, 11]. The bivariate interpolation formula for the antisymmetric sine functions, related to the discrete sine transform of type VII, is tested here. The construction algorithm for (anti)symmetric bivariate generalization of Chebyshev polynomials of the second and fourth kinds are calculated. The continuous and discrete orthogonality relations are presented. The corresponding bivariate cubature formulas are explicitly stated.

For the parameter $\lambda = (\lambda_1, \lambda_2) \in \mathbb{R}^2$ and the variable $x = (x_1, x_2) \in \mathbb{R}^2$, the antisymmetric sine functions $\sin^-_\lambda(x)$ and symmetric sine functions $\sin^+_\lambda(x)$ are explicitly given by the following formulas,

$$\begin{aligned}\sin^-_\lambda(x) &= \sin(\pi\lambda_1 x_1)\sin(\pi\lambda_2 x_2) - \sin(\pi\lambda_1 x_2)\sin(\pi\lambda_2 x_1),\\ \sin^+_\lambda(x) &= \sin(\pi\lambda_1 x_1)\sin(\pi\lambda_2 x_2) + \sin(\pi\lambda_1 x_2)\sin(\pi\lambda_2 x_1),\end{aligned} \tag{1}$$

and the (anti)symmetric cosine functions $\cos^{\pm}_{\lambda}(x)$ are specified similarly by

$$\begin{aligned}\cos^{-}_{\lambda}(x) &= \cos(\pi\lambda_1 x_1)\cos(\pi\lambda_2 x_2) - \cos(\pi\lambda_2 x_1)\cos(\pi\lambda_1 x_2),\\ \cos^{+}_{\lambda}(x) &= \cos(\pi\lambda_1 x_1)\cos(\pi\lambda_2 x_2) + \cos(\pi\lambda_2 x_1)\cos(\pi\lambda_1 x_2).\end{aligned} \tag{2}$$

For the purpose of this paper, the parameter λ labelling the generalized sine functions is considered only in the form $\lambda = k + \rho_i^{\pm}$, $i = 2,4$, where $k = (k_1, k_2) \in \mathbb{Z}^2$ satisfies $k_1 \geq k_2 \geq 0$ and $\rho_i^{\pm}$ are constant shifts given by

$$\rho_2^- = (2,1), \quad \rho_2^+ = (1,1), \quad \rho_4^- = \left(\frac{3}{2}, \frac{1}{2}\right), \quad \rho_4^+ = \left(\frac{1}{2}, \frac{1}{2}\right). \tag{3}$$

The functions $\sin^{\pm}_{k+\rho_i^{\pm}}$, $i = 2,4$, have several symmetries derived from the properties of determinants, permanents and classical trigonometric functions [1, 4, 5, 11]. Therefore, without loss of generality, these bivariate (anti)symmetric sine functions are evaluated only on the corresponding domain $F(\widetilde{S}_2^{\mathrm{aff}})$,

$$F(\widetilde{S}_2^{\mathrm{aff}}) = \left\{(x_1, x_2) \in \mathbb{R}^2 \,|\, 1 \geq x_1 \geq x_2 \geq 0\right\}.$$

Furthermore, the four fundamental functions $\sin^{\pm}_{\rho_i^{\pm}}$, $i = 2,4$ do not vanish [4, 5] in the interior of the domain $F(\widetilde{S}_2^{\mathrm{aff}})$.

2. Bivariate discrete trigonometric transforms

The classical discrete trigonometric transforms DCTs and DSTs [2] are generalized using the multivariate trigonometric functions in [4, 5] into 16 symmetric and 16 antisymmetric discrete transforms for any dimension. The form of the transformations is demonstrated on the following two–dimensional case of the antisymmetric discrete sine transform of type VII.

2.1. Bivariate transform AMDST–VII

For an arbitrary scaling factor $N \geq 2$, the following discrete set of points on which the functions $\sin^{-}_{k+\rho_4^-}(s)$ are restricted is considered,

$$F_N^{\mathrm{VII},-} = \left\{ \left(\frac{2r_1}{2N+1}, \frac{2r_2}{2N+1}\right) \,\middle|\, N \geq r_1 > r_2 \geq 1 \right\}. \tag{4}$$

Assuming additional conditions $N - 2 \geq k_1, k_1'$ on labels k and k', the functions $\sin^{-}_{k+\rho_4^-}(s)$ are pairwise discretely orthogonal,

$$\sum_{s \in F_N^{\mathrm{VII},-}} \sin^{-}_{k+\rho_4^-}(s) \sin^{-}_{k'+\rho_4^-}(s) = \left(\frac{2N+1}{4}\right)^2 \delta_{kk'}. \tag{5}$$

Relation (5) implies that any function f defined on the set $F_N^{\mathrm{VII},-}$ is expressed in terms of the bivariate antisymmetric sine functions as

$$f(s) = \sum_{\substack{k_1,k_2=0\\ k_1 \geq k_2}}^{N-2} A_k \sin^{-}_{k+\rho_4^-}(s), \tag{6}$$

with the coefficients A_k determined by

$$A_k = \left(\frac{4}{2N+1}\right)^2 \sum_{s\in F_N^{\mathrm{VII},-}} f(s)\,\mathrm{sin}^-_{k+\rho_4^-}(s). \tag{7}$$

The forward and backward transforms (6) and (7) correspond to the interpolation formula for functions that are defined on the fundamental domain $F(\widetilde{S}_2^{\mathrm{aff}})$ and sampled on the discrete point set $F_N^{\mathrm{VII},-}$.

2.2. Interpolation via AMDST–VII

To any function $f : F_N^{\mathrm{VII},-} \to \mathbb{C}$, transformed by (7), is assigned its AMDST–VII interpolation $I^{\mathrm{VII},-}[f] : F(\widetilde{S}_2^{\mathrm{aff}}) \to \mathbb{C}$ of the form

$$I_N^{\mathrm{VII},-}[f](x) = \sum_{\substack{k_1,k_2=0\\ k_1\geq k_2}}^{N-2} A_k\,\mathrm{sin}^-_{k+\rho_4^-}(x), \quad x\in F(\widetilde{S}_2^{\mathrm{aff}}). \tag{8}$$

As a 2D model function is taken the function f given by the following expression,

$$\begin{aligned} f(x,y) = {} & \frac{1}{3}\exp\left(\frac{(x-0,45)^2+(y-0,2)^2}{0,01}+3\right) + \frac{1}{6}\exp\left(\frac{(x-0,8)^2+(y-0,3)^2}{0,01}+3\right) \\ & + \frac{1}{2}\exp\left(\frac{(x-0,85)^2+(y-0,6)^2}{0,005}+3\right). \end{aligned} \tag{9}$$

The function f is plotted over the fundamental domain $F(\widetilde{S}_2^{\mathrm{aff}})$ in Figure 1. The AMDST–VII interpolations $I_N^{\mathrm{VII},-}[f]$ of the model function (9) are for $N = 8, 13, 18$ depicted in Figure 2.

Figure 1. The contour plot of the model function (9) plotted over the fundamental domain $F(\widetilde{S}_2^{\mathrm{aff}})$.

3. Generalized Chebyshev-like polynomials of bivariate sine functions

Recall that the classical Chebyshev polynomials U_n and W_n of degree $n \in \mathbb{N}$ of the second and fourth kind are defined using the standard sine functions as

$$U_n(x) = \mathcal{P}_n^{\mathrm{II}}(x) = \frac{\sin\left((n+1)\,\theta\right)}{\sin(\theta)}, \qquad W_n(x) = \mathcal{P}_n^{\mathrm{IV}}(x) = \frac{\sin\left(\left(n+\frac{1}{2}\right)\theta\right)}{\sin\left(\frac{1}{2}\theta\right)} \tag{10}$$

with variable $x = \cos(\theta)$. The Chebyshev-like polynomials arise when the classical trigonometric functions are replaced by the (anti)symmetric trigonometric functions [4, 5].

Figure 2. The AMDST–VII interpolations $I_N^{\mathrm{VII},-}[f]$ of the model function (9) with $N = 8, 13, 18$. The sets of points $F_N^{\mathrm{VII},-}$ are depicted as the black dots.

Thus, the bivariate (anti)symmetric sine polynomials of degree k_1 are induced by the formulas

$$\begin{aligned}
\mathcal{P}_k^{\mathrm{II},-}(X_1(x), X_2(x)) &= \frac{\sin^-_{k+\rho_2^-}(x)}{\sin^-_{\rho_2^-}(x)}, &\qquad \mathcal{P}_k^{\mathrm{II},+}(X_1(x), X_2(x)) &= \frac{\sin^+_{k+\rho_2^+}(x)}{\sin^+_{\rho_2^+}(x)}, \\
\mathcal{P}_k^{\mathrm{IV},-}(X_1(x), X_2(x)) &= \frac{\sin^-_{k+\rho_4^-}(x)}{\sin^-_{\rho_4^-}(x)}, &\qquad \mathcal{P}_k^{\mathrm{IV},+}(X_1(x), X_2(x)) &= \frac{\sin^+_{k+\rho_4^+}(x)}{\sin^+_{\rho_4^+}(x)},
\end{aligned} \tag{11}$$

where the variables X_1, X_2 are defined by

$$X_1 = \cos^+_{(1,0)}, \quad X_2 = \cos^+_{(1,1)}. \tag{12}$$

The general form of the recurrence relations of the polynomials defined above can be obtained using generalized trigonometric identities from [4]. Using these identities, the following recurrence relations for the bivariate case are obtained.

3.1. Polynomials $\mathcal{P}^{\mathrm{II},+}_{(k_1,k_2)}$

The first three polynomials, calculated directly from definition, are given by

$$\mathcal{P}^{\mathrm{II},+}_{(0,0)} = 1, \quad \mathcal{P}^{\mathrm{II},+}_{(1,0)} = X_1, \quad \mathcal{P}^{\mathrm{II},+}_{(1,1)} = 2X_2. \tag{13}$$

The higher degree polynomials can then be obtained via recurrence relations of the form

$$\begin{aligned}
k_1 \geq k_2 + 2 \geq 3: &\quad \mathcal{P}^{\mathrm{II},+}_{(k_1,k_2)} = 2\mathcal{P}^{\mathrm{II},+}_{(k_1-1,k_2)} X_1 - \mathcal{P}^{\mathrm{II},+}_{(k_1-1,k_2+1)} - \mathcal{P}^{\mathrm{II},+}_{(k_1-1,k_2-1)} - \mathcal{P}^{\mathrm{II},+}_{(k_1-2,k_2)}, \\
k_1 \geq k_2 + 2 = 2: &\quad \mathcal{P}^{\mathrm{II},+}_{(k_1,0)} = 2\mathcal{P}^{\mathrm{II},+}_{(k_1-1,0)} X_1 - \mathcal{P}^{\mathrm{II},+}_{(k_1-1,1)} - \mathcal{P}^{\mathrm{II},+}_{(k_1-2,0)}, \\
k_1 = k_2 + 1 \geq 4: &\quad \mathcal{P}^{\mathrm{II},+}_{(k_1,k_1-1)} = 2\mathcal{P}^{\mathrm{II},+}_{(k_1-1,k_1-2)} X_2 - 2\mathcal{P}^{\mathrm{II},+}_{(k_1-1,k_1-3)} X_1 + \mathcal{P}^{\mathrm{II},+}_{(k_1-1,k_1-4)}, \\
k_1 = k_2 + 1 = 3: &\quad \mathcal{P}^{\mathrm{II},+}_{(3,2)} = 2\mathcal{P}^{\mathrm{II},+}_{(2,1)} X_2 - 2\mathcal{P}^{\mathrm{II},+}_{(2,0)} X_1,
\end{aligned}$$

$$
\begin{aligned}
k_1 = k_2 + 1 = 2: \quad & \mathcal{P}^{\mathrm{II},+}_{(2,1)} = 2\mathcal{P}^{\mathrm{II},+}_{(1,0)} X_2 - \mathcal{P}^{\mathrm{II},+}_{(1,0)}, \\
k_1 = k_2 \geq 3: \quad & \mathcal{P}^{\mathrm{II},+}_{(k_1,k_1)} = 2\mathcal{P}^{\mathrm{II},+}_{(k_1-1,k_1-1)} X_2 - 4\mathcal{P}^{\mathrm{II},+}_{(k_1-1,k_1-2)} X_1 + 2\mathcal{P}^{\mathrm{II},+}_{(k_1-1,k_1-1)} \\
& \quad + 2\mathcal{P}^{\mathrm{II},+}_{(k_1-1,k_1-3)} + \mathcal{P}^{\mathrm{II},+}_{(k_1-2,k_1-2)}, \\
k_1 = k_2 = 2: \quad & \mathcal{P}^{\mathrm{II},+}_{(2,2)} = 2\mathcal{P}^{\mathrm{II},+}_{(1,1)} X_2 - 4\mathcal{P}^{\mathrm{II},+}_{(1,0)} X_1 + 2\mathcal{P}^{\mathrm{II},+}_{(1,1)} + \mathcal{P}^{\mathrm{II},+}_{(0,0)}.
\end{aligned}
$$

3.2. Polynomials $\mathcal{P}^{\mathrm{II},-}_{(k_1,k_2)}$

The first three polynomials, calculated directly from definition, are given by

$$
\mathcal{P}^{\mathrm{II},-}_{(0,0)} = 1, \quad \mathcal{P}^{\mathrm{II},-}_{(1,0)} = 2X_1, \quad \mathcal{P}^{\mathrm{II},-}_{(1,1)} = 2X_2 + 1. \tag{14}
$$

The higher degree polynomials can then be obtained via recurrence relations of the form

$$
\begin{aligned}
k_1 \geq k_2 + 2 \geq 3: \quad & \mathcal{P}^{\mathrm{II},-}_{(k_1,k_2)} = 2\mathcal{P}^{\mathrm{II},-}_{(k_1-1,k_2)} X_1 - \mathcal{P}^{\mathrm{II},-}_{(k_1-1,k_2+1)} - \mathcal{P}^{\mathrm{II},-}_{(k_1-1,k_2-1)} - \mathcal{P}^{\mathrm{II},-}_{(k_1-2,k_2)}, \\
k_1 \geq k_2 + 2 = 2: \quad & \mathcal{P}^{\mathrm{II},-}_{(k_1,0)} = 2\mathcal{P}^{\mathrm{II},-}_{(k_1-1,0)} X_1 - \mathcal{P}^{\mathrm{II},-}_{(k_1-1,1)} - \mathcal{P}^{\mathrm{II},-}_{(k_1-2,0)}, \\
k_1 = k_2 + 1 \geq 2: \quad & \mathcal{P}^{\mathrm{II},-}_{(k_1,k_1-1)} = 2\mathcal{P}^{\mathrm{II},-}_{(k_1-1,k_1-1)} X_1 - \mathcal{P}^{\mathrm{II},-}_{(k_1-1,k_1-2)}, \\
k_1 = k_2 \geq 3: \quad & \mathcal{P}^{\mathrm{II},-}_{(k_1,k_1)} = 2\mathcal{P}^{\mathrm{II},-}_{(k_1-1,k_1-1)} X_2 - 2\mathcal{P}^{\mathrm{II},-}_{(k_1-1,k_1-2)} X_1 + 2\mathcal{P}^{\mathrm{II},-}_{(k_1-1,k_1-1)} + \mathcal{P}^{\mathrm{II},-}_{(k_1-1,k_1-3)}, \\
k_1 = k_2 = 2: \quad & \mathcal{P}^{\mathrm{II},-}_{(2,2)} = 2\mathcal{P}^{\mathrm{II},-}_{(1,1)} X_2 - 2\mathcal{P}^{\mathrm{II},-}_{(1,0)} X_1 + 2\mathcal{P}^{\mathrm{II},-}_{(1,1)}.
\end{aligned}
$$

3.3. Polynomials $\mathcal{P}^{\mathrm{IV},+}_{(k_1,k_2)}$

The first three polynomials, calculated directly from definition, are given by

$$
\mathcal{P}^{\mathrm{IV},+}_{(0,0)} = 1, \quad \mathcal{P}^{\mathrm{IV},+}_{(1,0)} = X_1 + 1, \quad \mathcal{P}^{\mathrm{IV},+}_{(1,1)} = 2X_2 + 2X_1 + 1. \tag{15}
$$

The higher degree polynomials can then be obtained via recurrence relations of the form

$$
\begin{aligned}
k_1 \geq k_2 + 2 \geq 3: \quad & \mathcal{P}^{\mathrm{IV},+}_{(k_1,k_2)} = 2\mathcal{P}^{\mathrm{IV},+}_{(k_1-1,k_2)} X_1 - \mathcal{P}^{\mathrm{IV},+}_{(k_1-1,k_2+1)} - \mathcal{P}^{\mathrm{IV},+}_{(k_1-1,k_2-1)} - \mathcal{P}^{\mathrm{IV},+}_{(k_1-2,k_2)}, \\
k_1 \geq k_2 + 2 = 2: \quad & \mathcal{P}^{\mathrm{IV},+}_{(k_1,0)} = 2\mathcal{P}^{\mathrm{IV},+}_{(k_1-1,0)} X_1 - \mathcal{P}^{\mathrm{IV},+}_{(k_1-1,1)} + \mathcal{P}^{\mathrm{IV},+}_{(k_1-1,0)} - \mathcal{P}^{\mathrm{IV},+}_{(k_1-2,0)}, \\
k_1 = k_2 + 1 \geq 4: \quad & \mathcal{P}^{\mathrm{IV},+}_{(k_1,k_1-1)} = 2\mathcal{P}^{\mathrm{IV},+}_{(k_1-1,k_1-2)} X_2 - 2\mathcal{P}^{\mathrm{IV},+}_{(k_1-1,k_1-3)} X_1 + \mathcal{P}^{\mathrm{IV},+}_{(k_1-1,k_1-4)}, \\
k_1 = k_2 + 1 = 3: \quad & \mathcal{P}^{\mathrm{IV},+}_{(3,2)} = 2\mathcal{P}^{\mathrm{IV},+}_{(2,1)} X_2 - 2\mathcal{P}^{\mathrm{IV},+}_{(2,0)} X_1 - \mathcal{P}^{\mathrm{IV},+}_{(2,0)}, \\
k_1 = k_2 + 1 = 2: \quad & \mathcal{P}^{\mathrm{IV},+}_{(2,1)} = 2\mathcal{P}^{\mathrm{IV},+}_{(1,0)} X_2 + 2\mathcal{P}^{\mathrm{IV},+}_{(1,0)} X_1 - \mathcal{P}^{\mathrm{IV},+}_{(1,1)}, \\
k_1 = k_2 \geq 3: \quad & \mathcal{P}^{\mathrm{IV},+}_{(k_1,k_1)} = 2\mathcal{P}^{\mathrm{IV},+}_{(k_1-1,k_1-1)} X_2 - 4\mathcal{P}^{\mathrm{IV},+}_{(k_1-1,k_1-2)} X_1 + 2\mathcal{P}^{\mathrm{IV},+}_{(k_1-1,k_1-1)} \\
& \quad + 2\mathcal{P}^{\mathrm{IV},+}_{(k_1-1,k_1-3)} + \mathcal{P}^{\mathrm{IV},+}_{(k_1-2,k_1-2)}, \\
k_1 = k_2 = 2: \quad & \mathcal{P}^{\mathrm{IV},+}_{(2,2)} = 2\mathcal{P}^{\mathrm{IV},+}_{(1,1)} X_2 - 4\mathcal{P}^{\mathrm{IV},+}_{(1,0)} X_1 + 2\mathcal{P}^{\mathrm{IV},+}_{(1,1)} \\
& \quad - 2\mathcal{P}^{\mathrm{IV},+}_{(1,0)} + \mathcal{P}^{\mathrm{IV},+}_{(0,0)}.
\end{aligned}
$$

3.4. Polynomials $\mathcal{P}^{\mathrm{IV},-}_{(k_1,k_2)}$

The first three polynomials, calculated directly from definition, are given by

$$
\mathcal{P}^{\mathrm{IV},-}_{(0,0)} = 1, \quad \mathcal{P}^{\mathrm{IV},-}_{(1,0)} = 2X_1 + 1, \quad \mathcal{P}^{\mathrm{IV},-}_{(1,1)} = 2X_2 + 2X_1 + 2. \tag{16}
$$

The higher degree polynomials can then be obtained via recurrence relations of the form

$$\begin{aligned}
k_1 \geq k_2+2 \geq 3: &\quad \mathcal{P}^{\mathrm{IV},-}_{(k_1,k_2)} = 2\mathcal{P}^{\mathrm{IV},-}_{(k_1-1,k_2)}X_1 - \mathcal{P}^{\mathrm{IV},-}_{(k_1-1,k_2+1)} - \mathcal{P}^{\mathrm{IV},-}_{(k_1-1,k_2-1)} - \mathcal{P}^{\mathrm{IV},-}_{(k_1-2,k_2)},\\
k_1 \geq k_2+2 = 2, &\quad \mathcal{P}^{\mathrm{IV},-}_{(k_1,0)} = 2\mathcal{P}^{\mathrm{IV},-}_{(k_1-1,0)}X_1 - \mathcal{P}^{\mathrm{IV},-}_{(k_1-1,1)} + \mathcal{P}^{\mathrm{IV},-}_{(k_1-1,0)} - \mathcal{P}^{\mathrm{IV},-}_{(k_1-2,0)},\\
k_1 = k_2+1 \geq 2: &\quad \mathcal{P}^{\mathrm{IV},-}_{(k_1,k_1-1)} = 2\mathcal{P}^{\mathrm{IV},-}_{(k_1-1,k_1-1)}X_1 - \mathcal{P}^{\mathrm{IV},-}_{(k_1-1,k_1-2)},\\
k_1 = k_2 \geq 3: &\quad \mathcal{P}^{\mathrm{IV},-}_{(k_1,k_1)} = 2\mathcal{P}^{\mathrm{IV},-}_{(k_1-1,k_1-1)}X_2 - 2\mathcal{P}^{\mathrm{IV},-}_{(k_1-1,k_1-2)}X_1 + 2\mathcal{P}^{\mathrm{IV},-}_{(k_1-1,k_1-1)} + \mathcal{P}^{\mathrm{IV},-}_{(k_1-1,k_1-3)},\\
k_1 = k_2 = 3: &\quad \mathcal{P}^{\mathrm{IV},-}_{(2,2)} = 2\mathcal{P}^{\mathrm{IV},-}_{(1,1)}X_2 - 2\mathcal{P}^{\mathrm{IV},-}_{(1,0)}X_1 + 2\mathcal{P}^{\mathrm{IV},-}_{(1,1)} - \mathcal{P}^{\mathrm{IV},-}_{(1,0)}.
\end{aligned}$$

The sets of relations above allow to recursively construct polynomials for any parameter $k = (k_1, k_2) \in \mathbb{Z}^2$, $k_1 \geq k_2 \geq 0$. Furthermore, from these relations is obvious that the polynomials are of degree equal to k_1. For the constructive proof see [4].

4. Orthogonality and cubature formulas

The bivariate Chebyshev-like polynomials defined above fulfill both continuous and discrete orthogonality relations [4]. Furthermore, four cubature formulas of various precision and summation nodes for each bivariate Chebyshev-like polynomials are obtained.

4.1. Continuous orthogonality

The continuous orthogonality relations of the Chebyshev-like polynomials arise directly from the following orthogonality relations of (anti)symmetric bivariate sine functions,

$$\int_{F(\widetilde{S}_2^{\mathrm{aff}})} \sin^-_{k+\rho_i^-}(x)\sin^-_{k'+\rho_i^-}(x)\,dx = \frac{1}{4}\delta_{kk'}, \tag{17}$$

$$\int_{F(\widetilde{S}_2^{\mathrm{aff}})} \sin^+_{k+\rho_i^+}(x)\sin^+_{k'+\rho_i^+}(x)\,dx = \frac{1}{4}H_k\delta_{kk'}, \tag{18}$$

where H_k is the number of permutations stabilizing the parameter k. After the change of variables and rewriting in terms of the bivariate Chebyshev-like polynomials the orthogonality relations transform as

$$\int_{\mathfrak{F}(\widetilde{S}_2^{\mathrm{aff}})} \mathcal{P}^{\mathrm{j},-}_k(X)\mathcal{P}^{\mathrm{j},-}_{k'}(X)w^{\mathrm{j},-}(X)\,dX = \frac{1}{4}\delta_{kk'}, \tag{19}$$

$$\int_{\mathfrak{F}(\widetilde{S}_2^{\mathrm{aff}})} \mathcal{P}^{\mathrm{j},+}_k(X)\mathcal{P}^{\mathrm{j},+}_{k'}(X)w^{\mathrm{j},+}(X)\,dX = \frac{1}{4}H_k\delta_{kk'}, \tag{20}$$

where $j = \mathrm{II}, \mathrm{IV}$ and $\mathfrak{F}(\widetilde{S}_2^{\mathrm{aff}})$ is the fundamental domain of the bivariate sine functions after the change of variables. The weight functions are defined as

$$w^{\mathrm{II},\pm} = \frac{\mathcal{J}^{\mathrm{II},\pm}}{\mathcal{J}}, \qquad w^{\mathrm{IV},\pm} = \frac{\mathcal{J}^{\mathrm{IV},\pm}}{\mathcal{J}} \tag{21}$$

with

$$\begin{aligned}
\mathcal{J}^{\mathrm{II},\pm}(X_1(x), X_2(x)) &= \sin^{\pm}_{\rho_2^{\pm}}(x)\sin^{\pm}_{\rho_2^{\pm}}(x),\\
\mathcal{J}^{\mathrm{IV},\pm}(X_1(x), X_2(x)) &= \sin^{\pm}_{\rho_4^{\pm}}(x)\sin^{\pm}_{\rho_4^{\pm}}(x),
\end{aligned} \tag{22}$$

and

$$\mathcal{J}(X_1(x), X_2(x)) = \left|\det \frac{\partial(X_1(x), X_2(x))}{\partial(x_1, x_2)}\right| = \pi^2 \left|\sin^-_{\rho_2^-}(x)\right|. \tag{23}$$

Journal of Physics: Conference Series **1416** (2019) 012007 doi:10.1088/1742-6596/1416/1/012007

All weight functions are well defined [4] and do not vanish in the interior of the fundamental domain $\mathfrak{F}(\widetilde{S}_2^{\mathrm{aff}})$. When calculated in the variables X_1, X_2, the functions $\mathcal{J}^{\mathrm{II},\pm}$, $\mathcal{J}^{\mathrm{IV},\pm}$ and $\mathcal{J}$ are of the form

$$\begin{aligned}
\mathcal{J}^{\mathrm{II},-}(X_1,X_2) &= \left(X_1^2-2X_2\right)(X_2+2X_1+2)(X_2-2X_1+2),\\
\mathcal{J}^{\mathrm{II},+}(X_1,X_2) &= (X_2+2X_1+2)(X_2-2X_1+2),\\
\mathcal{J}^{\mathrm{IV},-}(X_1,X_2) &= \frac{1}{2}\left(X_1^2-2X_2\right)(X_2-2X_1+2),\\
\mathcal{J}^{\mathrm{IV},+}(X_1,X_2) &= \frac{1}{2}(X_2-2X_1+2),\\
\mathcal{J}(X_1,X_2) &= \pi^2\sqrt{\left(X_1^2-2X_2\right)(X_2+2X_1+2)(X_2-2X_1+2)}.
\end{aligned} \tag{24}$$

4.2. Bivariate cubature formulas

The cubature formulas are obtained when the continuous orthogonality relations above and the discrete bivariate sine transforms are combined. The cubature formulas arising from the bivariate polynomials $\mathcal{P}^{\mathrm{II},-}$ are exemplified bellow. The general cubature formulas for all cases of multivariate (anti)symmetric trigonometric Chebyshev-like polynomials [4, 5] are of similar form. According to their precision, the bivariate cubature formulas induced by the polynomials $\mathcal{P}^{\mathrm{II},-}$ are divided into the four following cases.

(i) The Gaussian cubature formula of precision $2N-1$ that holds exactly for any bivariate polynomial $f(Y)$ of degree at most $2N-1$,

$$\int_{\mathfrak{F}(\widetilde{S}_2^{\mathrm{aff}})} f(Y)w^{\mathrm{II},-}(Y)\,dY = \left(\frac{1}{N+1}\right)^2 \sum_{Y\in\mathfrak{F}_N^{\mathrm{I},-}} f(Y)\mathcal{J}^{\mathrm{II},-}(Y). \tag{25}$$

(ii) The cubature formulas of precision $2N-2$ that hold exactly for bivariate polynomials $f(Y)$ of degree at most $2N-2$,

$$\begin{aligned}
\int_{\mathfrak{F}(\widetilde{S}_2^{\mathrm{aff}})} f(Y)w^{\mathrm{II},-}(Y)\,dY &= \left(\frac{2}{2N+1}\right)^2 \sum_{Y\in\mathfrak{F}_N^{\mathrm{V},-}} f(Y)\mathcal{J}^{\mathrm{II},-}(Y),\\
\int_{\mathfrak{F}(\widetilde{S}_2^{\mathrm{aff}})} f(Y)w^{\mathrm{II},-}(Y)\,dY &= \left(\frac{2}{2N+1}\right)^2 \sum_{Y\in\mathfrak{F}_N^{\mathrm{VI},-}} f(Y)\mathcal{J}^{\mathrm{II},-}(Y).
\end{aligned} \tag{26}$$

(iii) The cubature formula of precision $2N-3$ that holds exactly for any bivariate polynomial $f(Y)$ of degree at most $2N-3$,

$$\int_{\mathfrak{F}(\widetilde{S}_2^{\mathrm{aff}})} f(Y)w^{\mathrm{II},-}(Y)\,dY = \left(\frac{1}{N}\right)^2 \sum_{Y\in\mathfrak{F}_N^{\mathrm{II},-}} f(Y)\mathcal{J}^{\mathrm{II},-}(Y). \tag{27}$$

The explicit forms of summation nodes $\mathfrak{F}_N^{*,-}$, $*=\mathrm{I},\mathrm{II},\mathrm{V},\mathrm{VI}$, which arise from the corresponding bivariate discrete sine transforms, are analogous to the transformed defining relation (4) and detailed in [4].

5. Concluding remarks

- Besides the recurrence relations algorithm, presented for the bivariate (anti)symmetric trigonometric polynomials in this paper, the generating functions method provides another efficient tool for construction of the multivariate polynomials [12, 13]. The calculation procedure and the resulting explicit form of the generating functions for the low-dimensional cases of the present polynomials poses an open problem.
- Criteria for convergence of the polynomial interpolating functions of type (8) and precision Lebesgue constant estimates for the cubature formulas for all cases deserve further study. Good behavior of the bivariate cases, demonstrated in Figure 2, indicates viability of the presented polynomial interpolation methods. Adaptation of the Clenshaw–Curtis [14, 15] methods of numerical integration to the present cases merits additional research.
- Relations between the character functions of representations of simple Lie algebras and the symmetric orbit functions are encompassed in the dominant weight multiplicities [16]. Analogous quantities to dominant weight multiplicities can be defined for the labels of the current (anti)symmetric trigonometric polynomials that are expressed in terms of the multivariate symmetric cosine functions. Theoretical properties as well as practical calculation algorithm for the label multiplicities deserve further study.

Acknowledgments
The authors acknowledge support from the Czech Science Foundation (GAČR), Grant No. 19-19535S. A.B. is also grateful for partial support by the student grant of the Grant Agency of the Czech Technical University in Prague, grant number SGS19/183/OHK4/3T/14.

References

[1] Klimyk A and Patera J 2007 (Anti)symmetric multivariate trigonometric functions and corresponding Fourier transforms *J. Math. Phys.* **48** 093504

[2] Britanak V, Rao K and Yip P 2007 *Discrete cosine and sine transforms: general properties, fast algorithms and integer approximation* (Amsterdam: Elsevier/Academic Press)

[3] Rivlin T J 1990 *The Chebyshev Polynomials* (New York: John Wiley & Sons)

[4] Brus A, Hrivnák J and Motlochová L 2018 Discrete Transforms and Orthogonal Polynomials of (Anti)symmetric Multivariate Sine Functions *Entropy* **20** 938

[5] Hrivnák J and Motlochová L 2010 Discrete transforms and orthogonal polynomials of (anti)symmetric multivariate cosine functions *SIAM J. Numer. Anal.* **51** (7) 073509

[6] Cools R 2003 An encyklopedia of cubature formulas *Journal of Complexity* **19** 445-453

[7] Háková L, Hrivnák J and Motlochová L 2016 On cubature rules associated to Weyl group orbit functions *Acta Polytechnica* **56** 202-213

[8] Hrivnák J, Motlochová L and Patera J 2016 Cubature formulas of multivariate polynomials arising from symmetric orbit functions *Symmetry* **8** (7) 63

[9] Moody R V, Motlochová L and Patera J 2014 Gaussian cubature arising from hybrid characters of simple lie groups *J. Fourier Anal. Appl.* **20** 1257-1290

[10] Hrivnák J and Patera J 2010 Two dimensional symmetric and antisymmetric generalization of exponential and cosine functions *J. Math. Phys.* **51** 023515

[11] Hrivnák J, Motlochová L and Patera J 2010 Two dimensional symmetric and antisymmetric generalization of sine functions *J. Math. Phys.* **51** 073509

[12] Czyżycki T, Hrivnák J and Patera J 2018 Generating functions for orthogonal polynomials of A_2, C_2 and G_2 *Symmetry* **10** 354

[13] Damaskinsky E V, Kulish P P and Sokolov M A, 2015 On calculation of generating functions of Chebyshev polynomials in several variables *J. Math. Phys.* **56** 063507

[14] Clenshaw C W and Curtis A R 1960 A method for numerical integration on an automatic computer *Numer. Math.* **2** 197-205

[15] Sommariva A, Vianello M and Zanovello R 2008 Nontensorial Clenshaw-Curtis cubature *Numer. Algorithms* **49** 409-427

[16] Bremner M, Moody R and Patera J 1985 *Tables of dominant weight multiplicities for representations of simple Lie algebras* (New York: Dekker)

Journal of Physics: Conference Series **1416** (2019) 012008 doi:10.1088/1742-6596/1416/1/012008

Generalized Four-Dimensional Effective Hadronic Supersymmetry based on QCD: New Results

Čestmir Burdik[1,2,†], Sultan Catto[3,4,††], Yasemin Gürcan[5], Amish Khalfan[6,†††], V. Kato La[7] and Enxi Yu[3]

[1]Bogoliubov Laboratory of Theoretical Physics, JINR, 141980 Dubna, Russia
[2]Department of Mathematics, Faculty of Nuclear Sciences and Physical Engineering, Czech Technical University, Prague, Trojanova 13, CZ-120 00 Czech Republic
[3]Physics Department, The Graduate School, City University of New York, New York, NY 10016-4309
[4]Theoretical Physics Group, Rockefeller University, 1230 York Avenue, New York, NY 10021-6399
[5] Department of Science, Borough of Manhattan CC, The City University of NY, New York, NY 10007
[6] Physics Department, LaGuardia CC, The City University of New York, LIC, NY 11101
[7] Columbia University, New York, NY 10027

[†]Grant No. SG15/215/OHK4/3T/14 Czech Technical University in Prague
[††] Work supported in part by the PSC-CUNY Research Awards, and DOE contracts No. DE-AC-0276 ER 03074 and 03075; and NSF Grant No. DMS-8917754.
[†††] Work supported in part by the PSC-CUNY Research Award No:69379-00 47

Abstract. New results are shown on the generalization of phenomenological hadronic supersymmetry based on QCD as an extension of spin-flavor $SU(6)$ symmetry. A more general supersymmetric extension of the group $SU(6/21)$ will be given with applications to pentaquark and tetraquark states for which there is a great deal of ongoing research.

1. Introduction

It was previously shown [1, 2] that the algebra of octonions neatly represented hadronic supersymmetry based on the extension of the spin-flavor group $SU(6)$ to the supergroup $SU(6/21)$. In considering a rotationally excited baryon (qqq), the most energetically stable configuration occurs when two of the quarks come together at a point to form a diquark $(D = qq)$ on one end of a spherical bag structure leaving the remaining quark (q) on the opposite end. Therefore, the baryonic system (qqq) can be effectively likened to a mesonic one $(q\bar{q})$ in which we identify the diquark (D) of the baryon with the antiquark $(\bar{q})$ of the meson. This symmetry between baryons and mesons served as a stepping stone into closely examining the Regge trajectories of various hadronic states. Upon doing so, one readily discovered a parallelism among all the Regge trajectories thereby extracting a universal slope parameter. The parallelism of Regge trajectories established a deep connection between fermionic and bosonic hadronic states. In the context of the work done by Catto and Gürsey [3, 4], baryonic states are fermionic and map into mesonic ones which are bosonic under the transformation of $D = qq \to \bar{q}$, or alternatively, $\bar{D} = \bar{q}\bar{q} \to q$. These symmetries were further bolstered by the use of the split octonionic algebra [5, 6]. In light of this symmetry formalism, mass calculations were carried out [7, 8] concerning low-lying baryonic states. The predicted masses obtained fell to within 5% accuracy of experimental

Journal of Physics: Conference Series **1416** (2019) 012008 doi:10.1088/1742-6596/1416/1/012008

observations. Furthermore, the predicted relationship between the mass differences among mesons and the mass differences among baryons was found to be in excellent agreement with experimental results. For example, the difference in π and ρ masses as it relates to the difference in the masses of N and Δ were within less than 1% accuracy.

The algebra of octonions is the only Heisenberg-like algebra in modern mathematics to incorporate color degrees of freedom that underlie hadronic phenomena. The use of octonions in hadronic physics gave rise to dynamical supersymmetry and provided the first description of existing quark models through the edifice of exceptional Lie algebras. Supergroups and infinite groups, like those generated by the Virasoro algebra, emerged as useful descriptions of certain properties regarding hadronic spectra. The exceptional Lie groups are G_2, F_4, E_6, E_7 and E_8.

Are supersymmetry and its accompanying algebraic structures effective properties of more complex structures? Are they to be found in an exact form only through an artificial extension of physical Hilbert spaces, or, are they really fundamental in terms of a new extension of quantum field theory? We may not be able to provide answers to such questions but our hope is to be able to shed light on these problems and define them more precisely.

The outstanding difficulty in applying supersymmetry to hadrons was that supersymmetry is badly broken, otherwise the proton and the pion would have the same mass. The arrival of broken supersymmetry stems from the underlying difference between quark and antidiquark masses, differences in spin, and also, variations in size.

It should be noted that the spin-dependence of forces in QCD plays an important role. This effect was minimized by performing an appropriate averaging over spin. The main assumption here was that the spin-dependent interaction energy between two quarks in a diquark is independent of the nature of the hadron in which the diquark is embedded. The spin-dependent contribution to the interaction energy was approximately extracted from the known values of baryon masses.

Presently in literature there exists no conclusive papers on the sizes of quarks or diquarks. Recently [9, 10] we have constructed octonionic geometries that tie together, in an algebraic way, many well-known and apparently disconnected projective geometries [11]. Through such octonionic geometries, there appears to be novel ways of relating diquark and antiquark sizes within a bag structure of a $q - D$ system. We present relevant theoretical work and await experimental results from LHC, SLAC and other labs for potential confirmations. In addition, herein we develop a broad scheme for calculating hadronic masses based on an extension of $SU(6/21)$.

2. Bilocal approximation to hadronic structure and inclusion of color.

It is known that the low-lying baryons occur in the symmetric 56 representation [12] of $SU(6)$, whereas the Pauli principle would have led to the antisymmetrical 20 representation. This fact was a crucial fact for the introduction of color degree of freedom [13] based on $SU(3)^c$. Hence, to obtain the correct representation of the $q - D$ system we must also include color degrees of freedom for the constituents. Since the quark field transforms like a color triplet and the diquark like a color antitriplet under $SU(3)^c$, the color degrees of freedom of the constituents must be included correctly in order to obtain a correct representation of the q-D system. Hadronic states must be color singlets. These are represented by bilocal operators $O(\mathbf{r}_1, \mathbf{r}_2)$ in the bilocal approximation (for (qD) system see [14]) that gives $\bar{q}(1)q(2)$ for mesons and $D(1)q(2)$ for baryons. Here $\bar{q}(1)$ represents the antiquark situated at $\mathbf{r}_1$, $q(2)$ the quark situated at $\mathbf{r}_2$, and $D(1) = q(1)q(1)$ the diquark situated at $\mathbf{r}_1$. If we denote the c.m. and the relative coordinates of the constituents by $\mathbf{R}$ and $\mathbf{r}$, where $\mathbf{r} = \mathbf{r}_2 - \mathbf{r}_1$ and

$$\mathbf{R} = \frac{(m_1\mathbf{r}_1 + m_2\mathbf{r}_2)}{(m_1 + m_2)} \tag{1}$$

with m_1 and m_2 being their masses, we can then write $O(\mathbf{R}, \mathbf{r})$ for the operator that creates hadrons out of the vacuum. The matrix element of this operator between the vacuum and the hadronic state h will be

Journal of Physics: Conference Series **1416** (2019) 012008 doi:10.1088/1742-6596/1416/1/012008

of the form

$$< h|O(\mathbf{R},\mathbf{r})|0 >= \chi(\mathbf{R})\psi(\mathbf{r}) \tag{2}$$

where $\chi(\mathbf{R})$ is the free wave function of the hadron as a function of the c.m. coordinate and $\psi(\mathbf{r})$ is the bound-state solution of the $U(6/21)$ invariant Hamiltonian describing the $q-\bar{q}$ mesons, $q-D$ baryons, $\bar{q}-\bar{D}$ antibaryons and $D-\bar{D}$ exotic mesons, given by

$$\begin{aligned} i\partial_t\psi_{\alpha\beta} &= [\sqrt{(m_\alpha+\frac{1}{2}V_s)^2+\mathbf{p}^2}+\sqrt{(m_\beta+\frac{1}{2}V_s)^2+\mathbf{p}^2} \\ &\quad -\frac{4}{3}\frac{\alpha_s}{r}+k\frac{\mathbf{s}_\alpha\cdot\mathbf{s}_\beta}{m_\alpha m_\beta}]\ \psi_{\alpha\beta} \end{aligned} \tag{3}$$

Here $\mathbf{p}=-i\nabla$ in the c.m. system and m and s denote the masses and spins of the constituents, α_s the strong-coupling constant, $V_s=br$ is the scalar potential with r being the distance between the constituents in the bilocal object, and $k=|\psi(0)|^2$.

The operator product expansion [15] will give a singular part depending only on $\mathbf{r}$ and proportional to the propagator of the field binding the two constituents. There will be a finite number of singular coefficients $c_n(\mathbf{r})$ depending on the dimensionality of the constituent fields. For example, for a meson, the singular term is proportional to the propagator of the gluon field binding the two constituents. Once we subtract the singular part, the remaining part $\tilde{O}(\mathbf{R},\mathbf{r})$ is analytic in r and thus we can write

$$\tilde{O}(\mathbf{R},\mathbf{r})=O_0(\mathbf{R})+\mathbf{r}\cdot\mathbf{O}_1(\mathbf{R})+o(r^2), \tag{4}$$

where the remainder term is of the order r^2.

Now $O_0(\mathbf{R})$ creates a hadron at its c.m. point $\mathbf{R}$ equivalent to a $\ell=0$, s-state of the two constituents. For a baryon this is a state associated with q and $D\sim qq$ at the same point $\mathbf{R}$, hence it is essentially a 3-quark state when the three quarks are at a common location. The $\mathbf{O}_1(\mathbf{R})$ can create three $\ell=1$ states with opposite parity to the state created by $O_0(\mathbf{R})$. Hence, if we denote the nonsingular parts of $\bar{q}(1)q(2)$ and $D(1)q(2)$ by $[\bar{q}(1)q(2)]$ and $[D(1)q(2)]$, respectively, we have

$$[\bar{q}(1)q(2)]|0>=|M(\mathbf{R})>+\mathbf{r}\cdot|\mathbf{M}'(\mathbf{R})>+o(r^2), \tag{5}$$

$$[D(1)q(2)]|0>=|B(\mathbf{R})>+\mathbf{r}\cdot|\mathbf{B}'(\mathbf{R})>+o(r^2), \tag{6}$$

and similarly for the exotic meson states $D(1)\bar{D}(2)$.

The meson states $|M(\mathbf{R})$ being $\ell=0$ bound state of a quark and antiquark will correspond to the singlet and 35-dimensional representations of $SU(6)$. The $M'(\mathbf{R})$ is an orbital excitation ($\ell=1$) of opposite parity, which are in the $(35+1,3)$ representation of the group $SU(6)\times O(3)$, $O(3)$ being associated with the relative angular momentum of the constituents. The M' states contain mesons like B, $\mathbf{A}_1$, $\mathbf{A}_2$, and scalar particles. On the whole, the $\ell=0$ and $\ell=1$ part $\bar{q}(1)q(2)$ contain $4\times(35+1)=144$ meson states.

Switching to the baryon states, the requirement of antisymmetry in color, and symmetry in spin-flavor indices gives the $(56)^+$ representation for $B(\mathbf{R})$. The $\ell=1$ multiplets have negative parity and have mixed spin-flavor symmetry. They belong to the representation $(70^-,3)$ of $SU(6)\times O(3)$ and are represented by the states $|\mathbf{B}'(\mathbf{R})>$ which are 210 in number. On the whole, these 266 states account for all the observed low-lying baryon states obtained form $56+3\times 70=266$. A similar analysis can be carried out for the exotic meson states $D(1)\bar{D}(2)$, where the diquark and the antidiquark can be bound in a $\ell=0$ or $\ell=1$ state with opposite parities.

Journal of Physics: Conference Series **1416** (2019) 012008 doi:10.1088/1742-6596/1416/1/012008

3. Algebraic treatment of color through Split Octonions.

The octonions have seven imaginary units e_α ($\alpha = 1, \dots, 7$) which obey the algebra

$$e_\alpha e_\beta = -\delta_{\alpha\beta} + \epsilon_{\alpha\beta\gamma} e_\gamma \tag{7}$$

Here the structure constants $\epsilon_{\alpha\beta\gamma}$ are antisymmetrical in all three indices. They are equal to one for $(\alpha\beta\gamma)$ being of the seven triplets (123), (246), (435), (367), (651), (572) and (714). They vanish otherwise.

The behavior of various states under the color group are best seen if we use split octonion units defined by

$$u_0 = \frac{1}{2}(1 + ie_7), \qquad u_0^* = \frac{1}{2}(1 - ie_7), \tag{8}$$

$$u_j = \frac{1}{2}(e_j + ie_{j+3}), \quad u_j^* = \frac{1}{2}(e_j - ie_{j+3}), \quad j = 1, 2, 3. \tag{9}$$

The automorphism group of the octonion algebra is the 14-parameter exceptional group G_2. The imaginary octonion units $e_\alpha(\alpha = 1, ..., 7)$ fall into its 7-dimensional representation.

Under the $SU(3)^c$ subgroup of G_2 that leaves e_7 invariant, u_0 and u_0^* are singlets, while u_j and u_j^* correspond, respectively, to the representations $\mathbf{3}$ and $\bar{\mathbf{3}}$. The multiplication table can now be written in a manifestly $SU(3)^c$ invariant manner (together with the complex conjugate equations):

$$u_0^2 = u_0, \qquad u_0 u_0^* = 0 \tag{10}$$

$$u_0 u_j = u_j u_0^* = u_j, \qquad u_0^* u_j = u_j u_0 = 0 \tag{11}$$

$$u_i u_j = -u_j u_i = \epsilon_{ijk} u_k^*, \tag{12}$$

$$u_i u_j^* = -\delta_{ij} u_0 \tag{13}$$

where ϵ_{ijk} is completely antisymmetric with $\epsilon_{ijk} = 1$ for ijk = 123, 246, 435, 651, 572, 714, 367. Here, one sees the virtue of octonion multiplication. If we consider the direct products

$$C: \quad \mathbf{3} \otimes \bar{\mathbf{3}} = \mathbf{1} + \mathbf{8} \tag{14}$$

$$G: \quad \mathbf{3} \otimes \mathbf{3} = \bar{\mathbf{3}} + \mathbf{6} \tag{15}$$

for $SU(3)^c$, then these equations show that octonion multiplication gets rid of $\mathbf{8}$ in $\mathbf{3} \otimes \bar{\mathbf{3}}$, while it gets rid of $\mathbf{6}$ in $\mathbf{3} \otimes \mathbf{3}$. Combining Eq.(12) and Eq.(13) we find

$$(u_i u_j) u_k = -\epsilon_{ijk} u_0^* \tag{16}$$

Thus the octonion product leaves only the color part in $\mathbf{3}\otimes\bar{\mathbf{3}}$ and $\mathbf{3}\otimes\mathbf{3}\otimes\mathbf{3}$, so that it is a natural algebra for colored quarks. We note that under the $SU(3)$ subgroup of G_2 (the 14 parameter automorphism group of octonions), the split units u_i transform like a triplet, u_i^* like an antitriplet, u_0 and u_0^* like singlets. From the octonion algebra we can extract a subalgebra satisfied by u_i and u_j^* only, i.e.

$$\{u_i, u_j\} = 0, \quad \{u_i^\dagger, u_j^\dagger\} = 0, \quad \{u_i, u_j^\dagger\} = \delta_{ij}, \tag{17}$$

where

$$u_i^\dagger = -u_i^* = \bar{u_i}^*, \tag{18}$$

the bar denoting octonionic conjugation. This tells us that the three split units u_i are Grassmann numbers. u_j^* form a conjugate set of Grassmann numbers and together they form a fermionic Heisenberg algebra. Note that the split units represent an exceptional Grassmann algebra because they are not associative. In fact their associator is

$$[u_i, u_j, u_k] - (u_i, u_j)u_k - u_i(u_j u_k) = \epsilon_{ijk}(u_0 - u_0^*) = i\epsilon_{ijk} e_7. \tag{19}$$

We also have

$$[u_i, u_j, u_k^*] = \delta_{ki} u_j - \delta_{kj} u_i. \tag{20}$$

As a result the split units, like the imaginary units e_α cannot be represented by matrices. Note that the split octonion algebra we gave above is such that the product of two triplets gives an antitriplet, while the product of a triplet and an antitriplet gives a singlet. Hence the dynamical suppression of octet and sexted states mentioned above is automatically achieved.

The quarks, being in the triplet representation of the color group $SU(3)^c$, they are represented by the local fields $q_\alpha^i(x)$, where $i = 1, 2, 3$ is the color index and α the combined spin-flavor index. Antiquarks at point y are color antitriplets $\bar{q}_\beta^i(y)$. Consider the two-body systems

$$C_{\alpha j}^{\beta i} = q_\alpha^i(x_1)\bar{q}_\beta^j(x_2) \tag{21}$$

$$G_{\alpha\beta}^{ij} = q_\alpha^i(x_1)q_\beta^j(x_2) \tag{22}$$

so that C is either a color singlet or color octet, while G is a color antitriplet or a color sextet. Now C contains meson states that are color singlets and hence observable. The octet $q - \bar{q}$ state is confined and not observed as a scattering state. In the case of two-body G states, the antitriplets are diquarks which, inside a hadron can be combined with another triplet quark to give observable, color singlet, three-quark baryon states. The color sextet part of G can only combine with a third quark to give unobservable color octet and color decuplet three-quark states. Hence the hadron dynamics is such that the **8** part of C and the **6** part of G are suppressed. This can best be achieved by the use of above split octonion algebra. The dynamical suppression of the octet and sextet states in Eq.(21) and Eq.(22) is therefore automatically achieved. The split octonion units can be contracted with color indices of triplet or antitriplet fields. For quarks and antiquarks we can define the "transverse" octonions (calling u_0 and u_0^* longitudinal units)

$$q_\alpha = u_i q_\alpha^i = \mathbf{u} \cdot \mathbf{q}_\alpha, \quad \bar{q}_\beta = u_i^\dagger \bar{q}_\beta^j = -\mathbf{u}^* \cdot \bar{\mathbf{q}}_\beta \tag{23}$$

We find

$$q_\alpha(1)\bar{q}_\beta(2) = u_0 \mathbf{q}_\alpha(1) \cdot \bar{\mathbf{q}}_\beta(2) \tag{24}$$

$$\bar{q}_\alpha(1)q_\beta(2) = u_0^* \bar{\mathbf{q}}_\alpha(1) \cdot \mathbf{q}_\beta(2) \tag{25}$$

$$G_{\alpha\beta}(12) = q_\alpha(1)q_\beta(2) = \mathbf{u}^* \cdot \mathbf{q}_\alpha(1) \times \mathbf{q}_\beta(2) \tag{26}$$

$$G_{\beta\alpha}(21) = q_\beta(2)q_\alpha(1) = \mathbf{u}^* \cdot \mathbf{q}_\beta(2) \times \mathbf{q}_\alpha(1) \tag{27}$$

Because of the anticommutativity of the quark fields, we have

$$G_{\alpha\beta}(12) = G_{\beta\alpha}(21) = \frac{1}{2}\{q_\alpha(1), q_\beta(2)\} \tag{28}$$

If the diquark forms a bound state represented by a field $D_{\alpha\beta}(x)$ at the center-of-mass location x

$$x = \frac{1}{2}(x_1 + x_2), \tag{29}$$

when x_2 tends to x_1 we can replace the argument by x, and we obtain

$$D_{\alpha\beta}(x) = D_{\beta\alpha}(x) \tag{30}$$

so that the local diquark field must be in a symmetric representation of the spin-flavor group. If the latter is taken to be $SU(6)$, then $D_{\alpha\beta}(x)$ is in the 21-dimensional symmetric representation, given by

$$(\mathbf{6} \otimes \mathbf{6})_s = \mathbf{21} \tag{31}$$

If we denote the antisymmetric 15 representation by $\Delta_{\alpha\beta}$, we see that the octonionic fields single out the 21 diquark representation at the expense of $\Delta_{\alpha\beta}$. We note that without this color algebra supersymmetry would give antisymmetric configurations as noted by Salam and Strathdee [16] in their possible supersymmetric generalization of hadronic supersymmetry. Using the nonsingular part of the operator product expansion we can write

$$\tilde{G}_{\alpha\beta}(\mathbf{x}_1, \mathbf{x}_2) = D_{\alpha\beta}(x) + \mathbf{r} \cdot \boldsymbol{\Delta}_{\alpha\beta}(x) \tag{32}$$

The fields $\Delta_{\alpha\beta}$ have opposite parity to $D_{\alpha\beta}$; $\mathbf{r}$ is the relative coordinate at time t if we take $t = t_1 = t_2$. They play no role in the excited baryon which becomes a bilocal system with the 21- dimensional diquark as one of its constituents.

Now consider a three-quark system at time t. The c.m. and relative coordinates are

$$\mathbf{R} = \frac{1}{\sqrt{3}}(\mathbf{r}_1 + \mathbf{r}_2 + \mathbf{r}_3) \tag{33}$$

$$\boldsymbol{\rho} = \frac{1}{\sqrt{6}}(2\mathbf{r}_3 - \mathbf{r}_1 - \mathbf{r}_2) \tag{34}$$

$$\mathbf{r} = \frac{1}{\sqrt{2}}(\mathbf{r}_1 - \mathbf{r}_2), \tag{35}$$

giving

$$\mathbf{r}_1 = \frac{1}{\sqrt{3}}\mathbf{R} - \frac{1}{\sqrt{6}}\boldsymbol{\rho} + \frac{1}{\sqrt{2}}\mathbf{r} \tag{36}$$

$$\mathbf{r}_2 = \frac{1}{\sqrt{3}}\mathbf{R} - \frac{1}{\sqrt{6}}\boldsymbol{\rho} - \frac{1}{\sqrt{2}}\mathbf{r} \tag{37}$$

$$\mathbf{r}_3 = \frac{1}{\sqrt{3}}\mathbf{R} + \frac{2}{\sqrt{6}}\boldsymbol{\rho} \tag{38}$$

The baryon state must be a color singlet, symmetric in the three pairs (α, x_1), (β, x_2), (γ, x_3). We find

$$(q_\alpha(1)q_\beta(2))q_\gamma(3) = -u_0^* F_{\alpha\beta\gamma}(123) \tag{39}$$

$$q_\gamma(3)(q_\alpha(1)q_\beta(2)) = -u_0 F_{\alpha\beta\gamma}(123) \tag{40}$$

so that

$$-\frac{1}{2}\{\{q_\alpha(1), q_\beta(2)\}, q_\gamma(3)\} = F_{\alpha\beta\gamma}(123) \tag{41}$$

The operator $F_{\alpha\beta\gamma}(123)$ is a color singlet and is symmetrical in the three pairs of coordinates. We have

$$F_{\alpha\beta\gamma}(123) = B_{\alpha\beta\gamma}(\mathbf{R}) + \boldsymbol{\rho} \cdot \mathbf{B}'(\mathbf{R}) + \mathbf{r} \cdot \mathbf{B}''(\mathbf{R}) + C \tag{42}$$

where C is of order two and higher in $\boldsymbol{\rho}$ and $\mathbf{r}$. Because $\mathbf{R}$ is symmetric in $\mathbf{r}_1$, $\mathbf{r}_2$ and $\mathbf{r}_3$, the operator $B_{\alpha\beta\gamma}$ that creates a baryon at $\mathbf{R}$ is totally symmetrical in its flavor-spin indices. In the $SU(6)$ scheme it belongs to the (56) representation. In the bilocal $q - D$ approximation we have $\mathbf{r} = 0$ so that $F_{\alpha\beta\gamma}$ is a function only of $\mathbf{R}$ and $\boldsymbol{\rho}$ which are both symmetrical in $\mathbf{r}_1$ and $\mathbf{r}_2$. As before, $\mathbf{B}'$ belongs to the orbitally excited 70^- representation of $SU(6)$. The totally antisymmetrical representation (20) is absent in the bilocal approximation. It would only appear in the trilocal treatment that would involve the 15-dimensional diquarks. Hence, if we use local fields, any product of two octonionic quark fields gives a (21) diquark

$$q_\alpha(\mathbf{R})q_\beta(\mathbf{R}) = D_{\alpha\beta}(\mathbf{R}), \tag{43}$$

and any nonassociative combination of three quarks, or a diquark and a quark at the same point give a baryon in the 56^+ representation:

$$(q_\alpha(\mathbf{R})q_\beta(\mathbf{R}))q_\gamma(\mathbf{R}) = -u_0^* B_{\alpha\beta\gamma}(\mathbf{R}) \tag{44}$$

$$q_\alpha(\mathbf{R})(q_\beta(\mathbf{R})q_\gamma(\mathbf{R})) = -u_0 B_{\alpha\beta\gamma}(\mathbf{R}) \tag{45}$$

$$q_\gamma(\mathbf{R})(q_\alpha(\mathbf{R})q_\beta(\mathbf{R})) = -u_0 B_{\alpha\beta\gamma}(\mathbf{R}) \tag{46}$$

$$(q_\gamma(\mathbf{R})q_\alpha(\mathbf{R}))q_\beta(\mathbf{R}) = -u_0^* B_{\alpha\beta\gamma}(\mathbf{R}) \tag{47}$$

The bilocal approximation gives the $(35 + 1)$ mesons and the 70^- baryons with $\ell = 1$ orbital excitation.

4. Algebraic Realization

Octonionic Supersymmetry of Catto and Gürsey (C-G) was based on the supergroup $SU(6/21)$ acts on a quark and antidiquark situated at the same point x_1. At the point x_2 we can consider the action of the supergroup with the same parameters, or one with different parameters. In the first case we have a global symmetry. In the second case, if we only deal with bilocal fields the symmetry will be represented by $SU(6/21) \times SU(6/21)$, doubling the C-G supergroup. On the other hand, if any number of points are considered, with different parameters attached to each point, we are led to introduce a local supersymmetry $SU(6/21)$ to which we should add the local color group $SU(3)^c$. Since it is not a fundamental symmetry, we shall not deal with the local $SU(6/21)$ group here. However, the double $SU(6/21)$ supergroup is useful for bilocal fields since the decomposition of the adjoint representation of the 728-dimensional group with respect to $SU(6) \times SU(21)$ gives

$$728 = (35, 1) + (1, 440) + (6, 21) + (\bar{6}, \bar{21}) + (1, 1) \tag{48}$$

A further decomposition of the double $SU(6/21)$ supergroup into its field with respect to its c.m. coordinates, as seen above (see Eq.(42)), leads to the decomposition of the 126-dimensional cosets $(6, 21)$ and $(21, 6)$ into $56^+ + 70^-$ of the diagonal $SU(6)$.

We would have a much tighter and more elegant scheme if we could perform such a decomposition from the start and be able to identify $(1, 21)$ part of the fundamental representation of $SU(6/21)$ with the 21-dimensional representation of the $SU(6)$ subgroup, which means going beyond the $SU(6/21)$ supersymmetry to a smaller supergroup having $SU(6)$ as a subgroup.

5. Beyond $SU(6/21)$– A minimal scheme

In order to go beyond the $SU(6/21)$ supersymmetry we propose a nonsimple super-Lie algebra that has parabolic structure with the odd part being represented by a symmetric third rank tensor $F_{\alpha\beta\gamma}$ with 56 components and the even part by the 36 $U(6)$ generators L^ρ_σ. The bracket relations are

$$[L^\rho_\sigma, L^\lambda_\mu] = \delta^\rho_\mu L^\lambda_\sigma - \delta^\lambda_\sigma L^\rho_\mu \tag{49}$$

$$\{F_{\alpha\beta\gamma}, F_{\rho\sigma\tau}\} = 0 \tag{50}$$

$$[L^\rho_\sigma, F_{\alpha\beta\gamma}] = \delta^\rho_\alpha F_{\beta\gamma\sigma} + \delta^\rho_\beta F_{\gamma\alpha\sigma} + \delta^\rho_\gamma F_{\alpha\beta\sigma} \tag{51}$$

There is a conjugate superalgebra generated by L^ρ_σ and $F^{\alpha\beta\gamma}$. Note that

$$N = \sum_\rho L^\rho_\rho \tag{52}$$

acts like a number operator and traceless part of L^ρ_σ coincides with the elements of the $SU(6)$ Lie algebra. Hence this minimal superalgebra S has $56 + 35 = 91$ generators. With the addition of color, the algebra becomes $SU(3)^c \times S$.

We can immediately construct a matrix representation of this algebra by means of fermionic operators f^i_α that transform like the $(3,6)$ representation of $SU(3)^c \times SU(6)$ and the conjugate operators $\bar{f}^\beta_j$ associated with the $(\bar{3},\bar{6})$ representation. We have

$$\{f^i_\alpha, f^j_\beta\} = 0, \qquad \{\bar{f}^\alpha_i, \bar{f}^\beta_j\} = 0, \tag{53}$$

$$\{f^i_\alpha, \bar{f}^\beta_j\} = \delta^i_j \delta^\beta_\alpha. \tag{54}$$

Since f^i_α are Grassmann numbers they can be represented by finite matrices belonging to a Clifford algebra. The conjugate operators $\bar{f}^i_\alpha$ are represented by the corresponding Hermitian conjugate matrices. Out of this fermionic Heisenberg basis we construct the following matrices:

$$\Phi_{\alpha\beta\gamma} = \epsilon_{ijk} f^i_\alpha f^j_\beta f^k_\gamma \tag{55}$$

$$M^\rho_\sigma = \bar{f}^\rho_i f^i_\sigma - \frac{1}{6}\delta^\rho_\sigma N \tag{56}$$

where

$$N = \bar{f}^\tau_j f^j_\tau \tag{57}$$

Then $\Phi_{\alpha\beta\gamma}$ and M^ρ_α are the representations of $F_{\alpha\beta\gamma}$ and L^ρ_σ since they satisfy the algebra of Eqs.(49-51). If f^i_α are identified with the colored-quark annihilation operators, the odd elements of Φ of the algebra correspond to baryons, while the even elements M correspond to mesons. N represents the number operator.

At this point let us introduce a module that consists of the representations $(\bar{3},\bar{6})$ and $(\bar{3},21)$ of $SU(3)^c \times SU(6)$. They correspond to the operators associated with the antiquarks and diquarks:

$$\bar{Q}^\alpha_j = \bar{f}^\alpha_j, \qquad D_{\alpha\beta k} = \epsilon_{ijk} f^i_\alpha f^j_\beta \tag{58}$$

This multiplet is closed under the group $SU(3)^c \times S$. Indeed we have

$$\{\Phi_{\alpha\beta\gamma}, \bar{Q}^\rho_k\} = \delta^\rho_\alpha D_{\beta\gamma k} + \delta^\rho_\beta D_{\gamma\alpha k} + \delta^\rho_\gamma D_{\alpha\beta k} \tag{59}$$

$$[\Phi_{\rho\sigma\tau}, D_{\alpha\beta k}] = 0 \tag{60}$$

$$[M^\gamma_\rho, \bar{Q}^\sigma_k] = \delta^\sigma_\rho \bar{Q}^\gamma_k \tag{61}$$

$$[M^\gamma_\rho, D_{\alpha\beta k}] = \delta^\gamma_\alpha D_{\beta\rho k} + \delta^\gamma_\beta D_{\alpha\rho k} \tag{62}$$

Eq.(61) and Eq.(62) show that $\bar{Q}$ and D transform like $(\bar{6})$ and (21) under $SU(6)$. Eq.(59) and Eq.(60) tell us that the odd part of the supergroup maps $\bar{Q}$ into D and D into zero. Letting the corresponding odd parameters be $\eta^{\alpha\beta\gamma}$, we now consider the superalgebra element

$$\Phi = \frac{1}{6}\eta^{\alpha\beta\gamma}\Phi_{\alpha\beta\gamma} \tag{63}$$

We find

$$\delta\bar{Q}^\rho_k = [\Phi, \bar{Q}^\rho_k] = \frac{1}{2}\eta^{\rho\alpha\beta} D_{\alpha\beta k} \tag{64}$$

$$\delta D_{\rho\sigma} = [\Phi, D_{\rho\sigma}] = 0 \tag{65}$$

Neglecting mass and spin differences, the Hamiltonian we wrote is invariant under the infinitesimal transformations given above.

The baryon-meson system transforms like the adjoint representation of the supergroup S. We find

$$\delta B_{\alpha\beta\gamma} = [\Phi, B_{\alpha\beta\gamma}] = 0 \tag{66}$$

$$\delta M^{\rho}_{\sigma} = [\Phi, M^{\rho}_{\sigma}] = -\frac{1}{2}\eta^{\rho\alpha\beta} B_{\alpha\beta\sigma} \tag{67}$$

Hence the odd part of S maps mesons into baryons and baryons into zero. If the conjugate supergroup is used, baryons and diquarks are replaced by antibaryons and antidiquarks, while the antiquarks are replaced by quarks.

Note that in this minimal scheme the exotic mesons do not appear. Nevertheless the transformations Eq.(66) and Eq.(67) are sufficient for establishing the equality of the Regge slopes for baryons and mesons.

Because of the form of the cubic function and the quadratic functions of the generators of S in terms of the fermionic annihilation and creation operators, it is natural to introduce the transverse octonion operators as in equation Eq.(23)

$$f_{\alpha} = u_i f^i_{\alpha}, \qquad \bar{f}^{\beta} = -u^{*j} \bar{f}^{\beta}_j \tag{68}$$

Then we obtain in a general fashion

$$\Phi_{\alpha\beta\gamma} = \frac{1}{2}\{\{f_{\alpha}, f_{\beta}\}, f_{\gamma}\} \tag{69}$$

$$\{\bar{f}^{\rho}, f_{\rho}\} = \frac{3}{2}\delta^{\rho}_{\sigma} + ie_7 M^{\rho}_{\sigma} \tag{70}$$

We also have

$$[f_{\alpha}, f_{\beta}] = 0, \quad [\bar{f}^{\alpha}, \bar{f}^{\beta}] = 0 \tag{71}$$

$$[f_{\alpha}, \bar{f}^{\beta}] = M^{\beta}_{\alpha} + \frac{3}{2} ie_7 \delta^{\beta}_{\alpha} \tag{72}$$

Eq.(71) and Eq.(72) together with the $SU(6)$ Lie algebra of the form Eq.(49) satisfied by the operators M^{β}_{α} suggest that M^{β}_{α}, f_{α}, $\bar{f}^{\beta}$, and $ie_7 = (u_0 - u^*_0)$ form an octonionic extension of $SU(6)$ that has $SU(6) \times SU(3)^c$ as a subgroup. It is related to the simple supergroup $SU(6/3)$, but not equivalent to it because of the suppression of the color octet and sextet states in the octonionic products $f_{\alpha}\bar{f}^{\beta}$ and $f_{\alpha}f_{\beta}$.

The bilocal treatment outlined earlier gets carried over unchanged to the minimal scheme. The $(\bar{q}, D)$ system forms one multiplet $R(1)$ at point x_1 under the supergroup S. The conjugate system $(q, \bar{D})$ forms a multiplet $R(2)$ at point x_2 under the conjugate group $\bar{S}$. Then the bilocal system $R(1) \times R(2)$ is expanded as in Eq.(32) and it transforms under the group $S \times \bar{S}$. Its local part with respect to c.m. coordinate x gives the meson-baryon and the meson-antibaryon multiplets $(35, 56^{+})$ and $(35, 56^{+})$, while the coefficients of the relative coordinate $\mathbf{r}$ give the $\ell = 1$ excitations of mesons in the $(35 + 1)$ representations and the $\ell = 1$ excited baryons in the 70^{-} representation of $SU(6)$. In the symmetric approximation, the masses of various states are proportional to the sum of the quark number plus the antiquark number.

6. Symmetries of the Three Quark System

For the three quark system we can write the generalized form of the two body Hamiltonian as

$$\begin{aligned}
\tilde{H}_{123} &= \tilde{V}_{12} + \tilde{V}_{23} + \tilde{V}_{31} + \gamma_4^{(1)}\sqrt{[m_1 + \frac{1}{2}S_{12}]^2 + \boldsymbol{p}_{12}^2} \\
&+\gamma_4^{(2)}\sqrt{[m_2 + \frac{1}{2}S_{12}]^2 + \boldsymbol{p}_{12}^2} + \gamma_4^{(2)}\sqrt{[m_2 + \frac{1}{2}S_{23}]^2 + \boldsymbol{p}_{23}^2} \\
&+\gamma_4^{(3)}\sqrt{[m_3 + \frac{1}{2}S_{23}]^2 + \boldsymbol{p}_{23}^2} + \gamma_4^{(3)}\sqrt{[m_3 + \frac{1}{2}S_{31}]^2 + \boldsymbol{p}_{31}^2} \\
&+\gamma_4^{(1)}\sqrt{[m_1 + \frac{1}{2}S_{31}]^2 + \boldsymbol{p}_{31}^2}
\end{aligned} \tag{73}$$

Journal of Physics: Conference Series **1416** (2019) 012008 doi:10.1088/1742-6596/1416/1/012008

where

$$\tilde{V}_{ij} = V_{ij} + \text{spin dependent terms} \tag{74}$$

with $i, j = 1, 2, 3$.

To study the symmetry of this system we introduce fermion annihilation and creation operators for colored quarks f^i_α with $i = 1, 2, 3$ the color index, and $\alpha = 1, \ldots, 6$ the flavor index, so that

$$\{f^i_\alpha, f^j_\beta\} = 0, \quad \{\bar{f}^\alpha_i, \bar{f}^\beta_j\} = 0, \quad \{f^i_\alpha, \bar{f}^\beta_j\} = \delta^i_j \delta^\beta_\alpha \tag{75}$$

The diquark transforms like

$$D_{\alpha\beta k} = \epsilon_{ijk} f^i_\alpha f^k_\beta \tag{76}$$

It is in $(\bar{3})$ representation for color, and in 21 representation for $SU(6)$. The baryons transform like

$$\Phi_{\alpha\beta\gamma} = \epsilon_{ijk} f^i_\alpha f^j_\beta f^k_\gamma \tag{77}$$

It is a color singlet belonging to the 56 representation of $SU(6)$. Color singlet mesons that are in the 35 representation of $SU(6)$ transform like

$$M^\rho_\sigma = \bar{f}^\rho_i f^i_\sigma - \frac{1}{6}\delta^\rho_\sigma N \tag{78}$$

with N being a color singlet and an $SU(6)$ singlet, and is given by

$$N = \bar{f}^\tau_j f^j_\tau \tag{79}$$

We have the relations

$$[M^\rho_\sigma, M^\lambda_\mu] = \delta^\rho_\mu M^\lambda_\sigma - \delta^\lambda_\sigma M^\rho_\mu \tag{80}$$

$$\{\Phi_{\alpha\beta\gamma}, \Phi_{\rho\sigma\tau}\} = 0 \tag{81}$$

$$[M^\rho_\sigma, \Phi_{\rho\sigma\tau}] = \delta^\rho_\alpha \Phi_{\beta\gamma\sigma} + \delta^\rho_\beta \Phi_{\gamma\alpha\sigma} + \delta^\rho_\gamma \Phi_{\alpha\beta\sigma} \tag{82}$$

Now $\bar{f}^\rho_j$ and $D_{\alpha\beta,k}$ form a multiplet that transforms under Eqs.(80 - 82) as follows:

$$\{\Phi_{\alpha\beta\gamma}, \bar{f}^\alpha_k\} = \delta^\rho_\alpha D_{\beta\gamma,k} + \delta^\rho_\beta D_{\gamma\alpha,k} + \delta^\rho_\gamma D_{\alpha\beta,k} \tag{83}$$

$$[\Phi_{\rho\sigma\tau}, D_{\alpha\beta,k}] = 0 \tag{84}$$

$$[M^\gamma_\rho, \bar{f}^\sigma_k] = \delta^\sigma_\rho \bar{f}^\gamma_k \tag{85}$$

$$[M^\gamma_\rho, D_{\alpha\beta,k}] = \delta^\gamma_\alpha D_{\beta\rho,k} + \delta^\gamma_\beta D_{\alpha\rho,k} \tag{86}$$

When the three quark system gets excited, it becomes a $q - D$ system, but retains its symmetry

$$B_{123} = \frac{1}{\sqrt{3}}(q_1 D_{23} + q_2 D_{31} + q_3 D_{12}) \tag{87}$$

and remains in the 56 representation. The 70 representation has opposite parity (for example in $\boldsymbol{r}_1 - \boldsymbol{r}_{23}$, etc.).

When in the excited baryon quarks 2 and 3 form a diquark, and we can put

$$\boldsymbol{p}_{12} = \boldsymbol{p}_{13} = \boldsymbol{p} \quad \boldsymbol{p}_{23} = 0 \tag{88}$$

which leads to

$$\tilde{H}_{123} \longrightarrow \tilde{H}_{q_1 D_{23}} \cong 2\tilde{V}_{12} + \gamma_4^{(1)} \sqrt{[m_1 + \frac{1}{2}S]^2 + \boldsymbol{p}^2}$$
$$+(\gamma_4^{(2)} + \gamma_4^{(3)}) \sqrt{[(m_2 + m_3 + \bar{S}_{23}) + \frac{1}{2}S]^2 + \boldsymbol{p}^2} \tag{89}$$

and similarly for the other two configurations obtained by cyclic permutations. Note here that $m_D = m_2 + m_3 + \bar{S}_{23}$ is the mass of the diquark.

Then the superalgebra given by Eqs.(80-82) which is symmetric in (123) becomes, for a given $q - D$ configuration the $SU(6/21)$ supersymmetry $1-(23)$.

When $\tilde{H}_{123}$ is symmetrised it becomes invariant under the discrete group S_3. Hence its eigenstates are also eigenstates of S_3, i.e. correspond to representations of S_3. The splitting of 3 quarks into q and D breaks the S_3 symmetry down to $Z_2 \sim S_2$ which has two eigenstates (associated with ± 1). As mentioned in our earlier papers, Iachello [17, 18] has shown that the 70 representation of $SU(6)$ associated with $\ell = 1$ is parity doubled while the ground state 56 ($\ell = 0$) is not. This is supported by experiment.

We can now define eight auxiliary octonions of quadratic norm $\frac{1}{2}$ constructed out of the seven imaginary units f_α ($\alpha = 1, \cdots 7$).

$$f_1 = \frac{1}{2}(e_1 + e_4), \qquad f_2 = \frac{1}{2}(e_2 + e_5), \tag{90}$$

$$f_3 = \frac{1}{2}(e_3 + e_6), \qquad f_4 = \frac{1}{2}(e_1 - e_4), \tag{91}$$

$$f_5 = \frac{1}{2}(e_2 - e_5), \qquad f_6 = \frac{1}{2}(e_3 - e_6), \tag{92}$$

$$f_7 = \frac{1}{2}(1 - e_7), \quad f_8 = \frac{1}{2}(1 + e_7) \tag{93}$$

These are intimately related to the eight split octonionic units and to their conjugates we discussed above. If we keep one of the f's fixed, say f_j, then the difference combination $(f_i f_j f_k - f_k f_j f_i)$ is always either zero or equal to another f. We have worked out isomorphisms for such combination rules and they seem to play a fundamental role in dealing with symmetries of three quark systems we discussed, with further applications in multiquark systems. These will be discussed in another publication.

Combinations of f's also appear naturally within the root systems of groups associated with a magic square and play a profound role in superstring theories and their compactifications.

7. Particle multiplets including a giant supermultiplet

The multiplet X that sits in the adjoint representation of $SU(6/21)$ is

$$X = \begin{pmatrix} M & B \\ \bar{B} & N \end{pmatrix} \tag{94}$$

where M and N are mesons and exotics, and B and $\bar{B}$ are fermions. The M and N are square matrices, and B is a rectangular matrix. Specifically, $M = 6 \times \bar{6}$, $B = 6 \times 21$, $\bar{B} = \bar{21} \times \bar{6}$, and $N = \bar{21} \times 21$. M and N are taken to be hermitian. If we have three flavors the $SU(6)$ content of these matrices are

Journal of Physics: Conference Series **1416** (2019) 012008 doi:10.1088/1742-6596/1416/1/012008

$M = 1 + 35$ (negative parity), $N = 1 + 35 + 405$ (positive parity), and $B = 56 + 70$ (positive parity). The fundamental representation **F** is the color triplet

$$\mathbf{F} = \begin{pmatrix} \mathbf{Q} \\ \bar{\mathbf{D}} \end{pmatrix} = \begin{pmatrix} \mathbf{q} \\ \bar{\mathbf{q}} \times \bar{\mathbf{q}} \end{pmatrix} \tag{95}$$

with $\mathbf{q} = 6 \times 1$ and $\bar{\mathbf{q}} \times \bar{\mathbf{q}} = \bar{21} \times 1$. Now let Ξ be the superalgebra element of $SU(6/21)$ ($SU(3)$ singlet). Ξ is a color singlet given by

$$\Xi = \begin{pmatrix} m & b \\ \bar{b} & n \end{pmatrix} \tag{96}$$

and the transformation laws for the fundamental representation $[3, (6 + \overline{21})]$ are

$$\begin{aligned} \delta\mathbf{F} &= \Xi\mathbf{F} = \begin{pmatrix} m & b \\ \bar{b} & n \end{pmatrix} \begin{pmatrix} \mathbf{Q} \\ \bar{\mathbf{D}} \end{pmatrix} \\ &= \begin{pmatrix} m\mathbf{Q} + b\bar{\mathbf{D}} \\ \bar{b}\mathbf{Q} + n\bar{\mathbf{D}} \end{pmatrix} = \begin{pmatrix} \delta\mathbf{Q} \\ \delta\bar{\mathbf{D}} \end{pmatrix} \end{aligned} \tag{97}$$

and

$$\overline{\delta\mathbf{F}} = \overline{\mathbf{F}}\Xi = (\bar{\mathbf{Q}}m + \mathbf{D}\bar{b} \quad \bar{\mathbf{Q}}b + \mathbf{D}n) = (\delta\bar{\mathbf{Q}} \quad \delta\mathbf{D}) \tag{98}$$

or in the index notation

$$\delta q^i_\alpha = m^\beta_\alpha q^i_\beta + b_{\alpha\beta\gamma}(\bar{D}^i)^{\beta\gamma}, \tag{99}$$

$$(\delta\bar{D}^i)^{\beta\gamma} = \bar{b}^{\alpha\beta\gamma} q^i_\alpha + n^{\beta\gamma}_{\rho\sigma}(\bar{D}^i)^{\rho\sigma} \tag{100}$$

On the other hand, the transformation law for the adjoint representation is

$$\delta\Xi = i[\Xi, X], \tag{101}$$

where

$$\begin{aligned} \Xi X &= \begin{pmatrix} m & b \\ \bar{b} & n \end{pmatrix} \begin{pmatrix} M & B \\ \bar{B} & N \end{pmatrix} \\ &= \begin{pmatrix} mM + b\bar{B} & mB + bN \\ \bar{b}M + n\bar{B} & \bar{b}B + nN \end{pmatrix} \end{aligned} \tag{102}$$

$$X\Xi = \begin{pmatrix} Mm + B\bar{b} & Mb + Bn \\ \bar{B}m + N\bar{b} & \bar{B}b + Nn \end{pmatrix} \tag{103}$$

so that

$$\begin{pmatrix} \delta M & \delta B \\ \delta\bar{B} & \delta N \end{pmatrix} = \\ i\begin{pmatrix} [M, m] + b\bar{B} - \bar{B}b & Mb - Bn + bN - Mb \\ -\bar{B}m + n\bar{B} - N\bar{b} + \bar{b}M & [n, N] + \bar{b}B - \bar{B}b \end{pmatrix} \tag{104}$$

Next we build a giant supermultiplet containing M, N, L, B, $\bar{B}$, Q, $\bar{Q}$, D, and $\bar{D}$. The fundamental representation of $SU(6/21) \times [SU(3)^c]_{triplet}$ and the adjoint representation of $U(6/21) \times [SU(3)^c]_{singlet}$

fits in the adjoint representation of an octonionic version of $SU(6/22)$ denoted by Z, given by

$$\begin{aligned} Z &= u_0 \begin{pmatrix} M & B & 0 \\ B^\dagger & N & 0 \\ 0 & 0 & 0 \end{pmatrix} + u_0^* \begin{pmatrix} 0 & 0 & 0 \\ 0 & 0 & 0 \\ 0 & 0 & L \end{pmatrix} \\ &+ \mathbf{u} \cdot \begin{pmatrix} 0 & 0 & \mathbf{Q} \\ 0 & 0 & \mathbf{D}^* \\ 0 & 0 & 0 \end{pmatrix} + \mathbf{u}^* \cdot \begin{pmatrix} 0 & 0 & 0 \\ 0 & 0 & 0 \\ \epsilon \mathbf{Q}^\dagger & \epsilon \mathbf{D}^T & 0 \end{pmatrix} \\ &= \begin{pmatrix} u_0 M & u_0 B & \mathbf{u} \cdot \mathbf{Q} \\ u_0 B^\dagger & u_0 N & \mathbf{u} \cdot \mathbf{D}^* \\ \epsilon \mathbf{u}^* \cdot \mathbf{Q}^\dagger & \epsilon \mathbf{u}^* \cdot \mathbf{D}^T & u_0^* L \end{pmatrix} \end{aligned} \tag{105}$$

where mesons M $(6 \times \bar{6})$ and exotics N $(\bar{21} \times 21)$ are Hermitian; B (6×21), $\bar{B}$ $(\bar{21} \times \bar{6})$, Q (6×1), $\bar{Q}$ $(1 \times \bar{6})$, D (1×21), $\bar{D}$ $(\bar{21} \times 1)$, and L (1×1); ϵ can be taken as 1 if $u^\dagger = \bar{u}^* = -u^*$, -1 if $u^\dagger = u^*$, and zero. Closure properties of Z matrices are such that

$$[Z, Z'] = iZ'', \quad \{Z, Z'\} = Z''' \tag{106}$$

and in general they are nonassociative (Jacobian $J = f(Q, D) \neq 0$), except in the case when $\epsilon = 0$ we have

$$[[Z, Z'], Z''] + [[Z', Z''], Z] + [[Z'', Z], Z'] = 0. \tag{107}$$

Then we have a true superalgebra (non-semisimple) which is a contraction of a simple algebra that closes but does not satisfy the Jacobi identity. In both cases we get an extension of $SU(6/21)$.

We now consider the element of the algebra

$$\Omega = \begin{pmatrix} u_0 m & u_0 b & \mathbf{u} \cdot \boldsymbol{\xi} \\ u_0 b^\dagger & u_0 n & \mathbf{u} \cdot \mathbf{d}^* \\ \epsilon \mathbf{u}^* \cdot \boldsymbol{\xi}^\dagger & \epsilon \mathbf{u}^* \cdot \mathbf{d}^T & u_0^* \ell \end{pmatrix} \tag{108}$$

where

$$\begin{pmatrix} u_0 m & u_0 b \\ u_0 b^\dagger & u_0 n \end{pmatrix} \tag{109}$$

are the $[SU(6/21)]$ color singlet parameters,

$$\begin{pmatrix} \mathbf{u} \cdot \boldsymbol{\xi} \\ \mathbf{u} \cdot \mathbf{d}^* \end{pmatrix} \tag{110}$$

and

$$(\epsilon \mathbf{u}^* \cdot \boldsymbol{\xi}^\dagger \quad \epsilon \mathbf{u}^* \cdot \mathbf{d}^T) \tag{111}$$

are the colored parameters,

$$(u_0 b \quad \mathbf{u} \cdot \boldsymbol{\xi}) \tag{112}$$

and

$$\begin{pmatrix} u_0 b^\dagger \\ \epsilon \mathbf{u}^* \cdot \boldsymbol{\xi}^\dagger \end{pmatrix} \tag{113}$$

are the fermionic parameters, and $m \in SU(6)$.

The change in Z is given by

$$\delta Z = [\Omega, Z] \tag{114}$$

which leads to

$$\delta\mathbf{Q} = m\mathbf{Q} - M\boldsymbol{\xi} + b\mathbf{D}^* - B\mathbf{d}^* + \boldsymbol{\xi}L - \mathbf{Q}\ell, \tag{115}$$

and

$$\delta\mathbf{D}^* = b^\dagger\mathbf{Q} - B^\dagger\boldsymbol{\xi} + n\mathbf{D}^* - N\mathbf{d}^* + \mathbf{d}^*L - \mathbf{D}^*\ell. \tag{116}$$

The $SU(6/21)$ subgroup is obtained by taking

$$\boldsymbol{\xi} = 0, \quad \mathbf{d} = 0 \quad \ell = 0 \tag{117}$$

so that

$$\delta\mathbf{Q} = m\mathbf{Q} + b\mathbf{D}^* \tag{118}$$

$$\delta\mathbf{D}^* = b^\dagger\mathbf{Q} + n\mathbf{D}^* \tag{119}$$

which, in index form, is equivalent to Eqs.(99 -100). This subgroup is valid for a Hamiltonian describing $q(x_1)$ and $\bar{q}(x_2)$, $q(x_1)$ and $D(x_2)$, $\bar{D}(x_1)$ and $q(x_2)$, $\bar{D}(x_1)$ and $D(x_2)$ interacting through a scalar potential $V = br$ as we have seen earlier.

In general for $\frac{m}{2}$ flavors and $n = \frac{1}{2}m(m+1)$, we have

$$Z = \begin{pmatrix} u_0 M & u_0 B & \mathbf{u}\cdot\mathbf{Q} \\ u_0 B^\dagger & u_0 N & \mathbf{u}\cdot\mathbf{D}^* \\ \epsilon\mathbf{u}^*\cdot\mathbf{Q}^\dagger & \epsilon\mathbf{u}^*\cdot\mathbf{D}^T & u_0^* L \end{pmatrix} = \begin{pmatrix} m\times m & m\times n & m\times 1 \\ n\times m & n\times n & n\times 1 \\ 1\times m & 1\times n & 1\times 1 \end{pmatrix} \tag{120}$$

As examples, for 2 flavors, $M = 4\times 4$, $N = 10\times 10$; for 6 flavors (icluding the top quark), $M = 12\times 12$, $N = 78\times 78$.

The automorhism group of this algebra includes $SU(m)\times SU(n)\times SU(3)^c$. If $m = 6$, it includes $SU(6)\times SU(3)^c$.

If $\mathbf{Q}$ is Majorana and $\mathbf{D}$ real, then the group becomes $Osp(n/m)\times SU(3)^c$, with subgroup $Sp(2n/R)\times O(m)\times SU(3)^c$. We shall explore quark models built by use of such groups as well as use of auxiliary octonions of quadratic norm $\frac{1}{2}$ which are related to the split octonion units we used in this paper in a subsequent publication.

We would also like to mention that we have worked out the relativistic formulation through the spin realization of the Wess-Zumino algebra in detail, but due to lack of space we leave the discussion for another publication at this point.

8. Future of Mass formulae and some prospects

Here we have shown that the quark model with potentials derived from QCD, including the quark-diquark model for excited hadrons gives mass formulae in very good agreement with experiment and goes a long way in explaining the approximate symmetries and supesymmetries of the hadronic spectrum, including the symmetry breaking mechanism. For heavy quarks the non relativistic approximation can be used so that the potential models for the spectra of charmonium [19] and the $b\bar{b}$ system [20] are even simpler. In this approach gluons are eliminated leaving quarks interacting through potentials.

It is also possible to take an opposite approach by eliminating quarks as well as gluons, leaving only an effective theory that involves mesons (quark bound states) and baryons as collective excitations (solitons) of the mesonic field. This is the way pioneered by Skyrme [21] and revived in recent years (for a review see [22]). The simplest model uses an $SU(2)\times SU(2)$ nonlinear chiral model involving pi-mesons as originally formulated by Skyrme in 1958. In 1960 Skyrme added a fourth order term to stabilize the soliton, so that it could be interpreted as a nucleon. The model was enriched by the Syracuse school

[23] and by Witten and his collaborators [24]. More recently vector mesons were added, improving the predictions (for review see [25]). In the original model, mass formulae were obtained only for a tower of $I = S$ excitations. Recently $SU(4)$ symmetry [26] and in principle $SU(6)$ symmetry were incorporated and through fermionic quantization of the soliton the tower excitations were chopped off, leading to a more realistic model [27], [28].

However, even in these improved versions the mass relations are not nearly as good as in the potential quark model when the correct pion decay constant is used [29]. There is also an overall positive contribution to hadronic masses that is not easily disposed of, although there are some recent attempts to deal with the problem [30]. In the absolute scale baryon masses are about 20% too high and vector mesons enter the theory on a very different footing than pseudoscalar mesons, making it very difficult to relate their masses.

A skyrme model that can compete with the potential quark model is still hidden in the future.

A better understanding of hadron masses through the QCD theory of quarks and gluons has shifted the mass problem from the hadronic level to the level of the elementary constituent quarks, leptons, weak bosons and Higgs bosons of the standard model. Since we now know that there are three generations of fundamental fermions with light left handed neutrinos, the development of a theory of masses for the finite system has become a burning fundamental problem. What is needed is the equivalent of a Gell-Mann-Okubo formula for the six quarks and for the leptons.

A group theoretical treatment would have to involve a horizontal group associated with the three generations, namely a new $SU(3)$. If E_6 is the right grand unification group for each generation [31] as also suggested by string theory [32] then one has to enlarge it to a group admitting $E_6 \times SU(3)$ as a subgroup. Such a group also arises in string theory is E_8, the ultimate exceptional group [33]. It will also accommodate the $O(10)$ GUT through its $O(10) \times O(6)$ or $O(10) \times SU(3) \times U(1)$ decomposition [11]. An unusual symmetry breaking pattern of E_8 through a subgroup $O(5)$ or $SU(5)$ might isolate the t-quark and give it a high mass while all the other fundamental fermions are massless in this approximation. In order to approach the fundamental mass problem in imitation of hadron masses we would have to introduce new primary particles (preons) that would yield quarks and leptons as bound states. But then it would be very difficult to understand why the fundamental fermions of the standard model are finite in number.

Acknowledgments

One of us (SC) would like to thank the organizers of ISQS-26 for the invited plenary talk given in Prague (July, 2019) and for their kind hospitality. We would like to thank Professors Vladimir Akulov, Adrian Dumitru, Borys Grinyuk, Francesco Iachello, Evgeny Ivanov, Ramzi Khuri, Sergey Krivonos, V.Parameswaran Nair, Michael Rios, Aldemaro Romero and Piero Truini for enlightening conversations.

References

[1] Burdik C et.al. 2017 *Phys. Part. Nuclei Lett.* **14** 390
[2] Catto S et.al. 2018 *Adv. Appl. Clifford Algebras* **28** 81
[3] Catto S and Gürsey F 1985 *Nuovo Cimento* A **86** 201
[4] Catto S and Gürsey F 1988 *Nuovo Cimento* A **99** 685
[5] Burdik C and Catto S 2018 *J. Physics* **965** 012009
[6] Catto S et.al. 2014 *J. Physics* **563** 012006
[7] Catto S et.al. 2019 *Uk. J. Phys.* **64** 8
[8] Catto S and Gürcan Y 2018 *J. Physics* **2150** 03004
[9] Catto S et.al. 2016 *J. Physics* **670** 012016
[10] Catto S et.al. 2016 *J. Physics* **766** 012001
[11] Burdik C et.al. 2017 *J. Physics* **804** 012016
[12] Gürsey F and Radicati L A 1964 *Phys. Rev. Lett.* **13** 172
Gürsey F, Pais A and Radicati L A 1964 *Phys. Rev. Lett.* **13** 299
Gürsey F 1965 *Phys. Lett.* **14** 330
[13] Greenberg O W 1964 *Phys. Rev. Lett.* **13** 598

Nambu Y 1966 *Preludes in Theoretical Physics* eds. A. de Shalit, H. Fesbach and L. van Hove (Amsterdam) pp 133
Fritzsch H, Gell-Mann M and Leutwyler H 1973 *Phys. Lett.* B **47** 365
[14] Takakura S et al. 1987 *Prog. Theor. Phys.* **77** 917
[15] Wilson K 1969 *Phys. Rev.* **179** 1499
Wilson K 1971 *Phys. Rev.* D **3** 1818
[16] Salam A and Strathdee J 1975 *Nucl. Phys.* B **87** 85
[17] Iachello F 1989 *Nucl. Phys.* A **497**
[18] Iachello F et.al. 1991 *Phys. Rev.* B **44** 898
[19] Appelquist T A and Politzer H D 1975 *Phys. Rev. Lett.* **34** 43
[20] Eichten E et.al. 1980 *Phys. Rev.* D **21** 203
[21] Skyrme, T R H 1961 *Proc.Roy.Soc.* A **260** 127
Skyrme T R H 1961 *Proc.Roy.Soc.* A **262** 237
Skyrme T R H 1962 *Nucl. Phys.* **31** 566
[22] Chodos A, E. Hadjimichael E and H.C. Tze H C 1984 *Solitons in Nuclear and Elementary Particle Physics* (World Scientific, Singapore)
[23] Balachandran A P, Nair V P, Rajeev S G and Stern A 1982 *Phys. Rev. Lett.* **49** 1124 Balachandran A P, Nair V P, Rajeev S G and Stern A 1983 *Phys. Rev. Lett.* **50** 1620(E)
[24] Adkins G S, Nappi C R and Witten E 1983 *Nucl. Phys.* **228** 522
[25] Bando M, Kugo T and Yamawaki K 1988 *Phys. Rep.* **164** 217
[26] Cheung H Y and Gürsey F 1989 *Phys. Lett.* B **219** 127
Meissner U G and Pasquier B 1990 *Phys. Lett.* **235** 153
[27] Cheung H Y 1991 *Ph.D. thesis* (Yale)
[28] Catto S, Cheung H Y and Gürsey F 1991 *Mod. Phys. Lett* A **6** 3485
[29] Ruiz Arriola E, Alberto P, Goeke K and Urbano J N 1990 *Phys. Lett.* **236** 381
[30] Islam M M 1992 *Z. Phys.* C **53** 253
[31] Gürsey F, Ramond P and Sikivie P 1976 *Phys. Lett.* B **60** 177
Gürsey F and Serdaroğlu M 1981 *Nuovo Cimento* A **65** 337
[32] Candelas P, Horowitz G, Strominger A and Witten E 1985 *Nucl. Phys.* B **258** 46
[33] Bars I and Günaydin M 1980 *Phys. Rev. Lett.* **45** 859
Gürsey F 1981 *Proc. of Workshop on Weak Interactions as Probes of Unification* eds G.B. Collins, L.N. Chang and J.R. Ficenec (AIP) p 635

Journal of Physics: Conference Series **1416** (2019) 012009 doi:10.1088/1742-6596/1416/1/012009

Operator-algebraic construction of the deformed Sine-Gordon model

Daniela Cadamuro

Universität Leipzig, Institut für Theoretische Physik, Brüderstraße 16, 04103 Leipzig, Germany

E-mail: daniela.cadamuro@itp.uni-leipzig.de

Abstract. We consider the construction of integrable quantum field theories in the operator-algebraic approach, which is based on quantum fields localized in infinitely extended wedge regions. This approach has been successful for the construction of a class of models with scalar S-matrices and without bound states. In extension of these results, we apply similar methods to S-matrices with poles in the physical strip ("bound states"). Specifically, we consider a deformed version of the Sine-Gordon model, containing only breathers. We exhibit wedge-local fields in this model, which differ from those in non-bound state models by an additive term, the so called "bound state operator".

1. Introduction

The construction of interacting quantum field theories is a difficult problem in Mathematical Physics. A usual approach is based on the perturbative expansion of the S-matrix constructed from a Lagrangian, but the appearance of infrared and ultraviolet divergences in the expansion, as well as the problem of convergence of the perturbation series, make expressions mathematically ill-defined.

There is however a class of quantum field theories that can be constructed with techniques which avoid perturbative expansions: the quantum integrable models. These are $1+1$-dimensional field theories where two-particle scattering processes characterize the theory completely. They are constructed as an inverse scattering problem: Given a function S as a mathematical input, one constructs the corresponding QFT having this two-particle scattering function. The interaction of these models is simple enough for certain methods familiar from free field theory to be carried over to the interacting situation.

Despite being restricted to low dimensions, there is physical interest in studying these models. In fact, they share interesting common features with interacting theories in higher dimensions. For example, the nonlinear $O(N)$ sigma models [1] are linked to experimentally realizable situations in condensed matter systems. They can also be regarded as simplified analogues of four-dimensional nonabelian gauge theories, inasmuch as they share crucial features with them, including renormalizability, asymptotic freedom, and the existence of instanton solutions. In this sense, integrable systems provide a "landscape" of possible interactions, where one may hope to test techniques that can be adapted to, or inspire methods for, systems in higher dimensions, and where one may obtain insight into the structure of local QFTs with a less simple S-matrix.

Here, we will focus on a particular model of the integrable class, the Sine-Gordon model, which has been treated with several methods before. In particular, the integrability of the massless Sine-Gordon model and its S-matrix have been conjectured in [2]. The corresponding inverse scattering problem was studied in the form factor programme [3, 4] where certain matrix elements of the pointlike local fields ("form factors") have been computed; the existence of a Wightman field remains unclear though due to convergence issues in the infinite expansion of its n-point functions in terms of form factors [5]. In a rigorous framework (pAQFT) it was shown [6, 7] that in the massless Sine-Gordon model in the ultraviolet-finite regime (hence, without need of renormalization), the expectation value of the Epstein-Glaser S-matrix given as formal power series converges in a Hadamard state. The authors also constructed the interacting currents and the interacting field as formal power series in the coupling constant, and showed convergence in expectation value in that state. Yet, the factorization of the S-matrix has not been proven there. We should also mention the conjectured equivalence of the massless Sine-Gordon model with the Thirring model (Coleman's equivalence), which has been proven by Benfatto, Falco and Mastropietro [8] in the case of finite volume interaction of the massless Sine-Gordon model and of the Thirring model with a finite volume mass term. A construction of the Thirring model by functional integral methods can be found in [9].

In this paper, we report on progress towards a construction of a deformed version of the Sine-Gordon model, which contains only "breather" particles (Sec. 3). We construct quantum fields associated with this model which are—in the sense of their commutation relations—localized in spacelike wedges (Sec. 4), as proposed by Schroer [10]. This extends our previous results for scalar models with bound states [11]. The construction of strictly local observables, which is ongoing work, would then use methods from the theory of C^*-algebras and the abstract framework of Tomita-Takesaki modular theory, as already known for the Sinh-Gordon model [12].

This proceeding is based on joint work with Y. Tanimoto [13].

2. Properties of the S-matrix

Quantum integrable models have the property that the S-matrix is factorizable into two-particle scattering matrices. This two-particle scattering matrix, a meromorphic matrix-valued function (in the following only referred to as S), needs to fulfill a number of axioms, including *unitarity*, *crossing symmetry*, the *Bootstrap equation* and, in the case of several particle species, the *Yang-Baxter equation*. In some cases, only matrix elements of the form $S^{\beta\alpha}_{\alpha\beta}$ are nonzero; in physical terms, two particles of type α, β can scatter with the exchange of a unitary factor, but the type of particles stays the same in scattering. In this case, the S-matrix is said to be "diagonal". There are also "non-diagonal" S-matrices where the type of particle can change due to scattering.

We are interested in the class of integrable models where bound states of particles are present. Mathematically, these bound states correspond to poles of S in the *physical strip* (the region $0 \leq \operatorname{Im}\zeta \leq \pi$ in complex rapidity space). We denote with θ_1, θ_2 the rapidities of the particles α, β, which parametrize their momenta. These two particles can fuse into a third particle of type γ in the following sense:

$$\boldsymbol{p}_\alpha(\theta + i\theta_{(\alpha\beta)}) + \boldsymbol{p}_\beta(\theta - i\theta_{(\beta\alpha)}) = \boldsymbol{p}_\gamma(\theta), \tag{1}$$

i.e., the momenta of two virtual particles add to the momentum of a third real particle (the bound particle) which lies on the mass shell. The numbers $\theta_{(\alpha\beta)}, \theta_{(\beta\alpha)}$ are determined by this equation, once the masses of the particles are fixed. The bound state formed would correspond to a pair of simple poles of the component $S^{\beta\alpha}_{\alpha\beta}(\zeta)$ in the physical strip: the s-channel pole, located at $\zeta = i\theta_{\alpha\beta} := i\theta_{(\alpha\beta)} + i\theta_{(\beta\alpha)}$, and the t-channel pole at $\zeta = i\pi - i\theta_{\alpha\beta}$, which arises due to crossing symmetry. The numbers $\theta_{\alpha\beta}$ are called *fusion angles*. The possible fusion processes, which we denote by the symbol $(\alpha\beta) \to \gamma$, are characteristic of each model, and listed in its fusion table.

processes	rapidities of particles	fusion angles
$(s\bar{s}) \longrightarrow b_k$	$\theta^{b_k}_{(s\bar{s})} = \frac{\pi}{2}(1-\nu k)$	$\theta^{b_k}_{s\bar{s}} = \pi(1-\nu k)$
$(sb_k) \longrightarrow s$	$\theta^{s}_{(sb_k)} = \pi\nu k, \theta^{s}_{(b_k s)} = \frac{\pi}{2}(1-\nu k)$	$\theta^{s}_{sb_k} = \frac{\pi}{2}(1+\nu k)$
$(\bar{s}b_k) \longrightarrow \bar{s}$	$\theta^{\bar{s}}_{(\bar{s}b_k)} = \pi\nu k, \theta^{b_k}_{(\bar{s}b_k)} = \frac{\pi}{2}(1-\nu k)$	$\theta^{\bar{s}}_{\bar{s}b_k} = \frac{\pi}{2}(1+\nu k)$
$(b_k b_\ell) \longrightarrow b_{k+\ell}$	$\theta^{b_{k+\ell}}_{(b_k b_\ell)} = \frac{\pi\nu}{2}\ell, \theta^{b_{k+\ell}}_{(b_\ell b_k)} = \frac{\pi\nu}{2}k$	$\theta^{b_{k+\ell}}_{b_k b_\ell} = \frac{\pi\nu}{2}(k+\ell)$
$(b_{k+\ell}b_k) \longrightarrow b_\ell$	$\theta^{b_\ell}_{(b_{k+\ell}b_k)} = \frac{\pi\nu}{2}k, \theta^{b_\ell}_{(b_k b_{k+\ell})} = \pi\left(1-\frac{\nu}{2}(k+\ell)\right)$	$\theta^{b_\ell}_{b_{k+\ell}b_k} = \pi\left(1-\frac{\nu}{2}\ell\right)$

Table 1. Fusion processes in the (usual) Sine-Gordon model

processes	rapidities of particles	fusion angles
$(b_1 b_1) \longrightarrow b_2$	$\theta^{b_2}_{(b_1 b_1)} = \frac{\pi\nu}{2}$	$\theta^{b_2}_{b_1 b_1} = \pi\nu$
$(b_2 b_1) \longrightarrow b_1, (b_1 b_2) \longrightarrow b_1$	$\theta^{b_1}_{(b_1 b_2)} = \pi(1-\nu), \theta^{b_1}_{(b_2 b_1)} = \frac{\pi\nu}{2}$	$\theta^{b_1}_{b_2 b_1} = \theta^{b_1}_{b_1 b_2} = \pi\left(1-\frac{\nu}{2}\right)$
$(b_2 b_2)$ not a fusion		

Table 2. Fusion processes in the deformed Sine-Gordon model

3. The deformed Sine-Gordon model

The particle spectrum of the Sine-Gordon model contains solitons (s), anti-solitons $(\bar{s})$ and several breathers $(b_k,\ 1 \le k \le K)$, which are neutral particles (i.e., $\overline{b_k} = b_k$). The number K of breathers is the largest integer such that $K\nu < 2$, where $0 < \nu < 1$ is a coupling constant. The fusion processes between these particles and corresponding fusion angles are shown in Table 1. The corresponding two-particle scattering matrix S_{SG} is of non-diagonal type or, more precisely, block-diagonal: On the soliton-antisoliton subspace, it has the form

$$S_{\mathrm{SG,soliton}}(\zeta) = \begin{pmatrix} a(\zeta) & 0 & 0 & 0 \\ 0 & t(\zeta) & r(\zeta) & 0 \\ 0 & r(\zeta) & t(\zeta) & 0 \\ 0 & 0 & 0 & a(\zeta) \end{pmatrix}, \tag{2}$$

where

$$a(\zeta) := S^{ss}_{ss}(\zeta) = S^{\bar{s}\bar{s}}_{\bar{s}\bar{s}}(\zeta), \quad t(\zeta) := S^{\bar{s}s}_{s\bar{s}}(\zeta) = S^{s\bar{s}}_{\bar{s}s}(\zeta), \quad r(\zeta) := S^{s\bar{s}}_{s\bar{s}}(\zeta) = S^{\bar{s}s}_{\bar{s}s}(\zeta); \tag{3}$$

the expressions $a(\zeta), t(\zeta), r(\zeta)$, depending on ν, are explicitly given in [14, Section 3.3]. On the breather subspace, S_{SG} is diagonal [14].

In this paper, we will not deal with the massless Sine-Gordon model itself but with a new model with the same fusion structure, which is derived as a deformation of the "breather-breather" S-matrix of the massless Sine-Gordon model. To that end, we restrict the value of the coupling constant to $\frac{2}{3} < \nu < \frac{4}{5}$, so that there are exactly two breathers, b_1, b_2. We now consider the two-particle scattering function

$$S^{b_\ell b_k}_{b_k b_\ell}(\zeta) = S_{\mathrm{SG}}{}^{b_\ell b_k}_{b_k b_\ell}(\zeta) S_{\mathrm{CDD}}{}^{b_\ell b_k}_{b_k b_\ell}(\zeta), \quad \ell, k = 1, 2, \tag{4}$$

where the "CDD factors" S_{CDD} are explicitly given in [13, Sec. 2.2]; they are so arranged that the required properties of the function S, such as unitarity and crossing symmetry, are retained. Thus we obtain an S-matrix of diagonal type with particle spectrum b_1, b_2; we do not consider (anti-)solitons.

Journal of Physics: Conference Series **1416** (2019) 012009 doi:10.1088/1742-6596/1416/1/012009

Range of ν	Residue of pole of $S_{b_1b_1}^{b_1b_1}$	Comment
$4/5 < \nu < 1$	$-i\mathbb{R}_+$	No adjusting CDD factors found.
$2/3 < \nu < 4/5$	$-i\mathbb{R}_+$	Adjusting CDD factors found.
$1/2 < \nu < 2/3$	$-i\mathbb{R}_+$	There are three breathers if one requires the maximal analyticity within breathers. No adjusting CDD factors found.
$0 < \nu < 1/2$	$i\mathbb{R}_+$	There is a breather b_k for which $\operatorname{res}_{\zeta=i\theta_{b_1b_k}^{b_{k+1}}} S_{b_1b_k}^{b_kb_1}(\zeta) \in -i\mathbb{R}_+$.

Table 3. Sign of the residue of $S_{b_1b_1}^{b_1b_1}$ for different values of ν

Each of the components $S_{b_1b_1}^{b_1b_1}(\zeta)$, $S_{b_2b_1}^{b_1b_2}(\zeta)$ and $S_{b_1b_2}^{b_2b_1}(\zeta)$ has only two simple poles in the physical strip (an s-channel pole as shown in Table 2, and a t-channel pole at $i\pi$ minus that value), and no higher-order poles in that strip. The corresponding fusion processes are shown in Table 2; note that the fusion table for b_1, b_2 is closed under fusions. We have *maximal analyticity* for the particle b_1, i.e., all above-mentioned poles of S correspond to fusion processes. The poles in the component $S_{b_2b_2}^{b_2b_2}(\zeta)$ will not matter in our construction of the model in Sec. 4.

The reason for including the CDD factor is that, for our proof of wedge locality of quantum fields (see Sec. 4), the residue of the poles in the physical strip of those S-components involving b_1 must have a certain sign, namely we need,

$$\operatorname{res}_{\zeta=i\theta_{b_1b_k}^{b_\ell}} S_{b_1b_k}^{b_kb_1}(\zeta) \in i\mathbb{R}_+ \tag{5}$$

(*positivity of residue* for b_1). For different values of the coupling constant $0 < \nu < 1$, it may or may not be possible to find such CDD factors that adjust the signs of the residues. We summarize the situation in Table 3. Due to the special properties of particle b_1 with respect to poles and residues of S, we call it an "elementary particle" (the precise definition can be found in [15, Section 2.1]); this again will play an important role in Sec. 4.

4. Wedge-local fields

We now aim to construct a quantum field theory that is associated with the the two-particle scattering matrix (4). We wish to bypass problems that arise when directly constructing point-local quantum fields, and hence focus on objects with weaker localization properties, so-called wedge local fields [10].

More specifically: Let W_L be the left and W_R the right wedge in 1+1-dimensional Minkowski space, i.e., $W_L = \{x : x^1 < -|x^0|\}$, $W_R = \{x : x^1 > |x^0|\}$. These regions are causal complements of each other. We are looking for fields ϕ, ϕ' associated with S, such that

$$[\phi(x), \phi'(y)] = 0 \quad \text{if } x \in W_L,\ y \in W_R. \tag{6}$$

This equation will need to be read in the sense of distributions, i.e., smeared with test functions supported in W_L and W_R respectively. Additionally, ϕ should be covariant under the proper orthochronous Poincaré group, and ϕ' should arise from ϕ by adjoint action of the PCT operator.

To find these fields, we follow the techniques presented in [12, 16] where the construction was established for scattering functions *without* poles in the physical strip. However, we shall see that certain modifications are required.

We start by specifying the underlying Hilbert space. In our case, the single-particle Hilbert space should accommodate the two species of particles, hence we consider

$$\mathcal{H}_1 = \bigoplus_{k=1,2} \mathcal{H}_{1,b_k}, \quad \mathcal{H}_{1,b_k} = L^2(\mathbb{R}, d\theta). \tag{7}$$

The full Hilbert space $\mathcal{H}$ is then constructed as an S-symmetric Fock space over $\mathcal{H}_1$. Similar to the free field, one introduces creation and annihilation operators on $\mathcal{H}$ (labeled by the particle species) which now fulfill twisted commutation relations involving the S-matrix:

$$z^\dagger_{b_k}(\theta) z^\dagger_{b_\ell}(\theta') = S^{b_\ell b_k}_{b_k b_\ell}(\theta - \theta') z^\dagger_{b_\ell}(\theta') z^\dagger_{b_k}(\theta), \tag{8}$$
$$z_{b_k}(\theta) z_{b_\ell}(\theta') = S^{b_\ell b_k}_{b_k b_\ell}(\theta - \theta') z_{b_\ell}(\theta') z_{b_k}(\theta), \tag{9}$$
$$z_{b_k}(\theta) z^\dagger_{b_\ell}(\theta') = S^{b_k b_\ell}_{b_\ell b_k}(\theta' - \theta) z^\dagger_{b_\ell}(\theta') z_{b_k}(\theta) + \delta_{b_k b_\ell} \delta(\theta - \theta') \mathbf{1}_{\mathcal{H}}. \tag{10}$$

$\mathcal{H}$ is equipped with a unitary representation of the proper orthochronous Poincaré group which preserves particle numbers, and the antiunitary CPT operator J; we denote by J_1 its restriction to $\mathcal{H}_1$.

Now we are ready to define our quantum field: For $f \in \bigoplus_{k=1,2} \mathcal{S}(\mathbb{R}^2)$, we set

$$\phi(f) := z^\dagger(f^+) + z(J_1 f^-) = \sum_{k=1,2} \int d\theta \left(f^+_{b_k}(\theta) z^\dagger_{b_k}(\theta) + (J_1 f^-)_{b_k}(\theta) z_{b_k}(\theta) \right), \tag{11}$$

where $f^\pm$ indicates the Fourier transform with positive and negative frequencies,

$$f^\pm_{b_k}(\theta) := \frac{1}{2\pi} \int d^2x \, e^{\pm i p_{b_k}(\theta)\cdot x} f_{b_k}(x). \tag{12}$$

Further, we define the *reflected field* as $\phi'(g) := J\phi(g_j)J$, where $(g_j)_{b_k}(x) := \overline{g_{b_k}(-x)}$.

These fields $\phi(f)$, $\phi'(g)$ are, as expected, not strictly local. The analogous fields for scalar S-matrices without bound states can however shown to be wedge-local [12]. To compare this with our case, let us compute their commutator: it turns out to be

$$\begin{aligned}
&([\phi'(g), \phi(f)]\Psi_n)^{b_{k_1},\dots,b_{k_n}}(\theta_1, \cdots, \theta_n) \\
&= \int d\theta' \left(g^-_{b_1}(\theta') \left(\prod_{p=1}^n S^{b_{k_p} b_1}_{b_1 b_{k_p}}(\theta' - \theta_p) \right) f^+_{b_1}(\theta') - g^+_{b_1}(\theta') \left(\prod_{p=1}^n \overline{S^{b_{k_p} b_1}_{b_1 b_{k_p}}(\theta' - \theta_p)} \right) f^-_{b_1}(\theta') \right) \\
&\quad \times (\Psi_n)^{b_{k_1},\dots,b_{k_n}}(\theta_1, \dots, \theta_n).
\end{aligned} \tag{13}$$

One now asks whether this expression vanishes when $\operatorname{supp} f \subset W_L$ and $\operatorname{supp} g \subset W_R$. In fact, if S were analytic in the physical strip, then the integration contour in one of the summands could be shifted by $i\pi$ using the Cauchy theorem, and the commutator would then vanish as a consequence of crossing symmetry. In the presence of poles of S in the physical strip, however, the Cauchy theorem yields contributions from the residues of the integrand. Hence to achieve wedge commutativity, we have to modify the field with an additional term whose commutator will cancel these contributions. To that end we introduce the bound state operator $\chi(f)$, whose action on a one-particle vector (in a suitable domain, see below) is defined as follows:

$$(\chi_1(f)\xi)_{b_k}(\theta) := \begin{cases} -i\eta^{b_1}_{b_1 b_2} f^+_{b_1}(\theta + i\theta^{b_1}_{(b_1 b_2)}) \xi_{b_2}(\theta - i\theta^{b_1}_{(b_2 b_1)}) & \text{if } k = 1, \\[2ex] -i\eta^{b_2}_{b_1 b_1} f^+_{b_1}(\theta + i\theta^{b_2}_{(b_1 b_1)}) \xi_{b_1}(\theta - i\theta^{b_2}_{(b_1 b_1)}) & \text{if } k = 2. \end{cases} \tag{14}$$

Here the matrix elements $\eta^{b_1}_{b_1 b_2}, \eta^{b_2}_{b_1 b_1}$ are related to the residues of S at the poles in the strip, see [13, Eq. (3)] for the precise expressions. It is at this point that the positivity of residue condition (5) is needed, since for the following construction to work, the η need to be defined as proportional to $(-i \operatorname{res} S)^{1/2}$. The field $\chi(f)$ is then defined by lifting to second quantization,

$$\chi_n(f) := nP_n(\chi_1(f) \otimes \mathbf{1} \otimes \cdots \otimes \mathbf{1})P_n, \quad \chi(f) := \bigoplus_{n=0}^{\infty} \chi_n(f), \tag{15}$$

where P_n is the projection onto the S-symmetric part of the Fock space. A corresponding "reflected" field is defined as $\chi'(g) := J\chi(g_j)J$.

The domain of the operators $\chi(f)$ and $\chi_1(f)$ is rather subtle. For the expression (14) to be in L^2, the vectors ξ_{b_2}, ξ_{b_1} must be analytic until $\mathbb{R} - i\theta^{b_1}_{(b_2 b_1)}$ and $\mathbb{R} - i\theta^{b_2}_{(b_1 b_1)}$, respectively, with their L^2-norm uniformly bounded in those strips. The domain of $\chi(f)$ is even more difficult to describe; it consists of n-particle vectors that have such analytic continuations in each variable, but also possess certain zeros that compensate poles of S in the physical strip, occurring in the operator P_n. (See [15, Proof of Prop. 3.1].)

Now the main result of [13] is as follows. For the field defined as

$$\tilde{\phi}(f) = \phi(f) + \chi(f) \tag{16}$$

and its reflected field $\tilde{\phi}'(g)$, we can show that those components corresponding to elementary particles (namely, of type b_1) commute weakly on a dense domain $D \subset \operatorname{dom}(\tilde{\phi}(f)) \cap \operatorname{dom}(\tilde{\phi}'(g))$. (Note that $\operatorname{Dom}(\tilde{\phi}(f)) = \operatorname{Dom}(\chi(f))$.) More precisely:

Theorem 1. *Let f and g be test functions supported in W_L and W_R, respectively, and with the property that $f = f^*$ and $g = g^*$. Furthermore, assume that f, g have components $f_{b_k} = 0$ and $g_{b_k} = 0$ for $k \neq 1$. Then, for each $\Phi, \Psi \in D$, we have*

$$\langle \tilde{\phi}(f)\Phi, \tilde{\phi}'(g)\Psi \rangle = \langle \tilde{\phi}'(g)\Phi, \tilde{\phi}(f)\Psi \rangle. \tag{17}$$

In the proof of this theorem, the commutator $[\tilde{\phi}(f), \tilde{\phi}'(g)]$ (meant always in a weak sense) gives the contributions $[\phi(f), \phi'(g)]$, $[\chi(f), \chi'(g)]$, $[\phi(f), \chi'(g)]$ and $[\chi(f), \phi'(g)]$. We can show that the last two are independently zero, while the first two cancel each other due to the chosen form of the $\chi(f)$.

5. Strictly local observables: Open questions

Having constructed a wedge-local field, one would now like to proceed to strictly localized observables, either algebras of bounded operators localized in the intersection of wedges, i.e., in double cones [12], or pointlike fields affiliated with these algebras [17]. The approach would be to define von Neumann algebras associated with W_L and W_R, generated by bounded functions of the $\tilde{\phi}(f)$ and $\tilde{\phi}'(g)$, respectively, and corresponding algebras for shifted wedges by covariance. Then one would use abstract methods of Tomita-Takesaki theory to show that the intersection of two such algebras is sufficiently large.

For the present model, however, these questions are not settled at this point. We highlight some of the difficulties.

First, the field $\tilde{\phi}(f)$ is densely defined and symmetric on a suitable domain of vectors, but it is not self-adjoint on a naive domain. For defining bounded functions of the fields, one needs to show the existence of self-adjoint extensions of $\tilde{\phi}(f)$ and $\tilde{\phi}'(g)$, and then select those that strongly commute. This is a highly non-trivial task, but some progress has been made in the Bullough-Dodd model [18].

Second, the nontriviality of double-cone algebras was shown for certain scalar models without bound states [12, 16] using the so-called "modular nuclearity condition" for the von Neumann algebras of wedges. Notwithstanding the construction of these algebras in our case, techniques for showing the nuclearity conditions are still being investigated.

As a final step, once local operators are fully constructed, it would be important to verify that the S-matrix which results by Haag-Ruelle scattering theory is indeed factorizing, and that the two-particle scattering matrix coincides with the function S in (4) that our construction started from. This is another point of our ongoing work.

6. Conclusions

We have investigated the construction of the Sine-Gordon model [3, 4] in the operator-algebraic framework. More specifically, we have considered a deformed version of the original S-matrix by multiplying with a CDD factor for a certain range of the coupling constant, not only limiting the number of particle species involved but also adjusting the signs of the residues of S. This allowed us to establish a pair of wedge-local quantum fields, which commute at least weakly (on a suitable dense domain) at spacelike distance of the wedges if their components are restricted to elementary particles (b_1). For a full construction, some steps are still missing, in particular strong commutativity of the fields.

An extension to other ranges of the coupling constant, as well as the construction of the usual Sine-Gordon model without modifications by CDD factors, have not been treated in the algebraic framework yet; this remains challenging due to the subtle pole structure of the S-matrix and the sign of its residues as mentioned above. We hope to return to this problem in the near future.

Acknowledgements

The author is supported by the Deutsche Forschungsgemeinschaft (DFG) within the Emmy Noether grant CA1850/1-1.

References

[1] Babujian H M, Foerster A and Karowski M 2013 *JHEP* **2013** 1–53
[2] Zamolodchikov A B and Zamolodchikov A B 1979 *Annals Phys.* **120** 253–91
[3] Babujian H M, Fring A, Karowski M and Zapletal A 1999 *Nucl. Phys.* **B538** 535–86
[4] Babujian H and Karowski M 2002 *J. Phys. A: Math. Gen.* **35** 9081–104
[5] Babujian H M and Karowski M 2004 *Int. J. Mod. Phys.* **A19S2** 34–49 (*Preprint* hep-th/0301088)
[6] Bahns D and Rejzner K 2018 *Commun. Math. Phys.* **357** 421–46 (*Preprint* 1609.08530)
[7] Bahns D, Fredenhagen K and Rejzner K 2017 Local nets of von Neumann algebras in the Sine-Gordon model (*Preprint* 1712.02844)
[8] Benfatto G, Falco P and Mastropietro V 2009 *Commun. Math. Phys.* **285** 713–62
[9] Benfatto G, Falco P and Mastropietro V 2007 *Commun. Math. Phys.* **273** 67–118
[10] Schroer B and Wiesbrock H W 2000 *Rev. Math. Phys.* **12** 301–26 (*Preprint* hep-th/9812251)
[11] Cadamuro D and Tanimoto Y 2015 *Commun. Math. Phys.* **340** 661–97
[12] Lechner G 2008 *Commun. Math. Phys.* **277** 821–60 (*Preprint* math-ph/0601022)
[13] Cadamuro D and Tanimoto Y 2018 *Rev. Math. Phys.* **30** 1850010
[14] Quella T 1999 *Formfactors and locality in integrable models of quantum field theory in 1+1 dimensions (in German)* Diploma thesis Freie Universität Berlin
[15] Cadamuro D and Tanimoto Y 2017 *Ann. H. Poincaré* **18** 233–79 (*Preprint* 1601.07092)
[16] Alazzawi S and Lechner G 2017 *Commun. Math. Phys.* **354** 913–56 (*Preprint* 1608.02359)
[17] Bostelmann H and Cadamuro D 2019 *Ann. H. Poincaré* (to appear) (*Preprint* arXiv:1806.00269)
[18] Tanimoto Y 2016 *SIGMA* **12** 100 (*Preprint* 1602.04696)

Journal of Physics: Conference Series **1416** (2019) 012010 doi:10.1088/1742-6596/1416/1/012010

On calculation of quadrupole operator in orthogonal Bargmann-Moshinsky basis of SU(3) group

A. Deveikis

Vytautas Magnus University, Kaunas, Lithuania

E-mail: algirdas.deveikis@vdu.lt

A.A. Gusev

Joint Institute for Nuclear Research, Dubna, Russia

E-mail: gooseff@jinr.ru

S.I. Vinitsky

Joint Institute for Nuclear Research, Dubna, Russia
RUDN University, Moscow, Russia, 6 Miklukho-Maklaya st, Moscow, 117198

A. Pędrak

National Centre for Nuclear Research, Warsaw, Poland

A. Góźdź

Institute of Physics, University of M. Curie-Skłodowska, Lublin, Poland

Č Burdík

Czech Technical University, Prague, Czech Republic
Joint Institute for Nuclear Research, Dubna, Russia

P.M. Krassovitskiy

Institute of Nuclear Physics, Almaty, Kazakhstan
Joint Institute for Nuclear Research, Dubna, Russia

Abstract. Construction of orthonormal states of the noncanonical Bargmann-Moshinsky basis in a nonmultiplicity-free case is presented. It is implemented by means of the both Gram-Schmidt procedure and solving eigenvalue problem of the Hermitian labeling operator of an envelope algebra of the SU(3) group. Calculations of the quadrupole and Bargmann-Moshinsky operators and its matrix elements needed for construction of the nuclear models are tested. Comparison of results in the integer and floating point calculations with help of the proposed procedures implemented in Wolfram Mathematica is given.

Journal of Physics: Conference Series **1416** (2019) 012010 doi:10.1088/1742-6596/1416/1/012010

1. Introduction

The quadrupole momentum operator is one of the most important quantities applied in nuclear physics modeling and large-scaled calculations and treated as generator of SU(3) group [1, 2, 3, 4]. It should be noted that the quadrupole momentum is an essential part in construction of the Hamiltonians for nuclear models with a tetrahedral and octahedral symmetries [5, 6], which are studied by modern experiments [7, 8]. However, in spite of a long history [9], there are still no sufficient efficient and cost effective algorithms and programs for constructing the required orthogonal SU(3) bases of the non-canonical group reduction chain of subgroups SU(3)$\supset$ SO(3) $\supset$ SO(2) and computing tensor operators in these bases similar as a conventional program implemented in the canonical Gel'fand-Tseitlin basis [10]. Another problem is that the above non-canonical group reduction chain is known of one label missing to characterize the states completely [11, 12, 13, 14]. Solving these problem using adapted versions of Gram-Schmidt procedure for construction of an orthogonal Bargmann-Moshinsky(BM) basis [15] has been subject of our study given in the recent papers [16, 17, 18]. Below they are labeled as I, II and III and their main results are summarized briefly.

The degeneracy problem for SO(3) states in the SU(3) representation can be also solved by introducing of an additional Hermitian operator $X^{(3)}$ proposed by Bargmann and Moshinsky [15] of an enveloping algebra of the SU(3) group [19]. The eigenvalues of the operator $X^{(3)}$ provide the missing label for the state vectors.

In this paper we developed effective procedures for computing matrices of the quadrupole operators as well as the matrices of the operator $X^{(3)}$ in the orthogonal BM basis. The developed procedure for calculation of quadrupole operators was applied to calculate the spectrum of the Hamiltonian with the quadrupole-quadrupole interaction exhibited a tetrahedral/octahedral symmetry. It should be noted that the developed procedures allow both numerical and exact symbolic implementation. All developed procedures were implemented in Wolfram Mathematica, which allows for a comprehensive study of their effectiveness.

The paper is organized as follows. In section 2 the procedure of calculations of the orthogonal BM basis is described. In section 3 the scheme of evaluation of matrix elements of tensor operator components is given. In section 4 the spectrum of the tetrahedral/octahedral quadrupole-quadrupole Hamiltonian is calculated. In Conclusions further applications of the elaborated procedures are outlined.

2. The orthogonal Bargmann-Moshinsky basis

2.1. Classification of Bargmann-Moshinsky states

The non-canonical BM basis corresponding to the subgroup chain SU(3)$\supset$ SO(3) $\supset$ SO(2) is indexed by the quantum numbers λ and μ, L and M of the respective group SU(3), SO(3) and SO(2). In the case of the BM basis, α is an additional index that is used to unambiguously distinguish the equivalent SO(3) irreps L in a given SU(3) irrep λ and μ. In the case of the Elliott basis, the additional index K is used instead of α for the same purpose. Note the set of quantum numbers λ, μ, L, α (or K) must be consistent according to the subgroup chain

$$|u_\alpha\rangle \equiv \left| \begin{matrix} (\lambda\mu)_B \\ \alpha LM \end{matrix} \right\rangle \quad \begin{matrix} \mathrm{SU}(3) & \supset & \mathrm{SO}(3) & \supset & \mathrm{SO}(2) \\ (\lambda\mu) & \alpha & L & & M \end{matrix} \tag{1}$$

The ranges of quantum numbers are determined by the following relations:

$$\lambda, \mu = 0, 1, 2, \ldots \quad \text{and} \quad \lambda \geq \mu; \tag{2}$$

$$K = \mu, \mu-2, \mu-4, ..., 1 \text{ or } 0; \tag{3}$$

$$\alpha = \frac{1}{2}(\mu - K); \quad \alpha = 0, 1, 2, \ldots \alpha_{\max}; \quad \alpha_{\max} = \begin{cases} \frac{\mu}{2}, & \mu \text{ even}, \\ \frac{\mu-1}{2}, & \mu \text{ odd}; \end{cases} \tag{4}$$

$$K = 0: \quad L = \lambda, \lambda - 2, \lambda - 4, ..., 1 \text{ or } 0; \tag{5}$$

$$K \neq 0: \quad L_{\min} = K; \quad L_{\max} = \mu - 2\alpha + \lambda - \beta; \quad \beta = \begin{cases} 0, & \lambda + \mu - L \text{ even}, \\ 1, & \lambda + \mu - L \text{ odd}; \end{cases} \tag{6}$$

$$M = -L, \ldots, +L. \tag{7}$$

2.2. Overlap Integrals of non-orthogonal BM Basis

The construction of a basis is mainly based on an efficient procedure for calculating the overlap integrals. A very effective procedure for the overlap integrals of the non-orthogonal basis of BM was proposed by [20]. We developed the procedure for calculating these overlap integrals and checked its results with the results provided by the authors [20] in I and II:

$$\begin{aligned} \langle u_{\alpha'}|u_\alpha\rangle \equiv \langle u_\alpha|u_{\alpha'}\rangle = \left\langle \begin{matrix} (\lambda,\mu)_B \\ \alpha, L, L \end{matrix} \middle| \begin{matrix} (\lambda,\mu)_B \\ \alpha', L, L \end{matrix} \right\rangle = C_1(\lambda, L, \Delta)(\lambda+2)^\beta (L-\mu+2\alpha)! \\ \times(\lambda - L + \mu - 2\alpha' - \beta)!!(\mu - 2\alpha' - \beta + \Delta - 1)!! \\ \times\sum_{l,z} \begin{pmatrix} \alpha' \\ \frac{1}{2}(l-\beta-\Delta) \end{pmatrix} (-1)^{(\mu+2\alpha-\Delta-\beta)/2+z} \begin{pmatrix} \frac{1}{2}(\mu-2\alpha-\Delta-\beta) \\ z \end{pmatrix} \\ \times\frac{(\mu-l)!!}{(\mu-l-2z)!!}\frac{(\mu+\beta+\Delta)!!}{(\mu-2\alpha'+l)!!}(l-\Delta+\beta-1)!!(\mu-\Delta-\beta-2z)!! \\ \times\frac{(\lambda-L+\mu-2\alpha-\beta)!!}{(\lambda-L+\Delta+2z)!!}\frac{(\lambda+L-\Delta+2)!!}{(\lambda+L-\mu+2\alpha+\beta+2z+2)!!}\frac{(L+l)!}{L!} \\ \times\frac{(\lambda+\mu+L+\beta+2)!!}{(\lambda+L+l+\beta+2z+2)!!}\frac{(\lambda+\beta+2z+1)!}{(\lambda+\beta+1)!}\frac{(\lambda+\mu-l-L+\Delta)!!}{(\lambda-L+\mu-2\alpha'-\beta)!!} \\ \times C_2(\lambda, L, \Delta, z). \end{aligned} \tag{8}$$

Here $\alpha \geq \alpha'$ and we use the following notations

$$\beta = \begin{cases} 0, & \lambda+\mu-L \text{ even}, \\ 1, & \lambda+\mu-L \text{ odd}, \end{cases} \qquad \Delta = \begin{cases} 0, & \lambda-L \text{ even}, \\ 1, & \lambda-L \text{ odd}, \end{cases} \qquad \begin{pmatrix} m \\ n \end{pmatrix} = \frac{m!}{n!(m-n)!}, \tag{9}$$

$$C_1(\lambda, L, \Delta) = \begin{cases} 1, & L > \lambda+\Delta, \\ \frac{(\lambda+L+\Delta+1)!!}{(2L+1)!!}, & L \leq \lambda+\Delta, \end{cases} \qquad C_2(\lambda, L, \Delta, z) = \begin{cases} \frac{(\lambda+L+\Delta+1+2z)!!}{(2L+1)!!}, & L > \lambda+\Delta, \\ \frac{(\lambda+L+\Delta+1+2z)!!}{(\lambda+L+\Delta+1)!!}, & L \leq \lambda+\Delta. \end{cases}$$

Note, the upper alternative in definition of the coefficients C_1 and C_2 corresponds to the overlap integrals which contain only λ in their final expression. The summation parameter z runs from 0 to $\frac{1}{2}(-2\alpha-\beta-\Delta+\mu)$ except when $z < \frac{1}{2}(-\Delta-\lambda+L)$. The summation parameter l runs from $\beta+\Delta$ to $2\alpha'+\beta+\Delta$ except when $\mu-l$ or $l-\Delta-\beta$ is odd. We check also applicability of the above expressions of the overlap $\langle u_\alpha|u_{\alpha'}\rangle$ in the case of values $\lambda < \mu$ in benchmark calculations of matrices of the BM operator $X^{(3)}$ and its eigenvalues given in section 3.

2.3. Adaptation of the Gram-Schmidt orthonormalization procedure

The developed universal algorithm for orthonormalizing the noncanonical basis SU(3) with vectors of highest weights of irreducible representations (irreps) of SO(3) is based on the Gram-Schmidt orthonormalization procedure adopted in accordance with I and II

$$|\phi_i\rangle \equiv \left| \begin{matrix} (\lambda,\mu) \\ i, L, L \end{matrix} \right\rangle = \sum_{\alpha=0}^{\alpha_{\max}} A_{i,\alpha}^{(\lambda,\mu)}(L) \left| \begin{matrix} (\lambda,\mu)_B \\ \alpha, L, L \end{matrix} \right\rangle, \quad A_{i,\alpha}^{(\lambda,\mu)}(L) = 0, \quad \text{if } i > \alpha. \tag{10}$$

Here, a multiplicity index i is introduced to differentiate the orthonormalized BM states and takes the same range of values as α. $A^{(\lambda,\mu)}_{i,\alpha}(L)$ denotes matrix elements of the upper triangular matrix $\mathcal{A}$ of the BM basis orthonormalization coefficients.

It should be noted that the developed symbolic orthonormalization algorithm can be called non-standard recursive, or actually iterative, since it traverses the computation graph not from top to bottom, but from bottom to top

$$|\psi_\alpha\rangle = -|u_\alpha\rangle + \sum_{\alpha'=\alpha+1}^{\alpha_{\max}} c_{\alpha\alpha'}|\phi_{\alpha'}\rangle, \text{ when } 0 \le \alpha < \alpha_{\max}. \tag{11}$$

Here $\alpha_{\max}$ is the number of linearly independent BM states $|u_\alpha\rangle$, $|\psi_\alpha\rangle$ is the orthogonalized BM states, $|\phi_\alpha\rangle$ is the orthonormalized BM states and $c_{\alpha\alpha'}$ are linear coefficients. The orthonormalization process (11) starts (somewhat deliberately) by taking $|\psi_{\alpha_{\max}}\rangle = |u_{\alpha_{\max}}\rangle$.

2.4. Calculation of $f^{(n)}_{\alpha,\alpha'}$, $n = 0, 1, \ldots, \alpha_{\max K} - 1$, and $\langle\psi_\alpha|\psi_\alpha\rangle$

In this algorithm, it is necessary to introduce and iteratively calculate the convenient intermediate values $f^{(n)}_{\alpha,\alpha'}$, where n denotes the iteration number. $\alpha_{\max K}$ is the maximal α value for which there is a BM state for this particular L. The iteration starts at $n = 0$ by calculation of all $f^{(n)}_{\alpha,\alpha'}$ which are determined by the overlap integrals $\langle u_{\alpha'}|u_\alpha\rangle$ of BM states

$$f^{(0)}_{\alpha,\alpha'} = \langle u_\alpha|u_{\alpha'}\rangle, \tag{12}$$

$$\alpha = \alpha_{\min K}, \alpha_{\min K}+1, \ldots, \alpha_{\max K}-1 \text{ and } \alpha < \alpha' \le \alpha_{\max K};$$

$$f^{(1)}_{\alpha,\alpha'} = -f^{(0)}_{\alpha,\alpha'} + \frac{f^{(0)}_{\alpha,\alpha_{\max K}} f^{(0)}_{\alpha',\alpha_{\max K}}}{\langle u_{\alpha_{\max K}}|u_{\alpha_{\max K}}\rangle}, \tag{13}$$

$$\alpha = \alpha_{\min K}, \alpha_{\min K}+1, \ldots, \alpha_{\max K}-2 \text{ and } \alpha < \alpha' \le \alpha_{\max K}-1;$$

$$f^{(n)}_{\alpha,\alpha'} = f^{(n-1)}_{\alpha,\alpha'} + \frac{f^{(n-1)}_{\alpha,\alpha_{\max K}-n+1} f^{(n-1)}_{\alpha',\alpha_{\max K}-n+1}}{\langle\psi_{\alpha_{\max K}-n+1}|\psi_{\alpha_{\max K}-n+1}\rangle}, \tag{14}$$

$$\alpha = \alpha_{\min K}, \alpha_{\min K}+1, \ldots, \alpha_{\max K}-n-1 \text{ and } \alpha < \alpha' \le \alpha_{\max K}-n;$$

$$\langle\psi_\alpha|\psi_\alpha\rangle = \langle u_\alpha|u_\alpha\rangle - \sum_{\alpha'=\alpha+1}^{\alpha_{\max K}} \frac{\left(f^{(\alpha_{\max K}-\alpha')}_{\alpha,\alpha'}\right)^2}{\langle\psi_{\alpha'}|\psi_{\alpha'}\rangle}. \tag{15}$$

Here the indices are in the indicated ranges. In the next step of the iteration $n = 1$, the calculation of $f^{(1)}_{\alpha,\alpha'}$ is determined by the formula (13). For all subsequent steps of the iteration $n > 1$, $f^{(n)}_{\alpha,\alpha'}$ is determined by the formula (14). Here, the normalizing integral is determined by the formula (15). One can see that to calculate in symbolic analytical form $f^{(n)}_{\alpha,\alpha'}$ the square root operation cannot be applied, see I and II. It should be noted that all the $f^{(n)}_{\alpha,\alpha'}$ values at step n of the iteration can be calculated solely from the values of $f^{(n-1)}_{\alpha,\alpha'}$ and normalization integrals obtained at the previous step $n-1$ of the iteration.

2.5. Orthonormalization of the basis of Bargmann-Moshinsky

Having calculated the normalization integrals by means of (8)–(9) and Eqs. (12)–(15), one may directly calculate the required coefficients of matrix $\mathcal{A}$. The formulas for calculating them are very simple

(i) $\alpha = \alpha' = \alpha_{\max K}$:

$$A_{\alpha_{\max K},\alpha_{\max K}} = \langle u_{\alpha_{\max K}}|u_{\alpha_{\max K}}\rangle^{-1/2}. \tag{16}$$

Journal of Physics: Conference Series **1416** (2019) 012010 doi:10.1088/1742-6596/1416/1/012010

(ii) $\alpha = \alpha'$ and $\alpha = \alpha_{\min K}, \alpha_{\min K}+1, \ldots, \alpha_{\max K}-1$:

$$A_{\alpha,\alpha} = -\langle \psi_\alpha | \psi_\alpha \rangle^{-1/2}, \tag{17}$$

(iii) $\alpha < \alpha' < \alpha_{\max K}$:

$$A_{\alpha,\alpha'} = \frac{1}{\langle \psi_{\alpha'} | \psi_{\alpha'} \rangle} \sum_{\alpha''=\alpha}^{\alpha'-1} A_{\alpha,\alpha''} f_{\alpha'',\alpha'}^{(\alpha_{\max K}-\alpha')}, \tag{18}$$

(iv) $\alpha = \alpha_{\min K}, \alpha_{\min K}+1, \ldots, \alpha_{\max K}-1$ and $\alpha' = \alpha_{\max K}$:

$$A_{\alpha,\alpha_{\max K}} = -\frac{1}{\langle \psi_{\alpha_{\max K}} | \psi_{\alpha_{\max K}} \rangle} \sum_{\alpha''=\alpha}^{\alpha_{\max K}-1} A_{\alpha,\alpha''} f_{\alpha'',\alpha_{\max K}}^{(0)}. \tag{19}$$

It should be noted that the square root operation is applied only at the last stage of calculating coefficient of matrix $\mathcal{A}$.

2.6. Test condition

The orthonormalization property of the matrix of orthonormalization coefficients of the matrix $\mathcal{A}$ has the following form. Product of the nonzero blocks of the indicated matrices is the unity matrix. The matrix elements of the matrix $\mathcal{U}$ are the overlap integrals of the BM states. The equality of the product of these matrices to the identity matrix is actually the final test for the correctness of orthonormalization. Because the BM vectors $|u_\alpha\rangle$ are linearly independent, one can require the orthonormalization properties for the vectors $|\phi_i\rangle$

$$\langle \phi_i | \phi_j \rangle = \delta_{ij}. \tag{20}$$

From Eq. (20) there may be derived the orthonormalization property of the orthonormalization coefficients $A_{\alpha,\alpha'}$ of matrix $\mathcal{A}$

$$\mathcal{A}\mathcal{U}\tilde{\mathcal{A}} = \mathcal{I}. \tag{21}$$

Here matrices $\mathcal{A}, \mathcal{U}$ and $\mathcal{I}$ have dimension $\alpha_{\max}+1$, the matrix $\tilde{\mathcal{A}}$ is transposed of $\mathcal{A}$. In general case, when $\alpha_{\max K} < \alpha_{\max}$ and $\alpha_{\min K} > 0$, these matrices have the following block structure

$$\mathcal{A} = \begin{pmatrix} 0 & 0 & 0 \\ 0 & \mathbf{A} & 0 \\ 0 & 0 & 0 \end{pmatrix}, \quad \mathcal{U} = \begin{pmatrix} 0 & 0 & 0 \\ 0 & \mathbf{U} & 0 \\ 0 & 0 & 0 \end{pmatrix}, \quad \mathcal{I} = \begin{pmatrix} 0 & 0 & 0 \\ 0 & \mathbf{I} & 0 \\ 0 & 0 & 0 \end{pmatrix}, \quad \mathbf{U} = \mathbf{A}^{-1}\tilde{\mathbf{A}}^{-1}. \tag{22}$$

Here the matrices $\mathbf{A}$, $\tilde{\mathbf{A}} = \mathbf{A}^T$ is transposed matrix $\mathbf{A}$, $\mathbf{U}$ and $\mathbf{I}$ have the dimension $\alpha_{\max K} - \alpha_{\min K} + 1$. The zeroes represents the sub-blocks with the appropriate dimensions that are filled with zeroes. The $\mathbf{I}$ is the unit matrix, and the entries of the matrix $\mathbf{U}$ are the overlap integrals $\langle u_\alpha | u_{\alpha'} \rangle$. The Eq. (21) may be used for control of consistency and accuracy of calculations.

3. The matrix elements of tensor operator components

3.1. The action of the zero component of the quadrupole operator onto the BM basis

The procedure for calculating the matrices of a quadrupole operator can be developed according to the method proposed in [21, 22] and recalculated in II and III. The action of the zero component of Q_0 onto nonorthogonal BM vectors reads as

$$Q_0 \left| \begin{matrix} (\lambda,\mu)_B \\ \alpha, L, L \end{matrix} \right\rangle = \sum_{\substack{k=0,1,2 \\ s=0,\pm 1}} a_s^{(k)} \left| \begin{matrix} (\lambda,\mu)_B \\ \alpha+s, L+k, L \end{matrix} \right\rangle, \tag{23}$$

Using the inverse transformation of the orthonormalization matrix $(A^{-1})^{(\lambda\mu)}_{\alpha,i}(L)$ connected nonorthogonal and orthogonal BM bases

$$\left| \begin{matrix} (\lambda,\mu)_B \\ \alpha, L, L \end{matrix} \right\rangle = \sum_{i=0}^{\alpha} (A^{-1})^{(\lambda\mu)}_{\alpha,i}(L) \left| \begin{matrix} (\lambda,\mu) \\ i, L, L \end{matrix} \right\rangle, \tag{24}$$

we have action of zero component of Q_0 on orthogonal BM basis vector

$$Q_0 \left| \begin{matrix} (\lambda,\mu) \\ i, L, L \end{matrix} \right\rangle = \sum_{\substack{j=0,\ldots,\alpha_{\max} \\ k=0,1,2}} q^{(\lambda\mu)}_{ijk}(L) \left| \begin{matrix} (\lambda,\mu) \\ j, L+k, L \end{matrix} \right\rangle. \tag{25}$$

Here matrix elements $q^{(\lambda\mu)}_{ijk}(L)$ are calculated

$$q^{(\lambda\mu)}_{ijk}(L) = \sum_{\substack{\alpha=0,\ldots,\alpha_{\max} \\ s=0,\pm1}} A^{(\lambda\mu)}_{i,\alpha}(L) a^{(k)}_s (A^{-1})^{(\lambda\mu)}_{(\alpha+s),j}(L+k) \tag{26}$$

via the coefficients $a^{(k)}_s$ determined by formulas

$$\begin{aligned}
a^{(2)}_0 &= \frac{6(\lambda+\mu-L-2\alpha-\beta)}{[(L+2)(2L+3)]^{1/2}}, \quad a^{(2)}_{-1} = \frac{12\alpha}{[(L+2)(2L+3)]^{1/2}}, \quad a^{(2)}_{-1}=0, \\
a^{(1)}_0 &= -6\frac{2\alpha\beta(L+2\alpha-\mu+1)+(\lambda+\mu-L-2\alpha-\beta)(\mu-2\alpha-\beta)}{(L+2)(L+1)^{1/2}} - \frac{6\beta}{(L+1)^{1/2}}, \\
a^{(1)}_{-1} &= \frac{12\alpha(\lambda-\mu+2\alpha)}{(L+2)(L+1)^{1/2}}, \quad a^{(1)}_1 = \frac{6\beta(\lambda+\mu-L-2\alpha-\beta)}{(L+2)(L+1)^{1/2}}, \\
a^{(0)}_0 &= 4\alpha\frac{L(L+1)-3(L+2\alpha-\mu+\beta)^2}{(L+1)(2L+3)} - 2(\lambda+\mu-L-\beta-2\alpha)\frac{L(L+1)-3(\mu-2\alpha)^2}{(L+1)(2L+3)} \\
&-(L-\mu+4\alpha+\beta)\left(1+\frac{3\beta}{L+1}\right), \\
a^{(0)}_{-1} &= \frac{6(\lambda+\mu-L-2\alpha-\beta)(L-\mu+2\alpha)(L-\mu+2\alpha-1)}{(L+1)(2L+3)}, \\
a^{(0)}_1 &= -\frac{6(\lambda+\mu-L-2\alpha-\beta)(\mu-2\alpha-\beta)(\mu-2\alpha-\beta-1)}{(L+1)(2L+3)}, \\
\beta &= \begin{cases} 0, & \lambda+\mu-L \text{ even}, \\ 1, & \lambda+\mu-L \text{ odd}. \end{cases}
\end{aligned}$$

The analytical expressions for the $q^{(\lambda\mu)}_{ijk}(L)$ coefficients for values up to $\mu=3$ were presented in [22] and were used to verify the procedures developed by us for $q^{(\lambda\mu)}_{ijk}(L)$ coefficients calculations in III.

3.2. The matrix elements of the quadrupole operator components

The reduced matrix elements of the quadrupole operator are related to the previously determined coefficients $q^{(\lambda\mu)}_{ijk}(L)$ by the following formula

$$\left\langle \begin{matrix} (\lambda\mu) \\ j, L+k \end{matrix} \right\| Q^{(2)} \left\| \begin{matrix} (\lambda\mu) \\ i, L \end{matrix} \right\rangle = (-1)^k \frac{\sqrt{2L+1}}{(L+k, L, 20|LL)} q^{(\lambda,\mu)}_{i,j,k}(L). \tag{27}$$

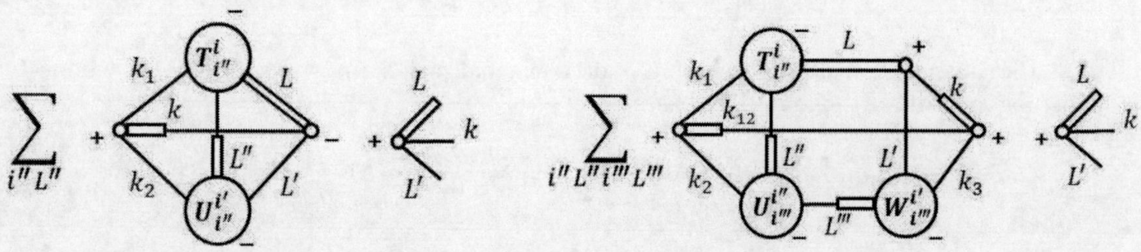

Figure 1. Calculations of tensor product of two and three tensor operators in graphical form.

The matrix elements of the quadrupole operator components can be obtained from the reduced matrix elements using the Wigner-Eckart theorem

$$\left\langle \begin{matrix} (\lambda\mu) \\ jL'M' \end{matrix} \middle| Q_p^{(2)} \middle| \begin{matrix} (\lambda\mu) \\ iLM \end{matrix} \right\rangle = \frac{(LM\,2p|L',M')}{\sqrt{2L'+1}} \left\langle \begin{matrix} (\lambda\mu) \\ j,L' \end{matrix} \middle\| Q^{(2)} \middle\| \begin{matrix} (\lambda\mu) \\ i,L \end{matrix} \right\rangle. \tag{28}$$

Here the conventional notation of $O(3)$ Clebsch-Gordan coefficients [23] is used. The dimension of the matrix of the quadrupole operator for given values of μ and λ can be calculated by the following formula

$$D_{\lambda\mu} = \frac{1}{2}(\lambda+1)(\mu+1)(\lambda+\mu+2). \tag{29}$$

3.3. $Q \cdot Q$ matrix

The quadratic Casimir operator of SU(3) group and its eigenvalues of a subspace for given λ, μ read as

$$C_2(SU(3)) = Q \cdot Q + 3L \cdot L = 4(\lambda^2 + \mu^2 + \lambda\mu + 3\lambda + 3\mu), \tag{30}$$

where the scalar product of the quadrupole operators is given by the formula

$$Q \cdot Q = \sum_{\eta=-2}^{2} (-1)^\eta Q_\eta Q_{-\eta}, \tag{31}$$

which was verified in integer arithmetic calculations.

3.4. Illustration of $\left(L^{(1)} \cdot L^{(1)}\right)$ calculation by graphical method

The effective expression for the matrix elements of the operator $X^{(3)}$ was derived by the graphical method [24](see fig. 1). The great advantage of the graphical method is that usually it can significantly simplify the momentum recoupling coefficients that are presented in formulas

$$\begin{aligned} \langle iLM| [T^{(k_1)} \otimes U^{(k_2)}]_q^{(k)} |i'L'M'\rangle = & \sum_{i''L''M''q_1,q_2} (k_1q_1\,k_2q_2|kq)\, \langle iLM| T_{q_1}^{(k_1)} |i''L''M''\rangle \\ & \times \langle i''L''M''| U_{q_2}^{(k_2)} |i'L'M'\rangle = (L'M'\,kq|LM) \sum_{i''L''} [(2L+1)(2L''+1)]^{-1/2} \\ & \times \langle iL\| T^{(k_1)} \|i''L''\rangle\, \langle i''L''\| U^{(k_2)} \|i'L'\rangle\, \langle (L',(k_1,k_2)k)L|((L',k_2)L'',k_1)L\rangle, \end{aligned} \tag{32}$$

where coefficients $\langle (L',(k_1,k_2)k)L|((L',k_2)L'',k_1)L\rangle$ read as

$$\begin{aligned} \langle (L',(k_1,k_2)k)L|((L',k_2)L'',k_1)L\rangle = & \frac{1}{2L+1} \sum_{q_1q_2qM'M''M} (k_1q_1\,k_2q_2|kq)(L'M'\,kq|LM) \\ & \times (L'M'\,k_2q_2|L''M'')(L''M''\,k_1q_1|LM) - (-1)^{k_1}(2k_1+1)^{-1/2}\,\delta_{k_1,k_2}\,\delta_{L,L'}\,\Delta(k_1L''L) \end{aligned} \tag{33}$$

3.5. Illustration of $\left(L^{(1)} \cdot L^{(1)}\right)$ calculation

Using this relation between the scalar product and the tensor product of zero rank and the expression for the reduced matrix element of the angular momentum, we obtain the correct formula for the scalar product of the angular momentum operators

$$[T^{(k)} \otimes U^{(k)}]_0^{(0)} = (-1)^k (2k+1)^{-1/2} (T^{(k)} \cdot U^{(k)}), \tag{34}$$

$$\langle iL \| \, L^{(1)} \, \| i''L'' \rangle = [L(L+1)(2L+1)]^{1/2} \, \delta_{iL,i''L''}, \tag{35}$$

$$\langle iLM | \, (L^{(1)} \cdot L^{(1)}) \, | i'L'M' \rangle = L(L+1) \, \delta_{iLM,i'L'M'}, \tag{36}$$

$$\left\langle \begin{matrix} (\lambda\mu) \\ i, L, L \end{matrix} \left| \left(L^{(1)} \cdot L^{(1)}\right) \right| \begin{matrix} (\lambda\mu) \\ i, L', L' \end{matrix} \right\rangle = -\sqrt{3}\delta_{LL'} \tag{37}$$

$$\times \sum_{m_1=-1}^{1} (1m_1 1-m_1|00) \left\langle \begin{matrix} (\lambda\mu) \\ i, L, L \end{matrix} \left| L^{(1)}_{m_1} \right| \begin{matrix} (\lambda\mu) \\ i, L, L-m_1 \end{matrix} \right\rangle \left\langle \begin{matrix} (\lambda\mu) \\ i, L, L-m_1 \end{matrix} \left| L^{(1)}_{-m_1} \right| \begin{matrix} (\lambda\mu) \\ i, L, L \end{matrix} \right\rangle .$$

It should be noted that the same numerical values gives and direct summation of angular momentum matrix elements. This type of procedure can be used for testing purposes.

3.6. Illustration of $X^{(3)} = \left([L^{(1)} \otimes L^{(1)}]^{(2)} \cdot Q^{(2)}\right)$ calculation by graphical method

In the case of three operators, their general tensor product can be expressed in graphical form as follows (see fig. 1). The appeared angular momentum recoupling coefficient can be significantly simplified by the graphical method

$$\begin{aligned}
&\langle iLM | \left[[T^{(k_1)} \otimes U^{(k_2)}]^{(k_{12})} \otimes W^{(k_3)} \right]^{(k)}_q | i'L'M' \rangle \\
&= \sum_{i''L''M''q_1,q_2} \; \sum_{i'''L'''M'''q_{12},q_3} (k_1 q_1 \, k_2 q_2 | k_{12} q_{12})(k_{12} q_{12} \, k_3 q_3 | kq) \\
&\times \langle iLM | \, T^{(k_1)}_{q_1} \, | i''L''M'' \rangle \, \langle i''L''M'' | \, U^{(k_2)}_{q_2} \, | i'''L'''M''' \rangle \, \langle i'''L'''M''' | \, W^{(k_3)}_{q_3} \, | i'L'M' \rangle \\
&= (L'M' \, kq | LM) \sum_{i''L''i'''L'''} [(2L+1)(2L''+1)(2L'''+1)]^{-1/2} \\
&\times \langle iL \| \, T^{(k_1)} \, \| i''L'' \rangle \, \langle i''L'' \| \, U^{(k_2)} \, \| i'''L''' \rangle \, \langle i'''L''' \| \, W^{(k_3)} \, \| i'L' \rangle \\
&\times \langle (L', ((k_1, k_2)k_{12}, k_3)k)L | (((L', k_3)L''', k_2)L'', k_1)L \rangle,
\end{aligned} \tag{38}$$

where coefficients $\langle (L', ((k_1, k_2)k_{12}, k_3)k)L | (((L', k_3)L''', k_2)L'', k_1)L \rangle$ are determined similar to (33) and calculated following the graphical method [24].

3.7. Matrix elements of $X^{(3)} = \left([L^{(1)} \otimes L^{(1)}]^{(2)} \cdot Q^{(2)}\right)$ operator

We present the result of our calculation for the operator $X^{(3)}$ that coincides with the expression of the operator $X^{(3)}$ given in the article [22] with the exception of the numerical factor $1/\sqrt{6}$

$$\left\langle \begin{matrix} (\lambda\mu) \\ j, L', M' \end{matrix} \left| X^{(3)} \right| \begin{matrix} (\lambda\mu) \\ i, L, M \end{matrix} \right\rangle = \frac{1}{\sqrt{6}} (L+1)(2L+3) \, q^{(\lambda\mu)}_{ij0}(L) \, \delta_{LL'} \, \delta_{MM'}. \tag{39}$$

It should be noted that some numerical factor is quite deliberately used by different authors. The obtained formula of the matrix elements of the operator $X^{(3)}$ was verified with results that give a direct summation of the matrix elements of these operators

$$\begin{aligned}\left\langle \begin{matrix}(\lambda\mu)\\ j,L',L'\end{matrix}\right| X^{(3)} \left| \begin{matrix}(\lambda\mu)\\ i,L,L\end{matrix}\right\rangle &= \sqrt{5}\delta_{LL'}\sum_{m_1=-2}^{2}\sum_{m_4=-1}^{1}(2m_1 2-m_1|00)\\ &\times(1m_1-m_4 1m_4|2m_1)\left\langle \begin{matrix}(\lambda\mu)\\ j,L,L\end{matrix}\right| L^{(1)}_{m_1-m_4} \left| \begin{matrix}(\lambda\mu)\\ j,L,L-m_1+m_4\end{matrix}\right\rangle\\ &\times\left\langle \begin{matrix}(\lambda\mu)\\ j,L,L-m_1+m_4\end{matrix}\right| L^{(1)}_{m_4} \left| \begin{matrix}(\lambda\mu)\\ j,L,L-m_1\end{matrix}\right\rangle \left\langle \begin{matrix}(\lambda\mu)\\ j,L,L-m_1\end{matrix}\right| Q^{(2)}_{-m_1} \left| \begin{matrix}(\lambda\mu)\\ i,L,L\end{matrix}\right\rangle .\end{aligned} \tag{40}$$

3.8. Matrix elements of $\tilde{X}^{(3)} = \left([L^{(1)}\otimes Q^{(2)}]^{(1)}\cdot L^{(1)}\right)$ *operator*

Another form of the operator $X^{(3)}$ was proposed in [19]. In this case, the expressions obtained by the graphical method (38) and with direct summation have the following form

$$\left\langle \begin{matrix}(\lambda\mu)\\ j,L',M'\end{matrix}\right| \tilde{X}^{(3)} \left| \begin{matrix}(\lambda\mu)\\ i,L,M\end{matrix}\right\rangle = \sqrt{\frac{5}{2}}\left(-\frac{1}{\sqrt{10}}\right)(L+1)(2L+3)\, q^{(\lambda\mu)}_{ij0}(L)\,\delta_{LL'}\,\delta_{MM'}, \tag{41}$$

$$\begin{aligned}\left\langle \begin{matrix}(\lambda\mu)\\ j,L',L'\end{matrix}\right| \tilde{X}^{(3)} \left| \begin{matrix}(\lambda\mu)\\ i,L,L\end{matrix}\right\rangle &= \sqrt{\frac{5}{2}}(-\sqrt{3})\,\delta_{LL'}\sum_{m_1=-2}^{2}\sum_{m_4=-1}^{1}(2m_1 2-m_1|00)\\ &\times(1m_1-m_4 1m_4|2m_1)\left\langle \begin{matrix}(\lambda\mu)\\ j,L,L\end{matrix}\right| L^{(1)}_{m_1-m_4} \left| \begin{matrix}(\lambda\mu)\\ j,L,L-m_1+m_4\end{matrix}\right\rangle\\ &\times\left\langle \begin{matrix}(\lambda\mu)\\ j,L,L-m_1+m_4\end{matrix}\right| L^{(1)}_{m_4} \left| \begin{matrix}(\lambda\mu)\\ j,L,L-m_1\end{matrix}\right\rangle \left\langle \begin{matrix}(\lambda\mu)\\ j,L,L-m_1\end{matrix}\right| Q^{(2)}_{-m_1} \left| \begin{matrix}(\lambda\mu)\\ i,L,L\end{matrix}\right\rangle .\end{aligned} \tag{42}$$

3.9. Calculation of the matrix $\tilde{X}^{(3)} = \left([L^{(1)}\otimes Q^{(2)}]^{(1)}\cdot L^{(1)}\right)$ *and its eigenvalues*

The resulting expression for the matrix elements of the operator $\tilde{X}^{(3)}$ gives the same numerical results as in [19]. We present example of calculation of the matrix $\tilde{X}^{(3)}$ with $\mu=6$, $\lambda=6$ and $L=6$ for comparison with result of [19] at $\lambda=k_1-k_2$ and $\mu=k_2$, i.e. $k_1=12$ and $k_2=6$. For example, here is the matrix $\tilde{X}^{(3)}$, calculated exactly and numerically, as well as the eigenvalues of the matrix:

$$\tilde{\mathbf{X}}^{(3)} = \begin{pmatrix} -\frac{2407509}{3455} & -\frac{18\sqrt{44049486}}{3455} & 0 & 0\\ -\frac{18\sqrt{44049486}}{3455} & -\frac{9677048733}{102692965} & \frac{720\sqrt{20733455}}{29723} & 0\\ 0 & -\frac{720\sqrt{20733455}}{29723} & \frac{1976289111}{10492219} & \frac{36\sqrt{4577342}}{353}\\ 0 & 0 & \frac{36\sqrt{4577342}}{353} & \frac{212751}{353}\end{pmatrix} \tag{43}$$

$$\tilde{\mathbf{X}}^{(3)} = \begin{pmatrix} -696.818813314038 & -34.5776023854303 & 0 & 0\\ -34.5776023854303 & -94.2328301943565 & -110.300046158617 & 0\\ 0 & -110.300046158617 & 188.357592516893 & 218.189816246354\\ 0 & 0 & 218.189816246354 & 602.694050991501\end{pmatrix}$$

Numerical eigenvalues E_n of eigenvalue problem $\tilde{\mathbf{X}}^{(3)}\mathbf{C} = E_n\mathbf{C}$ read as

$$E_n = \{698.844211758832,\ -698.844211758832,\ -138.429648887718,\ 138.429648887718\}.$$

Table 1. The example of spectrum $E_n \equiv E_{\lambda,\mu,\nu}$ of Hamiltonian (32) for $\gamma = 1.5$ and $h_{4Q} = 10$. The pair (λ, μ) labels the irreducible representations of the group SU(3) and the label $\nu_{\max}$ denote degeneration of eigenvalues due to the intrinsic tetrahedral/octahedral symmetry.

(λ, μ)	$D_{\lambda\mu}$	$\gamma C_2(SU(3))$	Exact eigenvalues	Numerical eigenvalues	$\nu_{\max}$
(1,0)	15	96	$\frac{12}{5}\left(1+\sqrt{89}\right)$	25.0415547	3
			$\frac{12}{5}\left(-1-\sqrt{65}\right)$	−21.7494186	3
			$\frac{12}{5}\left(1-\sqrt{89}\right)$	−20.2415547	3
			$\frac{12}{5}\left(-1+\sqrt{65}\right)$	16.9494186	3
			$\frac{48}{5}$	9.6000000	1
			$-\frac{24}{5}$	−4.8000000	2

We have also agreement of our calculations of eigenvalues of the operator $\tilde{X}^{(3)}$ with benchmark results presented in [10] up to a constant factor appearing in a definition of the operator $\tilde{X}^{(3)}$. Moreover, at the case of values $\mu > \lambda$ eigenvalues of the operator $X^{(3)}$ differ only sign if the values of $\lambda > \mu$ are only interchanged, as pointed out in [25, 26]. So, we are sure that we correctly calculated the matrix $\tilde{X}^{(3)}$ and its eigenvalues.

4. Spectrum of the tetrahedral/octahedral quadrupole-quadrupole Hamiltonian

We present a benchmark example that is related to the problem of existence of higher symmetries in nuclei. An influence of point symmetries within the nuclear SU(3) approximation has been proposed in [6] and investigated in II floating points arithmetic for example the low lying energy levels $E_n \equiv E_{\lambda,\mu,\nu}$ of the Hamiltonian:

$$H \equiv H/h_{4Q} = \gamma C_2(SU(3)) + H''_{4Q}/h_{4Q}, \; H|\lambda,\mu,\nu >= E_n|\lambda,\mu,\nu > . \tag{44}$$

Here we perform calculation of matrix of one of the quadrupole-quadrupole terms that can simulate both tetrahedral and octahedral nuclear symmetry with Hamiltonian H''_{4Q}

$$H''_{4Q} = \sqrt{\frac{14}{5}}(Q \otimes Q)^4_0 + (Q \otimes Q)^4_{-4} + (Q \otimes Q)^4_4.$$

To illustrate the effect of this term on the energy spectrum, the eigenvalues of H''_{4Q} for irr. $\mu = 0$, $\lambda = 1$ were calculated. The computational results for the example of spectrum of Hamiltonian (44) are presented in Table 1. The columns of the table are: (λ, μ) labels the irreducible representations of the group SU(3); $D_{\lambda\mu}$ is the dimension of the (λ, μ) irrep. from Eq.(29) that is determined a complexity of the above algorithm; $C_2(SU(3))$ marks the eigenvalues of the second order Casimir operator (30); E_n presents the energy levels that results after diagonalization of the Hamiltonian (44); $\nu = 1, ..., \nu_{\max}$ is the degeneration $\nu_{\max}$ of the corresponding energy levels E_n. The computations were done with integer arithmetic and numerically to 10-digit precision.

The obtained results of exact and numerical calculations are presented in Table 1 here in two columns respectively. These results indicate correct degeneracy of the eigenvalues expected be for interactions with tetrahedral symmetry.

5. Conclusion

We present the new, extremely fast and universal symbolic algorithm for construction of the orthonormalized non-canonical SU(3) basis with the highest weight vectors of SO(3) irreps. The proposed orthonormalization algorithm allows one to implement analytic, exact and numerical calculations. The high efficiency of the program implemented in the Wolfram Mathematica was illustrated by exact as well as numerical orthonormalisation of the BM basis up to extremely high quantum numbers λ and μ which is not given by other symbolic algorithms known in the literature [9]. The efficient exact as well as numerical procedures for calculation of the quadrupole operator as well as the BM operator $X^{(3)}$ matrices up to high λ and μ values were developed and implemented. The presented benchmark calculations of spectrum of the quadrupole-quadrupole interaction with Tetrahedral/Octahedral symmetry demonstrate efficiency of the proposed method for calculation of spectral characteristics (especially close to resonances) of quantum systems under consideration and to study their analytical properties for understanding the dominant symmetries.

Acknowledgments

The work was partially supported by the Bogoliubov-Infeld program, the Votruba-Blokhintsev program, the RUDN University Program 5-100 and grant of Plenipotentiary of the Republic of Kazakhstan in JINR. A. Deveikis is grateful to Prof. A. Góźdź for hospitality during visits in Institute of Physics, Maria Curie-Skłodowska University.

References

[1] Harvey M 1968 *The Nuclear SU_3 Model* Advances in Nuclear Physics Springer Eds: Baranger M and Vogt E (Boston, MA)
[2] Moshinsky M and Smirnov Yu F 1996 *The Harmonic Oscillator in Modern Physics* (Netherlands: HAP)
[3] Cseh J 2015 *Phys. Lett. B* **743** 213
[4] Dytrych T, et al 2016 *Comput. Phys. Commun.* **207** 202
[5] Dudek J, Góźdź A, Schunck N and Miśkiewicz M 2002 *Phys. Rev. Lett.* **88** 252502
[6] Góźdź A , Pędrak A, Gusev A A and Vinitsky S I 2018 *Acta Phys. Polonica B Proc. Suppl.* **11** 19
[7] Dudek J, Gurien D, Dedes, Mazurek K, et al 2018 *Phys. Rev. C* **97** 021302(R)
[8] Saha A, Bhattacharjee T, Curien D, Dudek J, Dedes I, Mazurek K, Góźdź A et al 2019 *J. Phys. G* **46** 055102
[9] Pan F, Yuan S, Launey K D and Draayer J P 2016 *Nucl. Phys A* **743** 70
[10] McKay W, Patera J and Sharp R T 1975 *Comp. Phys. Commun.* **10** 1
[11] Moshinsky M, Patera J, Sharp R T and Winternitz P 1975 *Ann. Phys. (NY)* **95** 139
[12] Campoamor-Stursberg R 2007 *J. Phys. A* **40** 14773
[13] Campoamor-Stursberg R 2011 *J. Phys. A* **44** 025204
[14] Campoamor-Stursberg R 2019 *J. Phys. Conf. Ser.* **1194** 012019
[15] Bargmann V and Moshinsky M 1961 *Nucl. Phys.* **23** 177
[16] Deveikis A, Gusev A A, Gerdt V P, Vinitsky S I, Góźdź A and Pędrak A 2018 *Lect. Notes Computer Sci.* **11077** 131
[17] Deveikis A, Gusev A A, Gerdt V P, Vinitsky S I, Góźdź A, Pędrak A and Burdik Č 2019 *Lect. Notes Computer Sci.* **11661** 91
[18] Vinitsky S, Burdik Č, Gusev A, Deveikis A, Góźdź A, Pędrak A, Krassovitskiy P M 2019 *J. Phys. Conf. Ser.* **1194** 012109
[19] Judd B R, Miller W , Patera J and Winternitz P 1974 *J. Math. Phys.* **15** 1787
[20] Alisauskas S, Raychev P and Roussev R 1981 *J. Phys. G* **7** 1213
[21] Afanasjev G N, Avramov S A and Raychev P P 1973 *Sov. J. Nucl. Phys.* **16** 53
[22] Raychev P and Roussev R 1981 *J. Phys. G* **7** 1227
[23] Varshalovitch D A, Moskalev A N and Hersonsky V K 1988 *Quantum Theory of Angular Momentum* (Singapore: World Sci.)
[24] Jucys A P and Bandzaitis A A, 1977 *Theory of Angular Momentum in Quantum Mechanics* (Vilnius: Mokslas)
[25] Asherova R M and Smirnov Yu F 1973 *Rept. Math. Phys.* **4** 83
[26] Pustovalov V V and Smorodinsky Ya A 1970 *Sov. J. Nucl. Phys.* **10** 729

Journal of Physics: Conference Series **1416** (2019) 012011 doi:10.1088/1742-6596/1416/1/012011

Two-qubit geometric discord. The solution

Andrzej Frydryszak[1], Piotr Ługiewicz, Lech Jakóbczyk
Institute of Theoretical Physics, University of Wroclaw, Poland
E-mail: andrzej.frydryszak@ift.uni.wroc.pl

Abstract. We report the analytical solution of the problem of description of quantum geometric discord for a two-qubit system.

1. Introduction

Contrary to the common belief not all important questions for compound multi qubit systems can be answered explicitly. Important characterization of apparently one of the simplest compound systems i.e. two-qubit system, known as the geometric discord, has been not known for a long time. Only recently has been found strict answer [1]. Explicit construction turns out to be rather involved, but can be brought to the compact form. Here we want to report the final result in a generic form. Partial answers for selected families of states were known before including: the Bell diagonal states (or states with maximally mixed marginals) (2013) [2], X-shaped states (2014) [3], states with the correlation matrix having only one nonzero singular value and arbitrary Bell vectors of the marginals and so-called quantum-classical states (2014) [3]. In the present contribution we provide brief description of the general solution for arbitrary type of mixed or pure state of two-qubit compound system

2. Notation and tools

The set $\mathcal{E}_2$ of all states of 2 - level system is conventionally parametrized in the following way

$$\rho = \frac{1}{2}\left(\mathbb{1}_2 + \langle n, \sigma\rangle\right), \quad n \in \mathbb{R}^3 \tag{1}$$

where $||n|| \leq 1$. The $\mathcal{E}_2$ forms a unit ball in $\mathbb{R}^3$ with the pure states laying on the unit sphere $||n|| = 1$.

The set of states of two-qubit compound system can be described by

$$\rho = \frac{1}{4}\left(\mathbb{1}_2 \otimes \mathbb{1}_2 + \langle x, \sigma\rangle \otimes \mathbb{1}_2 + \mathbb{1}_2 \otimes \langle y, \sigma\rangle + \sum_{j,k=1}^{3} K_{jk}\sigma_j \otimes \sigma_k\right) \tag{2}$$

where $x, y \in \mathbb{R}^3$ and $K = (K_{jk})$ is the correlation matrix. In the case of qubits we can use the group of orthogonal transformations to diagonalize correlation matrix, hence any two-qubit state is locally equivalent to the state with diagonal K.

Typical examples:

[1] Talk given by A.F.

Journal of Physics: Conference Series **1416** (2019) 012011 doi:10.1088/1742-6596/1416/1/012011

- States with maximally mixed marginals (MMM)

$$\mathrm{tr}_{\mathcal{A}}\rho = \frac{1}{2}\mathbb{1}_2 \quad \text{and} \quad \mathrm{tr}_{\mathcal{B}}\rho = \frac{1}{2}\mathbb{1}_2$$

- X-states

$$\rho_X = \begin{pmatrix} \rho_{11} & 0 & 0 & \rho_{14} \\ 0 & \rho_{22} & \rho_{23} & 0 \\ 0 & \rho_{32} & \rho_{33} & 0 \\ \rho_{41} & 0 & 0 & \rho_{44} \end{pmatrix}$$

- Bell diagonal states (special case of X-states)

$$\rho = \frac{1}{4}\left(\mathbb{1}_2 \otimes \mathbb{1}_2 + \sum_{j=1}^{3} c_j \sigma_j \otimes \sigma_j\right)$$

Geometric discord

Let us consider a state ρ of bipartite system $\mathcal{AB}$. When we perform local measurement on the subsystem $\mathcal{A}$, the state ρ may be disturbed due to such measurement. The trace-norm (one-sided) measurement induced geometric discord is defined as the minimal disturbance induced by projective measurement $\mathbb{P}_{\mathcal{A}}$ on subsystem $\mathcal{A}$, computed using the trace distance in the set of states. It can be compared with the standard geometric discord equal to the distance from a given state to the set of classical - quantum states [4, 5, 6, 7, 8]. As it was already stated in the Introduction, in the case of qubits these two notions coincide [7]. It is more convenient to use the quantity based on the disturbance induced by the measurement which will be simply called trace - norm geometric discord. The formal definition is as follows [2]

$$D_1(\rho) = \min_{\mathbb{P}_{\mathcal{A}}} ||\rho - \mathbb{P}_{\mathcal{A}}(\rho)||_1 \tag{3}$$

where $||A||_1 = \mathrm{tr}\,|A|$.

In the case of qubits, the local projective measurement $\mathbb{P}_{\mathcal{A}}$ is given by the one - dimensional projectors P_1, P_2 on $\mathbb{C}^2$, such that

$$P_1 + P_2 = \mathbb{1}_2, \quad P_j P_k = \delta_{jk} P_k$$

and $\mathbb{P}_{\mathcal{A}} = \mathbb{P} \otimes id$, where

$$\mathbb{P}(A) = P_1 A P_1 + P_2 A P_2 \tag{4}$$

One - dimensional projectors P_k can be always chosen as

$$P_k = u P_k^0 u^* \quad \text{for some} \quad u \in SU(2)$$

where

$$P_1^0 = \begin{pmatrix} 1 & 0 \\ 0 & 0 \end{pmatrix}, \quad P_2^0 = \begin{pmatrix} 0 & 0 \\ 0 & 1 \end{pmatrix}$$

A real orthogonal projector $\mathcal{P}$ on $\mathbb{R}^3$ can be defined by the following relation

$$\langle \mathcal{P}m, \sigma \rangle = \mathbb{P}(\langle m, \sigma \rangle), \quad m \in \mathbb{R}^3 \tag{5}$$

If $\mathcal{P}_0$ denotes such projector given by (4), where we take P_1^0 and P_2^0, then

$$\mathcal{P}_0 = diag(0,0,1)$$

and

$$\mathcal{P} = V \mathcal{P}_0 V^T, \quad V \in SO(3) \tag{6}$$

Journal of Physics: Conference Series **1416** (2019) 012011 doi:10.1088/1742-6596/1416/1/012011

It is convenient to use orthogonal complements to $\mathcal{P}_0$ and $\mathcal{P}$

$$\mathcal{M}_0 = \mathbb{1}_3 - \mathcal{P}_0, \quad \mathcal{M} = \mathbb{1}_3 - \mathcal{P} \tag{7}$$

Obviously $\mathcal{M}_0 = diag(1,1,0)$,

$$\mathcal{M} = V\mathcal{M}_0V^T, \quad V \in SO(3)$$

and

$$dim Ran\, \mathcal{M}_0 = dim Ran\, \mathcal{M} = 2$$

Such projectors $\mathcal{M}$ cover the whole set of projectors with dimension 2.

Disturbance
As a disturbance of the state (2) caused by measurement $\mathbb{P}_{\mathcal{A}}$ we shall understand

$$S(\mathcal{M}) = \rho - \mathbb{P}_{\mathcal{A}}(\rho) = \frac{1}{4}\left(\langle \mathcal{M}x, \sigma\rangle \otimes \mathbb{1}_2 + \sum_{k=1}^{3} \langle \mathcal{M}Ke_k, \sigma\rangle \otimes \langle e_k, \sigma\rangle\right) \tag{8}$$

where e_k, $k = 1,2,3$ are the vector of the canonical basis of $\mathbb{R}^3$. So

$$D_1(\rho) = \min_{\mathcal{M}} \mathrm{tr}\, |S(\mathcal{M})| = \min_{\mathcal{M}} \mathrm{tr}\, \sqrt{Q(\mathcal{M})} \tag{9}$$

where $Q(\mathcal{M}) = S(\mathcal{M})S(\mathcal{M})^*$ and the minimum is taken over all projectors $\mathcal{M}$ on two dimensional subspaces of $\mathbb{R}^3$.

3. Characterization of the disturbance

The first step in obtaining the analytical solution consists on a derivation of suitable expression of the trace norm of the disturbance allowing convenient study of behaviour of matrix dependent real function. One gets the following elegant result [1].

Theorem 1. *The trace norm of the disturbance $S(\mathcal{M})$ of the state (2) is given by the formula*

$$||S(\mathcal{M})||_1 =$$
$$\frac{1}{\sqrt{2}}\sqrt{||\mathcal{M}x||^2 + \mathrm{tr}\,(\mathcal{M}KK^T) + \sqrt{[||\mathcal{M}x||^2 + \mathrm{tr}\,(\mathcal{M}KK^T)]^2 - 4\,[||K^T\mathcal{M}x||^2 + \mathrm{tr}\,(E^TE - \mathcal{M}E^TE)]}}$$

where $||\cdot||$ denotes the Euclidean norm in $\mathbb{R}^3$ and $E = adjK$ is the adjunct matrix of the correlation matrix K (i. e. the transpose of its cofactor matrix).

Now, the question of finding a minimum of the trace norm of the $S(\mathcal{M})$ can be addressed. To find critical points of the mapping $\mathcal{M} \to ||S(\mathcal{M})||_1$. We use a unit vector $v \in \mathbb{R}^3$ to characterize $\mathcal{P} = P_v$, given by the third column of the matrix $V \in SO(3)$ relating $\mathcal{P}$ and $\mathcal{P}_0$. Introducing auxiliary function $\mathbf{g}$ on the unit sphere $\mathbb{S}^2 \subset \mathbb{R}^3$ with values in $\mathbb{R}^2$

$$\mathbf{g}(v) = (g_1(v), g_2(v)) \tag{10}$$

where

$$g_1(v) = ||(\mathbb{1}_3 - P_v)x||^2 + \mathrm{tr}\,((\mathbb{1}_3 - P_v)KK^T) \tag{11}$$

and

$$g_2(v) = 4\,\left(||K^T(\mathbb{1}_3 - P_v)x||^2 + \mathrm{tr}\,(E^TEP_v)\right) \tag{12}$$

Using the functions (11) and (12) we get

$$||S(\mathcal{M})||_1 = ||S(\mathbb{1}_3 - P_v)||_1 = \frac{1}{\sqrt{2}}\sqrt{g_1(v) + \sqrt{g_1(v)^2 - g_2(v)}}$$

The mapping **g** can be viewed as a function of two-dimensional projectors $\mathcal{M}$ or vectors $v \in \mathbb{S}^2$.

Introducing the following operators

$$W_y z = \langle z, y\rangle y, \quad y, z \in \mathbb{R}^3 \tag{13}$$

and

$$L_+ = KK^T + W_x \tag{14}$$

where x is a Bloch vector, we can obtain relations

$$g_1(v) = \mathrm{tr}\, L_+ - \langle v, L_+ v\rangle \tag{15}$$

and

$$g_2(v) = 4\left(||K^T(x - \langle x, v\rangle x||^2 + \langle v, E^T E v\rangle\right) \tag{16}$$

It turns out that there are basically two types of situations emerging in the search of a minimum, depending of the type of set of points on which it is achieved. We shall name them as: singular critical points, smooth critical points.

Singular critical points

When the minimum of the mapping $\mathcal{M} \to ||S(\mathcal{M})||_1$ is achieved on the set

$$\mathbb{D} = \{v \in \mathbb{S}^2 \,:\, g_1^2(v) = g_2(v)\} \tag{17}$$

a critical point $v_0 \in \mathbb{D}$ we call the singular critical point. In such a case the minimal value of the $||S(\mathcal{M})||_1$ is given by the minimum of the function g_1. New representations of the functions g_1 and g_2

$$g_1(v) = \mathrm{tr}\, L_- - \langle v, L_- v\rangle + 2\left(||x||^2 - \langle x, v\rangle^2\right) \tag{18}$$

and

$$g_2(v) = 4\left(||x||^2 - \langle x, v\rangle^2\right)\left[\mathrm{tr}\, L_- - \langle v, L_- v\rangle + ||x||^2 - \langle x, v\rangle^2\right] + 4\langle v, adj L_- v\rangle \tag{19}$$

where $L_- = KK^T - W_x$, allow simplification of analysis of properties of the set $\mathbb{D}$. One can show that the set (17) is discrete or is equal to the whole sphere $\mathbb{S}^2$. The characterization of the function $g_1(v)$ for such a case is following: for all vectors $v \in \mathbb{D}$

$$g_1(v) = 2\left(int\{\widetilde{\lambda}_1, \widetilde{\lambda}_2, \widetilde{\lambda}_3\} + ||x||^2 - \langle x, v\rangle^2\right) \tag{20}$$

where *int* denotes the intermediate value and $\widetilde{\lambda}_1, \widetilde{\lambda}_2, \widetilde{\lambda}_3$ are the eigenvalues of the matrix L_-.

Smooth critical points

By a smooth critical point we understand the critical point belonging to the set $\mathbb{S}^2 \setminus \mathbb{D}$, in such a case for $v_0 \in \mathbb{S}^2 \setminus \mathbb{D}$ we have

$$g_1^2(v_0) \neq g_2(v_0)$$

Any smooth critical point of the mapping $\mathcal{M} \to ||S(\mathcal{M})||_1$ is a critical point of the function **g** and satisfies equation

$$\left[W_x P_v KK^T + KK^T P_v W_x - KK^T W_x - W_x KK^T + E^T E + \mu L_+\right] v = \omega_0 v \tag{21}$$

where a vector $v \in \mathbb{S}^2$ is a solution of (21) if there exists a real number μ such that v is a solution of the above eigenvector problem.

4. The solution

In this section we describe the solution of the main problem, namely we find the value of geometric discord in general case. It is convenient to consider the matrix

$$G_\mu = -W_x \mathcal{M} K K^T - K K^T \mathcal{M} W_x + E^T E + \mu L_+ \tag{22}$$

represented in the ordered basis of orthonormal eigenvectors of the operator L_- assuming that $\widetilde{\lambda}_1 \geq \widetilde{\lambda}_2 \geq \widetilde{\lambda}_3$, and neglecting parts which commute with the projectors P_v, we obtain the matrix

$$\widetilde{G}_\mu = W_x P_v \Lambda + \Lambda P_v W_x + 2 P_x W_v P_x - (\widetilde{\lambda}_1 - \widetilde{\lambda}_2) W_x + \widetilde{\mu}(\Lambda + 2W_x) + (\widetilde{\lambda}_1 - \widetilde{\lambda}_2)(\widetilde{\lambda}_2 - \widetilde{\lambda}_3) P_{y_0}$$

where

$$\Lambda = \begin{pmatrix} \widetilde{\lambda}_1 - \widetilde{\lambda}_3 & 0 & 0 \\ 0 & \widetilde{\lambda}_2 - \widetilde{\lambda}_3 & 0 \\ 0 & 0 & 0 \end{pmatrix}, \quad y_0 = \begin{pmatrix} 0 \\ 1 \\ 0 \end{pmatrix}$$

and

$$\widetilde{\mu} = \mu - \widetilde{\lambda}_2 - ||x||^2$$

Let us observe that if L_- has a degenerate spectrum, the form of the operator $\widetilde{G}_\mu$ is particulary simple and the desired minimal value of $\mathcal{M} \to ||S(\mathcal{M})||_1$ can be found in a straightforward way.

Nondegenerate spectrum

This is a harder part of the problem under consideration. The proof is rather subtle, but final result can be formulated in the form of the following theorem

Theorem 2. *In the case of non-degenerate spectrum of the matrix $L_- = KK^T - W_x$, the value of trace - norm geometric discord is given by*

$$D_1(\rho) = \sqrt{\min\{d_1, d_2, d_3\}}$$

where

$$d_1 = \widetilde{\lambda}_1 + x_1^2,$$

$$d_2 = \frac{1}{2}\left(\widetilde{\lambda}_1 + \widetilde{\lambda}_2 + ||x||^2 + x_3^2 - \sqrt{(\widetilde{\lambda}_1 - \widetilde{\lambda}_2 + ||x||^2 - x_3^2)^2 - 4(\widetilde{\lambda}_1 - \widetilde{\lambda}_2)x_2^2}\right)$$

$$d_3 = \min\{\mu_*, \mu_{**}\}$$

with

$\mu_* = \frac{1}{2}\left(\widetilde{\lambda}_1 + \widetilde{\lambda}_3 + ||x||^2 + x_2^2 + N_*(1 - 2p(\theta_*)\right)$

$\mu_{**} = \widetilde{\lambda}_2 + ||x||^2 - (||x||^2 - x_2^2) r(\theta_{**})$ *and*

$$\cos\theta_* = \frac{\sqrt{2}|x_1 x_3|}{\sqrt{N_*(N_* - \widetilde{\lambda}_1 + \widetilde{\lambda}_3 - x_1^2 + x_3^2)}}, \quad \sin\theta_* = -\frac{\sqrt{2}x_1 x_3}{\sqrt{N_*(N_* + \widetilde{\lambda}_1 - \widetilde{\lambda}_3 + x_1^2 - x_3^2)}} \tag{23}$$

where

$$N_* = \sqrt{(\widetilde{\lambda}_1 - \widetilde{\lambda}_3 + ||x||^2 - x_2^2)^2 - 4(\widetilde{\lambda}_1 - \widetilde{\lambda}_3)x_3^2}$$

and

$$\cos\theta_{**} = \frac{\sqrt{2}|x_1 x_3|}{\sqrt{N_{**}(N_{**} - x_1^2 + x_3^2)}}, \quad \sin\theta_{**} = -\frac{\sqrt{2}x_1 x_3}{\sqrt{N_{**}(N_{**} + x_1^2 - x_3^2)}} \tag{24}$$

with

$$N_{**} = ||x||^2 - x_2^2.$$

Degenerate spectrum
In the case of degenerate spectrum the analysis definitely simplifies and final answer takes the form

Theorem 3. *1. Let* $\widetilde{\lambda}_1 = \widetilde{\lambda}_2 \geq \widetilde{\lambda}_3$, *then*

$$D_1(\rho) = \sqrt{\widetilde{\lambda}_2}.$$

2. Let $\widetilde{\lambda}_1 > \widetilde{\lambda}_2 = \widetilde{\lambda}_3$, *then*

$$D_1(\rho) = \frac{1}{\sqrt{2}}\sqrt{\widetilde{\lambda}_1 + \widetilde{\lambda}_2 + ||x||^2 - \sqrt{(\widetilde{\lambda}_1 - \widetilde{\lambda}_2 + ||x||^2)^2 + 4(\widetilde{\lambda}_1 - \widetilde{\lambda}_2)x_1^2}}.$$

5. Example: new class of states $\rho_{\gamma,a}$

Knowing the generic formulas which determine the explicit values of the geometric discord we can distinguish completely new two-parameter family of states

$$\rho_{\gamma,a} = \frac{1}{4}\begin{pmatrix} 1+a(1+w_2) & -iaz & -iaz & a(2-w_1) \\ iaz & 1+a(1-w_2) & a(2+w1) & iaz \\ iaz & a(2+w_1) & 1-a(1+w_2) & iaz \\ a(2-w_1) & -iaz & -iaz & 1-a(1-w_2) \end{pmatrix}, \tag{25}$$

where for $\widetilde{\gamma} = \sqrt{1+16\gamma^2}$ we have

$$w_1 = \frac{\sqrt{7-\widetilde{\gamma}}\,(\widetilde{\gamma}-1) + \sqrt{7+\widetilde{\gamma}}\,(\widetilde{\gamma}+1)}{2\sqrt{2}\widetilde{\gamma}}$$

$$w_2 = \frac{\sqrt{7-\widetilde{\gamma}}\,(\widetilde{\gamma}+1) + \sqrt{7+\widetilde{\gamma}}\,(\widetilde{\gamma}-1)}{2\sqrt{2}\widetilde{\gamma}}$$

$$z = \frac{\sqrt{7+\widetilde{\gamma}} - \sqrt{7-\widetilde{\gamma}}}{\widetilde{\gamma}}\sqrt{2}\gamma.$$

The (25) is a state if $|\gamma| \leq \sqrt{3}$, where a is the function of γ. The correlation matrix and Bloch vector are the following

$$K = a\begin{pmatrix} 2 & 0 & 0 \\ 0 & w_1 & z \\ 0 & z & w_2 \end{pmatrix}, \quad x = \begin{pmatrix} 0 \\ 0 \\ a \end{pmatrix}$$

therefore a is equal to the norm $||x||$ of the Bloch vector. The correlation matrix of the $\rho_{\gamma,a}$ yields the relation

$$KK^T = ||x||^2\begin{pmatrix} 4 & 0 & 0 \\ 0 & 4 & 2\gamma \\ 0 & 2\gamma & 3 \end{pmatrix}$$

providing that the L_- matrix has a simple form

$$L_- = ||x||^2\begin{pmatrix} 4 & 0 & 0 \\ 0 & 4 & 2\gamma \\ 0 & 2\gamma & 2 \end{pmatrix} \tag{26}$$

and the ordered eigenvalues are the following

$$\widetilde{\lambda}_1 = ||x||^2\left(3+\sqrt{1+4\gamma^2}\right), \quad \widetilde{\lambda}_2 = 4||x||^2, \quad \widetilde{\lambda}_3 = ||x||^2\left(3-\sqrt{1+4\gamma^2}\right). \tag{27}$$

The spectrum of L_- is non-degenerate. The value of $D_1(\rho_{\gamma,a})$ can be shown to be fully determined by the norm of the Bloch vector and to be equal

$$D_1(\rho_{\gamma a}) = 2||x||^2.$$

Conclusions

Strict general formulas characterising quantum correlations valid for all class of quantum states of compound systems in general were not known. In the present report we have sketched how the problem of finding values of quantum correlation measure - the geometric quantum discord - for two-qubit compound system can be solved completely. We have given the new compact formula for the trace norm of the disturbance $S(\mathcal{M})$ for the general two-qubit state.

Let us observe a remarkable effect, that the information about critical points is encoded in the spectrum of the matrix $L_- = KK^T - W_x$. Moreover, for non-degenerate spectrum critical points are located on big circles in $\mathbb{S}^2$. Therefore the problem of finding geometric discord could be reduced to minimization of the quadratic form restricted to big circles. The complete solution of this problem was rather involved and it simplified significantly, when any degeneracy of the spectrum of L_- was present. The full analysis can be found in the work [1] (see also Ref. [9]).

Our solution agrees with all known particular answers for selected families of states. Moreover, as a non-trivial illustration, we have discussed the totally new family of mixed states $\rho_{\gamma,a}$ and have found the value of its geometric discord.

References

[1] Ługiewicz P, Frydryszak A and Jakóbczyk L 2019 *Quantum Inf. Process.* **18** 185
[2] Paula F M, de Oliveira T R and Sarandy M S 2013 *Phys. Rev.* A **87** 064101
[3] Ciccarello F, Tufarelli T, Giovannetti V 2014 *New J. Phys* **16** 013038
[4] Modi K, Brodutch A, Cable H, Paterek T and Vedral V 2012 *Rev. Mod. Phys.* **84** 1655
[5] Streltsov A, Adesso G and Plenio M B 2017 *Rev. Mod. Phys.* **89** 041003
[6] Ming-Liang Hu, Xueyuan Hu, Jieci Wang, Yi Peng, Yu-Ran Zhang and Heng Fan 2018 *Phys. Rep.* **762-764** 1
[7] Ługiewicz P, Frydryszak A and Jakóbczyk L 2017 *J. Phys. A: Math. Theor.* **50** 245301
[8] Dakić B, Vedral V and Brukner C 2010 *Phys.Rev. Lett.* **105** 190502
[9] Jakóbczyk L, Frydryszak A and Ługiewicz P 2016 *Phys. Lett.* A **380** 1535

Journal of Physics: Conference Series **1416** (2019) 012012 doi:10.1088/1742-6596/1416/1/012012

Beyond coherent state quantization

Laure Gouba
Abdus Salam International Centre for Theoretical Physics, ICTP - Strada Costiera, 11, I-34151 Trieste Italy

E-mail: lgouba@ictp.it

Abstract. We present an original approach to quantization based on operator-valued measure that generalizes the so-called Berezin-Klauder-Toeplitz quantization, and more generally coherent state quantization approches.

1. Introduction

In Physics, quantization is generally understood as a correspondence between a classical and a quantum theory and dequantization is the opposite process by which one starts with a quantum theory and arrives back at its classical counterpart. The processes of quantization and dequantization have evolved into mathematical theories, affecting the areas of group representation theory and symplectic geometry. Quantization is not a method for deriving quantum mechanics but a way to understand the deeper physical reality which underlies the structure of both classical and quantum mechanics and which unifies the two from a geometrical point of view. There is a certain mathematical richness in the various theories of quantization where the procedure does make sense. However, not every quantum system has a classical counterpart and then for such a system a quantization method does not make sense, moreover different quantum systems may reduce to the same classical theory. Originally P. A. M. Dirac introduced the canonical quantization in his 1926 doctoral thesis, *The method of classical analogy for quantization* [1]. The canonical quantization or correspondence principle is an attempt to take a classical theory described by the phase space variables, let's say p and q, and a Hamiltonian $H(q,p)$ to define or construct its corresponding quantum theory. The following simple technique for quantizing a classical system is used. Let q^i, $p_i, i = 1, 2 \dots n$, be the canonical positions and momenta for a classical system with n degrees of freedom. Their quantized counterparts $\hat{q}^i, \hat{p}_i$ are to be realized as operators on the Hilbert space $\mathcal{H} = L^2(\mathbb{R}^n, dx)$ by the prescription

$$(\hat{q}^i\psi)(x) = x^i\psi(x); \quad (\hat{p}_i\psi)(x) = -i\hbar\frac{\partial}{\partial x^i}\psi(x); \quad i = 1, 2, \dots n, \quad x \in \mathbb{R}^n. \tag{1}$$

This procedure is known as canonical quantization and is the basic procedure of quantization of a classical mechanics model [2, 3, 4]. More general quantities, such as the Hamiltonians, become operators according to the rule

$$H(p,q) \to \mathcal{H} = H(\hat{p}, \hat{q}), \tag{2}$$

an expression that may have ordering ambiguities [5, 6].

In which canonical coordinates system does such a quantization procedure work?

(i) According to Dirac replacing classical canonical coordinates by corresponding operators is found in practice to be successful only when applied with the dynamical coordinates and momenta referring to a cartesian system of axes and not to more general curvilinear coordinates.

(ii) Cartesian coordinates can only exist on a flat space.

(iii) The canonical quantization seems to depend on the choice of coordinates.

(iv) Beyond the ordering problem, one should keep in mind that $[Q, P] = i\hbar I_d$ holds true with self adjoint operators Q, P, only if both have continuous spectrum $(-\infty, +\infty)$, and there is uniqueness of the solution, up to unitary equivalence (von Neumann).

There are two attitudes that may be taken towards this apparent dependence of the procedure of the canonical quantization procedure on the choice of coordinates. The first view would be to acknowledge the cartesian character that is seemingly part of the procedure. The second view would be to regard it as provisional and seek to find a quantization formulation that eliminates this apparently unphysical feature of the current approaches. The aim of eliminating the dependence on cartesian coordinates in the standard approaches is no doubt one of the motivations for several procedures such as the geometric quantization [7, 8, 9, 10], the path integral quantization [11], the deformation quantization [12, 13, 14, 15, 16, 17], the Klauder-Berezin-Toeplitz quantization [18, 19, 20, 21]. However, the canonical quantization is of the most famous and simplest procedure that is mostly accepted due to its experimental validations. In a recent review [22] by S. T. Ali and M. Engliš various techniques of quantization conceived to make the transition from classical mechanics to quantum mechanics in the last five decades are presented. The point of departure is always an analysis of the geometrical structure of either the classical phase space or the classical configuration space. There is no general theory of quantization presently available which is applicable in all cases, and indeed, often the techniques used to quantize has to be tailored to the problem in question. As we have already said, canonical quantization generally requires the use of cartesian coordinates and not more general coordinates. So for a dynamical system without any constraints, its phase space is assumed to be flat and admits a standard quantization of its canonical variables. In presence of constraints, all the variables of the system are not physical and the unphysical variables that may cause some little concerns at a classical level are highly unwelcome at the quantum level. It is then necessary to eliminate the unphysical variables and keep only the true physical degrees of freedom. The quantization of the physical degrees of freedom follows as in standard quantization. The study of constrained systems for the purpose of quantization was initiated by Dirac [23, 24, 25], where he developed a method for treating singular systems and constraints. While Dirac's method and subsequent developments can cope with most models of interest, there are some problems in the sense that the quantization of the remaining degrees of freedom after one eliminates the unphysical variables may not be straightforward because the reduced (physical) phase space is non-euclidean in the sense that an obstruction has arisen where none existed before and that obstruction precludes the existence of self-adjoint canonical operators satisfying the canonical commutation relations. Other approaches for quantizing constrained systems of classical theories have been subsequently developed [26, 27, 28, 29, 30]. Recently we have been interested in applying the Dirac's method in the study of the two-dimensional damped harmonic oscillator in the extended phase space [31] as some years ago, we have used the method to study nonperturbatively scalar and spinor abelian gauge theories in $1+1$ dimensions [32]. Inspired by Klauder's paper on coherent state quantization of constrained systems [33], we are interested in understanding a procedure of quantization beyond the coherent states quantization, generically called integral quantization mostly developed by J. P. Gazeau and al. [34, 35, 36, 37, 38], in order to better study some of the constrained systems where the method is applicable. The paper is organized as follows. In section 2, starting from the Klauder-Berezin-Toeplitz quantization we reach the coherent state or frame quantization. In section 3, we start by introducing first the sea star algebra and then we introduce the integral quantization. The perspectives are included in the later section.

2. Coherent states quantization

Quantization procedure is an important aspect of coherent states, for instance standard coherent states offer a classical-like representation of the evolution of quantum observables. In coherent states quantization, we are interested in quantization of sets, more precisely measure spaces through coherent states. For the normalized coherent states $|z\rangle$, the resolution of the unity

$$\frac{1}{\pi}\int_{\mathbb{C}} |z\rangle\langle z| d^2 z = I_d \tag{3}$$

is a crucial property for our purpose in setting the bridge between the classical and the quantum worlds.

2.1. Klauder-Berezin-Toeplitz quantization

The Klauder-Berezin coherent state quantization also named anti-Wick quantization or Toeplitz quantization by many authors, consists in associating with any classical operator f that is a function of phase space variable (q,p) or equivalently of $(z,\bar{z})$, the operator valued integral

$$\frac{1}{\pi}\int_{\mathbb{C}} f(z,\bar{z})|z\rangle\langle z| d^2 z = A_f. \tag{4}$$

The function f is usually supposed to be smooth, but we will not retain in the sequel this too restrictive attribute. The resulting operator A_f, if it exists, at least in a weak sense, acts on the Hilbert space $\mathcal{H}$ of quantum states for which the set of Fock (or number) states $|n\rangle$ is an orthogonal basis. It is worthwhile being more explicit about what we mean by *weak sense*: the integral

$$\langle\psi|A_f|\psi\rangle = \int_{\mathbb{C}} f(z,\bar{z})|\langle\psi|z\rangle|^2 \frac{d^2 z}{\pi} \tag{5}$$

should be finite for any $|\psi\rangle$ in some dense subset in $\mathcal{H}$. One should note that if $|\psi\rangle$ is normalized then the integral (5) represents the mean value of the function f with respect to the ψ -dependent probability distribution $z \to |\langle\psi|z\rangle|^2$ on the phase space. More mathematical rigor is necessary here, and we will adopt the following acceptance criteria for a function (or distribution) to belong to the class of quantizable classical observables. A function $f : \mathbb{C} \to \mathbb{C}, \quad z \mapsto f(z,\bar{z})$ (more generally a distribution $T \in \mathcal{D}'(\mathbb{R}^2)$) is a coherent state quantizable classical observable along the map $f \mapsto A_f$ defined by $A_f = \frac{1}{\pi}\int_{\mathbb{C}} f(z,\bar{z})|z\rangle\langle z| d^2 z$ (more generally $T \mapsto A_T$) :

(i) if the map $z \mapsto \langle z|A_f|z\rangle$ (or $z \mapsto \langle z|A_T|z\rangle$), $z = \frac{1}{\sqrt{2}}(q+ip) \equiv (q,p)$, is a smooth ($\in C^\infty$) function with respect to the (q,p) coordinates of the phase space,

(ii) and if we restore the dependence on $\hbar$ through $z \to \frac{z}{\sqrt{\hbar}}$, we must get the right semiclassical limit, which mean that $\langle\frac{z}{\sqrt{\hbar}}|A_f|\frac{z}{\sqrt{\hbar}}\rangle \to f(\frac{z}{\sqrt{\hbar}}, \frac{\bar{z}}{\sqrt{\hbar}})$,

the same asymptotic behavior must hold in a distributional sense if we are quantizing distributions.

2.2. Coherent states or frame quantization

We consider the Hilbert space $L^2_K(X,\mu)(K=\mathbb{R}$ or $K=\mathbb{C})$ of square-integrable real or complex functions $f(x)$ on the observation set $X : \int_X |f(x)|^2\mu(dx) < \infty$. A classical observable is a function $f(x)$ on X having specific properties with respect to some supplementary structure allocated to X, such as topology, geometry, or something else. Given a frame, more about the concept of frame can be read in references [39, 40], $\{|x\rangle, x\in X\}$, a function $f : x(\in X)\mapsto f(x)(\in K)$, possibly undertood in a distribution sense, is a coherent state quantizable classical observable along the map $f\mapsto A_f = \int_X f(x)|x\rangle\langle x|\nu(dx)$, if the map $x(\in X)\mapsto \langle x|A_f|x\rangle$ is a regular function with respect to some additional structures allocated to X. We must get the right classical limit, which means that $\langle x|A_f|x\rangle = f(x)$ as a certain parameter goes to 0. A quantization of the observation set X is in one-to-one correspondence with the choice of a frame in the sense of $\int_X |x\rangle\langle x|\nu(dx) = I_d$, where $\nu(dx) = \mathcal{N}(x)\mu(dx)$. This term of frame is more appropriate for designating the total family $\{|x\rangle\}_{x\in X}$. This frame can be discrete or continuous, depending on the topology additionnaly allocated to the set X, and it can be overcompleted. The validity of a precise frame choice is determined by comparing spectral characteristics of quantum observables A_f with experimental data. A quantization method associated with a specific frame is intrinsically limited to all those classical observables for which the expansion $A_f = \int_X f(x)|x\rangle\langle x|\nu(dx)$ is mathematically justified within the theory of operators in Hilbert space (weak convergence). The coherent state quantization in this context requires a minimal significant structure on X, namely, the existence of a measure $\mu(dx)$, together with a σ-algebra of measurable subsets, and some additional structure to be defined depending on the context. The construction of the Hilbert space $\mathcal{H}$ is equivalent to the choice of a class of eligible quantum states, together with a technical condition of continuity. A correspondence between classical and quantum observables is then provided through a suitable generalization of the standard coherent states.

3. Beyond coherent state quantization: integral quantization

Let us brieffly recall the following. We consider an orthonormal basis (or frame) of the Euclidean plane $\mathbb{R}^2$ defined by the two vectors (in Dirac ket notations)

$$\vec{i} \equiv |0\rangle \equiv \begin{pmatrix} 1 \\ 0 \end{pmatrix} \quad \text{and} \quad \vec{j} \equiv |\frac{\pi}{2}\rangle \equiv \begin{pmatrix} 0 \\ 1 \end{pmatrix}, \tag{6}$$

and $|\theta\rangle$ denotes the unit vector with polar angle $\theta \in [0,\ 2\pi)$. This basis or frame is such that $\langle 0|0\rangle = 1 = \langle\frac{\pi}{2}|\frac{\pi}{2}\rangle$ and $\langle 0|\frac{\pi}{2}\rangle = 0$. The sum of their corresponding orthogonal projectors resolves the unit $I_d = |0\rangle\langle 0| + |\frac{\pi}{2}\rangle\langle\frac{\pi}{2}|$, that is equivalent to

$$\begin{pmatrix} 1 & 0 \\ 0 & 1 \end{pmatrix} = \begin{pmatrix} 1 & 0 \\ 0 & 0 \end{pmatrix} + \begin{pmatrix} 0 & 0 \\ 0 & 1 \end{pmatrix}. \tag{7}$$

To the unit vector $|\theta\rangle = \cos\theta|0\rangle + \sin\theta|\frac{\pi}{2}\rangle$ corresponds the orthogonal projector P_θ given by $P_\theta = |\theta\rangle\langle\theta|$ and

$$|\theta\rangle\langle\theta| = \begin{pmatrix} \cos\theta \\ \sin\theta \end{pmatrix} (\cos\theta\ \ \sin\theta) = \begin{pmatrix} \cos^2\theta & \cos\theta\sin\theta \\ \cos\theta\sin\theta & \sin^2\theta \end{pmatrix}, \tag{8}$$

then $P_\theta = \mathcal{R}(\theta)|\theta\rangle\langle\theta|\mathcal{R}(-\theta)$, where

$$\mathcal{R}(\theta) = \begin{pmatrix} \cos\theta & -\sin\theta \\ \sin\theta & \cos\theta \end{pmatrix} \in SO(2). \tag{9}$$

3.1. Sea star algebra

The five-fold symmetry of sea star (starfish) motivated the development in a comprehensive way, the quantization of functions on a set with five elements which yields in particular the notion of a quantum angle. The purpose of this introductory model is to enhance our intuition that quantization can be viewed as the analysis of a set from the point of view of a family of coherent states.

Figure 1. A sea star as a five-fold frame for the plane

(i) sea stars have five arms, or rays, connected to a small round body;
(ii) sea stars detect light with five purple eyespots at the end of each arm;
(iii) sea stars typically show pentameral symmetry.

How the ocean floor is "five-fold orientationally ' explored by a sea star?

A sea star possibly do this in a noncommutative way through the pentagonal set of unit vectors (the arms) forming a five-fold frame in the plane. The unit vectors (arms) are determined by $|\frac{2n\pi}{5}\rangle = \mathcal{R}(\frac{2n\pi}{5})|0\rangle \equiv$ "coherent" state $n = 0, 1, 2, 3, 4 \bmod 5$. We get the resolution of the identity:

$$\frac{2}{5}\sum_{n=0}^{4} |\frac{2\pi n}{5}\rangle\langle\frac{2\pi n}{5}| = \begin{pmatrix} 1 & 0 \\ 0 & 1 \end{pmatrix} \equiv I_d. \tag{10}$$

The property holds for any regular N-fold polygon in the plane

$$\frac{2}{N}\sum_{n=0}^{N-1} |\frac{2\pi n}{N}\rangle\langle\frac{2\pi n}{N}| = \begin{pmatrix} 1 & 0 \\ 0 & 1 \end{pmatrix}. \tag{11}$$

Let's make more precise the set X explored by the sea star, determine a probabilistic construction of the frame by using the Hilbert space structure of the plane and an exact way of exploring X through functions on this finite set. The sea star senses its possible orientations via the five angles $\frac{2\pi n}{5}$ and $X = \{0, 1, 2, 3, 4\}$ the set of orientations. It is equipped with discrete measure with uniform weight

$$\int_x f(x)d\mu(x) = \frac{2}{5}\sum_{n=0}^{4} f(n). \tag{12}$$

Let us choose two orthonormal elements, $\phi_0(n) = \cos\frac{2\pi n}{5}$ and $\phi_1(n) = \sin\frac{2\pi n}{5}$ in the Hilbert space $L^2(X, \mu)$, and build the five unit vectors in the real two-dimensional Hilbert space, that is the Euclidean plane $\mathbb{R}^2 = \mathcal{H}$, with the usual orthonormal basis $|0\rangle$, $|\frac{\pi}{2}\rangle$:

$$n \in X, \quad n \mapsto |n\rangle \equiv |\frac{2\pi n}{5}\rangle = \phi_0(n)|0\rangle + \phi_1(n)|\frac{\pi}{2}\rangle. \tag{13}$$

The operators $P_n = |\frac{2\pi n}{5}\rangle\langle\frac{2\pi n}{5}|$ acts on $\mathcal{H} = \mathbb{R}^2$. Given a $n_0 \in \{0,1,2,3,4\}$, one derives the probability distribution on $X = \{0,1,2,3,4\}$

$$\mathrm{tr}(P_{n_0}P_n) = |\langle\frac{2\pi n_0}{5}|\frac{2\pi n}{5}\rangle|^2 = \cos^2\left(\frac{2\pi(n_0-n)}{5}\right). \tag{14}$$

The five unit vectors $|n\rangle \equiv |\frac{2\pi n}{5}\rangle$ resolve the identity in $\mathcal{H}$. They form a finite unit frame for analysing the complex-valued functions $n \mapsto f(n)$ on X through what we call a coherent state (CS) quantization:

$$f(n) \mapsto \int_X |x\rangle\langle x| f(x) d\mu(x) = \frac{2}{5}\sum_{n=0}^{4} f(n)|\frac{2n\pi}{5}\rangle\langle\frac{2n\pi}{5}| \equiv A_f. \tag{15}$$

If we had chosen instead as a finite frame the orthonormal basis $|0\rangle, |\frac{\pi}{2}\rangle$, in $\mathbb{R}^2$, we would have obtained the trivial commutative quantization:

$$(f(0), f(1)) \mapsto A_f = \mathrm{diag}(f(0), f(1)). \tag{16}$$

This map should be analysed more precisely through spectral values of the 2×2 symmetric matrix A_f, and also through coherent states mean values of A_f:

$$n \mapsto \check{f}(n) := \langle\frac{2n\pi}{5}|A_f|\frac{2n\pi}{5}\rangle = \frac{2}{5}\sum_{m=0}^{4} f(m)\cos^2\frac{2(n-m)\pi}{5}, \quad n = 0,1,2,3,4. \tag{17}$$

3.2. Integral quantization

An original approach to quantization based on operator valued measures is named generically integral quantization. The approach generalizes the coherent state quantization. The probabilistic aspects appearing is highlighted at each stage of the quantization procedure. To be more mathematically precise and still remaining at a basic level, quantization is a linear map $\mathcal{Q} : \mathcal{C}(X) \to \mathcal{A}(\mathcal{H})$ from a vector space $\mathcal{C}(X)$ of complex valued function $f(x)$ on a set X to a vector space $\mathcal{A}(\mathcal{H})$ of linear operator $\mathcal{Q}(f) \equiv A_f$ in some complex Hilbert space $\mathcal{H}$, such that

(i) to the function $f = 1$ there corresponds the identity operator I_d on $\mathcal{H}$;

(ii) to a real function $f \in \mathcal{C}(X)$ there corresponds a (an essentially) self adjoint operator A_f in $\mathcal{H}$.

The above conditions may be easily fulfilled if one uses the resources offered by the pair (measure, integration). Let (X,ν) be a measure space. Let $\mathcal{H}$ be a complex Hilbert space and $x \in X$, $x \mapsto M(x) \in \mathcal{L}(\mathcal{H})$ an X-labelled family of bounded operators on $\mathcal{H}$ resolving the identity:

$$\int_X M(x) d\nu(x) = I_d, \tag{18}$$

provided that the equality is valid in a weak sense, which implies ν integrability for the family $M(x)$. If the operators $M(x)$ are positive and have unit trace, they will be preferentially denoted by $M(x) = \rho(x)$ in order to comply with the usual notation for a density matrix in quantum mechanics. The corresponding quantization of complex-valued functions $f(x)$ on X is then defined by the linear map:

$$f \mapsto A_f = \int_X M(x) f(x) d\nu(x). \tag{19}$$

This operator-valued integral is again understood in the weak sense, as the sesquilinear form,

$$B_f(\psi_1, \psi_2) = \int_X \langle \psi_1 | M(x) | \psi_2 \rangle f(x) d\nu(x). \tag{20}$$

The form B_f is assumed to be defined on a dense subspace of $\mathcal{H}$. If f is real and at least semi-bounded, the Friedrichs extension of B_f univocally defines a self-adjoint operator. If f is not semi-bounded, there is no natural choice of a self-adjoint operator associated with B_f, in this case, we can consider directly the symmetric operator A_f, enabling us to obtain a self-adjoint extension (unique for particular operators). The situation when the operators $M(x)$ are negative has not been discussed here and may be an interesting aspect to check. The spectral properties of the operators A_f may be analysed. An interesting aspect of this integral quantization method is the possibility to quantize constraints.

Acknowledgments

This paper is a slightly enlarged version of the talk given at the XXVIth International Conference on Integrable Systems and Quantum symmetries (ISQS-26), Prague, Czech Republic, 08 -12 July 2019. L. Gouba would like to acknowledge support from ICTP/Trieste and to thank the organizing committee of the ISQS-26.

References

[1] Dirac P A M 1925 *Proc. of the Royal Society of London Series A* Vol 109 pp 642-653
[2] Dirac P A M 1947 *The Principles of Quantum Mechanics* (London: Oxford University Press)
[3] von Neumann J 1955 *Mathematical Foundations of Quantum Mechanics* (Princeton: Princeton University Press)
[4] Weyl H 1950 *The Theory of Groups and Quantum Mechanics* (New York: Dover Publications)
[5] Born M and Jordan P 1925 On quantum mechanics *Zs f. Phys.* **34** 858
[6] Agarwal B S and Wolf E 1970 *Phys. Rev.* D **2** 2206
[7] Weyl H 1927 Quantenmechanik und gruppentheorie *Z. Phys.* **46** 1-46
[8] Groenewold H J 1946 *Physica* **12** 405-60
[9] Souriau J M 1966 *Commun. Math. Phys.***1** 374-98
[10] Kostant B 1970 *Lecture Notes in Mathematics* Vol 170 (Springer, Berlin, Heidelberg)
[11] Feynman R P 1942 *Principles of Least Action in Quantum Mechanics* (Thesis Feynman)
[12] Groenewold H J 1946 On the principles of elementary quantum mechanics *Physica* **12** 405-60
[13] Bayen F, Flato M, Fronsdal C, Lichnerowicz A and Sternheimer D 1978 *Annals Phys.* **111** 61
[14] Kontsevich M 1997 Deformation quantization of Poisson manifolds *Preprint q-alg/9709040*
[15] Bordemann M and Waldmann S 1998 *Commun. Math. Phys.* **195** 549-83
[16] Bordemann M, Neumaier N and Waldmann S 1998 *Comm. Math. Phys.* **198** 363-96
[17] Dito G and Sternheimer D 2002 Deformation quantization: genesis, developments and metamorphoses *Preprint math.QA/0201168*
[18] Klauder J R 1967 Weak correspondence principle *J. Math. Phys.* **8** 2392
[19] Berezin F A 1974 Quantization *Izv. Akad. Nauk SSSR Ser. Mat.* **38** 1116-1175
[20] Berezin F A 1975 General concept of quantization *Comm. Math. Phys.* **40** 153-174.
[21] Coburn L A 1994 *Algebraic Methods in Operator Theory* (Boston: Birkhäuser Springer) pp 101-108
[22] Ali S T and Engliš M 2004 *Review in Mathematical Physics* Vol 17 No 4 pp 391-490
[23] Dirac P A M 1950 Generalized hamiltonian dynamics *Canad. J. Math.* **2** 129
[24] Dirac P A M 1959 Fixation of coordinates in the Hamiltonian theory of gravitation *Phys. Rev.* **114** 924
[25] Dirac P A M 1964 *Lectures on Quantum Mechanics* (New York: Belfer Graduate School of Science, Yeshiva University)
[26] Faddeev L D 1969 The Feynman integral for singular Lagrangians *theor. math. phys.* **1** 1-13.
[27] Faddeev L D and Jackiw R W 1988 Hamiltonian reduction of unconstrained and constrained systems *Phys. Rev. lett.* **60** 1692
[28] Govaerts J 1991 Hamiltonian Quantisation and Constrained Dynamics (Louvain: Leuven University Press)
[29] Jackiw R W 1993 (Constrained) quantization without tears *Preprint hep-th/9306075*
[30] Klauder J R 2001 Quantization of constrained systems *Lect. Notes Phys.* **572** 143-182

[31] Gouba L 2018 Dirac's method for the two-dimensional damped harmonic oscillator in the extended phase space *Mathematics* **6** (10) 180
[32] Gouba L 2005 *Théories de Jauge Abéliennes Scalaire et Spinorielle à 1+1 Dimensions: une Etude Nonperturbative* PhD Thesis (Porto-Novo: Rep of Benin)
[33] Klauder J R 1997 Coherent state quantization of constrained systems *Annals of physics* **254** 419-453
[34] Bergeron H and Gazeau J P 2014 Integral quantizations with two basic examples *Annals of Physics* **344** 43-68
[35] Ali S T, Antoine J P and Gazeau J P 2014 *Coherent States, Wavelets and their Generalizations* (Springer)
[36] Bergeron H Curado E M F, Gazeau J P and Rodrigues M C S L 2014 *J. Phys.: Conf. Series* **512** 012032
[37] Gazeau J P 2016 Covariant integral quantizations and their applications to quantum cosmology *Acta polytechnica* **56**(3) 173-79
[38] Gazeau J P and Murenzi R 2016 *J. Math. Phys.* **57** 052102
[39] Ali S T, Antoine J P and Gazeau J P 1993 Continuous frames in Hilbert space *Annals of Physics* **222** 1-37
[40] Gazeau J P 2009 *Coherent States in Quantum Physics* (WileyVCH Verlag GmbH and Co. KGaA)

Journal of Physics: Conference Series **1416** (2019) 012013 doi:10.1088/1742-6596/1416/1/012013

Coherent states for rational extensions and ladder operators related to infinite-dimensional representations

Scott E. Hoffmann[1], Véronique Hussin[2], Ian Marquette[1] and Yao-Zhong Zhang[1]

[1]School of Mathematics and Physics, The University of Queensland, Brisbane, QLD 4072, Australia

[2]Département de Mathématiques et de Statistique, Université de Montréal, Montréal, Québec, H3C 3J7, Canada

E-mail: scott.hoffmann@uqconnect.edu.au

Abstract.
The systems we consider are rational extensions of the harmonic oscillator, the truncated oscillator and the radial oscillator. The wavefunctions for the extended states involve exceptional Hermite polynomials for the oscillator and truncated oscillator and exceptional Laguerre polynomials for the radial oscillator. In all cases it is possible to construct ladder operators that have infinite-dimensional representations of their polynomial Heisenberg algebras and couple all levels of the systems. We construct Barut-Girardello coherent states in all cases, eigenvectors of the respective annihilation operators with complex eigenvalues. Then we calculate their physical properties to look for classical or non-classical behaviour.

1. Introduction

This conference paper is a selection of results from three published papers [1, 2, 3] and a work in progress. It is concerned with supersymmetric rational extensions of the harmonic oscillator, the truncated oscillator and the radial oscillator. In addition, we construct special ladder operators associated with infinite-dimensional representations of their polynomial Heisenberg algebras. Then we form the coherent states associated with the annihilation operators and investigate their physical properties.

After a brief summary of supersymmetric quantum mechanics, in Section 2, we present results from [2, 3] on one-step rational extensions of the harmonic oscillator, in Section 3. Then, in Section 4, we present results from [1] on a multi-step rational extension of the truncated oscillator. In Section 5, we present unpublished results on a one-step rational extension of the radial oscillator.

In these papers, we constructed coherent states for special ladder operators that have infinite-dimensional representations. Then we investigated the physical properties of these coherent states using a variety of measures. In all cases, we looked for evidence of classical or non-classical behaviour.

Conclusions follow in Section 6.

2. Supersymmetric quantum mechanics

Supersymmetric quantum mechanics (SUSY QM) [4, 5, 6, 7, 8, 9, 10, 11, 12, 13, 14, 15, 16, 17, 18, 19, 20, 21] has been widely used to create partner Hamiltonians for a given, exactly solvable, Hamiltonian that have almost all of their spectrum in common with the original. Typically there are one or more energy levels of the partner Hamiltonian below the ground state energy of the original Hamiltonian. We consider only one-dimensional bound-state systems with discrete spectra. The methods can be applied to higher-dimensional systems [17, 22].

Suppose a Hamiltonian with an exact solution can be written in factorized form, $H^{(+)} = A^\dagger A$, in terms of the supercharge, A, and its Hermitian conjugate. It is is Hermitian and positive semi-definite. Then $H^{(-)} = AA^\dagger$ is another candidate Hamiltonian, Hermitian and positive semi-definite. The result of supersymmetric quantum mechanics is that these two Hamiltonians have the same energy spectra, except for additional eigenvector(s), $|\,0\,\rangle$, of $H^{(-)}$ satisfying $A\,|\,0\,\rangle = 0$.

Notice that the partner Hamiltonians satisfy the intertwining relation

$$A\,H^{(+)} = H^{(-)}\,A. \tag{1}$$

In higher-order supersymmetry [23, 22] the A operator is a product of first-order supercharges, such as

$$A = A^{(4)} A^{(3)} A^{(2)} A^{(1)} \tag{2}$$

that we will use in Section 4, with the $A^{(i)}$ defined in Eqs. (18,19). Then the simple forms for $H^{(+)}$ and $H^{(-)}$ do not hold. Instead, one requires an intertwining relation of the form of Eq. (1) to hold. It is often solved using symbolic manipulation software. The partial isospectrality of the partner Hamiltonians is then guaranteed.

3. One-step rational extensions of the harmonic oscillator

These papers [2, 3] are concerned with rational extensions of the harmonic oscillator. These have been studied by many authors [24, 25, 10, 21].

Of interest is the choice of ladder operators used to connect the basis states. If A and $A^\dagger$ are the supercharges used in the supersymmetric transformation and a and $a^\dagger$ are the ladder operators for the unmodified harmonic oscillator, one class of ladder operators is the *natural* [26],

$$b = A\,a\,A^\dagger, \quad b^\dagger = A\,a^\dagger A^\dagger. \tag{3}$$

The second class is the *intrinsic* class [26], that is a linearized version of the natural, with ladder operators of the form

$$\tilde{b} = f(H^{(-)})b \quad \tilde{b}^\dagger = b^\dagger f(H^{(-)}), \tag{4}$$

where f is chosen so that these ladder operators then satisfy the Heisenberg algebra, like the operators a and $a^\dagger$. Here $H^{(-)}$ is the supersymmetric partner Hamiltonian of the harmonic oscillator Hamiltonian $H^{(+)}$.

In general, the natural and intrinsic ladder operators will have finite-dimensional as well as infinite-dimensional representations in the state space of $H^{(-)}$. Ladder operators have been constructed [27, 22, 28, 29] with no finite-dimensional representations of their polynomial Heisenberg algebras. This is done using combinations of supercharges of Darboux-Crum and Krein-Adler type, allowing more than one possible path between the $H^{(+)}$ and $H^{(-)}$ states. Such ladder operators have been studied by the present authors [3, 2, 1] and by other authors [30, 29].

Polynomial Heisenberg algebras arise in many systems, such as the harmonic oscillator [24, 3, 2, 1], the Morse potential [31], the Scarf potential [9] and the infinite well [32]. These have been used in the study of superintegrable systems, Painlevé transcendents and coherent states [33, 16, 34, 35].

We note that we will find that the position wavefunctions for the modified states will involve exceptional Hermite orthogonal polynomials. These are defined in [36, 22, 23]. Their use has been connected with new potentials, shape invariance and position-dependent mass [37, 38, 39, 40].

For a one-step rational extension of the harmonic oscillator of order m (even), the supercharges are [22]

$$A = -\frac{d}{dx} + x + \frac{\mathcal{H}'_m}{\mathcal{H}_m}, \quad A^\dagger = +\frac{d}{dx} + x + \frac{\mathcal{H}'_m}{\mathcal{H}_m}. \tag{5}$$

The modified Hermite polynomials are $\mathcal{H}_m(x) = (-i)^m H_m(ix)$, which are positive definite for even m and satisfy the second-order differential equation

$$\mathcal{H}''_m + 2x\mathcal{H}'_m - 2m\mathcal{H}_m = 0. \tag{6}$$

Then we find

$$A^\dagger A = -\frac{d^2}{dx^2} + x^2 + 2m + 1 = H^{(+)} \quad \text{and} \quad AA^\dagger = -\frac{d^2}{dx^2} + V_m^{(-)} = H^{(-)}, \tag{7}$$

with partner potential

$$V_m^{(-)} = x^2 - 1 - \frac{\mathcal{H}_m\mathcal{H}''_m - 2x\mathcal{H}_m\mathcal{H}'_m - 2\mathcal{H}'^2_m}{\mathcal{H}^2_m}. \tag{8}$$

For $m = 2$, $\mathcal{H}_2(x) = 4x^2 + 2$, so

$$V_2^{(-)} = x^2 + 3 + \frac{16(4x^2 - 2)}{(4x^2 + 2)^2}. \tag{9}$$

The spectra of the partner Hamiltonian and the harmonic oscillator Hamiltonian are the same except for the unpartnered ground state of the partner Hamiltonian

$$\begin{aligned} E_\nu^{(+,m)} &= 2(\nu + m + 1) \quad \text{for } \nu = 0, 1, 2, \ldots \\ E_\nu^{(-,m)} &= 2(\nu + m + 1) \quad \text{for } \nu = -m - 1, 0, 1, 2, \ldots. \end{aligned} \tag{10}$$

The ladder operators, $c(m)$, are constructed and their matrix elements derived in [22], with

$$c(m) = A_m \ldots A_2 A_1 A^\dagger, \quad c^\dagger(m) = A A_1 A_2 \ldots A_m, \tag{11}$$

and

$$A_i = +\frac{d}{dx} + x + \frac{\mathcal{H}'_{i-1}}{\mathcal{H}_{i-1}} - \frac{\mathcal{H}'_i}{\mathcal{H}_i}, \quad A_i^\dagger = -\frac{d}{dx} + x + \frac{\mathcal{H}'_{i-1}}{\mathcal{H}_{i-1}} - \frac{\mathcal{H}'_i}{\mathcal{H}_i}. \tag{12}$$

We see that $c(m)$ involves a path from H^- to H^+ through the supercharge $A^\dagger$ then a return to H^- in m steps through intermediate Hamiltonians that may be singular.

The matrix elements of $c(m)$ are

$$\langle\, \nu - m - 1(-)\,|\, c(m)\,|\, \nu\,(-)\,\rangle = -[2^{m+1}(\nu - 1)(\nu - 2)\ldots(\nu - m)(\nu + m + 1)]^{\frac{1}{2}} \tag{13}$$

for all $\nu = -m - 1, 0, 1, 2, \ldots$. The operator $c(m)$ lowers the index by $m + 1$ and separates the spectrum into $m + 1$ distinct ladders with lowest weights $\mu = -(m + 1), 1, 2, 3, \ldots m$.

The operators $H^{(-,m)}, c(m)$ and $c^\dagger(m)$ satisfy a polynomial Heisenberg algebra [22], with polynomials of order $m+1$ in H^-.

The position wavefunctions involve type III Hermite exceptional orthogonal polynomials [37, 38, 39, 40].

Then we construct Barut-Girardello [41] coherent states as eigenvectors of $c(m)$ with complex eigenvalue z,

$$c(m)\,|\,z, c, m, \mu\,\rangle = z\,|\,z, c, m, \mu\,\rangle. \tag{14}$$

The time-dependent position probability densities for these coherent states involve the exceptional Hermite polynomials. For $m = 2, \mu = -3$ and $z = 15$, we construct the cat state superpositions

$$|\pm\rangle = \frac{1}{\sqrt{2}}\{|+z\rangle \pm |-z\rangle\}. \tag{15}$$

Their time-dependent position probability densities (Figure 1) show distinct wavepackets, oscillating and interfering semi-classically.

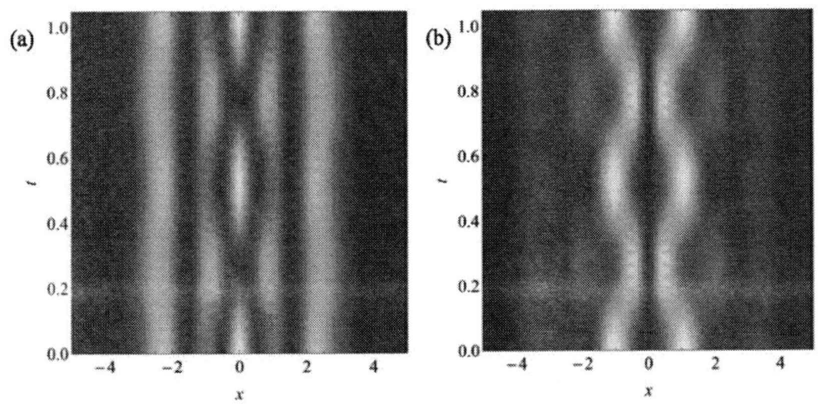

Figure 1. Cat state density for $c(2)$, $\mu = -3$ coherent states, (a) even and (b) odd, for $z = 15$.

Then for $m = 6$ and a large value of z, we see a great deal of wavepacket structure in the cat state densities (Figure 2), again semi-classical.

4. A multi-step rational extension of the truncated oscillator

The truncated oscillator has the Hamiltonian

$$\tilde{H}^{(+)} = \begin{cases} -\frac{d^2}{dx^2} + x^2 & 0 \le x < \infty, \\ \infty & x < 0. \end{cases} \tag{16}$$

The wavefunctions must vanish at the origin, which selects only the odd states of the usual harmonic oscillator. Again, these involve exceptional Hermite polynomials.

We choose a particular model, a four-step supersymmetric transformation with steps $\{m_1, m_2, m_3, m_4\} = \{1, 2, 3, 4\}$ [28, 2]. Then the special ladder operators are defined by

$$\tilde{C} = \bar{A}^{(5)}\bar{A}^{(4)}A^{(1)\dagger}A^{(2)\dagger}A^{(3)\dagger}A^{(4)\dagger}, \quad \tilde{C}^\dagger = A^{(1)}A^{(2)}A^{(3)}A^{(4)}\bar{A}^{(4)\dagger}\bar{A}^{(5)\dagger}. \tag{17}$$

Here the supercharges for the state adding approach are

$$A^{(i)} = +\frac{d}{dx} - \frac{d}{dx}\ln\varphi^{(i)}(x), \quad A^{(i)\dagger} = -\frac{d}{dx} - \frac{d}{dx}\ln\varphi^{(i)}(x), \tag{18}$$

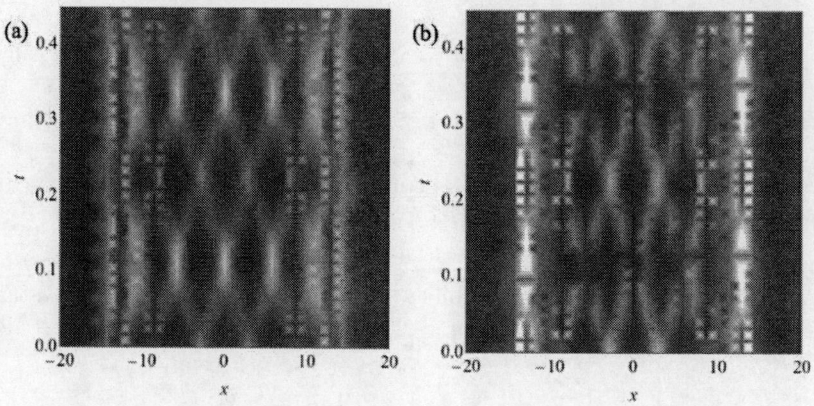

Figure 2. Time-dependent probability densities for (a) even and (b) odd cat states, for $c(6)$, $\mu = -7$ and $z = 10^8$.

with

$$\varphi^{(1)}(x) = \phi_1(x), \quad \varphi^{(i)}(x) = \frac{\mathcal{W}(\phi_1, \phi_2, \ldots, \phi_i)}{\mathcal{W}(\phi_1, \phi_2, \ldots, \phi_{i-1})}, \tag{19}$$

and $\phi_m(x) = \mathcal{H}_m(x)\, e^{x^2/2}$. The $\mathcal{W}(\phi_1, \phi_2, \ldots, \phi_n)$ are Wronskian determinants. The supercharges for the state deleting approach are

$$\bar{A}^{(i)} = +\frac{d}{dx} - \frac{d}{dx} \ln \bar{Q}^{(i)}, \quad \bar{A}^{(i)\dagger} = -\frac{d}{dx} - \frac{d}{dx} \ln \bar{Q}^{(i)}, \tag{20}$$

with

$$\bar{Q}^{(4)} = \psi_4, \quad \bar{Q}^{(5)} = \frac{\mathcal{W}(\psi_4, \psi_5)}{\psi_4} \tag{21}$$

and $\psi_\nu(x) = H_\nu(x)\, e^{-x^2/2}$.

The action of these operators separates the spectrum into three infinite ladders with lowest weights $\mu = -5, -3, 5$.

The matrix elements of $\tilde{C}$, as given in [28, 2], are labelled

$$\langle\, \mu + 6i - 6 \,|\, \tilde{C} \,|\, \mu + 6i \,\rangle = a_{\mu+6i}. \tag{22}$$

There are two particular cases

$$a_{-5+6\cdot 1} = [645\,120]^{\frac{1}{2}} \quad \text{and} \quad a_{-3+6\cdot 1} = [387\,072]^{\frac{1}{2}}. \tag{23}$$

The remaining matrix elements are

$$a_\nu = [2^6(\nu+6)\frac{(\nu-1)!}{(\nu-6)!}\frac{\nu+3}{\nu-3}\frac{\nu+4}{\nu-2}\frac{\nu+5}{\nu-1}]^{\frac{1}{2}}, \quad \text{for } \nu - 6 = 1, 3, 5, \ldots. \tag{24}$$

We construct the coherent states associated with these ladder operators for the three lowest weights.

We model placing a coherent state on one arm of an optical beamsplitter [2], with the excitations corresponding to photons. The beamsplitter entangles the state with the vacuum from the other arm. Then we calculate the linear entropy of the output state, an approximation to the von Neumann entropy. Small values indicate a low degree of entanglement. The maximum possible value is one. We see a moderate degree of entanglement, $S \sim 0.4$, a nonclassical feature, for $\mu = -5, -3$.

Journal of Physics: Conference Series **1416** (2019) 012013 doi:10.1088/1742-6596/1416/1/012013

5. One-step rational extension of the radial oscillator

For the three-dimensional spherically symmetric harmonic oscillator, the radial equation is (with $\hbar = 1, m = \frac{1}{2}, \omega = 1$)

$$(-\frac{d^2}{dx^2} + \frac{1}{4}x^2 + \frac{l(l+1)}{x^2})y_l(x) = E\, y_l(x). \tag{25}$$

We considered an $m = 2$ rational extension as in Section 3 [28, 3, 2], with the choice $l = 2$. The position wavefunctions of the partner Hamiltonian involve exceptional Laguerre polynomials [28].

The special ladder operators are defined by

$$c = \bar{\mathcal{A}}\tilde{\mathcal{A}}\mathcal{A}^\dagger \quad \text{and} \quad c^\dagger = \mathcal{A}\tilde{\mathcal{A}}^\dagger\bar{\mathcal{A}}^\dagger,$$

where $\mathcal{A}$ and $\mathcal{A}^\dagger$ use a state adding approach and $\bar{\mathcal{A}}$ and $\bar{\mathcal{A}}^\dagger$ use a state deleting approach. An intermediate transformation, enacted by $\tilde{\mathcal{A}}$ and $\tilde{\mathcal{A}}^\dagger$, is needed to shift the spectrum [28].

The lowest weights for the special ladder operators are $\mu = -3, 1, 2$. The matrix elements of c are

$$\langle -3 \,|\, c \,|\, 0 \rangle = 8[6\frac{\Gamma(\alpha+2)}{\Gamma(\alpha-1)}]^{\frac{1}{2}} \tag{26}$$

and

$$\langle \nu - 3 \,|\, c \,|\, \nu \rangle = 2^3 \left[(\nu+3)\frac{(\nu-1)!\,\Gamma(\nu+\alpha+2)}{(\nu-3)!\,\Gamma(\nu+\alpha-1)}\right]^{\frac{1}{2}}, \quad \text{for } \nu \geq 3, \tag{27}$$

with $\alpha = l + 1/2$.

The Wigner function for a coherent state is defined by

$$W(x,p;z) = \frac{1}{\pi}\int_{-\infty}^{\infty} dy \,\langle z \,|\, x - y \rangle \langle x + y \,|\, z \rangle e^{-i2py}, \tag{28}$$

is everywhere real and acts as a distribution in phase space. For the harmonic oscillator, it is everywhere positive and peaked at $x = \mathrm{Re}\, z, p = \mathrm{Im}\, z$.

We calculated the Wigner function for our three coherent states and found areas of negativity (the white areas in Figure 3) in all three cases, considered evidence of non-classicality.

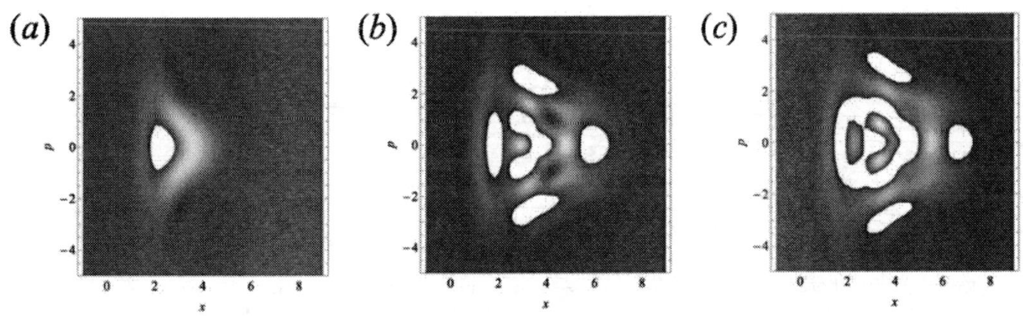

Figure 3. Wigner functions for $z = 500, l = 2$ and (a) $\mu = -3$ (b) $\mu = 1$ and (c) $\mu = 2$.

6. Conclusions

The coherent states of the harmonic oscillator have essentially classical behaviour [42]. The mean values of position and momentum follow classical trajectories. For rational extensions of the harmonic oscillator, much of this classical behaviour is lost. Instead, we find distinctly quantum behaviour such as wavepacket interference, significant entropy in the beamsplitter model and Wigner functions taking on negative values.

We have seen exceptional orthogonal polynomials in the position wavefunctions of the extended states. We have constructed special ladder operators with only infinite-dimensional representations of their polynomial Heisenberg algebras.

Future work will involve more results on the rational extensions of the radial oscillator and investigations of other potentials such as the Morse potential [31], the Scarf potential [9] and the infinite well [32].

Acknowledgements

IM was supported by Australian Research Council Discovery Project DP 160101376. YZZ was supported by Australian Research Council DP190101529. VH acknowledges the support of research grants from NSERC of Canada. SH receives financial support from a UQ Research Scholarship.

References

[1] Hoffmann S E, Hussin V, Marquette I and Zhang Y Z 2019 *J. Math. Phys.* **60** 052105
[2] Hoffmann S E, Hussin V, Marquette I and Zhang Y Z 2018 *J. Phys. A: Math. Theor.* **51** 315203
[3] Hoffmann S E, Hussin V, Marquette I and Zhang Y Z 2018 *J. Phys. A: Math. Theor.* **51** 085202
[4] Witten E 1981 *Nucl. Phys. B* **188** 513
[5] Witten E 1982 *Nucl. Phys. B* **202** 253
[6] Mielnik B 1984 *J. Math. Phys.* **25** 3387
[7] Andrianov A A, Ioffe M V and Spiridonov V P 1993 *Phys. Lett. A* **174** 273
[8] Andrianov A A, Ioffe M V, Cannata F and Dedonder J P 1995 *Int. J. Mod. Phys. A* **10** 2683
[9] Cooper F, Khare A and Sukhatme U 1995 *Phys. Rep.* **251** 267
[10] Fernández C D J, Glasser M L and Nieto L M 1998 *Phys. Lett. A* **240** 15
[11] Fernández C D J, Hussin V and Mielnik B 1998 *Phys. Lett. A* **244** 309
[12] Samsonov B F 1999 *Phys. Lett. A* **263** 274
[13] Mielnik B, Nieto L M and Rosas-Ortiz O 2000 *Phys. Lett. A* **269** 70
[14] Cariñena J F, Ramos A and Fernández C D J 2001 *Ann. Phys. (N.Y.)* **292** 42
[15] Aoyama H, Sato M and Tanaka T 2001 *Nucl. Phys. B* **619** 105
[16] Carballo J M, Fernández C D J, Negro J and Nieto L M 2004 *J. Phys. A. Math. Gen.* **37** 10349
[17] Marquette I 2009 *J. Math. Phys.* **50** 095202
[18] Fernández C D J 2010 Supersymmetric Quantum Mechanics *AIP Conf. Ser.* vol 1287 ed Garcia Rocha M, Lopez Fernandez R, Rojas Ochoa L F and Torres Vega G p 3
[19] Quesne C 2011 *Mod. Phys. Lett. A.* **26** 1843
[20] Marquette I 2012 *J. Math. Phys.* **53** 012901
[21] Gómez-Ullate D, Grandati Y and Milson R 2014 *J. Phys. A. Math. Gen.* **47** 015203
[22] Marquette I and Quesne C 2013 *J. Math. Phys.* **54** 102102
[23] Marquette I and Quesne C 2013 *J. Phys. A: Math. Theor.* **46** 155201
[24] Fernández D J, Hussin V and Nieto L M 1994 *J. Phys. A: Math. Gen.* **27** 3547
[25] Fernández D J, Nieto L M and Rosas-Ortiz O 1995 *J. Phys. A: Math. Gen.* **28** 2693
[26] Cariñena J F and Plyushchay M S 2017 *J. Phys. A: Math. Theor.* **50** 275202
[27] Odake S and Sasaki R 2013 *J. Phys. A: Math. Theor.* **46** 245201
[28] Marquette I and Quesne C 2014 *J. Math. Phys.* **55** 112103
[29] Marquette I and Quesne C 2016 *J. Math. Phys.* **57** 052101
[30] Quesne C 2015 Ladder operators for solvable potentials connected with exceptional orthogonal polynomials *J. Phys.: Conf. Ser.* vol 597 (IOP Publishing) p 012064
[31] Angelova M, Hertz A and Hussin V 2012 *J. Phys. A: Math. Theor.* **45** 244007
[32] Fiset M A and Hussin V 2015 *J. Phys.: Conf. Ser.* **624** 012016
[33] Marquette I 2010 *J. Phys. A: Math. Theor.* **43** 135203
[34] Fernández D J and Hussin V 1999 *J. Phys. A: Math. Gen.* **32** 3603
[35] Gelfand I and Fairlie D B 1991 *Comm. Math. Phys.* **136** 487
[36] Gómez-Ullate D, Kamran N and Milson R 2010 *J. Approx. Th.* **162** 987
[37] Dutta D and Roy P 2010 *J. Math. Phys.* **51** 042101
[38] Gómez-Ullate D, Kamran N and Milson R 2010 *J. Phys. A: Math. Theor.* **43** 434016

[39] Midya B and Roy B 2009 *Phys. Lett. A* **373** 4117
[40] Quesne C 2008 *J. Phys. A: Math. Theor.* **41** 392001
[41] Barut A O and Girardello L 1971 *Commun. Math. Phys.* **21** 41
[42] Cohen-Tannoudji C, Diu B and Laloë F 1977 *Quantum Mechanics* vol I and II (Hermann and John Wiley and Sons, Inc.)

Confluent hypergeometric expansions of the confluent Heun function governed by two-term recurrence relations

T A Ishkhanyan[1,2], V P Krainov[3] and A M Ishkhanyan[1,2]

[1]Russian-Armenian University, Yerevan, 0051 Armenia

[2]Institute for Physical Research, NAS of Armenia, Ashtarak, 0203 Armenia

[3]Moscow Institute of Physics and Technology, Dolgoprudny, 141700 Russian Federation

E-mail: aishkhanyan@gmail.com

Abstract. We show that there exist infinitely many nontrivial choices of parameters of the single confluent Heun equation for which the three-term recurrence relations governing the expansions of the solutions in terms of the confluent hypergeometric functions ${}_1F_1$ and ${}_0F_1$ are reduced to two-term ones. In such cases the expansion coefficients are explicitly calculated in terms of the Euler gamma functions.

PACS numbers: 02.30.Gp Special functions, 02.30.Hq Ordinary differential equations, 02.30.Mv Approximations and expansions
Keywords: Confluent Heun equation, series expansion, recurrence relation

1. Introduction

The single confluent Heun equation is a remarkable equation that generalizes both the Gauss ordinary and the Kummer confluent hypergeometric equations [1]-[3]. Though this equation has a wide coverage in contemporary physics and mathematics (see, e.g., [1]-[4] and references therein) a rather large (yet, not exhaustive) list concerning mainly the general relativity and cosmology is discussed in [5]), and for this reason it has extensively been studied during the recent years (see, e.g., [6]-[18]), however, the theory of the equation is currently far from being satisfactory for the most of applications. A reason for this is that the power-series solutions [1]-[3] as well as the expansions of the solutions in terms of simpler special functions (such as the incomplete beta, incomplete gamma, Bessel, Kummer, Coulomb wave, Goursat, Appell functions, etc., see, e.g., [8]-[12],[18]-[23]) are governed by three- or more-term recurrence relations for successive coefficients, so that apart from a few trivial cases the expansion coefficients are not calculated explicitly. However, we have recently shown that there exist infinitely many non-trivial particular choices of the involved parameters for which the three-term recurrence relations for power-series solutions become two-term [18]. In the present paper, we show that for the expansions of the solutions in terms of the hypergeometric functions ${}_1F_1$ and ${}_0F_1$ there also exist infinitely many particular non-trivial choices of parameters for which the recurrence relations reduce to two-term ones so that the expansion coefficients are explicitly written in terms of the Euler gamma functions.

2. Reductions

The confluent Heun equation written in its canonical form is given as [3]

$$\frac{d^2u}{dz^2}+\left(\frac{\gamma}{z}+\frac{\delta}{z-1}+\varepsilon\right)\frac{du}{dz}+\frac{\alpha z-q}{z(z-1)}u=0. \tag{1}$$

2.1 Let $\varepsilon\neq 0$. An expansion of the solution of this equation in terms of the Kummer confluent hypergeometric functions is written as [10]

$$u=\sum_{n=0}^{\infty}c_n\cdot {}_1F_1(\alpha/\varepsilon+n;\gamma+\delta+n;-\varepsilon z), \tag{2}$$

where the expansion coefficients obey the three-term recurrence relation

$$R_nc_n+Q_{n-1}c_{n-1}+P_{n-2}c_{n-2}=0 \tag{3}$$

with

$$R_n=-n(\gamma+\delta+n-1), \tag{4}$$

$$Q_n=-q+\alpha+(\gamma+\delta+\varepsilon+n-1)n, \tag{5}$$

$$P_n=-\frac{(\delta+n)(\varepsilon n+\alpha)}{\gamma+\delta+n}. \tag{6}$$

The expansion applies if ε is not zero, $\gamma+\delta$ is not zero or a negative integer and $\alpha^2+(\gamma+\delta)^2\neq 0$. The last condition assures that the involved confluent hypergeometric functions do not degenerate to ${}_0F_0(;-\varepsilon z)=e^{-\varepsilon z}$ for all n.

Our result is that this recurrence admits two-term reductions for infinitely many particular choices of the involved parameters. These reductions are achieved by the following ansatz guessed from the results of [18]:

$$c_n=\left(\frac{1}{n}\frac{\prod_{k=1}^{N+2}(a_k-1+n)}{\prod_{k=1}^{N+1}(b_k-1+n)}\right)c_{n-1}, \tag{7}$$

where we put

$$a_1,\dots,a_N,a_{N+1},a_{N+2}=1+e_1,\dots,1+e_N,\alpha/\varepsilon,\delta, \tag{8}$$

$$b_1,\dots,b_N,b_{N+1}=e_1,\dots,e_N,\gamma+\delta \tag{9}$$

with parameters $e_1,\dots,e_N$ to be defined later ($e_1,\dots,e_N$ are not zero or negative integers). We note that the form of this ansatz resembles that for the coefficients of the generalized hypergeometric series ${}_{N+2}F_{N+1}(a_1,\dots,a_{N+2};b_1,\dots,b_{N+1};z)$ [24],[25].

To derive the result, we note that the ratio c_n/c_{n-1} is explicitly written as

$$\frac{c_n}{c_{n-1}}=\frac{(\alpha/\varepsilon-1+n)(\delta-1+n)}{(\gamma+\delta-1+n)n}\prod_{k=1}^{N}\frac{e_k+n}{e_k-1+n}. \tag{10}$$

With this, the recurrence relation (3) is rewritten as

$$R_n\frac{(\alpha/\varepsilon-1+n)(\delta-1+n)}{(\gamma+\delta-1+n)n}\prod_{k=1}^{N}\frac{e_k+n}{e_k-1+n}+Q_{n-1}+P_{n-2}\frac{(\gamma+\delta-2+n)(n-1)}{(\alpha/\varepsilon-2+n)(\delta-2+n)}\prod_{k=1}^{N}\frac{e_k-2+n}{e_k-1+n}=0. \tag{11}$$

Substituting now R_n and P_n from equations (4),(6) and cancelling the common denominator, this equation becomes

$$-(\alpha/\varepsilon-1+n)(\delta-1+n)\prod_{k=1}^{N}(e_k+n)+Q_{n-1}\prod_{k=1}^{N}(e_k-1+n)-\varepsilon(n-1)\prod_{k=1}^{N}(e_k-2+n)=0. \tag{12}$$

This is a polynomial equation in n. Since the terms proportional to n^{N+2} cancel, it is of degree $N+1$. Hence, we have an equation of the form

$$\sum_{m=0}^{N+1} A_m(\gamma,\delta,\varepsilon,\alpha,q;e_1,\ldots,e_N)\,n^m = 0 . \tag{13}$$

Then, equating to zero the coefficients A_m warrants the satisfaction of the three-term recurrence relation (3) for all n. We thus have $N+2$ equations $A_m = 0$, $m = 0,1,..,N+1$, of which N equations serve for determination of the parameters $e_{1,2,\ldots,N}$ and the remaining two impose restrictions on the parameters of the confluent Heun equation (1).

One of the latter restrictions is derived by calculating the coefficient A_{N+1} of the term proportional to n^{N+1} which is readily shown to be $\gamma - N - 1 - \alpha/\varepsilon$. Hence,

$$\gamma = \alpha/\varepsilon + 1 + N . \tag{14}$$

The second restriction imposed on the parameters of the Heun equation is checked to be a polynomial equation of the degree $N+1$ for the accessory parameter q.

Resolving the recurrence (10), the coefficients of expansion (2) are explicitly written in terms of the gamma functions as ($c_0 = 1$)

$$c_n = \frac{\Gamma(n+\delta)\Gamma\left(n+\frac{\alpha}{\varepsilon}\right)\Gamma\left(\frac{\alpha}{\varepsilon}+\delta+1+N\right)}{n!\Gamma(\delta)\Gamma\left(\frac{\alpha}{\varepsilon}\right)\Gamma\left(n+\frac{\alpha}{\varepsilon}+\delta+1+N\right)} \prod_{k=1}^{N} \frac{e_k+n}{e_k}, \quad n = 0,1,2,\ldots . \tag{15}$$

Here are the explicit solutions of the recurrence relation (3)-(6) for $N = 0,1,2$.

$N = 0$:

$$\gamma = \alpha/\varepsilon + 1, \tag{16}$$

$$q = \alpha(1-\delta/\varepsilon), \tag{17}$$

$$c_n = \frac{\Gamma(n+\delta)\Gamma\left(n+\frac{\alpha}{\varepsilon}\right)\Gamma\left(\frac{\alpha}{\varepsilon}+\delta+1\right)}{n!\Gamma(\delta)\Gamma\left(\frac{\alpha}{\varepsilon}\right)\Gamma\left(n+\frac{\alpha}{\varepsilon}+\delta+1\right)} . \tag{18}$$

$N = 1$:

$$\gamma = \alpha/\varepsilon + 2, \tag{19}$$

$$q^2 + q\left(\frac{\alpha}{\varepsilon}(1+2\delta) - 2\alpha - \varepsilon + \delta\right) + \frac{\alpha}{\varepsilon}\left(\frac{\alpha}{\varepsilon}(\delta-\varepsilon)(1+\delta-\varepsilon) + \left(\delta+\delta^2-2\delta\varepsilon+\varepsilon^2\right)\right) = 0, \tag{20}$$

$$e_1 = q - \alpha + \delta - \varepsilon + \frac{\alpha(1+\delta)}{\varepsilon}, \tag{21}$$

$$c_n = \frac{\Gamma(n+\delta)\Gamma\left(n+\frac{\alpha}{\varepsilon}\right)\Gamma\left(\frac{\alpha}{\varepsilon}+\delta+2\right)}{n!\Gamma(\delta)\Gamma\left(\frac{\alpha}{\varepsilon}\right)\Gamma\left(n+\frac{\alpha}{\varepsilon}+\delta+2\right)} \frac{e_1+n}{e_1} . \tag{22}$$

$N = 2$:

$$\gamma = \alpha/\varepsilon + 3, \tag{23}$$

$$q^3+q^2\left(1+3\delta-3\varepsilon-3\alpha+\frac{3\alpha(1+\delta)}{\varepsilon}\right)+q\left(\frac{\alpha^2}{\varepsilon^2}\left(2+3\delta^2-6\delta(\varepsilon-1)+3\varepsilon(\varepsilon-2)\right)\right.$$
$$\left.+2\frac{\alpha}{\varepsilon}\left(1+3\delta^2+\delta(5-6\varepsilon)+3\varepsilon(\varepsilon-1)\right)+2\left(\delta+\delta^2-2\delta\varepsilon+\varepsilon^2\right)\right) \tag{24}$$
$$+\frac{\alpha}{\varepsilon}\left(\frac{\alpha^2}{\varepsilon^2}(\delta-\varepsilon)(1+\delta-\varepsilon)(2+\delta-\varepsilon)+\frac{\alpha}{\varepsilon}(1+\delta-\varepsilon)\left(3\delta(2+\delta)-2(1+3\delta)\varepsilon+3\varepsilon^2\right)\right.$$
$$\left.+2\left(\delta(1+\delta)(2+\delta)-3\delta(1+\delta)\varepsilon+3\delta\varepsilon^2-\varepsilon^3\right)\right)=0,$$

$$e_1+e_2=q+1-\alpha+2\delta-2\varepsilon+\frac{\alpha(2+\delta)}{\varepsilon}, \tag{25}$$

$$2e_1e_2+\left(\varepsilon-\delta-q-\frac{\alpha}{\varepsilon}(1+\delta-\varepsilon)\right)(e_1+e_2)-\frac{\alpha+2\alpha\delta+\varepsilon(\delta+\varepsilon)}{\varepsilon}=0, \tag{26}$$

$$c_n=\frac{\Gamma(n+\delta)\Gamma\left(n+\frac{\alpha}{\varepsilon}\right)\Gamma\left(\frac{\alpha}{\varepsilon}+\delta+3\right)}{n!\Gamma(\delta)\Gamma\left(\frac{\alpha}{\varepsilon}\right)\Gamma\left(n+\frac{\alpha}{\varepsilon}+\delta+3\right)}\frac{(e_1+n)(e_2+n)}{e_1e_2}. \tag{27}$$

These results can readily be checked by direct verification of the recurrence (3)-(6).

Since $e_1,\ldots,e_N$ are not zero or negative integers, it follows from equation (10) that c_n vanishes only if α/ε or δ is zero or a negative integer. The first case is not much interesting because then the resulting solution (2) is a polynomial in z. As regards the second choice, it can be checked that for integer $\delta<-N$ the solution identically vanishes, hence, the above recurrence generates non-zero finite-sum expansions of the solutions of the confluent Heun equation if

$$-N\le\delta\le 0. \tag{28}$$

It can further be checked that in these cases expansion (2) reduces to particular cases (because of the additional restriction $\gamma=\alpha/\varepsilon+1+N$, see equation (14)) of the ${}_{N+1}F_{1+N}$ solutions of the confluent Heun equation presented in [18].

2.2 There exist expansions of the solutions of the confluent Heun equation in terms of the Kummer confluent hypergeometric functions the form of which differs from those applied in expansion (2). It can be checked, however, that not for all such expansions two-term reductions of the governing recurrences are possible. For instance, consider the following expansion (compare with (2)-(6)) [10]:

$$u=\sum_{n=0}^{\infty}c_n\cdot{}_1F_1(\alpha/\varepsilon+n;\gamma;-\varepsilon z), \tag{29}$$

where the expansion coefficients obey the three-term recurrence relation

$$R_nc_n+Q_{n-1}c_{n-1}+P_{n-2}c_{n-2}=0 \tag{30}$$

with

$$R_n=n(\alpha/\varepsilon-\gamma+n), \tag{31}$$
$$Q_n=-q+n\gamma+(\alpha/\varepsilon+n)(\varepsilon-\delta-2n), \tag{32}$$
$$P_n=(\alpha/\varepsilon+n)(\delta+n). \tag{33}$$

As it is seen, this expansion applies if $\varepsilon\ne 0$ and γ is not zero or a negative integer. The possible two-term reduction of the recurrence (30) through ansatz (7) in this case reads

$$\frac{c_n}{c_{n-1}}=\frac{(\alpha/\varepsilon-1+n)(\delta-1+n)}{(\alpha/\varepsilon-\gamma+n)n}\prod_{k=1}^{N}\frac{e_k+n}{e_k-1+n}. \tag{34}$$

Hence, equation (30) is rewritten as

Journal of Physics: Conference Series **1416** (2019) 012014 doi:10.1088/1742-6596/1416/1/012014

$$R_n \frac{(\alpha/\varepsilon-1+n)(\delta-1+n)}{(\alpha/\varepsilon-\gamma+n)n}\prod_{k=1}^{N}\frac{e_k+n}{e_k-1+n}+Q_{n-1}+P_{n-2}\frac{(\alpha/\varepsilon-\gamma-1+n)(n-1)}{(\alpha/\varepsilon-2+n)(\delta-2+n)}\prod_{k=1}^{N}\frac{e_k-2+n}{e_k-1+n}=0. \quad (35)$$

Substituting R_n and P_{n-2} from equations (31),(33) and cancelling the common denominator, we obtain the following polynomial equation in n:

$$\begin{aligned}&(\alpha/\varepsilon-1+n)(\delta-1+n)\prod_{k=1}^{N}(e_k+n)\\ &+Q_{n-1}\prod_{k=1}^{N}(e_k-1+n)+(\alpha/\varepsilon-\gamma-1+n)(n-1)\prod_{k=1}^{N}(e_k-2+n)=0.\end{aligned} \quad (36)$$

Since the terms proportional to n^{N+2} cancel, this equation is of degree $N+1$. The point now is that the coefficient of the remaining highest-order term proportional to n^{N+1} is checked to be equal to ε. Since ε cannot be zero, this term cannot be canceled. Thus, the two-term reduction of the recurrence (30)-(33) via ansatz (7) is not possible.

2.3 Here is another expansion of the solution of the confluent Heun equation [10] for which the two-term reductions are possible:

$$u=\sum_{n=0}^{\infty}c_n\cdot{}_1F_1(\alpha/\varepsilon;\gamma+n;-\varepsilon z), \quad (37)$$

where the expansion coefficients obey the three-term recurrence relation

$$R_n c_n+Q_{n-1}c_{n-1}+P_{n-2}c_{n-2}=0 \quad (38)$$

with

$$R_n=-n(\gamma+\delta-1+n), \quad (39)$$

$$Q_n=-q+\alpha-\delta\varepsilon-n(1-\gamma-\delta+\varepsilon)+n^2, \quad (40)$$

$$P_n=\varepsilon\frac{(-\alpha/\varepsilon+\gamma+\delta+n)(\delta+n)}{\gamma+\delta+n}. \quad (41)$$

This expansion applies if $\varepsilon\neq 0$ and γ is not zero or a negative integer. Acting essentially as above, we arrive at the following explicit solution of this recurrence:

$$c_n=\frac{\Gamma(n+\delta)\Gamma(n+\gamma-1-N)\Gamma(\gamma+\delta)}{n!\Gamma(\delta)\Gamma(\gamma-1-N)\Gamma(n+\gamma+\delta)}\prod_{k=1}^{N}\frac{e_k+n}{e_k}, \quad n=0,1,2,\dots. \quad (42)$$

with

$$\delta=\alpha/\varepsilon-1-N. \quad (43)$$

The equations for the accessory parameter q and auxiliary parameters e_k for $N=0,1,2$ are as follows.

$N=0$:

$$q=(1-\gamma)\delta+\varepsilon, \quad (44)$$

$N=1$:

$$e_1=q-2+\gamma-\delta+\gamma\delta-\varepsilon, \quad (45)$$

$$\begin{aligned}&q^2+q(\gamma-2-3\delta+2\gamma\delta-3\varepsilon)\\ &+(2-3\gamma+\gamma^2)\delta^2+2\varepsilon(2-\gamma+\varepsilon)+\delta(2+\gamma^2+4\varepsilon-3\gamma-3\gamma\varepsilon)=0,\end{aligned} \quad (46)$$

$N=2$:

$$e_1+e_2+2q-7+3\gamma+3\delta-\varepsilon=0, \quad (47)$$

$$2e_1e_2+(1-q-\gamma-\gamma\delta+2\delta+2\varepsilon)(e_1+e_2)+2q-7+3\gamma+3\delta-\varepsilon=0, \quad (48)$$

$$\begin{aligned}&q^3+q^2\left(3\gamma(1+\delta)-2(4+3\delta+3\varepsilon)\right)\\&+q(12+26\delta+\gamma^2\left(2+3\delta(2+\delta)\right)+36\varepsilon+11(\delta+\varepsilon)^2\\&-2\gamma\left(5+7\varepsilon+\delta(13+6\delta+6\varepsilon)\right))+(\gamma-1)(\gamma-2)(\gamma-3)\delta(1+\delta)(2+\delta)\\&-2\left(9(1+\delta)(2+\delta)+\gamma^2\left(3+\delta(7+3\delta)\right)-\gamma\left(15+\delta(29+11\delta)\right)\right)\varepsilon\\&+\left(-18(2+\delta)+\gamma(15+11\delta)\right)\varepsilon^2-6\varepsilon^3.\end{aligned} \tag{49}$$

2.4 Consider now the case $\varepsilon=0$. This case should be examined separately because the nature of the irregular singularity of the confluent Heun equation at infinity is changed [26]. To start the discussion for this case, we note that if additionally $\alpha=0$ the confluent Heun equation is reduced to the Gauss ordinary hypergeometric equation. Therefore, since this specification is not interesting, we suppose $\alpha\neq 0$.

An appropriate expansion of the solution of the confluent Heun equation for $\varepsilon=0$ is constructed in terms of the hypergeometric functions ${}_0F_1$:

$$u=\sum_{n=0}^{\infty}c_n\cdot {}_0F_1(;\gamma+\delta+n;-\alpha z), \tag{50}$$

where the expansion coefficients obey the three-term recurrence relation

$$R_nc_n+Q_{n-1}c_{n-1}+P_{n-2}c_{n-2}=0 \tag{51}$$

with

$$R_n=-n(\gamma+\delta+n-1), \tag{52}$$

$$Q_n=-q+\alpha+n(\gamma+\delta+n-1), \tag{53}$$

$$P_n=-\frac{(\delta+n)\alpha}{\gamma+\delta+n}. \tag{54}$$

The expansion is valid if $\alpha\neq 0$ and $\gamma+\delta$ is not zero or a negative integer.

We note that this recurrence is a particular case of the recurrence relation (3)-(6) achieved by putting $\varepsilon=0$. Therefore, the above technique applies for this case as well (just, we keep in mind that the expansion functions are now the hypergeometric functions ${}_0F_1$). The resulting two-term reduction reads ($c_0=1$)

$$\gamma=-N,\quad N=0,1,2,\ldots, \tag{55}$$

$$c_n=\frac{\alpha^n\Gamma(\delta-N)}{n!\,\Gamma(n+\delta-N)}\prod_{k=1}^{N}\frac{e_k+n}{e_k}, \tag{56}$$

For the simplest case $N=0$ this solution is valid if $q=0$. For the next two cases $N=1$ and $N=2$ the parameters should obey the following equations:

$N=1$: $$qe_1+\alpha=0,\quad q^2+q(1-\delta)+\alpha=0, \tag{57}$$

$N=2$: $$e_1+e_2+q+3-2\delta=0,\quad q(e_1e_2-\alpha)+2\alpha(\delta-1)=0, \tag{58}$$

$$q^3+q^2(4-3\delta)+q\left(4-4\alpha-6\delta+2\delta^2\right)+4\alpha(\delta-1). \tag{59}$$

These solutions are readily checked by direct verification of the recurrence (51)-(54).

3. Discussion

Thus, we have shown that for the expansions of the solutions of the confluent Heun equation in terms of the confluent hypergeometric functions ${}_0F_1$ and ${}_1F_1$ there exist infinitely many particular choices of equation parameters for which the three-term recurrence relations for expansion coefficients become two-term. For these cases, we have presented the explicit solutions for expansion coefficients written in terms of the Euler gamma functions. Since in all cases the computing algorithms are regular and systematic, these solutions can easily be incorporated into a computer algebra system.

The approach we applied to derive the results is essentially based on the ansatz (7). This ansatz is guessed based on the pioneering observations by Letessier [27]-[29] and further notable discussions by Letessier et al. [30], Maier [31] and Takemura [32]. We have already applied the ansatz to reveal several two-term reductions of the three-term recurrence relations governing the series solutions of the general and single-confluent Heun equations in terms of powers or hypergeometric functions ${}_0F_1$, ${}_1F_1$ and ${}_2F_1$ [27],[33]. An observation here is that in all these cases the coefficients of the recurrence relations are (or can be rewritten as) polynomials, in the summation index n, of the degree at most 3. However, there exist other three-term expansions of the solutions of the Heun equations for which the coefficients of the recurrence relations are polynomials of higher degree (e.g., the expansions by Leaver [20], Mano et al. [21], El-Jaick and Figueiredo [8], or the eirlier Gauss-hypergeometric expansions of the general Heun functions by Svartholm [34] and Erdélyi [35]). It is interesting if the ansatz (7) works for these cases too. As we have seen in section **2.2**, not all three-term recurrences admit two-term reductions. Hence, the answer is not a priori evident and these recurrences need to be examined case by case. We conclude by noting that another appealing quest is if the ansatz is capable to produce non-trivial reductions for four- and more-term recurrences. We hope to explore these possibilities in the near future.

Acknowledgments

The work was supported by the Armenian Science Committee (SC Grants No. 18RF-139 and 18T-1C276), the Armenian National Science and Education Fund (ANSEF Grant No. PS-4986), the Russian-Armenian (Slavonic) University at the expense of the Ministry of Education and Science of the Russian Federation, the Russian Foundation for Basic Research (project No. 18-52-05006), and the Ministry of Education and Science of Russian Federation (Grant No. 3.873.2017/4.6). Tigran Ishkhanyan acknowledges the support from the French Embassy in Armenia for a doctoral grant as well as the Agence universitaire de la Francophonie and the Armenian Science Committee for a Scientific Mobility grant.

References

[1] Ronveaux A (ed.) 1995 *Heun's Differential Equations* (London: Oxford University Press)
[2] Slavyanov S Yu and Lay W 2000 *Special functions* (Oxford: Oxford University Press)
[3] Olver F W J, Lozier D W, Boisvert R F and Clark C W (eds.) 2010 *NIST Handbook of Mathematical Functions* (New York: Cambridge University Press)
[4] Erdélyi A, Oberhettinger F, Magnus W and Tricomi F 1953 *Higher Transcendental functions* (New York: McGraw Hill)
[5] Hortaçsu M 2018 Heun functions and some of their applications in physics *Advances in High Energy Physics* **2018** 8621573
[6] Fiziev P P 2010 Novel relations and new properties of confluent Heun's functions and their derivatives of arbitrary order *J. Phys. A* **43** 035203
[7] Kazakov A Ya and Slavyanov S Yu 2014 Euler integral symmetries for the confluent Heun equation and symmetries of the Painlevé equation PV *Theor. Math. Phys.* **179** 543-549
[8] El-Jaick L J and Figueiredo B D B 2013 Confluent Heun equations: convergence of solutions in series of Coulomb wave functions *J. Phys. A* **46** 085203
[9] El-Jaick L J and Figueiredo B D B 2015 Integral relations for solutions of the confluent Heun equation *Applied Mathematics and Computation* **256** 885–904
[10] Ishkhanyan T A and Ishkhanyan A M 2014 Expansions of the solutions to the confluent Heun equation in terms of the Kummer confluent hypergeometric functions *AIP Advances* **4** 087132
[11] Leroy C and Ishkhanyan A M 2015 Expansions of the solutions of the confluent Heun equation in terms of the incomplete Beta and the Appell generalized hypergeometric functions

Integral Transforms and Special Functions **26** 451-459

[12] El-Jaick L J and Figueiredo B D B 2016 Convergence and applications of some solutions of the confluent Heun equation *Applied Mathematics and Computation* **284** 234-259

[13] Hounkonnou M N and Ronveaux A 2015 On some differential transformations ofhypergeometric equations *J. Phys.: Conf. Series* **597** 012044

[14] Kazakov A Ya 2016 Integral symmetry for the confluent Heun equation with an added apparent singularity *J. Math. Sci.* **214** 268-276

[15] Motygin O V 2018 On evaluation of the confluent Heun functions, arXiv:1804.01007

[16] Bărar A, Mocanu G and Raşa I 2018 Heun functions related to entropies *RACSAM* **2018** 1-12

[17] Slavyanov S Yu and Salatich A A 2018 Confluent Heun equation and confluent hypergeometric equation *J. Math. Sci.* **232** 157-163

[18] Ishkhanyan T A and Ishkhanyan A M 2018 Generalized confluent hypergeometric solutions of the Heun confluent equation *Applied Mathematics and Computation* **338** 624-630

[19] Kurth Th. and Schmidt D 1986 On the global representation of the solutions of second-order linear differential equations having an irregular singularity of rank one in ∞ by series in terms of confluent hypergeometric functions *SIAM J. Math. Anal.* **17** 1086-1103

[20] Leaver E W 1986 Solutions to a generalized spheroidal wave equation: Teukolsky equations in general relativity, and the two-center problem in molecular quantum mechanics *J. Math. Phys.* **27** 1238-1265

[21] Mano S, Suzuki H and Takasugi E 1996 Analytic solutions of the Teukolsky equation and their low frequency expansions *Prog. Theor. Phys.* **95** 1079-1096

[22] Figueiredo B D B 2002 On some solutions to generalized spheroidal wave equations and applications *J. Phys. A* **35** 2877-2906

[23] Ishkhanyan A 2005 Incomplete Beta-function expansions of the solutions to the confluent Heun equation *J. Phys. A* **38** L491

[24] Slater L J 1966 *Generalized hypergeometric functions* (Cambridge: Cambridge University Press)

[25] Bailey W N 1964 *Generalized Hypergeometric Series* (New York: Stechert-Hafner Service Agency)

[26] Olver F W J 1974 *Asymptotics and Special Functions* (New York: Academic Press)

[27] Letessier J 1995 Co-recursive associated Jacobi polynomials *J. Comp. Appl. Math.* **5** 203-213

[28] Letessier J 1993 Co-recursive associated Laguerre polynomials *J. Comp. Appl. Math.* **49** 127-136

[29] Letessier J 1994 Some results on co-recursive associated Laguerre and Jacobi polynomials *SIAM J. Math. Anal.* **25** 528-548

[30] Letessier J, Valent G and Wimp J 1994 Some differential equations satisfied by hypergeometric functions *Internat. Ser. Numer. Math.* **119** 371-381

[31] Maier R S 2008 P-symbols, Heun identities, and ${}_3F_2$ identities *Contemporary Mathematics* **471** 139-159

[32] Takemura K 2012 Heun's equation, generalized hypergeometric function and exceptional Jacobi polynomial *J. Phys. A* **45** 085211

[33] Ishkhanyan T A, Shahverdyan T A and Ishkhanyan A M 2018 Expansions of the solutions of the general Heun equation governed by two-term recurrence relations for coefficients *Advances in High Energy Physics* **2018** 4263678

[34] Svartholm N 1939 Die Lösung der Fuchs'schen Differentialgleichung zweiter Ordnung durch Hypergeometrische Polynome *Math. Ann.* **116** 413-421

[35] Erdélyi A 1944 Certain expansions of solutions of the Heun equation *Q. J. Math. (Oxford)* **15** 62-69

Supersymmetric Calogero and Calogero-Sutherland models from gauging

Evgeny Ivanov
Bogoliubov Laboratory of Theoretical Physics, JINR, 141980 Dubna, Moscow Region, Russia
E-mail: eivanov@theor.jinr.ru

Abstract. We describe how the new kinds of $\mathcal{N}=2$ and $\mathcal{N}=4$ supersymmetric extensions of the rational and hyperbolic Calogero models can be derived by gauging $U(n)$ symmetry of the appropriate superfield matrix models. These systems feature non-standard numbers $\mathcal{N}n^2$ of physical fermionic variables as compared with $\mathcal{N}n$ in the standard case. An essential ingredient of $\mathcal{N}=4$ models is the necessary presence of semi-dynamical spin variables described by $d=1$ Wess-Zumino terms. The bosonic cores of $\mathcal{N}=4$ models are $U(2)$ spin Calogero and Calogero-Sutherland models. In the hyperbolic case two non-equivalent $\mathcal{N}=4$ extensions exist, with and without the interacting center-of-mass coordinate in the bosonic sector. The talk is based on joint works with Sergey Fedoruk and Olaf Lechtenfeld.

1. Motivations

Calogero-type models (CM) [1] (see also review [2]) provide nice examples of integrable $d=1$ systems. Simplest is the so-called rational n-particle Calogero model

$$S^c = \int dt \Big[\sum_a \dot{x}_a \dot{x}_a - \sum_{a\neq b} \frac{c^2}{4(x_a - x_b)^2} \Big]. \tag{1}$$

The Calogero-Moser system is obtained by adding the oscillator-type term $\sim \sum_{a\neq b}(x_a - x_b)^2$. The rational CM models are conformal: the action (1) is invariant under $d=1$ conformal group $SO(1,2)$

$$\delta t = \alpha,\ \delta x_a = \frac{1}{2}\dot{\alpha} x_a\,,\ \ \partial_t^3 \alpha = 0\,.$$

Conformal CM models, being interesting on their own, can be closely related to superstrings and M-theory [3]. Besides conformal models, there are known other integrable CM-type models, e.g., the trigonometric and hyperbolic Calogero-Sutherland systems,

$$S^{cs} = \int dt \Big[\sum_a \dot{x}_a \dot{x}_a - \sum_{a\neq b} \frac{c^2}{4\sinh^2 \frac{x_a - x_b}{2}} \Big]. \tag{2}$$

The first example of $\mathcal{N}=2$ superextended CM model was given in [4]: each bosonic coordinate x_a was enlarged to the multiplet $(\mathbf{1},\mathbf{2},\mathbf{1})$, i.e. n bosonic particles were completed by $2n$ fermionic $d=1$ fields. The appropriate $\mathcal{N}=2, d=1$ superfield action can be easily constructed and shown to yield the rational CM model in the limit of vanishing fermions. Analogously, $\mathcal{N}=2$ extension of the Calogero-Sutherland models can be constructed. An important role in gaining the correct pairwise potential terms is played by auxiliary fields of the supermultiplet.

There are some problems with higher $\mathcal{N}$ extensions [5], [6], [7]. For $\mathcal{N}=4$ case, x_a should be extended to the supermultiplets $(\mathbf{1},\mathbf{4},\mathbf{3})$, i.e. one deals with n bosonic and $4n$ fermionic fields. It is very difficult to construct the appropriate superfield action yielding n-particle Calogero potential in the bosonic sector. There arise a few functions of x_a related by the complicated WDVV [8], [9] equations the explicit solutions of which are known only for a few lowest values of n. This is in contrast to the standard one-particle $\mathcal{N}=4$ supersymmetric mechanics where no similar difficulties appear.

Thus, no universal convenient method was invented so far for constructing the "standard" $\mathcal{N}n$- extended supersymmetric CM systems.

Fortunately, another type of supersymmetrization is possible, such that no the above-mentioned problems arise, though the models appearing in this way are "non-minimal": they contain $\mathcal{N}n^2$ fermions for n bosonic coordinates [10], [11], [12]. It proceeds from the gauge approach to bosonic CM models developed earlier in [13] [14], [15]. One starts from some simple free matrix model and gauge the appropriate linear isometries by non-propagating $d=1$ gauge fields. After eliminating gauge fields by their algebraic equations of motion one or another CM model is reproduced.

Generalization to the supersymmetry case is straightforward: one starts from some free superfield matrix model action and gauge the isometry by a gauge superfield. After passing to components and gauge-fixing, the relevant supersymmetric CM model is recovered.

I will apply this approach to $\mathcal{N}=2$ and $\mathcal{N}=4$ superfield matrix models and show that some new versions of the corresponding supersymmetric CM models can be discovered in this way.

The talk is based mainly on the papers [10], [11], [12].

2. Toy example: conformal mechanics by gauging

The renowned conformal mechanics model by [16] is described by the action :

$$S_0 = \int dt\,\left(\dot{x}^2 - \gamma^2 x^{-2}\right) \equiv \int dt\,\mathcal{L}_0\,.$$

Let us obtain it from another $d=1$ action. As an input we take the Lagrangian of free complex field $z(t)$

$$L_z = \dot{z}\,\dot{\bar{z}} + im\,(\dot{z}\bar{z} - z\dot{\bar{z}})\,.$$

It is obviously invariant under rigid phase transformations

$$z' = e^{-i\lambda}z\,, \quad \bar{z}' = e^{i\lambda}\bar{z}\,.$$

Then we gauge this abelian symmetry, $\lambda \to \lambda(t)$, and introduce $d=1$ gauge field $A(t)$, so that $\dot{z} \to \dot{z} + iAz$:

$$\begin{aligned} L_z \to L^g_z &= (\dot{z} + iAz)\,(\dot{\bar{z}} - iA\bar{z}) + im\,(\dot{z}\bar{z} - z\dot{\bar{z}} + 2iAz\bar{z}) + 2\gamma\,A\,, \\ A' &= A + \dot{\lambda}\,. \end{aligned}$$

Here, a "Fayet-Iliopoulos term" $\propto \gamma$ has been also added. It is gauge invariant (up to a total derivative) by itself.

The next step is to effect the appropriate gauge in L^g:

$$z = \bar{z} \equiv q(t)\,.$$

We plug it back into L^g and obtain:

$$L^g_z = (\dot q + iAq)\,(\dot q - iAq) + 2im\,Aq^2 + 2\gamma\,A = (\dot q)^2 + A^2q^2 - 2mAq^2 + 2\gamma A\,.$$

The field $A(t)$ can be eliminated by its algebraic equation of motion:

$$\delta A: \qquad A = m - \gamma\, q^{-2}\,.$$

The final form of the gauge-fixed Lagrangian is

$$L^g \Rightarrow (\dot q)^2 - \left(mq - \gamma q^{-1}\right)^2 = (\dot q)^2 - m^2q^2 - \gamma^2 q^{-2} + 2\gamma m\,.$$

Up to an additive constant, this Lagrangian coincides with the mass-modified AFF. At $m = 0$, one recovers the standard conformal mechanics:

$$L^g_{(m=0)} = (\dot q)^2 - \gamma^2\, q^{-2}\,.$$

The initial action $S_z = \int dt L_z$ at $m = 0$ is invariant under the conformal $SO(1,2)$ transformations $\delta t = f(t), \delta z = \frac{1}{2}\dot f\, z\,,\ (\partial_t)^3 f = 0\,.$ The conformal invariance is preserved by the gauging procedure, provided that the gauge field $A(t)$ transform as ∂_t, *i.e.* $\delta A(t) = -\dot f\, A(t)\,.$

3. Calogero and Calogero-Sutherland by gauging

The generalization of this setting to the CM case goes as follows.

We start from the $U(n)$ invariant free action of the $n \times n$ hermitian matrix field X_a^b and complex U(n)-spinor field $Z_a(t)$, $\bar Z^a = (\overline{Z_a})$, $a, b = 1, \ldots, n$. We gauge $U(n)$ symmetry by n^2 hermitian gauge fields A_a^b. The resulting gauge invariant action reads

$$S_C = \int dt \left[\mathrm{tr}\,(\nabla X \nabla X) + \tfrac{i}{2}\,(\bar Z \nabla Z - \nabla \bar Z Z) + c\,\mathrm{tr}A \right], \tag{3}$$

$$\nabla X = \dot X + i[A, X]\,, \qquad \nabla Z = \dot Z + iAZ\,, \qquad \nabla \bar Z = \dot{\bar Z} - i\bar Z A\,.$$

The last term (Fayet-Iliopoulos term) includes only $U(1)$ gauge field, c is a real constant. As the next step we fix $U(n)$ gauge so as to kill all non-diagonal components of X_a^b

$$X_a{}^b = x_a \delta_a{}^b\,, \Rightarrow [X, A]_a{}^b = (x_a - x_b) A_a{}^b\,.$$

We have used just $n^2 - n$ gauge parameters, but there remains the residual abelian gauge subgroup $[U(1)]^n$, with local parameters $\varphi_a(t)$:

$$Z_a \to e^{i\varphi_a} Z_a\,,\ \bar Z^a \to e^{-i\varphi_a} \bar Z^a,\ A_a{}^a \to A_a{}^a - \dot\varphi_a \quad \text{(no sum with respect to } a)\,.$$

Then the next gauge-fixing can be effected:

$$\bar Z^a = Z_a\,,$$

which leads to the gauge-fixed action in the form

$$S_C = \int dt \sum_{a,b} \left[\dot x_a \dot x_a + (x_a - x_b)^2 A_a{}^b A_b{}^a - Z_a Z_b A_a{}^b + c\, A_a{}^a \right]$$

Varying it with respect to the non-propagating gauge fields, we obtain

$$A_a{}^b = \frac{Z_a Z_b}{2(x_a - x_b)^2} \qquad \text{for } a \neq b\,,$$
$$Z_a Z_a = c \qquad \forall\, a \quad \text{(no sum with respect to a).}$$

The diagonal entries A^a_a drop out from the action and, after substituting the explicit expressions for the rest of A^b_a, we obtain

$$S_C = \tfrac{1}{2}\int dt \Big[\sum_a \dot{x}_a \dot{x}_a - \sum_{a\neq b} \frac{c^2}{(x_a - x_b)^2}\Big],$$

that is just the rational Calogero model action. The action we started from is conformal, so the final Calogero action is conformal as well.

The Calogero-Sutherland model can be recovered by the same techniques, the only difference is that the initial action includes nonlinear sigma-model type kinetic term for the matrix X^b_a:

$$S_C = \int dt \left[\operatorname{tr}\left(X^{-1}\nabla X X^{-1}\nabla X\right) + \tfrac{i}{2}(\bar{Z}\nabla Z - \nabla\bar{Z} Z) + c\operatorname{tr}A \right]. \tag{4}$$

After passing through the same steps as in the rational case, we obtain the gauge-fixed action in the form

$$S_C = \tfrac{1}{2}\int dt \Big[\sum_a \frac{\dot{x}_a \dot{x}_a}{(x_a)^2} - \sum_{a\neq b} \frac{x_a x_b c^2}{(x_a - x_b)^2}\Big].$$

Introducing the new variables as $x_a = e^{q_a}$ brings this action to the standard Calogero-Sutherland form

$$S^{cs} = \int dt \Big[\sum_a \dot{q}_a \dot{q}_a - \sum_{a\neq b} \frac{c^2}{4\sinh^2 \frac{q_a - q_b}{2}}\Big]$$

This action is not conformal, since the initial action is not conformal.

4. $\mathcal{N}=2$ Calogero and Calogero-Sutherland

Once again, both cases follow the same strategy, are defined on the same set of $d=1$ superfields and differ only in the choice of the initial $\mathcal{N}=2$ matrix model action.

The starting point in the first case is the free $\mathcal{N}=2, d=1$ action of the $n\times n$ matrix hermitian superfield $\mathcal{X}_a{}^b(t,\theta,\bar\theta)$, $a,b=1,\ldots,n$, with each entry carrying $(\mathbf{1},\mathbf{2},\mathbf{1})$ multiplet, and of chiral $U(n)$-spinor superfield $\mathcal{Z}_a(t_R,\bar\theta)$, $\bar{\mathcal{Z}}^a(t_L,\theta)$, $\quad D\mathcal{Z}_a = 0\,, \quad \bar{D}\bar{\mathcal{Z}}^a = 0$,

$$S^{(N=2)}_{sC} = \int dt d\theta d\bar\theta \left[\operatorname{tr}\left(\bar{D}\mathcal{X}\, D\mathcal{X}\right) + \tfrac{1}{2}\bar{\mathcal{Z}}\,\mathcal{Z}\right].$$

This action is evidently invariant under rigid $U(n)$ transformations acting as rotations of the fundamental and co-fundamental indices a,b.

Next, in order to preserve the chiralities of $\mathcal{Z}_a, \bar{\mathcal{Z}}^a$ we gauge this global symmetry by chiral and anti-chiral superfield parameters λ and $\bar\lambda$,

$$\mathcal{Z}' = e^{i\lambda}\mathcal{Z}\,, \quad \bar{\mathcal{Z}}' = \bar{\mathcal{Z}}\,e^{-i\bar\lambda}\,, \quad \mathcal{X}' = e^{i\lambda}\,\mathcal{X}\,e^{-i\bar\lambda}$$

To construct the gauge invariant action, one introduces the hermitian gauge superfield V,

$$e^{2V'} = e^{i\bar{\lambda}} e^{2V} e^{-i\lambda}.$$

Then, the gauge-covariantized action reads

$$\tilde{S}_{sC}^{(N=2)} = \int dt d^2\theta \left[\operatorname{tr}\left(\bar{\mathcal{D}}\mathcal{X}\, e^{2V}\mathcal{D}\mathcal{X}\, e^{2V}\right) + \tfrac{1}{2}\,\bar{\mathcal{Z}}\, e^{2V}\mathcal{Z} - c\,\operatorname{tr}V \right], \tag{5}$$

where

$$\mathcal{D}\mathcal{X} = D\mathcal{X} + e^{-2V}(De^{2V})\,\mathcal{X}\,, \qquad \bar{\mathcal{D}}\mathcal{X} = \bar{D}\mathcal{X} - \mathcal{X}\, e^{2V}(\bar{D}e^{-2V})\,.$$

It can be shown that the original rigid matrix action, as well as the final gauge-covariantized one, are invariant under the $\mathcal{N}=2$ superconformal symmetry $SU(1,1|1)$.

In the component expansion of the involved superfields

$$\mathcal{X} = X + \theta\Psi - \bar{\theta}\bar{\Psi} + \theta\bar{\theta}Y\,,\ \mathcal{Z} = Z + 2i\theta\Upsilon + i\theta\bar{\theta}\dot{Z}\,,\ \bar{\mathcal{Z}} = \bar{Z} + 2i\bar{\theta}\bar{\Upsilon} - i\theta\bar{\theta}\dot{\bar{Z}}\,,$$

all the fields, besides $2n^2$ fermionic ones,

$$\Psi_a^b, \quad \bar{\Psi}_b^a,$$

are auxiliary and can be eliminated by their equations of motion.

Upon choosing the standard Wess-Zumino gauge $V = \theta\bar{\theta}A(t)$, the component action acquires the form

$$S_{sC}^{WZ} = \int dt \left[\operatorname{tr}\nabla X\,\nabla X + \tfrac{i}{2}\,(\bar{Z}\nabla Z - \nabla\bar{Z}Z) - c\,\operatorname{tr}A \,+\, i\,\operatorname{tr}(\bar{\Psi}\nabla\Psi - \nabla\bar{\Psi}\Psi)\right],$$

where

$$\nabla\Psi = \dot{\Psi} + i[\Psi, A]\,, \quad \nabla\bar{\Psi} = \dot{\bar{\Psi}} + i[\bar{\Psi}, A]$$

and $\nabla Z, \nabla X$ coincide with the purely bosonic expressions presented earlier. In the limit of zero fermions, the standard input gauge-invariant action of the rational Calogero is recovered.

So we have obtained a new $\mathcal{N}$=2 extension of the n-particle Calogero model with n bosons and $2n^2$ fermions, as opposed to the standard $\mathcal{N}$=2 super Calogero system of Freedman and Mende with only $2n$ fermions.

In terms of the physical variables, the component action reads

$$S_{sC}^{(N=2)} = \int dt \,[\, \sum_a \dot{x}_a\dot{x}_a + i\,(\bar{\Psi}_a{}^b\dot{\Psi}_b{}^a - \dot{\bar{\Psi}}_a{}^b\Psi_b{}^a) - V]\,, \tag{6}$$

$$V = \sum_{a\neq b} \frac{4}{(x_a - x_b)^2}\Big(Z_a\bar{Z}^a Z_b\bar{Z}^b + 2\bar{Z}^a\{\Psi,\bar{\Psi}\}_a{}^b Z_b + \{\Psi,\bar{\Psi}\}_a{}^b\{\Psi,\bar{\Psi}\}_b{}^a\Big).$$

The constraint on Z (after fixing the gauge $Z_a = \bar{Z}^a$) also essentially involves fermions:

$$(Z_a)^2 = c - R_a\,, \quad R_a \equiv \{\Psi,\bar{\Psi}\}_a{}^a = \sum_b (\Psi_a{}^b\bar{\Psi}_b{}^a + \bar{\Psi}_a{}^b\Psi_b{}^a)\,,$$

$$(R_a)^{2n-1} \equiv 0 \ (\text{nilpotency})$$

To date, it is unclear how to treat this huge amount of fermionic fields, and whether this number could be somehow reduced by imposing some extra (perhaps, fermionic) gauge invariance. More detailed study of the fermionic sectors of such models was recently undertaken in [17], [18].

Like in the bosonic case, the $\mathcal{N}=2$ CS system is obtained by starting from a sigma-model type gauged action

$$S_{cs}^{N=2} = \frac{1}{2}\int dt d^2\theta \left[\mathrm{tr}(\mathcal{X}^{-1}\bar{\mathcal{D}}\mathcal{X}\,\mathcal{X}^{-1}\mathcal{D}\mathcal{X}) - \bar{\mathcal{Z}}\,e^{2V}\,\mathcal{Z} + 2c\,\mathrm{tr}V\right] \tag{7}$$

After passing through the same steps as before, we obtain the component action

$$\begin{aligned} S^{N=2} &= \int dt\, L^{N=2}, \\ L^{N=2} &= \frac{1}{2}\mathrm{tr}(X^{-1}\nabla X\, X^{-1}\nabla X) + \frac{i}{2}(\bar{Z}\nabla Z - \nabla\bar{Z} Z) + c\,\mathrm{tr}A \\ &+ \frac{i}{2}\mathrm{tr}(X^{-1}\bar{\Psi}X^{-1}\nabla\Psi - X^{-1}\nabla\bar{\Psi}X^{-1}\Psi) \\ &- \frac{1}{4}\mathrm{tr}(X^{-1}\bar{\Psi}X^{-1}\bar{\Psi}X^{-1}\Psi X^{-1}\Psi) \end{aligned}$$

In the bosonic limit the gauge-invariant CS action is recovered. A new interesting feature is the appearance of the quartic fermionic term as compared to $\mathcal{N}=2$ rational Calogero.

5. Towards higher $\mathcal{N}$: $\mathcal{N}=4$ Calogero

The $\mathcal{N}=4$ case is of special interest because the same gauge procedure applied to it results in the $U(2)$ spin Calogero systems. The reason is that the additional multiplets $Z, \bar{Z}$ now cannot be entirely eliminated by using the gauge freedom and/or the constraints as earlier. Their remnants are just a sort of the target space $U(2)$ harmonics.

The universal superfield approach to $\mathcal{N}=4$ mechanics, both one-particle and multi-particle, is provided by the $\mathcal{N}=4, d=1$ harmonic superspace (HSS) [19] which is the $d=1$ version of the renowned $\mathcal{N}=2, d=4$ HSS [20].

The $d=1$ HSS is an extension of the standard $\mathcal{N}=4, d=1$ superspace $(t, \theta_i, \bar{\theta}^k)$ by the harmonic coordinates $u_i^{\pm}$:

$$(t, \theta^{\pm}, \bar{\theta}^{\pm}, u_i^{\pm}), \quad \theta^{\pm} = \theta^i u_i^{\pm}, \; \bar{\theta}^{\pm} = \bar{\theta}^i u_i^{\pm}, \; u^{+i}u_i^{-} = 1.$$

The commuting $SU(2)$ variables $u_i^{\pm}$ parametrize the 2-sphere $S^2 \sim SU(2)_R/U(1)_R$. The salient property of HSS is the presence of the harmonic analytic superspace in it, with half of the original Grassmann coordinates

$$(\zeta, u) = (t_A, \theta^+, \bar{\theta}^+, u_i^{\pm}), \quad t_A = t + i(\theta^+\bar{\theta}^- + \theta^-\bar{\theta}^+).$$

It is closed under $\mathcal{N}=4$ supersymmetry transformations.

Most off-shell $\mathcal{N}=4, d=1$ multiplets are described by analytic superfields "living" on this subspace.

The direct analog of the $\mathcal{N}=2$ multiplet $(\mathbf{1},\mathbf{2},\mathbf{1})$ is the multiplet $(\mathbf{1},\mathbf{4},\mathbf{3})$ represented by a general superfield $\mathcal{X}(t, \theta^{\pm}, \bar{\theta}^{\pm}, u)$ subjected to the constraints

$$\begin{aligned} &D^{++}\mathcal{X} = 0, \quad D^{++} = u^{+i}\frac{\partial}{\partial u^{-i}} + 2i\theta^+\bar{\theta}^+\partial_{t_A}, \\ &D^+D^-\mathcal{X} = 0, \; \bar{D}^+\bar{D}^-\mathcal{X} = 0, \; (D^+\bar{D}^- + \bar{D}^+D^-)\mathcal{X} = 0. \end{aligned}$$

Journal of Physics: Conference Series **1416** (2019) 012015 doi:10.1088/1742-6596/1416/1/012015

These constraints can be solved for by the analytic prepotential $\mathcal{V}$,

$$\mathcal{X}(t,\theta_i,\bar\theta^i) = \int du\, \mathcal{V}(t_A,\theta^+,\bar\theta^+,u)\Big|_{\theta^\pm=\theta^i u_i^\pm,\ \bar\theta^\pm=\bar\theta^i u_i^\pm}.$$

The needed field content is ensured by the invariance under gauge analytic transformations

$$\delta\mathcal{V} = D^{++}\lambda^{--}, \quad \lambda^{--} = \lambda^{--}(\zeta,u)$$

and is recovered in the appropriate WZ gauge for $\mathcal{V}$.

The $\mathcal{N}=4$ analogs of the chiral $\mathcal{N}=2$ multiplet $Z_a, \bar Z^a$ are the complex analytic superfields $\mathcal{Z}^+$, $\bar{\mathcal{Z}}^+$, subjected to the additional constraint

$$D^{++}\mathcal{Z}^+ = 0,$$

which ensures the off-shell content $(\mathbf{4},\mathbf{4},\mathbf{0})$ for these superfields:

$$\mathcal{Z}^+ = z^i u_i^+ + \theta^+\varphi + \bar\theta^+\phi - 2i\,\theta^+\bar\theta^+\partial_{t_A} z^i u_i^-.$$

Finally, the gauge field $A(t)$ is accommodated by the analytic unconstrained gauge prepotential V^{++},

$$V^{++\,\prime} = e^{i\lambda}\,V^{++}\,e^{-i\lambda} - i\,e^{i\lambda}(D^{++}e^{-i\lambda}),$$

where $\lambda_a^b(\zeta,u^\pm)\in u(n)$ is the hermitian analytic matrix parameter. Using this gauge freedom we can choose the WZ gauge

$$V^{++} = 2i\,\theta^+\bar\theta^+ A(t_A).$$

Now we have all ingredients required for constructing $\mathcal{N}=4$ Calogero and Calogero-Sutherland models. We will firstly discuss the first class of systems.

Our guiding principle will be invariance under the most general $\mathcal{N}=4, d=1$ conformal supergroup $D(2,1;\alpha)$. The appropriate matrix superfield action reads

$$S = -\tfrac{1}{4(1+\alpha)}\int \mu_H (\mathrm{tr}\mathcal{X}^2)^{-1/2\alpha} - \tfrac{1}{2}\int \mu_A^{(-2)}\mathcal{V}_0\,\widetilde{\mathcal{Z}}^{a+}\mathcal{Z}_a^+ - \tfrac{i}{2}c\int \mu_A^{(-2)}\,\mathrm{tr}\,V^{++} \qquad (8)$$

Here, all the superfields defined above are involved, with all derivatives properly covariantized with respect to local $U(n)$ group which acts as

$$\mathcal{X}' = e^{i\lambda}\mathcal{X}e^{-i\lambda}, \quad \mathcal{Z}^{+\prime} = e^{i\lambda}\mathcal{Z}^+,\ \bar{\mathcal{Z}}^{+\prime} = \bar{\mathcal{Z}}^+e^{-i\lambda}$$

E.g., $D^{++}\mathcal{Z}^+ \to \mathcal{D}^{++}\mathcal{Z}^+ = D^{++}\mathcal{Z}^+ + iV^{++}\mathcal{Z}^+$. The object $\mathcal{V}_0$ is a real analytic gauge prepotential for the U(n) singlet $(\mathbf{1},\mathbf{4},\mathbf{3})$ superfield $\mathcal{X}_0 \equiv \mathrm{tr}\,(\mathcal{X})$. It is defined by the integral transform

$$\mathcal{X}_0(t,\theta_i,\bar\theta^i) = \int du\mathcal{V}_0\left(t_A,\theta^+,\bar\theta^+,u^\pm\right)\Big|_{\theta^\pm=\theta^i u_i^\pm,\ \bar\theta^\pm=\bar\theta^i u_i^\pm}.$$

In what follows, we will be interested in the choice $\alpha=-1/2$, which corresponds to the free superfield Lagrangian $\sim \mathrm{tr}\mathcal{X}^2$ for the multiplet $(\mathbf{1},\mathbf{4},\mathbf{3})$, and at which $D(2,1;\alpha)\sim osp(4|2)$.

In WZ gauge, and with auxiliary fields eliminated, we end up with:

$$\begin{aligned}
S_4 &= S_b + S_f, \\
S_b &= \int dt\big[\mathrm{tr}(\nabla X \nabla X + c\,A) + \tfrac{n}{8}(\bar Z^{(i} Z^{k)})(\bar Z_i Z_k) + \tfrac{i}{2} X_0(\bar Z_k \nabla Z^k - \nabla \bar Z_k\, Z^k)\big], \\
S_f &= i\,\mathrm{tr}\int dt(\bar\Psi_k \nabla \Psi^k - \nabla \bar\Psi_k \Psi^k) - \int dt\,\frac{\Psi_0^{(i}\bar\Psi_0^{k)}(\bar Z_i Z_k)}{X_0}, \\
X_0 &:= \mathrm{tr}(X), \quad \Psi_0^i := \mathrm{tr}(\Psi^i), \quad \bar\Psi_0^i := \mathrm{tr}(\bar\Psi^i).
\end{aligned}$$

After fixing gauges with respect to the residual gauge group, elimination of $A_a^b, a \neq b$, and some redefinitions, the bosonic core of this action is found to be as follows

$$S_b = \int dt\Big\{\sum_a \dot x_a \dot x_a + \tfrac{i}{2}\sum_a(\bar Z_k^a \dot Z_a^k - \dot{\bar Z}{}_k^a Z_a^k) + \sum_{a\neq b}\frac{\mathrm{tr}(S_a S_b)}{4(x_a - x_b)^2} - \frac{n\,\mathrm{tr}(\hat S \hat S)}{2(X_0)^2}\Big\}. \tag{9}$$

Here, the fields Z_a^k are subject to the constraints

$$\bar Z_i^a Z_a^i = c \qquad \forall\, a,$$

and

$$(S_a)_i{}^j := \bar Z_i^a Z_a^j, \qquad (\hat S)_i{}^j := \sum_a \Big[(S_a)_i{}^j - \tfrac{1}{2}\delta_i^j (S_a)_k{}^k\Big].$$

To reveal the meaning of these composite objects, we note that in the Hamiltonian approach, the kinetic WZ term for Z gives rise to the following Dirac brackets:

$$[\bar Z_i^a, Z_b^j]_D = i\delta_b^a \delta_i^j,$$

that implies

$$[(S_a)_i{}^j, (S_b)_k{}^l]_D = i\delta_{ab}\Big\{\delta_i^l (S_a)_k{}^j - \delta_k^j (S_a)_i{}^l\Big\},$$

i.e., for each index a the quantities S_a form mutually commuting $u(2)$ algebras. The object $\hat S$ is just the conserved Noether $SU(2)$ current for the total system. Thus, the new feature of the $\mathcal{N}=4$ case is that not all out of the bosonic variables Z_a^i are eliminated by fixing gauges and solving the constraint; there survives a non-vanishing WZ term for them. After quantization these variables become purely internal (U(2)-spin) degrees of freedom. Since $tr\hat S\hat S$ is a constant of motion, we have the conformal inverse-square potential even in the sector of the center-of-mass coordinate X_0. This is the essential difference of the $\mathcal{N}=4$ case from the $\mathcal{N}=1,2$ cases where this coordinate decouples. Modulo this extra conformal potential, the bosonic limit of the $\mathcal{N}=4$ system constructed is none other than the integrable U(2)-spin Calogero model as it was formulated in [22], [2].

6. $\mathcal{N}=4$ Calogero-Sutherland

Like in the previous case, the input superfield action in the hyperbolic case is the sum of three parts

$$S^{N=4} = S_X + S_{WZ} + S_{FI}. \tag{10}$$

The basic distinguishing feature of this system is the choice of $\mathcal{X}$ action

$$S_{\mathcal{X}} = \frac{1}{2}\int \mu_H \operatorname{tr}\left(\ln \mathcal{X}\right), \tag{11}$$

while the structure of the remaining two pieces is the same, as well as the form of the superfield constraints.

The full structure of the component action is restored by passing through the same steps as in the rational Calogero case, i.e. imposing various gauges, elimination of the auxiliary fields, etc. It is rather involved, especially in the fermionic sector. In particular, it contains a few terms quartic in fermions. The number of physical fermions is the same, just $4n^2$. The actions, both the superfield and the component ones, lack superconformal symmetry, only "flat" $\mathcal{N}=4, d=1$ supersymmetry and $SU(2)$ R-symmetry are present.

The bosonic sector is described by the action

$$\begin{aligned} S^{N=4}_{bose} &= \frac{1}{2}\int dt\Big[\operatorname{tr}(X^{-1}\nabla X X^{-1}\nabla X + 2c\,A) + i(\bar{Z}_k\nabla Z^k - \nabla\bar{Z}_k\, Z^k) \\ &\qquad + \frac{(\bar{Z}^{(i}Z^{k)})(\bar{Z}_i Z_k)\operatorname{tr}(X^2)}{2(X_0)^2}\Big]. \end{aligned}$$

After fixing $U(n)$ gauge, $X_a{}^b = 0, \quad a \neq b$, eliminating fields A^c_b, using the constraint $\bar{Z}^a_i Z^i_a = c\,\forall\, a$, and passing to $x_a = e^{q_a}$, it becomes

$$\begin{aligned} S^{N=4}_{bose} &= \frac{1}{2}\int dt\Big\{\sum_a \left[\dot{q}_a\dot{q}_a + i(\bar{Z}^a_k\dot{Z}^k_a - \dot{\bar{Z}}^a_k Z^k_a)\right] - \sum_{a\neq b}\frac{(S_a)_i{}^k(S_b)_k{}^i}{4\sinh^2\frac{q_a-q_b}{2}} \\ &\qquad + \sum_{a,b}\frac{(S_a)^{(ik)}\,(S_b)_{(ik)}\operatorname{tr}(X^2)}{2(X_0)^2}\Big\}, \\ &\operatorname{Tr}\left(X^2\right) = \sum_c e^{2q_c}, \quad X_0 = \sum_c e^{q_c}, \quad (S_a)_i{}^k := \bar{Z}^a_i Z^k_a. \end{aligned} \tag{12}$$

The quantities $(S_a)_i{}^k$ generate n copies of $U(2)$ algebra. As a result, the above action, up to the last term, describes the hyperbolic $U(2)$-spin Calogero-Sutherland system [14], [2].

The choice of the $\mathcal{Z}^+$ action in the $\mathcal{N}=4$ rational Calogero model was mainly dictated by the requirement of superconformal invariance. In the hyperbolic case, no such a symmetry is present from the very beginning. In particular, the $\mathcal{X}$ part of the action already lacks such an invariance. So, there are no reasons to require it in other pieces.

Then it is natural to try the simplest action for the $(\mathbf{4},\mathbf{4},\mathbf{0})$ multiplets, just $-1/2\int \mu_A^{(-2)}\,\bar{\mathcal{Z}}^{+a}\mathcal{Z}^+_a$. In this case all steps are radically simplified, in particular, all fermionic auxiliary fields of $(\mathbf{4},\mathbf{4},\mathbf{0})$ multiplets are zero on shell. The bosonic sector of the component action reads

$$\tilde{S}^{N=4}_{bose} = \frac{1}{2}\int dt\Big\{\sum_a \left[\dot{q}_a\dot{q}_a + i(\bar{Z}^a_k\dot{Z}^k_a - \dot{\bar{Z}}^a_k Z^k_a)\right] - \sum_{a\neq b}\frac{(S_a)_i{}^k(S_b)_k{}^i}{4\sinh^2\frac{q_a-q_b}{2}}\Big\}. \tag{13}$$

The previous bosonic action involved $\operatorname{tr}(X^2)$ and X_0. The latter coordinate (the center-of-mass coordinate) decouples only for the trivial cases $n=1,2$. In contrast, the new action yields in the bosonic sector the pure hyperbolic U(2)-spin Calogero-Sutherland system for any n, without any additional interaction. The center-of-mass coordinate is fully detached and described by the free action in this sector.

While for $n = 1$ in the system obtained the $\mathcal{X}$ sector is fully decoupled from the $\mathcal{Z}$ sector and describes a free dynamics, at $n = 2$ one is left with a non-trivial system. The relative coordinate $\phi := \frac{1}{2}(q_1 - q_2)$ involves non-trivial interaction with the spin variables Z_1^k and Z_2^k

$$\sim \quad \frac{\operatorname{tr}(S_1 S_2)}{4 \sinh^2 \phi}\,. \tag{14}$$

So in the bosonic sector we obtain an extension of the standard hyperbolic Pöschl-Teller mechanics [23] by the spin variables. This new $\mathcal{N} = 4$ superextended Pöschl-Teller system certainly deserves a careful analysis. In the $\mathcal{N} = 1, 2$ cases analogous Pöschl-Teller potential without any additional variables is recovered (the known versions of supersymmetric Pöschl-Teller mechanics were given, e.g., in [24]).

7. Summary and Outlook

- We have outlined a universal method of constructing supersymmetric extensions of the Calogero-type models by the superfield gauging procedure directly generalizing the similar one for the n-particle bosonic Calogero systems. This method yields non-standard supersymmetrization, with $\mathcal{N}n^2$ physical fermionic fields, in contrast to $\mathcal{N}n$ such fields in the standard supersymmetrization.

- In this way, we have explicitly constructed new $\mathcal{N} = 2$ and $\mathcal{N} = 4$ superfield systems containing the rational Calogero and hyperbolic Calogero-Sutherland models as the bosonic cores. In the hyperbolic models, new superxtensions of the Pöschl-Teller mechanics for the relative coordinate arise at $n = 2$, with the extra spin variables in the $\mathcal{N} = 4$ case.

- Recently, it was shown, in the on-shell Hamiltonian approach, that similar systems, at least in the rational case, can be formulated for arbitrary $\mathcal{N}$ [17], [25]. It is unclear whether such systems can be re-obtained within the superfield gauging since the *off-shell* superfield matrix models (the starting point of the gauging procedure) are known up to now only until $\mathcal{N} = 8$.

Some further lines of development:

(a) An intriguing question is as to whether the new super Calogero models preserve the remarkable classical and quantum integrability properties of the bosonic models.

(b) It would be interesting to inquire whether other $\mathcal{N} = 4, d = 1$ multiplets (e.g., the multiplets $(\mathbf{2}, \mathbf{4}, \mathbf{2})$ or $(\mathbf{3}, \mathbf{4}, \mathbf{1})$) can be used to represent spin variables in various $\mathcal{N} = 4$ Calogero systems.

(c) To generalize new $\mathcal{N}$ super Calogero to the case of "weak" $\mathcal{N} = 4$ ($SU(2|1)$) supersymmetry [26], [27], [28], as well as to analogous deformed versions of $\mathcal{N} = 8$ supersymmetry [29], [30]. Such generalizations involve an intrinsic mass-dimension parameter m which would deform the Calogero models to a kind of Calogero-Moser systems, with extra oscillator-type terms.

(d) Quantizing all these models. In fact, the quantization of the deformed $SU(2|1)$ Calogero-Moser systems was recently done in [31]. The quantization of the $\mathcal{N} = 2$ models considered here was undertaken in a recent paper [32].

(e) At last, there are known other integrable Calogero-like multiparticle models [33], [34], e.g. the trigonometric Calogero-Sutherland models, the elliptic models, etc. All of them still wait their supersymmetrization.

Acknowledgments
The author thanks the organizers of the "XXVIth International Conference on Integrable Systems and Quantum symmetries" for the kind hospitality in Prague and Sergey Fedoruk, Olaf Lechtenfeld, and Stepan Sidorov for successful collaboration.

References
[1] Calogero F 1969 Solution of a three-body problem in one-dimension *J. Math. Phys.* **10** 2191
[2] Polychronakos A P 2006 Physics and mathematics of Calogero particles *J. Phys.* A **39** 12793 (*Preprint* arXiv:hep-th/0607033)
[3] Gibbons G W and Townsend P K 1999 Black holes and Calogero models *Phys. Lett.* B **454** 187 (*Preprint* arXiv:hep-th/9812034)
[4] Freedman D Z and Mende P F 1990 An exactly solvable N-particle system in supersymmetric quantum mechanics *Nucl. Phys.* B **344** 317.
[5] Wyllard N 2000 (Super)conformal many body quantum mechanics with extended supersymmetry *J. Math. Phys.* **41** 2826 (*Preprint* arXiv:hep-th/9910160)
[6] Bellucci S Galajinsky A and Krivonos S 2003 New many-body superconformal models as reductions of simple composite systems *Phys. Rev.* D **68** 064010 (*Preprint* arXiv:hep-th/0304087)
[7] Bellucci S Galajinsky A and Latini E 2005 New insight into WDVV equation *Phys. Rev.* D **71** 044023 (*Preprint* arXiv:hep-th/0411232)
[8] Witten E 1990 On the structure of the topological phase of two-dimensional gravity *Nucl. Phys.* B **340** 281
[9] Dijkgraaf R Verlinde H L and Verlinde E P 1991 Topological strings in $d < 1$ *Nucl. Phys.* B **352** 59
[10] Fedoruk S Ivanov E and Lechtenfeld O 2009 Supersymmetric Calogero models by gauging *Phys. Rev* D **79** 105015 (*Preprint* arXiv:0812.4276 [hep-th])
[11] Fedoruk S Ivanov E and Lechtenfeld O 2012 Superconformal mechanics *J. Phys.* A **45** 173001 (*Preprint* arXiv:0112.1947 [hep-th])
[12] Fedoruk S Ivanov E and Lechtenfeld O 2019 Supersymmetric hyperbolic Calogero-Sutherland modelsa by gauging *Nucl. Phys.* B **944** 114633 (*Preprint* arXiv:1902.08023 [hep-th])
[13] Polychronakos A P 1991 Integrable systems from gauged matrix models *Phys. Lett.* B **266** 29
[14] Polychronakos A P 2001 Quantum Hall states as matrix Chern-Simons theory *JHEP* **0104** 011 (*Preprint* arXiv:hep-th/0103013)
[15] Gorsky A and Nekrasov N 1995 Relativistic Calogero-Moser model as gauged WZW theory *Nucl. Phys.* B **436** 582 (*Preprint* arXiv:hep-th/9401017)
[16] de Alfaro V Fubini S and Furlan G 1976 Conformal invariance in quantum mechanics *Nuovo Cim.* A **34** 569
[17] Krivonos S Lechtenfeld O and Sutulin A 2018 $\mathcal{N}$-extended supersymmetric Calogero models *Phys. Lett.* B **784** 137 (*Preprint* arXiv:1804.10825 [hep-th])
[18] Krivonos S Lechtenfeld O and Sutulin A 2019 Supersymmetric many-body Euler-Calogero-Moser model *Phys. Lett.* B **790** 191 (*Preprint* arXiv:1812.03530 [hep-th])
[19] Ivanov E and Lechtenfeld O 2003 $\mathcal{N} = 4$ Supersymmetric Mechanics in Harmonic Superspace *JHEP* **0309** 073 (*Preprint* arXiv:hep-th/0307111)
[20] Galperin A S Ivanov E A Kalitzin S Ogievetsky V I and Sokatchev E S 1984 Unconstrained $\mathcal{N} = 2$ matter, Yang-Mills and supergravity theories in harmonic superspace *Class. Quant. Grav.* **1** 469
[21] Galperin A S Ivanov E A Ogievetsky V I and Sokatchev E S 2001 *Harmonic Superspace* (Cambridge University Press 306 p.)
[22] Polychronakos A P 2002 Calogero-Moser models with noncommutative spin interactions *Phys.Rev.Lett.* **89** 126403
[23] Pöschl G and Teller E 1933 Bemerkungen zur Quantenmechanik des anharmonischen Oszillators *Zeitschrift für Physik* **83** 143
[24] Galajinsky A Nersessian A and Saghatelian A 2013 Superintegrable models related to near horizon extremal Myers-Perry black hole in arbitrary dimension *JHEP* **1306** 002 (*Preprint* arXiv:1303.4901 [hep-th])
[25] Krivonos S Lechtenfeld O Provorov A and Sutulin A 2019 Extended supersymmetric Calogero model *Phys. Lett.* B **791** 385 (*Preprint* arXiv:1812.10168 [hep-th])
[26] Bellucci S and Nersessian A 2003 (Super)Oscillator on $CP(N)$ and Constant Magnetic Field *Phys. Rev.* D **67** 065013 (*Preprint* arXiv:hep-th/0211070)
[27] Smilga A V 2004 Weak supersymmetry *Phys. Lett.* B **585** 173 (*Preprint* arXiv:hep-th/0311023)
[28] Ivanov E and Sidorov S 2014 Deformed Supersymmetric Mechanics *Class. Quant. Grav.* **31** 075013 (*Preprint* arXiv:1307.7690 [hep-th])
[29] Ivanov E Lechtenfeld O Sidorov S 2016 SU(2|2) supersymmetric mechanics JHEP **1611** 031 (*Preprint* arXiv:1609.00490 [hep-th])

Journal of Physics: Conference Series **1416** (2019) 012015 doi:10.1088/1742-6596/1416/1/012015

[30] Ivanov E Lechtenfeld O and Sidorov S 2018 Deformed $\mathcal{N} = 8$ mechanics of (8, 8, 0) multiplets *JHEP* **1808** 193 (*Preprint* arXiv:1807.11804 [hep-th])
[31] Fedoruk S Ivanov E Lechtenfeld O and Sidorov S 2018 Quantum $SU(2|1)$ supersymmetric Calogero-Moser spinning systems *JHEP* **043** 1804, (*Preprint* arXiv:1801.00206)
[32] Fedoruk S 2019 $\mathcal{N} = 2$ supersymmetric hyperbolic Calogero-Sutherland model (*Preprint* arXiv:1910.07348 [hep-th])
[33] Olshanetsky M A and Perelomov A M 1981 Classical integrable finite dimensional systems related to Lie algebras *Phys. Rept.* **71** 313
[34] Olshanetsky M A and Perelomov A M 1983 Quantum integrable systems related to Lie algebras *Phys. Rept.* **94** 313

Journal of Physics: Conference Series **1416** (2019) 012016 doi:10.1088/1742-6596/1416/1/012016

Towards a Z_3-graded approach to quarks' symmetries

Richard Kerner[a] **and Jerzy Lukierski**[b,c]

a : Laboratoire de Physique Théorique de la Matière Condensée - CNRS URA 7600
Sorbonne-Université, B.C. 121, 4 Place Jussieu 75005 Paris, France
b: Institute of Theoretical Physics, Wrocław University, Plac Maxa Borna 9, Wrocław, Poland
c: Bogolubov Laboratory of Theoretical Physics, Joint Inst. for Nuclear Research, 141980 Dubna (Moscow Region), Russia

E-mail: a: richard.kerner@upmc.fr

E-mail: b: jerzy.lukierski@ift.uni.wroc.pl

Abstract.
Colour $SU(3)$ group is an exact symmetry of Quantum Chromodynamics, which describes strong interactions between quarks and gluons. Supplemented by two internal symmetries, $SU(2)$ and $U(1)$, it serves as the internal symmetry of the Standard Model, describing as well the electroweak interactions of quarks and leptons. The colour $SU(3)$ symmetry is exact, while two other symmetries are broken by means of the Higgs-Kibble mechanism. The three colours and fractional quarks charges with values $1/3$ and $2/3$ suggest that the cyclic group Z_3 may play a crucial role in quark field dynamics.

In this paper we consequently apply the Z_3 symmetry to field multiplets describing colour quark fields. Generalized Dirac equation for coloured 12-component spinors is introduced and its properties are discussed. Imposing Z_3-graded Lorentz and Poincaré covariance leads to enlargement of quark fields multiplets and incorporates additional $Z_2 \times Z_3$ symmetry which leads to the appearance of three generations (families) of distinct quark doublets.

1. Introduction

According to present knowledge, the ultimate undivisible and undestructible constituents of matter, called atoms by ancient Greeks, are in fact the elementary particles named *quarks*, carrying fractional electric and baryonic charges which are conserved in Standard Model under any circumstances.

What can be regarded as a mysterious feature is the fact that *freely propagating* quarks have never been directly observed. However, the *deep inelastic scattering* experiments in which the electrons with extremely high energies penetrate inside the nucleons and scatter on quarks, reveal the presence of almost point-like constituents which seem to be almost free as long as they are confined inside the protons or neutrons. When very high energy nucleons collide, they merge for a short while, and quarks interact via *gluon* exchange. The gluons are supposed to be the quanta of a *color* $SU(3)$ gauge field. Besides, quarks interact also via electroweak gauge fields.

In Quantum Chromodynamics (QCD) quarks are considered as fermions, endowed with spin $\frac{1}{2}$. Only *three quarks* or *three anti-quarks* can coexist inside a fermionic baryon (respectively, anti-baryon), and a pair *quark-antiquark* can form a meson with integer spin. Besides, baryons are composed of three quarks (anti-quarks) each with different colour.

Journal of Physics: Conference Series **1416** (2019) 012016 doi:10.1088/1742-6596/1416/1/012016

Six doublets of quarks are organized in three generations. Two quarks in the first generation, u and d ("up" and "down"), can be considered as two states of a more general object, just like proton and neutron are two isospin states of a nucleon. While in baryons there is place for *two* quarks in the same u-state or d-state, three u or d quarks in composite hadron are not allowed.

The gauge gluon fields that intermediate strong interactions between quarks should carry colors. In fact, the totally "colorless" quadratic combinations of quarks do not interact strongly with quarks, what means that the condition $R\bar{R} + B\bar{B} + G\bar{G} = 0$, where R(red), B(blue) and G(green) denote coloured quarks, does not affect the QCD strong forces. Eight bilinear combinations listed below describe the octet of gluons

$$\frac{1}{\sqrt{2}}(R\bar{B} + B\bar{R}), \quad \frac{1}{i\sqrt{2}}(R\bar{B} - B\bar{R}), \quad \frac{1}{\sqrt{2}}(R\bar{G} + G\bar{R}), \quad \frac{1}{i\sqrt{2}}(R\bar{G} - G\bar{R}),$$

$$\frac{1}{\sqrt{2}}(B\bar{G} + G\bar{B}), \quad \frac{1}{i\sqrt{2}}(B\bar{G} - G\bar{B}), \quad \frac{1}{\sqrt{2}}(R\bar{R} - B\bar{B}), \quad \frac{1}{\sqrt{6}}(R\bar{R} + B\bar{B} - 2G\bar{G}),$$

The above combinations can be linked with the choice of eight traceless Gell-Mann 3×3 matrices, spanning the $SU(3)$ Lie algebra.

In these lecture notes we shall study a way to introduce quark dynamics with incorporated Z_3-covariance and consider the Z_3-graded generalization of relativistic symmetries. As basic field equation describing coloured quarks we introduce 12-component colour Dirac equations, describing in alternative way the quarks with three colours. The proposed Z_3-graded free quark equations imply that coloured quarks do not appear asymptotically; one observes asymptotically only colourless composites, three-linear in the case of baryons and bilinear for mesons. In colour Dirac equation the Lorentz and colour symmetries are entangled, not described by single tensor products of Lorentz spin and SU(3) colour matrices.

The plan of our presentation is the following. In Sect. 2 we recall the way of obtaining Dirac equation as Z_2-covariant doublet of two-component Pauli equations. In Sect. 3 we replace discrete Z_2 maps by discrete $Z_2 \times Z_3$ transformations acting on indecomposable sextet of Pauli equations describing our basic colour Dirac equations. The Z_3-covariance modifies the dynamics of quark fields components - any component of coloured Dirac equation satisfy modified Klein-Gordon equations, which are of the sixth order. In Sect. 4 we demonstrate how such new quarks dynamics permits to explain algebraically the asymptotic nonobservability of coloured quarks. In order to incorporate fully Z_3-covariance of quark dynamics recently there were proposed the Z_3-graded generalization of Lorentz algebra [6] and its extension to Z_3-graded Poincar 'e algebra [17]. In Sect. 6 we show that the formulation covariant under standard Lorentz transformations requires the consideration of doublets of colour fields called Lorentz doublets [17]; we show also how one can introduce the action principle for the pairs of colour Dirac fields describing these Lorentz doublets. Last Sect. 7 contains final remarks and outlook. In particular we discuss there the option that coloured quark fields satisfy ternary generalization of canonical commutation relations, what leads to the Z_3-graded exclusion principle for coloured quarks (maximally only two identical colour quark states can coexist in physical quark composites).

2. From Pauli spinors to Dirac's equation

After the discovery of spin of the electron (the Stern-Gerlach experiment, Pauli understood that a Schroedinger equation involving only one complex-valued wave function is not enough to take into account this new degree of freedom. He proposed then to describe the dichotomic spin variable by introducing a two-component function forming a column on which hermitian matrices can act as linear operators.

The basis of complex traceless 2 × 2 hermitian matrices contains just three elements, since then known as *Pauli matrices* that can be arranged in a matrix-valued 3-vector $\boldsymbol{\sigma} = [\sigma_1, \sigma_2, \sigma_3]$:

$$\sigma_1 = \begin{pmatrix} 0 & 1 \\ 1 & 0 \end{pmatrix}, \quad \sigma_2 = \begin{pmatrix} 0 & -i \\ i & 0 \end{pmatrix}, \quad \sigma_3 = \begin{pmatrix} 1 & 0 \\ 0 & -1 \end{pmatrix}$$

The three Pauli matrices multiplied by $\frac{i}{2}$ span the three dimensional Lie algebra: let $\tau_k = \frac{i}{2}\sigma_k$, then

$$[\tau_1, \tau_2] = \tau_3, \quad [\tau_2, \tau_3] = \tau_1, \quad [\tau_3, \tau_1] = \tau_2.$$

They also form the Clifford algebra related to the Euclidean 3-dimensional metric:

$$\sigma_i \sigma_k + \sigma_k \sigma_i = 2\delta_{ik} \mathbb{1}_2$$

The simplest linear relation between the operators of energy, mass and momentum acting on a column vector (called *a Pauli spinor*) would read then:

$$\begin{pmatrix} E & 0 \\ 0 & E \end{pmatrix} \begin{pmatrix} \psi^1 \\ \psi^2 \end{pmatrix} = \begin{pmatrix} mc^2 & 0 \\ 0 & mc^2 \end{pmatrix} \begin{pmatrix} \psi^1 \\ \psi^2 \end{pmatrix} + c\, \boldsymbol{\sigma} \cdot \mathbf{p} \begin{pmatrix} \psi^1 \\ \psi^2 \end{pmatrix}, \tag{1}$$

where $\boldsymbol{\sigma} \cdot \mathbf{p} = \sigma_1 \, p^1 + \sigma_2 \, p^2 + \sigma_3 \, p^3 = \begin{pmatrix} p^3 & p^1 - i\, p^2 \\ p^1 + i\, p^2 & -p^3 \end{pmatrix}$. We can write (1) in a simplified manner, denoting the Pauli spinor by one letter ψ and treating the unit matrix symbolically like a number:

$$E\, \psi = mc^2\, \psi + c\, \boldsymbol{\sigma} \cdot \mathbf{p}\, \psi. \tag{2}$$

This equation is not invariant under Lorentz transformations. Indeed, by iterating, i.e. taking the square of this operator, we arrive at the following relation between the operators of energy and momentum and the mass of the particle:

$$E^2 = m^2 c^4 + 2\, mc^3 \mid \mathbf{p} \mid^2 \boldsymbol{\sigma} \cdot \mathbf{p} + c^2 \mathbf{p}^2, \tag{3}$$

instead of the relativistic relation

$$E^2 - c^2\, \mathbf{p}^2 = m^2 c^4. \tag{4}$$

The double product in the expression for the energy squared can be removed if one introduces a second Pauli spinor satisfying a smilar equation, and intertwining the two spinors. So let us denote the first Pauli spinor by ψ_+ and the second one by ψ_-, and let them satisfy the following system of coupled equations:

$$E\, \psi_+ = mc^2\, \psi_+ + \boldsymbol{\sigma} \cdot \mathbf{p}\, \psi_-, \quad E\, \psi_- = -mc^2\, \psi_- + \boldsymbol{\sigma} \cdot \mathbf{p}\, \psi_+, \tag{5}$$

(by the way, here $-1 = e^{i\pi}$, a complex number! which coincides with the relativistic equation for the electron found by Dirac a few years later.

The relativistic invariance is now manifest: due to the negative mass term in the second equation, the iteration leads to the separation of variables, and all the components satisfy the desired relation

$$[E^2 - c^2\, \mathbf{p}^2]\psi_+ = m^2 c^4\, \psi_+, \quad [E^2 - c^2\, \mathbf{p}^2]\, \psi_- = m^2 c^4\, \psi_-. \tag{6}$$

In a more appropriate basis the Dirac equation becomes manifestly relativistic: $[\gamma^\mu p_\mu - mc]\ \psi = 0$, with $p_0 = \frac{E}{c}$,

$$\gamma^0 = \sigma_3 \otimes \mathbb{1}_2 = \begin{pmatrix} \mathbb{1}_2 & 0 \\ 0 & -\mathbb{1}_2 \end{pmatrix}, \quad \gamma^k = (i\sigma_2) \otimes \sigma^k = \begin{pmatrix} 0 & \sigma^k \\ -\sigma^k & 0 \end{pmatrix}. \tag{7}$$

The so defined Dirac's gamma-matrices span the Clifford algebra related to the Minkowskian metric tensor:

$$\gamma^\mu \gamma^\nu + \gamma^\nu \gamma^\mu = 2 g^{\mu\nu} \mathbb{1}_4$$

with $g^{\mu\nu} = diag[1, -1, -1, -1]$, while the six anti-symmetric matrices

$$\Sigma^{\mu\nu} = \frac{1}{4} [\gamma^\mu \gamma^\nu - \gamma^\nu \gamma^\mu]$$

satisfy commutation relations characterizing the Lorentz group.

The side effect of this modification was the presence of solutions with negative mass, which led Dirac to the conclusion that the "holes" in the sea of such solutions could be interpreted as electrons with the same mass as the usual ones, but with opposite charge.

Now we have clearly the invariance under the product group $Z_2 \times Z_2$. The first Z_2 factor is relative to the half-integer dichotomic spin variable, the second Z_2 group is related to this new universal symmetry, called charge conjugation, stipulating the existence of anti-particles. The four-component complex function ψ is a column composed of two two-component spinors, ξ_α and $\chi_{\dot\beta}$, which are supposed to transform under two non-equivalent representations of the $SL(2, \mathbf{C})$ group:

$$\xi_{\alpha'} = S^\alpha_{\alpha'} \xi_\alpha, \quad \chi_{\dot\beta'} = S^{\dot\beta}_{\dot\beta'} \chi_{\dot\beta'}, \tag{8}$$

The electric charge conservation is equivalent to the vanishing of the four-divergence of j^μ:

$$\partial_\mu j^\mu = \left(\partial_\mu \psi^\dagger \gamma^\mu\right) \psi + \psi^\dagger \left(\gamma^\mu \partial_\mu \psi\right) = 0, \tag{9}$$

from which we infer that this condition will be satisfied if we have

$$\partial_\mu \psi^\dagger \gamma^\mu = -m\psi^\dagger \ \text{ and } \ \gamma^\mu \partial_\mu \psi = m\psi, \tag{10}$$

which coincides with the usual Dirac equation.

The two coupled Pauli equations can be re-written using a matrix notation:

$$\begin{pmatrix} E & 0 \\ 0 & E \end{pmatrix} \begin{pmatrix} \psi_+ \\ \psi_- \end{pmatrix} = \begin{pmatrix} mc^2 & 0 \\ 0 & -mc^2 \end{pmatrix} \begin{pmatrix} \psi_+ \\ \psi_- \end{pmatrix} + \begin{pmatrix} 0 & c\,\boldsymbol{\sigma}\mathbf{p} \\ c\,\boldsymbol{\sigma}\mathbf{p} & 0 \end{pmatrix} \begin{pmatrix} \psi_+ \\ \psi_- \end{pmatrix}, \tag{11}$$

where the entries in the energy operator and the mass matrix are in fact 2×2 identity matrices, as well as the σ-matrices appearing in the last matrix, so that in reality the above equation represents the 4×4 Dirac equation, only in a different basis.

The system of linear equations (11) displays two discrete Z_2 symmetries: the space reflection consisting in simultaneous change of directions of spin and momentum, $\boldsymbol{\sigma} \to -\boldsymbol{\sigma}, \quad \mathbf{p} \to -\mathbf{p}$, and the particle-antiparticle symmetry realized by the transformation

$$m \to -m, \ \psi_+ \to \psi_-, \quad \psi_- \to \psi_+.$$

Our next aim is to extend the $Z_2 \times Z_2$ symmetry by including the Z_3 group which will mix not only the two spin states and particles with anti-particles, but also the three colors.

3. Colour Dirac equation and free quark dynamics

Recently in [7], [8], [9], [10] a generalization of the Dirac equation for quarks was proposed, incorporating the color degrees of freedom via extending the discrete symmetry of the system to the $Z_3 \times Z_2 \times Z_2$ group.

The cyclic group Z_3 is generated by the third root of unity, denoted by $j = e^{\frac{2\pi i}{3}}$, with $j^2 = e^{\frac{4\pi i}{3}}$, $j^3 = 1$, and $1 + j + j^2 = 0$.

Just as taking into account the dichotomic half-integer spin variable, the introduction of color degrees of freedom requires additional Z_3 symmetry acting on a new discrete variable taking three possible (and exclusive) values, named symbolically "red", "blue" and "green". The 3×3 matrices had to be introduced, all representing *third roots* of the 3×3 unit matrix. Six Pauli spinors represent three colors and three anti-colors:

$$\varphi_+ = \begin{pmatrix} \varphi_+^1 \\ \varphi_+^2 \end{pmatrix}, \; \chi_+ = \begin{pmatrix} \chi_+^1 \\ \chi_+^2 \end{pmatrix}, \; \psi_+ = \begin{pmatrix} \psi_+^1 \\ \psi_+^2 \end{pmatrix}, \; \varphi_- = \begin{pmatrix} \varphi_-^1 \\ \varphi_-^2 \end{pmatrix}, \; \chi_- = \begin{pmatrix} \chi_-^1 \\ \chi_-^2 \end{pmatrix}, \; \psi_- = \begin{pmatrix} \psi_-^1 \\ \psi_-^2 \end{pmatrix}, \tag{12}$$

on which Pauli sigma-matrices act in a natural way, as on two-component column spinors.

Let us follow the logic that led from Pauli's to Dirac's equation extending it to the colors acted upon by the Z_3-group. In the expression for the energy operator (i.e. the Hamiltonian), mass terms is positive when acting on particles, and acquires negative sign acting on anti-particles, i.e. it changes sign while intertwining particle-antiparticle components. We shall also assume that the mass term acquires the factor j when we switch from the red component φ to the blue component χ, and j^2 for the green component ψ. The momentum operator will be non-diagonal, as in the Dirac equation, systematically intertwining not only particles with antiparticles, but also colors with anti-colors.

By analogy with the pair of equations (6) in which multiplying the mass by -1 led to the anti-particle appearance, now the mass term is multiplied by the generator of the Z_3 group, j, each time the colour changes. This yields the following set of what may be called the "colour Dirac equation":

$$\begin{aligned} E\,\varphi_+ &= mc^2\,\varphi_+ + c\,\boldsymbol{\sigma}\cdot\mathbf{p}\,\chi_- \\ E\,\chi_- &= -j\,mc^2\,\chi_- + c\,\boldsymbol{\sigma}\cdot\mathbf{p}\,\psi_+ \\ E\,\psi_+ &= j^2\,mc^2\,\psi_+ + c\,\boldsymbol{\sigma}\cdot\mathbf{p}\,\varphi_- \\ E\,\varphi_- &= -mc^2\,\varphi_- + c\,\boldsymbol{\sigma}\cdot\mathbf{p}\,\chi_+ \\ E\,\chi_+ &= j\,mc^2\,\chi_+ + c\,\boldsymbol{\sigma}\cdot\mathbf{p}\,\psi_- \\ E\,\psi_- &= -j^2\,mc^2\,\psi_- + c\,\boldsymbol{\sigma}\cdot\mathbf{p}\,\varphi_+ \end{aligned} \tag{13}$$

The energy operator is obviously diagonal, and its action on the spinor-valued column-vector can be represented as a 6×6 operator valued unit matrix. The mass operator is diagonal, too, but its elements represent all powers of the sixth root of unity $q = e^{\frac{2\pi i}{6}}$, which are

$$q = -j^2, \; q^2 = j, \; q^3 = -1, \; q^2 = j^2, \; q^5 = -j \text{ and } q^6 = 1.$$

The system (13) was formulated in a basis in which the "coloured" Pauli spinors alternate with their antiparticles; however, if we want to put forward the colour content, it is better to choose an alternative basis in the space of spinors. From now on, we choose the basis in which the column of six "colored" Pauli spinors is arranged as follows:

$$(\varphi_+, \varphi_-, \chi_+, \chi_-, \psi_+, \chi_-, \psi_-)^T. \tag{14}$$

Then the mass and momentum operators take on the following form:

$$M = \begin{pmatrix} m & 0 & 0 & 0 & 0 & 0 \\ 0 & -m & 0 & 0 & 0 & 0 \\ 0 & 0 & jm & 0 & 0 & 0 \\ 0 & 0 & 0 & -jm & 0 & 0 \\ 0 & 0 & 0 & 0 & j^2 m & 0 \\ 0 & 0 & 0 & 0 & 0 & -j^2 m \end{pmatrix}, \quad P = \begin{pmatrix} 0 & 0 & 0 & \boldsymbol{\sigma}\cdot\mathbf{p} & 0 & 0 \\ 0 & 0 & \boldsymbol{\sigma}\cdot\mathbf{p} & 0 & 0 & 0 \\ 0 & 0 & 0 & 0 & 0 & \boldsymbol{\sigma}\cdot\mathbf{p} \\ 0 & 0 & 0 & 0 & \boldsymbol{\sigma}\cdot\mathbf{p} & 0 \\ 0 & \boldsymbol{\sigma}\cdot\mathbf{p} & 0 & 0 & 0 & 0 \\ \boldsymbol{\sigma}\cdot\mathbf{p} & 0 & 0 & 0 & 0 & 0 \end{pmatrix}$$

In fact, the dimension of the two matrices M and P displayed above is 12×12: all the entries in the first one are proportional to the 2×2 identity matrix, so that in the definition one should read $\begin{pmatrix} m & 0 \\ 0 & m \end{pmatrix}$ instead of m, $\begin{pmatrix} jm & 0 \\ 0 & jm \end{pmatrix}$ instead of $j\, m$, etc. The entries in the second matrix P contain 2×2 Pauli's sigma-matrices, so that P is also a 12×12 matrix. The energy operator E is proportional to the 12×12 identity matrix.

Due to the fact that only even powers of σ-matrices are proportional to $\mathbb{1}_2$, and only the powers of circulant 3×3 circulant matrix that are multiplicities of 3 are proportional to $\mathbb{1}_3$, the diagonalization of the system is achieved only at the sixth iteration. The final result is extremely simple: all the components satisfy the same sixth-order equation,

$$E^6 \, \varphi_+ = m^6 c^{12} \, \varphi_+ + c^6 \mid \mathbf{p} \mid^6 \, \varphi_+, \quad E^6 \, \varphi_- = m^6 c^{12} \, \varphi_- + c^6 \mid \mathbf{p} \mid^6 \, \varphi_-. \tag{15}$$

and similarly for all other components.

Using a more rigorous approach the three operators can be expressed in terms of tensor products of matrices of lower dimensions. Let us introduce two following 3×3 matrices:

$$B = \begin{pmatrix} 1 & 0 & 0 \\ 0 & j & 0 \\ 0 & 0 & j^2 \end{pmatrix} \quad \text{and} \quad Q_3 = \begin{pmatrix} 0 & 1 & 0 \\ 0 & 0 & 1 \\ 1 & 0 & 0 \end{pmatrix} \tag{16}$$

whose products and powers generate the $U(3)$ Lie group algebra, or the $SU(3)$ algebra if we remove the unit matrix. Then the 12×12 matrices M and P can be represented as the following tensor products:

$$M = m \, B \otimes \sigma_3 \otimes \mathbb{1}_2, \qquad P = Q_3 \otimes \sigma_1 \otimes (\boldsymbol{\sigma} \cdot \mathbf{p}) \tag{17}$$

with as usual, $\mathbb{1}_2 = \begin{pmatrix} 1 & 0 \\ 0 & 1 \end{pmatrix}$, $\sigma_1 = \begin{pmatrix} 0 & 1 \\ 1 & 0 \end{pmatrix}$, $\sigma_3 = \begin{pmatrix} 1 & 0 \\ 0 & -1 \end{pmatrix}$.

Let us rewrite the matrix operator generating the system (13) when it acts on the column vector containing twelve components of three "colour" fields, in the basis (14) $[\varphi_+, \varphi_-, \chi_+, \chi_-, \psi_+, \psi_-]$:

$$E \mathbb{1}_3 \otimes \mathbb{1}_2 \otimes \mathbb{1}_2 = mc^2 \, B \otimes \sigma_3 \otimes \mathbb{1}_2 + Q_3 \otimes \sigma_1 \otimes c \, \boldsymbol{\sigma} \cdot \mathbf{p}$$

with energy and momentum operators on the left hand side, and the mass operator on the right hand side:

$$E \mathbb{1}_2 \otimes \mathbb{1}_3 \otimes \mathbb{1}_2 - Q_3 \otimes \sigma_1 \otimes c \, \boldsymbol{\sigma} \cdot \mathbf{p} = mc^2 \, B \otimes \sigma_3 \otimes \mathbb{1}_2 \tag{18}$$

Like with the standard Dirac equation, let us transform this equation so that the mass operator becomes proportional the the unit matrix. To do so, we multiply the equation (18) from the left by the matrix $B^\dagger \otimes \sigma_3 \otimes \mathbb{1}_2$.

Now we get the following equation which enables us to interpret the energy and the momentum as the components of a Minkowskian four-vector $c \, p^\mu = [E, \, c\mathbf{p}]$:

$$E \, B^\dagger \otimes \sigma_3 \otimes \mathbb{1}_2 - Q_2 \otimes (i\sigma_2) \otimes c \, \boldsymbol{\sigma} \cdot \mathbf{p} = mc^2 \mathbb{1}_3 \otimes \mathbb{1}_2 \otimes \mathbb{1}_2, \tag{19}$$

where we used the fact that under matrix multiplication, $\sigma_3 \sigma^3 = \mathbb{1}_2$, $B^\dagger B = \mathbb{1}_3$ and $B^\dagger Q_3 = Q_2$.

The sixth power of this operator gives the same result as before,

$$\left[E \, \sigma_3 \otimes B^\dagger \otimes \mathbb{1}_2 - i\sigma_2 \otimes Q_2 \otimes c \, \boldsymbol{\sigma} \cdot \mathbf{p} \right]^6 = \left[E^6 - c^6 \mathbf{p}^6 \right] \mathbb{1}_{12} = m^6 c^{12} \mathbb{1}_{12} \tag{20}$$

The equation (19) can be written in a concise manner using the Minkowskian indices and the usual pseudo-scalar product of two four-vectors as follows:

$$\Gamma^{\mu} p_{\mu} \, \Psi = mc \mathbb{1}_{12} \, \Psi, \quad \text{with } p^0 = \frac{E}{c}, \;\; p^k = [\, p^x, p^y, p^z \,]. \tag{21}$$

with 12×12 matrices Γ^{μ}, $(\mu = 0, 1, 2, 3)$ defined as follows:

$$\Gamma^0 = \sigma_3 \otimes B^{\dagger} \otimes \mathbb{1}_2, \quad \Gamma^k = i\sigma_2 \otimes Q_2 \otimes \sigma^k \tag{22}$$

In an appropriate basis, the system (13) can be represented in a Dirac-like form as follows:

$$\Gamma^{\mu} p_{\mu} \Psi = mc \mathbb{1}_{12} \, \Psi, \tag{23}$$

where Ψ is the generalized 12-component spinor made of 6 Pauli spinors (12), and the generalized 12×12 Dirac matrices Γ^{μ} are constructed as follows:

$$\Gamma^0 = B^{\dagger} \otimes \sigma_3 \otimes \mathbb{1}_2, \quad \Gamma^i = Q_2 \otimes (i\sigma_2) \otimes \sigma^i, \tag{24}$$

where

$$B^{\dagger} = \begin{pmatrix} 1 & 0 & 0 \\ 0 & j^2 & 0 \\ 0 & 0 & j \end{pmatrix}, \quad B = \begin{pmatrix} 1 & 0 & 0 \\ 0 & j & 0 \\ 0 & 0 & j^2 \end{pmatrix}, \quad Q_2 = \begin{pmatrix} 0 & 1 & 0 \\ 0 & 0 & j^2 \\ j & 0 & 0 \end{pmatrix}, \tag{25}$$

The two traceless matrices B and Q_2 are both cubic roots of unit 3×3 matrix. They generate the entire Lie algebra of the $SU(3)$ group.

The full set of six matrices Q_A and $Q^{\dagger}_B$, $A, B = 1, 2, 3$, together with two diagonal traceless matrices B and $B^{\dagger}$ generated by B and Q_3 form a special basis of the $SU(3)$ algebra [18]. They can be obtained by iteration, using the following multiplication table:

$$\begin{gathered} BQ_A{=}j^2 Q_A B{=}Q_{A+1}, \;\; B^{\dagger} Q_A{=}jQ_A B^{\dagger}{=}Q_{A-1}, \\ Q^{\dagger} B{=}j^2 B Q^{\dagger}_A{=}Q^{\dagger}_{A-1}, \;\; Q^{\dagger}_A B^{\dagger}{=}B^{\dagger} Q^{\dagger}_A{=}Q^{\dagger}_{A+1}, \\ Q_A Q_{A-1}{=}jQ^{\dagger}_{A+1}, \;\; Q^{\dagger}_{A-1} Q^{\dagger}_A{=}j^2 Q_{A+1}, \\ Q_A Q^{\dagger}_{A+1}{=}B^{\dagger}, \;\; Q_A Q^{\dagger}_{A-1}{=}B, \;\; Q^{\dagger}_A Q_{A-1}{=}jB^{\dagger}, \;\; Q^{\dagger}_A Q_{A+1}{=}j^2 B. \end{gathered} \tag{26}$$

and of course $Q_A Q^{\dagger}_A = Q^{\dagger}_a Q_A = \mathbb{1}_3$. where the indices A, $A+1$, $A-1$ are always taken modulo 3, so that e.g. $3 + 1 \,|_{modulo\ 3} = 4 \,|_{modulo\ 3} = 1$, etc., and the cube of each of the eight matrices in (26) is the unit 3×3 matrix.

In terms of their Fourier transforms, linear differential operators of any order are represented by corresponding algebraical expressions multiplying the Fourier transform of the unknown function. The Fourier transform of the Green function is then given by the inverse of this expression, for example, the Fourier transform of the Green function of the Klein-Gordon operator is defined as

$$\hat{G}(k^{\mu}) = \frac{1}{k_0^2 - \mathbf{k}^2 - mu^2},$$

(with $\mu = \frac{mc}{\hbar}$). The Fourier transform of Green's function for the Dirac equation is a 4×4 matrix:

$$\hat{D}(k^{\mu}) = \frac{\gamma^{\mu} k_{\mu} + m \mathbb{1}_4}{-k_0^2 + \mathbf{k}^2 - m^2},$$

because quite obviously one has

$$(\gamma^{\mu} k_{\mu} + m \mathbb{1}_4)(\gamma^{\mu} k_{\mu} - m \mathbb{1}_4) = k_0^2 + \mathbf{k}^2 - m^2 \mathbb{1}_4.$$

The ternary generalization of Dirac's equation being written in the most compact form as in (23), in terms of Fourier transforms it becomes

$$(\Gamma^\mu\, k_\mu - m\mathbb{1}_{12})\ \hat{\Psi}(k) = 0. \tag{27}$$

The sixth power of the matrix $\Gamma^\mu k_\mu$ is diagonal and proportional to m^6, so that we have

$$(\Gamma^\mu k_\mu)^6 - m^6 \mathbb{1}_{12} = \left(k_0^6 - \mid \mathbf{k} \mid^6 - m^6\right) \mathbb{1}_{12} = 0. \tag{28}$$

Now we have to find the inverse of the matrix $(\Gamma^\mu\, k_\mu - m\mathbb{1}_{12})$. To this effect, let us note that the sixth-order expression on the left-hand side in (28) can be factorized as follows:

$$(\Gamma^\mu k_\mu)^6 - m^6 = \left((\Gamma^\mu k_\mu)^2 - m^2\right)\ \left((\Gamma^\mu k_\mu)^2 - j\, m^2\right)\ \left((\Gamma^\mu k_\mu)^2 - j^2\, m^2\right). \tag{29}$$

The first factor is in turn the product of two linear expressions, one of which is the ternary Dirac operator:

$$(\Gamma^\mu k_\mu)^6 - m^6 = (\Gamma^\mu k_\mu - m)\ (\Gamma^\mu k_\mu + m)\ \left((\Gamma^\mu k_\mu)^2 - j\, m^2\right)\ \left((\Gamma^\mu k_\mu)^2 - j^2\, m^2\right). \tag{30}$$

Therefore the inverse of the Fourier transform of the ternary Dirac operator is given by the following matrix:

$$\left[(\Gamma^\mu k_\mu)^6 - m^6\right]^{-1} = \frac{(\Gamma^\mu k_\mu + m)\ \left((\Gamma^\mu k_\mu)^2 - j\, m^2\right)\ \left((\Gamma^\mu k_\mu)^2 - j^2\, m^2\right)}{\left(k_0^6 - \mid \mathbf{k} \mid^6 - m^6\right)}. \tag{31}$$

It takes almost no effort to prove that the numerator can be given a more symmetric form. Taking into account that

$$\left((\Gamma^\mu k_\mu)^2 - j\, m^2\right)\left((\Gamma^\mu k_\mu)^2 - j^2\, m^2\right) = (\Gamma^\mu k_\mu)^4 + m^2\ (\Gamma^\mu k_\mu)^2 + m^4,$$

we find that

$$(\Gamma^\mu k_\mu + m)\left((\Gamma^\mu k_\mu)^2 - j\, m^2\right)\left((\Gamma^\mu k_\mu)^2 - j^2\, m^2\right) =$$

$$(\Gamma^\mu k_\mu)^5 + m\ (\Gamma^\mu k_\mu)^4 + m^2\ (\Gamma^\mu k_\mu)^3 + m^3\ (\Gamma^\mu k_\mu)^2 + m^4\ (\Gamma^\mu k_\mu) + m^5,$$

so that the final expression can be written in a concise form as

$$\left[(\Gamma^\mu k_\mu)^6 - m^6\right]^{-1} = \frac{\sum_{s=0}^{5} m^s\ (\Gamma^\mu k_\mu)^{(5-s)}}{\left(k_0^6 - \mid \mathbf{k} \mid^6 - m^6\right)}. \tag{32}$$

In the massless case, the operator equation whose Green's function we want to evaluate, reduces to

$$\left[\frac{1}{c^6}\frac{\partial^6}{\partial t^6} - \left(\frac{\partial^2}{\partial x^2} + \frac{\partial^2}{\partial y^2} + \frac{\partial^2}{\partial z^2}\right)^3\right]\ G(t, \mathbf{r}) = \delta^4(x) = \delta(ct)\delta(x)\delta(y)\delta(z).$$

Using the Fourier transformation method, we can write:

$$\left[\frac{\omega^6}{c^6} - \mid \mathbf{k} \mid^6\right]\ \hat{G}(k_0, \mathbf{k}) = 1, \quad \text{where } k_0 = \frac{\omega}{c}, \tag{33}$$

from which we get

$$\hat{G}(k_0, \mathbf{k}) = \frac{1}{k_0^6 - \mid \mathbf{k} \mid^6} + \Phi(k_0, \mathbf{k}), \tag{34}$$

where $\Phi(k_0, \mathbf{k})$ is a solution of the homogeneous equation,

$$\left[k_0^6 - \mid \mathbf{k} \mid^6\right] \; \Phi(k_0, \mathbf{k}) = 0 \;\; \rightarrow \;\; \Phi(k_0, \mathbf{k}) = \delta(k_0^6 - \mid \mathbf{k} \mid^6). \tag{35}$$

The sixth-order polynomial $k_0^6 - \mid \mathbf{k} \mid^6$ can be split into the product of three second-order factors as follows:

$$k_0^6 - \mid \mathbf{k} \mid^6 = (k_0^2 - \mid \mathbf{k} \mid^2)\,(k_0^2 - j \mid \mathbf{k} \mid^2)\,(k_0^6 - j^2 \mid \mathbf{k} \mid^2), \tag{36}$$

each of which being a product of two linear expressions with opposite signs of $\mid \mathbf{k} \mid$:

$$(k_0^2 - \mid \mathbf{k} \mid^2) = (k_0 + \mid \mathbf{k} \mid)\,(k_0 - \mid \mathbf{k} \mid),$$

$$(k_0^2 - j \mid \mathbf{k} \mid^2) = (k_0 + j^2 \mid \mathbf{k} \mid)\,(k_0 - j^2 \mid \mathbf{k} \mid),$$

$$(k_0^2 - j^2 \mid \mathbf{k} \mid^2) = (k_0 + j \mid \mathbf{k} \mid)\,(k_0 - j \mid \mathbf{k} \mid),$$

so that the sixth-order expression appearing in (33) can be decomposed into a product of six linear terms. Let us represent the inverse of this expression appearing in (34) as a sum of three fractions with second-order expressions in their denominators:

$$\frac{1}{k_0^6 - \mid \mathbf{k} \mid^6} = \frac{1}{3 \mid \mathbf{k} \mid^4} \left[\frac{1}{k_0^2 - \mid \mathbf{k} \mid^2} + \frac{j}{k_0^2 - j \mid \mathbf{k} \mid^2} + \frac{j^2}{k_0^2 - j^2 \mid \mathbf{k} \mid^2} \right], \tag{37}$$

which is to be compared with the usual Fourier inverse of the d'Alembert operator:

$$\frac{1}{k_0^2 - \mid \mathbf{k} \mid^2} = \frac{1}{2 \mid \mathbf{k} \mid^2} \left[\frac{1}{k_0 - \mid \mathbf{k} \mid} - \frac{1}{k_0 + \mid \mathbf{k} \mid} \right] \tag{38}$$

The difference in the order of the equation leads to the difference in the algebraic structure of the polynomial representig the equation for the Fourier transform. Its inverse displays not just two, but as much as *six* simple poles displayed in the following figure:

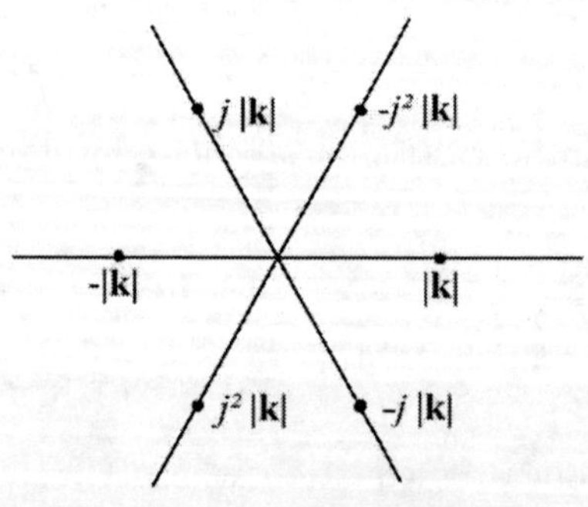

Figure 1. The six simple poles of the integral representation of zero-mass propagator of the sixth-order equation

Such field theories of higher order were considered by T.D.Lee and G.Wick [12] and were recently an object of renewed interest [13], providing arguments showing how QFT with Lee-Wick complex poles might lead to construction of unitary S-matrix.

Journal of Physics: Conference Series **1416** (2019) 012016 doi:10.1088/1742-6596/1416/1/012016

4. Plane wave solutions and algebraic confinement

Before discussing the properties of general solutions of our system, let us draw attention to its similarity with Maxwell and Dirac equations *in vacuo*, i.e. describing free fields with no interactions. The Maxwell equations, which after gauge fixing on the 4-potential A_μ reduce to a system of four linear differential equations for four functions of t and $\mathbf{r}$. In general, the characteristic equation of such a system should be of the fourth order; however, due to the particular symmetry of Maxwell's equations, the characteristic equation is of the second order (in fact, it is a fourth-order equation which is a square of the second-order equation). This ensures the existence of a *single light cone* and the absence of bi-refringence in vacuum.

The same phenomenon occurs for the characteristic equation od the Dirac equation, which is also a system of four linear differential equations for the four components of a Dirac spinor. Like in the case of Maxwell's equations, the diagonalization of the Dirac system occurs already at the level of the second order Klein-Gordon equation, while the characteristic equation of fourth order obtained from the determinant of the system is the square of the second-order expression. The four independent solutions are obtained by taking two possible signs of the square root of the determinant expression.

In our case we have as much as twelve linear equations imposed on twelve independent functions of time and space; therefore, one could expect a twelvth-order characteristic equation. However, due to the particular properties of Pauli's matrices and the symmetry of our system (23), the diagonalization occurs already at sixth iteration:

$$(\Gamma^\mu p_\mu)^6 = (p_0^6 - \mathbf{p}^6)\mathbb{1}_{12} = m^6 c^6 \mathbb{1}_{12} \tag{39}$$

This expression on the left-hand side is not manifestly relativistic invariant, but it represents a unique light cone multiplied by a positive form-factor:

$$(p_0^6 - \mathbf{p}^6) = (p_0^2 - \mathbf{p}^2)(jp_0^2 - \mathbf{p}^2)(j^2 p_0^2 - \mathbf{p}^2) = (p_0^2 - \mathbf{p}^2)(p_0^4 + p_0^2\mathbf{p}^2 + \mid \mathbf{p} \mid^4). \tag{40}$$

The characteristic equation of this system is obtained by taking the formal determinant of the 12×12 matrix operator; we get

$$\det(\Gamma^\mu p_\mu) = (p_0^6 - \mid \mathbf{p} \mid^6)^2 = m^{12} c^{12}, \tag{41}$$

which is the square of the dispersion relation (40).

In coordinate representation $\Psi = \Psi(x^\mu)$, and the four-momentum operator p_μ becomes $p_\mu = -i\hbar\partial_\mu$, and the system of equations satisfied by the 12-dimensional wave function $\Psi(x^\mu)$ is

$$-i\hbar\, \Gamma^\mu\, \partial_\mu\, \Psi = mc\Psi \tag{42}$$

As any linear system with constant coefficients, (13) representing twelve coupled linear equations, must have a basis of twelve independent solutions. General solutions are then given by their linear combinations whose coefficients depend on the choice of initial conditions.

As usually in the case of linear systems, we expect the solutions to be of general form

$$f(x^\mu) = e^{k_\mu x^\mu} \quad \text{with } k_\mu \text{ satisfying } k_0^6 - \mid \mathbf{k} \mid^6 = -\hbar^6 m^6 \tag{43}$$

Denoting $k^0 = \omega/c$, general solution can be written as $e^{\omega t - \mathbf{k}\cdot\mathbf{r}}$, $\mathbf{r} = [x, y, z]$:

The dispersion relation in (43) is invariant under the action of $Z_2 \times Z_3 = Z_6$ symmetry, because to any solution with given real ω and $\mathbf{k}$ one can add solutions with ω replaced by $j\omega$ or $j^2\omega$, $j\mathbf{k}$ or $j^2\mathbf{k}$, as well as $-\omega$; there is no need to introduce also $-\mathbf{k}$ instead of $\mathbf{k}$ because the vector $\mathbf{k}$ can take on all possible directions covering the unit sphere.

The nine complex solutions with positive frequency ω as well as with $j\,\omega$ and $j^2\,\omega$ obtained by the action of the Z_3-group can be displayed in a compact manner in form of a 3×3 matrix. The inclusion of the essential Z_2-symmetry ensuring the existence of *anti-particles* leads to the nine similar solutions with negative ω. This however will produce 18 solutions, of whiich only 12 can be linearly independent. We choose them in a symmetrical manner, displayed in two 3×3 matrices with vanishing diagonal terms:

$$\begin{pmatrix} 0 & e^{\omega t + j\mathbf{k}\cdot\mathbf{r}} & e^{\omega t + j^2\mathbf{k}\cdot\mathbf{r}} \\ e^{j\omega t + \mathbf{k}\cdot\mathbf{r}} & 0 & e^{j\omega t + j^2\mathbf{k}\cdot\mathbf{r}} \\ e^{j^2\omega t + \mathbf{k}\cdot\mathbf{r}} & e^{j^2\omega t + j\mathbf{k}\cdot\mathbf{r}} & 0 \end{pmatrix}, \quad \begin{pmatrix} 0 & e^{-\omega t - j\mathbf{k}\cdot\mathbf{r}} & e^{-\omega t - j^2\mathbf{k}\cdot\mathbf{r}} \\ e^{-j\omega t - \mathbf{k}\cdot\mathbf{r}} & 0 & e^{-j\omega t - j^2\mathbf{k}\cdot\mathbf{r}} \\ e^{-j^2\omega t - \mathbf{k}\cdot\mathbf{r}} & e^{-j^2\omega t - j\mathbf{k}\cdot\mathbf{r}} & 0 \end{pmatrix} \tag{44}$$

We can also produce similar matrix with six real functions, taking appropriate linear combinations as follows:

$$\begin{pmatrix} 0 & e^{\omega t - \frac{\mathbf{k}\cdot\mathbf{r}}{2}}\cos(\mathbf{K}\cdot\mathbf{r}) & e^{\omega t - \frac{\mathbf{k}\cdot\mathbf{r}}{2}}\sin(\mathbf{K}\cdot\mathbf{r}) \\ e^{-\frac{\omega t}{2} + \mathbf{k}\cdot\mathbf{r}}\cos\Omega t & 0 & e^{-\frac{\omega t}{2} - \frac{\mathbf{k}\cdot\mathbf{r}}{2}}\cos(\Omega t - \mathbf{K}\cdot\mathbf{r}) \\ e^{-\frac{\omega t}{2} + \mathbf{k}\cdot\mathbf{r}}\sin\Omega t & e^{-\frac{\omega t}{2} - \frac{\mathbf{k}\cdot\mathbf{r}}{2}}\sin(\Omega t - \mathbf{K}\cdot\mathbf{r}) & 0 \end{pmatrix}, \tag{45}$$

where $\Omega = \frac{\sqrt{3}}{2}\,\omega$ and $\mathbf{K} = \frac{\sqrt{3}}{2}\mathbf{k}$; a similar matrix, of course, can be produced for the alternative choice with negative ω .

All these solutions contain damping or exploding exponential factors, and cannot propagate farther than a few wavelengths. However certain quadratic or cubic products can be chosen in such a way that these nasty factors will mutually cancel leaving a freely propagating sinusoidal wave. Foir example, choosing two solutions from basis different than the one displayed in (44), namely

$$e^{j\omega t + j^2 kx} \cdot e^{-j^2\omega t - jkx} = e^{(j-j^2)\,\omega t - (j^2 - j)\,kx} = e^{i(\Omega t - Kx)}, \tag{46}$$

with $\Omega = \frac{\sqrt{3}}{2}\omega, \quad K = \frac{\sqrt{3}}{2}k$. An alternative choice of two elementary solutions leads to a free wave running in the opposite direction:

$$e^{-j^2\omega t - j^2 kx} \cdot e^{j\omega t + jkx} = e^{i(\Omega t + Kx)}.$$

This are the unique quadratic combinations yelding a running wave in both directions. This suggests that quark-anti-quark states behave as Lorentz scalars.

We face different situation if we consider possibilities of producing a running wave with cubic expressions involving solutions displayed in (45) and their conjugates with negative ω's. Let us produce the following basis of real solutions of (13):

$$F_1 = e^{-\frac{\omega t}{2} + kx}\,\sin\Omega t, \qquad F_2 = e^{-\frac{\omega t}{2} + kx}\,\cos\Omega t,$$

$$G_1 = e^{\omega t - \frac{kx}{2}}\,\sin Kx, \qquad G_2 = e^{\omega t - \frac{kx}{2}}\,\cos Kx,$$

$$H_1 = e^{-\frac{\omega t}{2} - \frac{kx}{2}}\,\sin(\Omega t - Kx), \qquad H_2 = e^{-\frac{\omega t}{2} - \frac{kx}{2}}\,\cos(\Omega t - Kx). \tag{47}$$

Neither of the six functions above can represent a freely propagating wave: even the last two functions, H_1 and H_2 contain, besides the running sinusoidal waves, the real exponentials which have a damping effect. (The wave cannot penetrate distances greater than a few wavelengths, and can last only for times comparable with few oscillations). However, we shall show that certain *cubic* expressions can represent a freely propagating wave, without any damping factors. Taking a closer look at the six solutions displayed in (47), we see that the only way to get rid of tho real exponents present in all those functions, but different damping factors, is to form cubic expressions constructed with three functions labelled with three *different* letters. Here is the exhaustive list of *eight* admissible cubic combinations:

$$F_1\, G_1\, H_1,\quad F_1\, G_1\, H_2,\quad F_1\, G_2\, H_1,\quad F_1\, G_2\, H_2;$$

$$F_2\, G_1\, H_1,\quad F_2\, G_1\, H_2,\quad F_2\, G_2\, H_1,\quad F_2\, G_2\, H_2. \tag{48}$$

But these expressions still contain, besides running waves with double frequency 2Ω, undesirable functions like $\sin \Omega t$ or $\cos Kx$. To take an example, we have

$$F_1\, G_2\, H_2 = \sin \Omega t\ \cos Kx\ \cos(\Omega t - Kx) =$$

$$\frac{1}{2}\ [\sin(\Omega t + Kx) + \sin(\Omega t - Kx)]\ \cos(\Omega t - Kx) =$$

$$\frac{1}{4}\ \sin(2\Omega t - 2Kx) + \frac{1}{4}\ \sin(2\Omega t) + \frac{1}{4}\ \sin(2Kx).$$

The explicit expressions, in terms of the trigonometric functions, of the eight independent cubic expressions displayed above, are quite cumbersome. Here we give the final result, showing that there are only *two* combinations of cubic products of solutions of the generalized ternary Dirac equation that represent running waves, which are the following:

$$F_1G_2H_2 + F_1G_1H_1 - F_2G_1H_2 + F_2G_2H_1 = \sin(2\Omega t - 2Kx), \tag{49}$$

$$F_2G_2H_2 + F_2G_1H_1 + F_1G_1H_2 - F_1G_2H_1 = \cos(2\Omega t - 2Kx). \tag{50}$$

The symmetry of these expressions appears better when grouped as follows:

$$F_1(G_2H_2 + G_1H_1) + F_2(G_2H_1 - G_1H_2) = \sin(2\Omega t - 2Kx), \tag{51}$$

$$F_2(G_2H_2 + G_1H_1) + F_1(G_1H_2 - G_2H_1) = \cos(2\Omega t - 2Kx). \tag{52}$$

Similar two running waves are produced by forming corresponding cubic combinations of negative frequency solutions obtained by substituting $-\omega$ instead of ω. We shall get then four independent running waves, two in one direction, two in the opposite direction. This circumstance suggests that the three-quark combnations may behave like four-component standard Dirac spinors.

5. Z_3-graded extension of $D = 4$ Lorentz and Poincaré algebra

It is usually taken for granted that there is no other way of combining internal symmetries of interactions between elementary particles and gauge fields with the Lorentz-Poincaré invariance interpreted as the fundamental symmetry of Minkowskian space-time, than simple cartesian product of these groups. In the case of the Standard Model the overall symmetry is $G \times SO(1,3)$, where $G = SU(3)_{color} \times SU(2) \times U(1)$ and $SO(1,3)$ is the Lorentz group. The invariance with respect to space-time translations P_μ extends $SO(1,3)$ to the full Poincaré group.

This statement results from the "no-go theorem" by Coleman and Mandula, published in 1967 [14]. However, it was shown since then that one can escape from the Coleman-Mandula theorem's constraints by changing the paradigm. One of the ways of weakening one or more of its assumptions. it is possible to allow anticommuting generators as well as commuting ones, which leads to the possibility of supersymmetry. Supersymmetry consists in introduction of anticommuting symmetry generators which transform in the (12,0) and (0,12) (i.e. spinor) representations of the Lorentz group. The new symmetry generators are spinors, not scalars; therefore supersymmetry is not an internal symmetry. It is rather an extension of the Poincaré space-time symmetries, and as such, has an obvious extension to space-time

Journal of Physics: Conference Series **1416** (2019) 012016 doi:10.1088/1742-6596/1416/1/012016

dimensions other than four, while the Coleman-Mandula theorem has no obvious extension beyond four dimensions. More details can be found in the review paper by J.D. Lykken [15]

As already observed in the previous section, the 12×12 matrices Γ^μ appearing in the coloured Dirac equation (84) do not satisfy the anti-commutation relations defining a 4-dimensional Clifford algebra. as it was the case with the 4×4 Dirac matrices γ^μ. What can be easily seen is that their *third powers* are of the form $\mathbb{1}_3\otimes\gamma^\mu$, and therefore can be used to construct the usual 4-dimensional Clifford algebra tensorized with trivial 3×3 unit matrix $\mathbb{1}_3$, like in the modified version of the Standard Model proposed recently by I. Sogami [16].

Here we shall construct a spinorial representation of the Z_3-graded extension of the Lorentz group following the method presented in [6], in which the usual Lorentz algebra $\overset{(0)}{\mathcal{L}}$ was embedded in the SZ_3-graded Lorentz algebra containing three subspaces,

$$\mathcal{L}_{Z_3}=\overset{(0)}{\mathcal{L}}\oplus\overset{(1)}{\mathcal{L}}\oplus\overset{(2)}{\mathcal{L}} \tag{53}$$

The $Z_3\otimes Z_2$ structure of Γ^μ-matrices implies that only their sixth powers are proportional to the unit matrix $\mathbb{1}_{12}$ (see also (41)). Thus, in order to obtain the realization of $D=4$ Lorentz algebra generators one can not use just two standard commutators

$$J_i=\frac{i}{2}\,\epsilon_{ijk}\left[\Gamma^j,\Gamma^k\right],\quad K_l=\frac{1}{2}\,\left[\Gamma_l,\Gamma_0\right]\,. \tag{54}$$

Nevertheless, the generators $(\overset{(0)}{J_i},\ \overset{(0)}{K_l})$ satisfying the standard Lorentz algebra relations (see also (59) for $r=0, s=0$) can be defined by triple commutators:

$$[K_i,[K_j,K_k]]=(\delta_{ij}\delta_{kl}-\delta_{ik}\delta_{jl})\ K_l^{(0)}\,. \tag{55}$$

Substituting in (55) the explicit form of Γ^μ given in (24), we get

$$\begin{aligned} J_i&=-\frac{i}{2}\,Q_2^\dagger\otimes\mathbb{1}_2\otimes\sigma_i,\quad K_l=-\frac{1}{2}\,Q_1\otimes\sigma_1\otimes\sigma_l,\\ \overset{(0)}{J_i}&=-\frac{i}{2}\mathbb{1}_3\otimes\mathbb{1}_2\otimes\sigma_i,\quad \overset{(0)}{K_l}=-\frac{1}{2}\mathbb{1}_3\otimes\sigma_1\otimes\sigma_l. \end{aligned} \tag{56}$$

In order to close the generalized Lorentz algebra (53) where $\overset{(0)}{L}=(\overset{(0)}{J_i},\overset{(0)}{K_j})$, $\overset{(1)}{L}=(\overset{(1)}{J_i},\overset{(1)}{K_j})$, $\overset{(2)}{L}=(\overset{(2)}{J_i},\overset{(2)}{K_j})$, one should supplement (55) by two missing triple commutators:

$$\begin{aligned} [J_i,[J_j,K_k]]&=(\delta_{ij}\delta_{kl}-\delta_{ik}\delta_{jl})\ K_l^{(2)},\\ [K_i,[K_j,J_k]]&=(\delta_{ij}\delta_{kl}-\delta_{ik}\delta_{jl})\ J_l^{(1)}, \end{aligned} \tag{57}$$

where using the representation (56) we get

$$\overset{(1)}{J_l}=-\frac{i}{2}\,Q_3\otimes\mathbb{1}_2\otimes\sigma_l,\quad \overset{(2)}{K_i}=-\frac{1}{2}\,Q_3^\dagger\otimes\sigma_1\otimes\sigma_i. \tag{58}$$

The full set of Z_3-graded relations defining the algebra (53) ($r, s,\ r+s$ are modulo 3):

$$\begin{aligned} &[\overset{(r)}{J_i},\overset{(s)}{J_k}]=\epsilon_{ikl}\ \overset{(r+s)}{J_l}\ ,\quad [\overset{(r)}{J_i},\overset{(s)}{K_k}]=\epsilon_{ikl}\ \overset{(r+s)}{K_l}\ ,\\ &[\overset{(r)}{K_i},\overset{(s)}{K_k}]=-\epsilon_{ikl}\ \overset{(r+s)}{J_l}\ . \end{aligned} \tag{59}$$

The remaining generators of $\mathcal{L}$ are obtained from commutators
$[\overset{(1)}{K},\overset{(1)}{K}] \simeq \overset{(2)}{J}$ and $[\overset{(1)}{J},\overset{(1)}{J}] \simeq \overset{(2)}{J}$:

$$\overset{(2)}{J_i} = -\frac{i}{2}\, Q_3^\dagger \otimes \mathbb{1}_2 \otimes \sigma_i, \quad \overset{(1)}{K_m} = -\frac{1}{2}\, Q_3 \otimes \sigma_1 \otimes \sigma_m. \tag{60}$$

The formulae (56, 58) and (60) provide the realization of $\mathcal{L}$ which follows from the choice (24) of matrices Γ^μ.

Before considering standard and generalized Lorentz covariance let us introduce the following notation:

$$\Gamma^\mu_{(A;\alpha)} = I_A \otimes \sigma_\alpha \otimes \sigma^\mu, \; A = 0,1,..,8; \; \alpha = 2,3; \; \mu = 0,1,2,3\,. \tag{61}$$

Let the 3×3 "colour matrices" I_A appearing as the first factor in (61) be defined as follows: $I_0 = \mathbb{1}_3$, $I_r = Q_r$, $I_{r+3} = Q_r^\dagger$, $I_7 = B$, $I_8 = B^\dagger$. Then the original Γ-matrices given by (24) are encoded as $\Gamma^0_{(8,3)} = B^\dagger \otimes \sigma_3 \otimes \mathbb{1}_2$ and $\Gamma^i_{(2;2)} = Q_2 \otimes (i\sigma_2) \otimes \sigma^i$. The eight matrices with $A = 1,2,...8$ with the multiplication rules given in (26) span the ternary basis, generated by the cyclic Z_3-automorphism of the $SU(3)$ algebra ([18], Sect. 8).

In order to get the closed formula for the action $\mathcal{S}^{(0)}\Gamma^\mu[\mathcal{S}^{(0)}]^{-1}$ of classical spinorial Lorentz symmetries generated by $L^{(0)}$, we should introduce the pairs of Γ^μ-matrices $\Gamma^\mu = (\Gamma^i_{(A;2)}, \Gamma^0_{(B;3)})$ and $\tilde{\Gamma}^\mu = (\Gamma^i_{(B;2)}, \Gamma^0_{(A;3)})$, $A \neq B$. For any choice of Γ^μ's in (61) we get:

$$[\overset{(0)}{J_i}, \Gamma^j_{(A;\alpha)}] = \epsilon_{ijk}\Gamma^k_{(A;\alpha)}, \quad [\overset{(0)}{J_i}, \Gamma^0_{(A;\alpha)}] = 0, \tag{62}$$

and the boosts $\overset{(0)}{K_i}$ act covariantly on doublets $\left(\Gamma^\mu, \tilde{\Gamma}^\mu\right)$ as follows:

$$[\overset{(0)}{K_i}, \Gamma^j_{(A;2)}] = \delta_i^j\, \Gamma^0_{(A;3)}, \quad [\overset{(0)}{K_i}, \Gamma^0_{(B;3)}] = \Gamma^i_{(B;2)},$$

$$[\overset{(0)}{K_i}, \Gamma^j_{(B;2)}] = \delta_i^j\, \Gamma^0_{(B;3)}, \; [\overset{(0)}{K_i}, \Gamma^0_{(A;3)}] = \Gamma^i_{(A;2)}, \tag{63}$$

(with $A \neq B$), i.e. the standard Lorentz covariance requires the *doublet* of coloured Dirac spinors; In particular, the Γ^μ matrices (24) should be supplemented by:

$$\tilde{\Gamma}^0 = \Gamma^0_{(2;3)} = Q_2 \otimes (\sigma_3) \otimes \mathbb{1}_2, \quad \tilde{\Gamma}^i = \Gamma^k_{(8;2)} = B^\dagger \otimes i\sigma_2 \otimes \sigma^k. \tag{64}$$

One can conjecture that the pairs of Γ-matrices generated by the standard Lorentz covariance requirement can be used for the introduction of weak isospin doublets of the $SU(2)\times U(1)$ electroweak symmetry. In such a way one can conclude that the internal symmetries $SU(3)\times SU(2)\times U(1)$ of Standard Model follow from the imposition of standard Lorentz covariance on colour Dirac multiplets.

Our next goal is to study the generalized Lorentz covariance of coloured Dirac equations, by generalization of standard invariance condition (65)

$$\tilde{\psi}(\Lambda x) = S(\Lambda)\, \psi(x)\, S^{-1}(\Lambda) \tag{65}$$

and incorporating the standard Γ^μ-matrices (24) into an irreducible representation of $\mathcal{L}$. For this purpose, we should study the 18-parameter symmetry transformation $\Gamma^\mu \to \mathcal{S}\Gamma^\mu\mathcal{S}^{-1}$, where

$$\mathcal{S} = \prod_{r=0}^{2} e^{i\,[\alpha^k_{(r)}\overset{(r)}{J_k} + \beta^m_{(r)}\overset{(r)}{K_m}]}, \tag{66}$$

with $\alpha^k_{(0)}$, $\beta^k_{(0)}$ real, $(\alpha^k_{(1)})^* = \alpha^k_{(2)}$, $(\beta^k_{(1)})^* = \beta^k_{(2)}$, $[\overset{(1)}{J_k}]^\dagger = \overset{(2)}{J_k}$ and $[\overset{(1)}{K_m}]^\dagger = \overset{(2)}{K_m}$. It follows that in order to obtain the closure of the faithful action of generators $(J^{(s)}_k,\ K^{(s)}_m)$ $(s = 0, 1, 2)$ on matrices Γ^μ, one should introduce two sets $\Gamma^\mu_{(a)}, \Gamma^\mu_{\dot a} = (\Gamma^\mu_{(a)})^\dagger$ $(a = 1, 2, ...6)$ of coloured 12×12 Dirac matrices supplemented by Lorentz doublet partners $(\tilde\Gamma^\mu_{(a)}, \tilde\Gamma^\mu_{(\dot a)})$. If we choose $(\overset{(1)}{J_k}, \overset{(1)}{K_m})$ as given by Eqs. (58), (60), and assume that $\Gamma^\mu_{(1)}$ is given by the formula (24), by calculating the multicommutators of $(\overset{(1)}{J_i}, \overset{(1)}{K_l})$ with the set $\Gamma^\mu_{(a)}$, $(a = 1, 2...6)$, we get the following sextet of Γ-matrices closed under the action of $\overset{(1)}{L}$:

$$\begin{aligned}
\Gamma^\mu_{(1)} &= \left(\Gamma^0_{(8;3)},\, \Gamma^i_{(2;2)}\right);\ \Gamma^\mu_{(4)} = \left(\Gamma^0_{(8;2)},\, \Gamma^i_{(2;3)}\right);\\
\Gamma^\mu_{(2)} &= \left(\Gamma^0_{(2;2)},\, \Gamma^i_{(4;3)}\right);\ \Gamma^\mu_{(5)} = \left(\Gamma^0_{(2;3)},\, \Gamma^i_{(4;2)}\right);\\
\Gamma^\mu_{(3)} &= \left(\Gamma^0_{(4;3)},\, \Gamma^i_{(8;2)}\right);\ \Gamma^\mu_{(6)} = \left(\Gamma^0_{(4;2)},\, \Gamma^i_{(8;3)}\right).
\end{aligned} \tag{67}$$

The six matrices (67) form three pairs, each of which transforms in itself under the action of the 0-grade subalgebra $(\overset{(0)}{J_i}, \overset{(0)}{K_l})$. The Z_3-graded components of $\mathcal{L}_{Z_3}$, $\overset{(1)}{\mathcal{L}}$ and $\overset{(2)}{\mathcal{L}}$, act on these pairs transforming them into other pairs, with conjugate Q and B matrices.

The realization of $\overset{(2)}{\mathcal{L}}$ sector is obtained by introducing the Hermitean-conjugate sextet $\Gamma^\mu_{(\dot a)} = (\Gamma^\mu_{(a)})^\dagger$; further one should add $\tilde\Gamma^\mu_{(\dot a)} = (\tilde\Gamma^\mu_{(a)})^\dagger$ due to standard Lorentz covariance. The generalized Lorentz transformations of 24 matrices $\Gamma^\mu_{(F)} = (\Gamma^\mu_{(a)}, \Gamma^\mu_{(\dot a)};\ \tilde\Gamma^\mu_{(a)}, \tilde\Gamma^\mu_{(\dot a)})$ will be expressed by the following generalization of the formula (65)

$$\mathcal{S}\Gamma^\mu_{(F)}\mathcal{S}^{-1} = \Lambda^{\mu\,(G)}_{\ \nu\,(F)}\ \Gamma^\nu_{(G)}, \quad \mu, \nu = 0, 1, 2, 3; \quad F, G = 1, 2, ..., 24. \tag{68}$$

It should be recalled that the Z_3-grading of Lorentz algebra was constructed explicitly in [6] only for spinorial (matrix) representations. For obtaining full Z_3-graded counterpart of relativistic symmetries described by Poincaré algebra one should introduce the orbital realization of Z_3-graded generalized Lorentz algebra (59), in order e.g. to apply to field-theoretic models. Such Z_3-graded extension of D=4 Poincaré algebra and its orbital realizations were constructed and discussed recently in [17] by one of the present authors (RK); here we give the outline of the main ideas expressed therein.

As one can extrapolate from the realizations of standard relativistic symmetries, in construction of Z_3-graded orbital realizations the basic tool is the introduction of 12-dimensional Z_3-graded triplets $(\overset{(0)}{p_\mu}, \overset{(1)}{p_\mu}, \overset{(2)}{p_\mu})$ of 4-momenta and dual triplets $(x^\mu_{(0)}, x^\mu_{(1)}, x^\mu_{(2)})$ of generalized Minkowski coordinates. Following earlier considerations [10] one can introduce these triplets of 4-momenta as describing three mass shells related by discrete Z_3-transformations. In this way we supplement the standard real mass shell $p_\mu p^\mu = m^2$ $(p_\mu \equiv \overset{(0)}{p_\mu})$ by a pair of complex-conjugated Lee-Wick mass shells [12, 13]

$$\overset{(1)}{p_\mu}\,\overset{(1)}{p^\mu} = j\,m^2, \qquad \overset{(2)}{p_\mu}\,\overset{(2)}{p^\mu} = j^2\,m^2, \overset{(1)}{p_\mu} = (\overset{(2)}{p_\mu})^\star, \tag{69}$$

where

$$\overset{(1)}{p}_\mu = (p_0^{(1)}, \overset{(1)}{\mathbf{p}}) \qquad \overset{(2)}{p}_\mu = (p_0^{(2)}, \overset{(2)}{\mathbf{p}}) \tag{70}$$

$$\overset{(1)}{p_0} = \pm j\sqrt{\overset{(1)}{\mathbf{p}^2} + m^2}, \quad \overset{(2)}{p_0} = \pm j^2\sqrt{\overset{(2)}{\mathbf{p}^2} + m^2}, \tag{71}$$

Further one can introduce as well three versions of 4-dimensional complexified space-times connected by discrete Z_3-transformations

$$z^\mu_{(s)} = (z^0_{(s)} = j^s\, x^0_{(s)}, x^i_{(s)}) \qquad\qquad s = 0, 1, 2 \tag{72}$$

where $x^\mu_{(s)}$ are real, but $z^\mu_{(r)}$ $(r = 1, 2)$ are complex-valued and $({z^\mu}_{(1)})^\star = {z^\mu}_{(2)}$. For s=0,1,2 one can introduce three real Minkowski scalar products in three 8-dimensional sectors $(\overset{(s)}{p}_\mu, z^\mu_{(s)})$ of 24-dimensional generalized phase space

$$(p, z)_{(s)} = \overset{(s)}{p}_\mu\, z^\mu_{(s)} = \overset{(s)}{\omega}\, x^0_{(s)} - \overset{(s)}{p}_i\, x^i_{(s)} \tag{73}$$

what allows to introduce the respective Fourier transforms.

Further, it has been shown in [17] that one can introduce the Z_3-graded extension of familiar orbital realizations $\overset{(0)}{M}_{\mu\nu}$ of standard Lorentz algebra

$$M_{\mu\nu} \equiv \overset{(0)}{M}_{\mu\nu} = x_{[\mu} p_{\nu]} \;\rightarrow\; (\overset{(0)}{M}_{\mu\nu}, \overset{(1)}{M}_{\mu\nu}, \overset{(2)}{M}_{\mu\nu}) \tag{74}$$

where all generators $\overset{(s)}{M}_{\mu\nu}$ are constructed as bilinears of Z_3-graded generalized phase space coordinates $(\overset{(s)}{p}_\mu, z^\mu_{(s)})$. The formula for total (orbital + spin) Z_3-graded relativistic angular momentum $\overset{(r)}{\mathcal{M}}_{\mu\nu} = (\overset{(r)}{\mathcal{J}_i}, \overset{(r)}{\mathcal{K}_l})$ is given by the formula

$$\overset{(r)}{\mathcal{J}_i} = \overset{(r)}{J}{}^{orb}_i + \overset{(r)}{J}_i \qquad \overset{(r)}{\mathcal{K}_i} = \overset{(r)}{K}{}^{orb}_i + \overset{(r)}{K}_i \tag{75}$$

where the matrix-like "spin" part of generators $(\overset{(r)}{J}_i, \overset{(r)}{K}_i)$ were studied in [6]. The orbital and spin parts of the realizations (75) commute with each other and $(\overset{(r)}{\mathcal{J}_i}, \overset{(r)}{\mathcal{K}_i})$ satisfy the Z_3-graded generalized Lorentz algebra relations (see (59)).

Further one can extend the Z_3-grade relations (59) to Z_3-graded Poincaré algebra, firstly presented in [17], with 18 generators $\overset{(s)}{M}_{\mu\nu}$ and 12 generators $\overset{(s)}{P}_\mu$:

$$[\,\overset{(r)}{\mathcal{P}_0}, \overset{(s)}{\mathcal{P}_k}\,] = 0; \qquad [\,\overset{(r)}{\mathcal{P}_i}, \overset{(s)}{\mathcal{P}_j}\,] = 0, \tag{76}$$

$$[\,\overset{(r)}{\mathcal{J}_k}, \overset{(s)}{\mathcal{P}_0}\,] = 0; \qquad [\,\overset{(r)}{\mathcal{J}_i}, \overset{(s)}{\mathcal{P}_k}\,] = \epsilon_{ikl}\, \overset{(r+s)}{\mathcal{P}_l}\,, \tag{77}$$

$$[\,\overset{(r)}{\mathcal{K}_i}, \overset{(s)}{\mathcal{P}_0}\,] = \overset{(r+s)}{\mathcal{P}_i}\,, \qquad [\,\overset{(r)}{\mathcal{K}_i}, \overset{(s)}{\mathcal{P}_k}\,] = -\delta_{ik}\, \overset{(r+s)}{\mathcal{P}_0}\,. \tag{78}$$

At present we study the possibilities of incorporating the Z_3-graded Poincaré symmetries (76-78) as a guiding line for the introduction of Z_3-graded generalization of field-theoretic QCD model, which incorporates complete quark sector of the Standard Model.

6. The Lorentz-covariant Lagrangean for quark doublets

In this Section we shall show how the free Lagrangean for the pair of colour Dirac spinors can be made covariant under the standard Lorentz transformations. Let us start by recalling the properties of color-spinor representation of the Lorentz group transforming the generalized colour Dirac matrices Γ^μ. According to the notation proposed in [6], we have:

$$\Gamma^\mu =: \left(\Gamma^0, \, \Gamma^k\right) = \left(B^\dagger \otimes \sigma_3 \otimes \mathbb{1}_2, \;\; Q_2 \otimes (i\sigma_2) \otimes \sigma^k\right), \tag{79}$$

while the generators of the colour Lorentz group are given by the following six matrices, three space rotations and three independent Lorentz boosts:

$$\overset{(0)}{J_i} = -\frac{i}{2} \mathbb{1}_3 \otimes \mathbb{1}_2 \otimes \sigma_i, \qquad \overset{(0)}{K_l} = -\frac{1}{2} \mathbb{1}_3 \otimes \sigma_1 \otimes \sigma_l. \tag{80}$$

Taking the commutators of these generators (56) we discover that the matrices Γ^μ are transformed into another set which we shall denote by $\tilde{\Gamma}^\mu$:

$$\left[\overset{(0)}{J}_i, \, \Gamma^0\right] = 0, \quad [\, \overset{(0)}{J}_i, \, \Gamma^k \,] = \epsilon_{ikm} \Gamma^m;$$

$$\left[\overset{(0)}{K}_i, \, \Gamma^0\right] = \tilde{\Gamma}^i, \quad [\, \overset{(0)}{K}_i, \, \Gamma^k \,] = \delta_i^k \tilde{\Gamma}^0; \tag{81}$$

where $\tilde{\Gamma}^\mu$ are given by the following tensor products:

$$\tilde{\Gamma}^\mu =: \left(\tilde{\Gamma}^0, \, \tilde{\Gamma}^k\right) = \left(Q_2 \otimes \sigma_3 \otimes \mathbb{1}_2, \;\; B^\dagger \otimes (i\sigma_2) \otimes \sigma^k\right), \tag{82}$$

Quite obviously, when acting with the same generators on $\tilde{\Gamma}^\mu$, we come back to the initial set Γ^μ, with exactly the same relations as (81).

It is also easy to check that the operator $\tilde{\Gamma}^\mu \, p_\mu$ satisfies the same fundamental relations as the original operator $\Gamma^\mu \, p_\mu$, namely

$$\left(\tilde{\Gamma}^\mu p_\mu\right)^6 = (p_0^6 - \mid \mathbf{p} \mid^6) \mathbb{1}_{12}, \quad \det\left(\tilde{\Gamma}^\mu p_\mu\right) = (p_0^6 - \mid \mathbf{p} \mid^6)^2. \tag{83}$$

The matrix-valued four-vector Γ^μ enter into the definition of the generalized ("colour") Dirac equation:

$$\Gamma^\mu \partial_\mu \, \Psi = m\Psi. \tag{84}$$

Now we clearly see that this equation cannot be covariant under the Lorentz transformations (81), because it does not contain the other member of the doublet, $\tilde{\Gamma}^\mu$. To restore the symmetry, we must introduce another coloured spinor, to be associated with the $\tilde{\Gamma}^\mu$ which form a Lorentz doublet of coloured spinors . In order to make the Lorentz doublet equations covariant under classical Lorentz transformations we should introduce mass terms as introducing the mixing between Ψ and $\tilde{\Psi}$:

$$\Gamma^\mu \partial_\mu \, \Psi = m\tilde{\Psi}, \quad \tilde{\Gamma}^\mu \partial_\mu \, \tilde{\Psi} = m\Psi, \tag{85}$$

The covariance is ensured because one can show that under the same Lorentz action Ψ becomes $\tilde{\Psi}'$ and $\tilde{\Psi}$ becomes Ψ'.

Journal of Physics: Conference Series **1416** (2019) 012016 doi:10.1088/1742-6596/1416/1/012016

In order to produce the Lagrangian action leading to the pair of coupled colour Dirac equations (85) we shall follow as faithfully as possible the classical Dirac scheme. The Dirac lagrangean was defined as follows:

$$\mathcal{L}_{Dirac} = \frac{i}{2}\left[\left(\partial_\mu \bar{\psi}\right)\gamma^\mu \psi - \bar{\psi}\gamma^\mu \partial_\mu \psi\right] + m\bar{\psi}\psi. \tag{86}$$

with Dirac conjugation defined as follows (in order to guarantee the positivity of the energy):

$$\bar{\psi} = \psi^\dagger \gamma^5, \qquad \gamma^5 = \sigma_1 \otimes \mathbb{1}_2 \tag{87}$$

The Lorentz-covariant generalization of the Lagrangean (86) looks as follows

$$\mathcal{L}_0 = \frac{i}{2}\left[\partial_\mu \bar{\tilde{\Psi}}\Gamma^\mu \Psi - \bar{\tilde{\Psi}}\Gamma^\mu \partial_\mu \Psi + \partial_\mu \bar{\Psi}\tilde{\Gamma}^\mu \tilde{\Psi} - \bar{\Psi}\tilde{\Gamma}^\mu \partial_\mu \tilde{\Psi}\right] + m\bar{\tilde{\Psi}}\tilde{\Psi} + m\bar{\Psi}\Psi. \tag{88}$$

One can easily check that variation with respect to Ψ or $\tilde{\Psi}$ lead respectively to the equations (85) and provides the conjugate equations if the variation is made with respect to $\bar{\Psi}$ and $\bar{\tilde{\Psi}}$.

What remains to be defined is the generalization of Dirac's conjugation $\bar{\Psi} = \Psi^+ \gamma_0$. In order to ensure the positivity of the energy, these conjugations should be defined as follows:

$$\bar{\Psi} = \Psi^\dagger (B \otimes \sigma_3 \otimes \mathbb{1}_2,) \qquad \bar{\tilde{\Psi}} = \tilde{\Psi}^\dagger (Q_2^\dagger \otimes \sigma_3 \otimes \mathbb{1}_2). \tag{89}$$

This choice of generalized Dirac conjugation for coloured spinors yields the desired result, namely:

$$\bar{\Psi}\Gamma^0 \Psi = \Psi^\dagger (B \otimes \sigma_3 \otimes \mathbb{1}_2)(B^\dagger \otimes \sigma_3 \otimes \mathbb{1}_2)\Psi = \Psi^\dagger \Psi, \tag{90}$$

and

$$\bar{\tilde{\Psi}}\tilde{\Gamma}^0 \tilde{\Psi} = \tilde{\Psi}^\dagger (Q_2^\dagger \otimes \sigma_3 \otimes \mathbb{1}_2)(Q_2 \otimes \sigma_3 \otimes \mathbb{1}_2)\tilde{\Psi} = \tilde{\Psi}^\dagger \tilde{\Psi}. \tag{91}$$

Similar construction can be performed with two remaining doublets of Γ-matrices introduced in (67), and two analogous Lagrangeans involving the pairs $(\Gamma^\mu_{(2)}, \Gamma^\mu_{(5)})$ and $(\Gamma^\mu_{(3)}, \Gamma^\mu_{(6)})$ should be added to the action given by (88).

7. Final remarks and outlook

The colour generalization of Dirac's equation proposed here with mixing of colour and spin degrees of freedom and incorporating particle-antiparticle duality displays natural $Z_3 \times Z_2 \times Z_2$ symmetry. Imposing the Z_3-graded Lorentz and Poincaré covariance required the enlagement of the usual 10-dimensional algebra to its 30-dimensional Z_3-graded extension, acting on two conjugate sextuplets of generalized 12-dimensional coloured spinors.

Although the characteristic equation of the colour Dirac operator is of sixth order, this operator will display the same ultraviolet behavior of perturbative Feynman integrals as the usual Dirac equation, because the inverse operator's Fourier transform shows that it behaves as k^{-1} in high energy limit. The complex-valued poles lead to the exponentially damped solutions which cannot propagate at distances larger than a few characteristic wavelenghts and other ones exponentially blowing up. However, as we did show in Sect. 4 the products of three suitable solutions of colour Dirac equations can produce freely propagating oscillating waves.

Such a phenomenon seems to be analogous to the properties of electrons' with opposite wave vectors $\mathbf{k}$ and $-\mathbf{k}$ couplings in Cooper pairs behaving like bosons and are the source of supraconductivity phenomenon. By analogy, the Z_3-symmetry creates the situation in which the damped quark states obeying the coloured Dirac equation can form composite states when in momentum space their wave vectors are $\mathbf{k}$, $j\mathbf{k}$ and $j^2\mathbf{k}$ forming a propagating fermion, or coupled in pairs with wave vectors $j\mathbf{k}$ and $-j^2\mathbf{k}$ forming a propagating boson.

Another remarkable property of Z_3 symmetry to be exploited in quark model consists in a generalization of Pauli's exclusion principle [4], [8]. Although quarks behave like Pauli spinors, obeying the Fermi-Dirac statistics, they possess, beside the two-valued spin and three-valued colour also another two-valued characteristics, labeled u and d in the first family, c and s in the second, and t and b in the third one. But this new attribute is not exclusive like the half-integer spin or colour: it can accomodate in composite state two particles labeled identically, but not three. Such a behavior can be modelled by a Z_3-graded exclusion principle making use of j-skew symmetry. In such a case the quark wave functions' algebra should take values in an associative Z_3-graded algebra whose generators obey the following *ternary commutation rules*;

$$\theta^A\theta^B\theta^C = j\,\theta^B\theta^C\theta^A = j^2\,\theta^C\theta^A\theta^B, \quad A,B,.. = 1,2,...N \tag{92}$$

along with their hermitian conjugated ternary rules for conjugate generators $\bar{\theta}^{\dot{B}}$.

The generators θ^A are of Z_3 grade 1, their hermitian conjugates $\bar{\theta}^{\dot{B}}$ of Z_3 grade 2, and to physical states we assign the products of generators whose Z_3 grade is 0. Briefly, the squares of generators are not zero, but their cubes vanish identically, $(\theta^A)^3 = 0$.

In such a framework, in the case of two generators, $A, B, .. = 1, 2,\ \dot{A}, \dot{B}, .. = \dot{1}, \dot{2}$, we can choose as the only non-vanishing independent cubic combinations of Z_3 grade 0 the following ones:

$$\theta^1\theta^1\theta^2, \quad \theta^1\theta^2\theta^2, \quad \bar{\theta}^{\dot{1}}\bar{\theta}^{\dot{1}}\bar{\theta}^{\dot{2}}, \quad \bar{\theta}^{\dot{1}}\bar{\theta}^{\dot{2}}\bar{\theta}^{\dot{2}}, \tag{93}$$

which can describe stable fermionic quark composites uud, udd, $\bar{u}\bar{u}\bar{d}$, $\bar{u}\bar{d}\bar{d}$;. Analogously, the independent quadratic monomials of Z_3 grade zero

$$\theta^1\bar{\theta}^{\dot{2}}, \quad \theta^2\bar{\theta}^{\dot{1}}, \quad \theta^1\bar{\theta}^{\dot{1}} - \theta^2\bar{\theta}^{\dot{2}}, \tag{94}$$

correspond to bosonic quark composites $u\bar{d}$, $\bar{u}d$, $u\bar{u} - d\bar{d}$.

Our next aim will be to investigate the quark fields endowed with the Z_3-graded exclusion principle and Z_3-graded extended relativistic symmetries as describing the dynamics of suitably modified Standard Model, in particular modified colour sector of strong interactions (chromodynamics). This stage of development of our framework, with the discussion of its applicability to particle physics is now under consideration.

Acknowledgement

One of the authors (JL) would like to express his gratitude to Bogolubov Institute of Theoretical Physics (JINR, Dubna) for warm hospitality extended to him during his stay there in Summer of 2019.

References

[1] Greensite J 2011 *An introduction to the confinement problem* Springer Lecture Notes in Physics **821**
[2] Wu T Y and Hwang W Y P 1991 *Relativistic Quantum Mechanics and Quantum Fields* World Scientific.
[3] Dirac P A M 1958 *The Principles of Quantum Mechanics Oxford at the Clarendon Press.*
[4] Pauli W 1926 *Zeitschrift für Physik* **26** (5) 336
[5] Dirac P A M 1928 *The Quantum Theory of the Electron Proc. Royal Soc. A* **117** (778) 610 *ibid* **118** (779) 351
[6] Kerner R and Lukierski J 2019 Z_3-graded colour Dirac equation for quarks, confinementt and generalized Lorentz symmetries *Phys. Letters B* **792** 233 also arXiv:1901.10936 [hep-th]
[7] Kerner R and Suzuki O 2014 The discrete quantum origin of the Lorentz group and the Z_3-graded ternary algebras Proceedings of RIMS Conference on Mathematical Physics Kyoto 2013 see also https://ci.nii.ac.jp/naid/110009863886

[8] Kerner R 2017 Ternary generalization of Pauli's principle and the Z_6-graded algebras *Physics of Atomic Nuclei*, **80** (3) 529 also arXiv:1111.0518, arXiv:0901.3961
[9] Kerner R 2018 *Mathematical Structures and Applications* Springer 311
[10] Kerner R 2018 Ternary $Z_2 \times Z_3$ graded algebras and ternary Dirac equation *Physics of Atomic Nuclei* **81** (6) 871 also: arXiv:1801.01403
[11] Kerner R 2019 The Quantum nature of Lorentz invariance *Universe*, **5** (1) 1 https://doi.org/10.3390/universe5010001
[12] Lee T D and Wick G C 1970 Finite Theory of Quantum Electrodynamics *Phys. Rev. D*, **2** 1033
[13] Anselmi D and Piva M 2017 Perturbative Unitarity of Lee-Wick Quantum Field Theory *Phys. Rev. D* **96** 045009
[14] Coleman S and Mandula J 1967 All possible symmetries of the S-Matrix *Physical Review* **159** 1251
[15] Lykken J D 1996 *Introduction to Supersymmetry*, arXiv:hep-th/9612114v1
[16] Sogami I 2018 arXiv:1512.09283 see also I. Sogami and K. Kozumi to appear
[17] Kerner R 2019 The Z_3-graded extension of the Poincaré algebra submitted to JMP see also arXiv:1908.02594
[18] Kac V G 1984 *Infinite-Dimensional Lie Algebras* Cambridge Univ.Press
[19] Hall B C 2015 *Lie groups, Lie algebras and representations* Graduate Texts in Mathematics **222** Springer (see Proposition 3.35).

Journal of Physics: Conference Series **1416** (2019) 012017 doi:10.1088/1742-6596/1416/1/012017

Worldsheet Instanton Corrections to Five-branes and Waves in Double Field Theory

Tetsuji Kimura[a], **Shin Sasaki**[b] **and Kenta Shiozawa**[c]

[a] Center for Physics and Mathematics, Institute for Liberal Arts and Sciences, Osaka Electro-Communication University, Neyagawa, Osaka 572-8530, Japan
[b,c] Department of Physics, Kitasato University, Sagamihara 252-0373, Japan

E-mail: [a] t-kimura@osakac.ac.jp
[b] shin-s@kitasato-u.ac.jp
[c] k.shiozawa@sci.kitasato-u.ac.jp

Abstract. We study string winding corrections to supergravity solutions in double field theory (DFT). We find five-brane and wave solutions of diverse codimensions where the winding coordinates dependence is interpreted as string worldsheet instanton corrections. In order to show this fact, we give a brief discussion on the origin of these winding corrections in gauged linear sigma models (GLSMs). We propose a GLSM that provides a string sigma model whose target spaces are a defect NS5-brane, a Kaluza-Klein vortex and an exotic 5^2_2-brane. They are codimension two objects and are related by T-duality. This contribution is based on the works [1, 2].

1. Introduction – double field theory

Double field theory (DFT) is an effective (or truncated) theory of superstring theory where T-duality is realized as a manifest symmetry [3]. In order to realize the T-duality symmetry, DFT is defined in a $2d$-dimensional doubled space whose coordinates are denoted by $X^M = (\tilde{x}_\mu, x^\mu)$, where $M = 1, \cdots, 2d$, $\mu = 1, \cdots, d$. Since the dimensionality of the space defined in DFT is doubled, all the dynamical fields and gauge parameters should satisfy a constraint. This is known as the strong constraint. The strong constraint is necessary for the gauge invariance of DFT.

A half of X^M for the doubled space are interpreted as the physical geometrical coordinates of space-time while the other half correspond to the *winding* coordinates of string. The T-duality is manifestly realized as an $O(d,d)$ transformation in DFT. By solving the strong constraint by dropping the winding coordinates dependence on the fields in a specific T-duality frame, we recover supergravity theory with the NSNS massless modes.

Classical solutions to DFT have been studied in various contexts. The solutions include those in the ordinary supergravity theory. For example, NS5-branes, fundamental strings, waves, Kaluza-Klein (KK) five-branes and so on are solutions to DFT. In addition to these standard branes, there are solutions of codimensions less than or equal to two which are known as the exotic branes [4]. The exotic branes are characterized by U-duality monodromy groups.

On the other hand, there are relatively little known objects in DFT which are not solutions to the ordinary supergravity theory [5, 6, 7]. These non-standard solutions depend not only on the space-time coordinates x^μ but also on the string winding coordinates $\tilde{x}_\mu$. Such a kind

Journal of Physics: Conference Series **1416** (2019) 012017 doi:10.1088/1742-6596/1416/1/012017

of winding coordinates dependence on geometries has been studied in the contexts of string worldsheet instanton corrections [8]. It is known that the worldsheet instanton effects are well understood in terms of two-dimensional gauged linear sigma models (GLSMs) [9]. Some GLSMs are UV completions of string worldsheet non-linear sigma models (NLSMs) in specific target spaces. It was shown that the gauge instantons in a GLSM is reduced to the string worldsheet instantons in the IR limit [10]. A GLSM which flows into a NLSM whose target space is the exotic 5^2_2-brane geometry has been proposed in [11]. Based on this GLSM, the instanton corrections to the 5^2_2-brane geometry has been studied in [12]. The corrections to the geometry are characterized by the winding coordinates dependence on the harmonic function that governs the solution. In this contribution, we introduce five-brane solutions in DFT where string winding corrections are included. We show that the winding coordinates dependence on the solutions can be interpreted as string worldsheet instanton corrections.

The organization of this contribution is as follows. In the next section, we briefly introduce DFT. We then write down five-brane solutions of codimension four. Parts of these solutions have winding coordinates dependence. In section 3, we show that the winding coordinates dependence can be interpreted as worldsheet instanton corrections. In section 4, we discuss GLSMs where instanton corrections are analyzed. Section 5 is the conclusion.

2. Brief introduction of DFT and its solutions

2.1. Double Field Theory (DFT)

The dynamical fields in DFT are the generalized metric $\mathcal{H}_{MN}$ and the generalized dilaton d. They are fields in the doubled space-time and parametrized by

$$\mathcal{H}_{MN} = \begin{pmatrix} g^{\mu\nu} & -g^{\mu\rho}B_{\rho\nu} \\ B_{\mu\rho}g^{\rho\nu} & g_{\mu\nu} - B_{\mu\rho}g^{\rho\sigma}B_{\sigma\nu} \end{pmatrix}, \qquad e^{-2d} = \sqrt{-g}\, e^{-2\phi}, \tag{1}$$

where $g_{\mu\nu}$, $B_{\mu\nu}$ are the metric and the NSNS B-fields and ϕ is the dilaton. The DFT action is given by [13]:

$$S_{\text{DFT}} = \int d^{2d}X\, e^{-2d}\, \mathcal{R}(\mathcal{H}, d). \tag{2}$$

Here $\mathcal{R}$ is called the generalized Ricci scalar defined by

$$\begin{aligned}\mathcal{R} = {} & 4\mathcal{H}^{MN}\partial_M\partial_N d - \partial_M\partial_N\mathcal{H}^{MN} - 4\mathcal{H}^{MN}\partial_M d\partial_N d + 4\partial_M\mathcal{H}^{MN}\partial_N d \\ & + \frac{1}{8}\mathcal{H}^{MN}\partial_M\mathcal{H}^{KL}\partial_N\mathcal{H}_{KL} - \frac{1}{2}\mathcal{H}^{MN}\partial_M\mathcal{H}^{KL}\partial_K\mathcal{H}_{NL}.\end{aligned} \tag{3}$$

The indices M, N are raised and lowered by the $O(d,d)$ invariant metric η_{MN} and its inverse η^{MN} defined by

$$\eta_{MN} = \begin{pmatrix} 0 & \delta^\mu{}_\nu \\ \delta_\mu{}^\nu & 0 \end{pmatrix}, \qquad \eta^{MN} = \begin{pmatrix} 0 & \delta_\mu{}^\nu \\ \delta^\mu{}_\nu & 0 \end{pmatrix}. \tag{4}$$

The equations of motion are derived from the variations of d and $\mathcal{H}_{MN}$:

$$\delta d: \ \mathcal{R} = 0, \qquad \delta\mathcal{H}^{MN}: \ P_{MN}{}^{KL}\mathcal{K}_{KL} = 0. \tag{5}$$

Here $\mathcal{K}_{MN}$ is defined by

$$\begin{aligned}\mathcal{K}_{MN} = {} & \frac{1}{8}\partial_M\mathcal{H}^{KL}\partial_N\mathcal{H}_{KL} - \frac{1}{4}(\partial_L - 2\partial_L d)\mathcal{H}^{KL}\partial_K\mathcal{H}_{MN} + 2\partial_M\partial_N d \\ & - \frac{1}{2}\partial_{(M}\mathcal{H}^{KL}\partial_L\mathcal{H}_{N)K} + \frac{1}{2}(\partial_L - 2\partial_L d)\left[\mathcal{H}^{KL}\partial_{(M}\mathcal{H}_{N)K} + \eta^{KP}\eta^{LQ}\mathcal{H}_{P(M}\partial_K\mathcal{H}_{N)Q}\right].\end{aligned} \tag{6}$$

The symbol P denotes a projection operator given by

$$P_{MN}{}^{KL} = \frac{1}{2}\left[\delta_M{}^{(K}\delta_N{}^{L)} - \mathcal{H}_{MP}\eta^{P(K}\eta_{NQ}\mathcal{H}^{L)Q}\right], \tag{7}$$

where the symmetrization is defined by $A_{(M}B_{N)} = (A_M B_N + A_N B_M)/2$.

2.2. Five-brane solutions in DFT

In the following, we introduce five-brane solutions of codimension four. Solutions of other codimensions of five-branes and wave can be found in our paper [1].

We split the generalized coordinates to $X^M = (\bar{x}_m, \bar{y}_a; x^m, y^a)$. Here x^m $(m = 0, \ldots, 5)$ and y^a $(a = 6, \ldots, 9)$ are the coordinates for the worldvolume of the five-branes and the transverse space, respectively. Their T-dual coordinates are denoted by $(\bar{x}_m, \bar{y}_a)$. Note that (x^m, y^a), $(\bar{x}_m, \bar{y}_a)$ are not necessarily the geometrical and winding coordinates. We will assign their roles when we specify the solutions.

We consider the following ansatz:

$$\begin{aligned} ds^2_{\mathrm{DFT}} &= H(\delta_{ab} - H^{-2}b_{ac}b^c{}_b)dy^a dy^b + H^{-1}\delta^{ab}d\bar{y}_a d\bar{y}_b + 2H^{-1}b_a{}^b dy^a d\bar{y}_b \\ &\quad + \eta_{mn}dx^m dx^n + \eta^{mn}d\bar{x}_m d\bar{x}_n, \\ d &= \text{const.} - \frac{1}{2}\log H. \end{aligned} \tag{8}$$

Here the line element in DFT is defined by $ds^2_{\mathrm{DFT}} = \mathcal{H}_{MN}dX^M dX^N$. $H = H(y^a)$ is a harmonic function satisfying $\Box H = 0$. Here $\Box$ is the four-dimensional Laplacian defined in the y^a-directions and b_{ab} satisfies the following condition:

$$3\partial_{[a}b_{bc]} = \varepsilon_{abcd}\partial^d H(y). \tag{9}$$

Here ε_{abcd} is the totally anti-symmetric symbol in the four transverse dimensions. This ansatz is nothing but the one for the localized DFT monopole solution [6].

We first choose the geometrical coordinates $x^\mu = (x^m, y^i, y^9)$ $(i = 6, 7, 8)$. Then we obtain the following solution:

$$\begin{aligned} ds^2 &= \eta_{mn}dx^m dx^n + H(\delta_{ij}dy^i dy^j + (dy^9)^2), \\ B &= (b_{ij}dy^i - b_{j9}dy^9) \wedge dy^j, \\ e^{2\phi} &= H, \end{aligned} \tag{10}$$

where $ds^2 = g_{\mu\nu}dx^\mu dx^\nu$, $H(y^i, y^9) = c + \frac{Q}{r^2}$, $r^2 = (y^i)^2 + (y^9)^2$ and c, Q are constants. B and ϕ are the NSNS B-field and the dilaton in supergravity. The solution (10) is nothing but the NS5-brane in type II supergravity.

Since the $O(d,d)$ T-duality symmetry is manifestly realized in DFT, we can obtain another solution by applying an $O(d,d)$ transformation to (10). An $O(d,d)$ transformation changes the role of the geometrical and winding coordinates. For example, by choosing a different set of the coordinates $x^\mu = (x^m, y^i, \bar{y}_9)$ as the geometrical ones, we obtain the following new solution:

$$\begin{aligned} ds^2 &= \eta_{mn}dx^m dx^n + H^{-1}(d\bar{y}_9 + b_{i9}dy^i)^2 + H\delta_{ij}dy^i dy^j, \\ B &= b_{ij}dy^i \wedge dy^j, \\ e^{2\phi} &= \text{const.} \end{aligned} \tag{11}$$

Here, again the harmonic function $H = H(y^i, y^9)$ is given by the one in the NS5-brane but now one of the coordinates y^9 is *not* geometrical but replaced by the T-dualized *winding* coordinate.

We therefore denote it as $y^9 = \tilde{y}_9$. Since the solution (11) depends on the winding coordinate, it is not a solution to supergravity but it is still a solution to DFT. The solution (11) is first obtained in [6]. When the harmonic function is smeared along the $\tilde{y}_9$-direction, it becomes $H(y^i, \tilde{y}_9) \to H(y^i) = c + \frac{Q'}{r'}$ and the solution is reduced to the ordinary KK-monopole (Taub-NUT). Here Q' is a constant and $r'^2 = (y^i)^2$. Therefore the solution (11) is interpreted as the KK-monopole with *winding corrections.*

We further proceed the above procedure. Choosing another set of the geometrical coordinates $x^\mu = (x^m, y^\alpha, \tilde{y}_8, \tilde{y}_9)$, we obtain the following solution:

$$\begin{aligned}
ds^2 &= \eta_{mn}\, dx^m dx^n + H\delta_{\alpha\beta}\, dy^\alpha dy^\beta \\
&\quad + \frac{H}{H^2 + b_{89}^2}\left[(d\tilde{y}_9 + b_{\alpha 9}dy^\alpha)^2 + (d\tilde{y}_8 + b_{\alpha 8}dy^\alpha)^2\right], \\
B &= b_{\alpha\beta}\, dy^\alpha \wedge dy^\beta - \frac{b_{89}}{H^2 + b_{89}^2}\Big[(d\tilde{y}_8 + b_{\alpha 8}dy^\alpha) \wedge (d\tilde{y}_9 + b_{\beta 9}dy^\beta)\Big], \\
e^{2\phi} &= \frac{H}{H^2 + b_{89}^2},
\end{aligned} \tag{12}$$

where $\alpha = 6, 7$. Once again, the harmonic function $H = H(y^\alpha, y^8, y^9)$ is given by the one in the NS5-brane but two of the coordinates ($y^8 = \tilde{y}_8, y^9 = \tilde{y}_9$) are replaced by the winding coordinates. If the solution is smeared along the $\tilde{y}_8, \tilde{y}_9$-directions, this is reduced to that of the exotic 5_2^2-brane. Therefore the solution (12) is the exotic 5_2^2-brane with winding corrections along T^2. This solution is first obtained in our paper [1].

Now it is straightforward to proceed our discussion further. By choosing $x^\mu = (x^m, y^6, \tilde{y}_{\hat{\imath}})$ ($\hat{\imath} = 7, 8, 9$) as the geometrical coordinates, we obtain the following solution:

$$\begin{aligned}
ds^2 &= \eta_{mn}\, dx^m dx^n + H(dy^6)^2 \\
&\quad + K_2^{-1}\left[H^2(d\tilde{y}_{\hat{k}} + b_{6\hat{k}}dy^6)^2 + \left(\frac{1}{2}\varepsilon^{\hat{\imath}\hat{\jmath}\hat{k}} b_{\hat{\imath}\hat{\jmath}}(d\tilde{y}_{\hat{k}} + b_{6\hat{k}}dy^6)\right)^2\right], \\
B &= -\frac{Hb_{\hat{\imath}\hat{\jmath}}}{K_2}(d\tilde{y}_{\hat{\imath}} + b_{6\hat{\imath}}dy^6) \wedge d\tilde{y}_{\hat{\jmath}}, \\
e^{2\phi} &= HK_2^{-1}, \\
K_2 &\equiv H(H^2 + b_{89}^2 + b_{79}^2 + b_{78}^2).
\end{aligned} \tag{13}$$

The harmonic function in this solution is $H = H(y^6, \tilde{y}^7, \tilde{y}^8, \tilde{y}^9)$ where only y^6 is the geometric coordinate while $\tilde{y}_7, \tilde{y}_8, \tilde{y}_9$ are the winding coordinates. When two of the three winding directions are smeared, this is denoted as the 5_2^3-brane and commonly known as the R-brane. We note that it is not possible to obtain a geometric five-brane in this frame by smearing all the winding directions $\tilde{y}_7, \tilde{y}_8, \tilde{y}_9$. Even though the smearing procedure affects the harmonic function, the condition (9) implies that the solution becomes trivial. Therefore the R-brane is a truely non-geometric object.

We now consider the final possibility. By choosing yet another different set of the geometrical coordinates $x^\mu = (x^m, \tilde{y}_a)$ $(a = 6, 7, 8, 9)$, we obtain

$$\begin{aligned}
ds^2 &= \eta_{mn}\, dx^m dx^n + \frac{H}{K_3}\left[\left(H^2 + \frac{1}{2}b_{cd}b^{cd}\right)\delta^{ab} - \delta_{cd}\, b^{ca}b^{db}\right] d\tilde{y}_a d\tilde{y}_b, \\
B &= K_3^{-1}\left[b^{cd}b_{ca}b_{db} - \left(H^2 + \frac{1}{2}b_{cd}b^{cd}\right) b_{ab}\right] d\tilde{y}_a \wedge d\tilde{y}_b, \\
e^{2\phi} &= HK_3^{-1}, \\
K_3 &\equiv H^4 + \frac{1}{2}H^2\, b_{ab}b^{ab} + \left(\frac{1}{8}\varepsilon^{abcd}b_{ab}b_{cd}\right)^2. \qquad (14)
\end{aligned}$$

This is denoted as the 5_2^4-brane solution. The harmonic function in this solution is $H = H(\tilde{y}_6, \tilde{y}_7, \tilde{y}_8, \tilde{y}_9)$ where all the coordinates represent the winding directions. If we smear all these directions, the solution becomes that of codimension zero, a space-filling five-brane in six dimensions.

3. Worldsheet instanton effects

In this section, we discuss a physical interpretation of the winding coordinates dependence on the DFT solutions. To this end, we focus on the harmonic function H appearing in the solutions. The harmonic function of the NS5-brane solution is

$$H(r) = c + \frac{Q}{r^2}, \qquad r^2 = (x^6)^2 + (x^7)^2 + (x^8)^2 + (x^9)^2. \qquad (15)$$

Since the T-duality transformation is performed in a compact space direction, we compactify the x^9-direction to a circle S^1 whose radius is R_9. Then the solution becomes a periodic array of the NS5-branes. The harmonic function becomes

$$H(r', x^9) = c + \sum_{n=-\infty}^{\infty} \frac{Q}{(r')^2 + (x^9 - 2\pi R_9 n)^2}, \qquad (r')^2 = (x^6)^2 + (x^7)^2 + (x^8)^2. \qquad (16)$$

By using the Poisson resummation formula

$$\sum_{n=-\infty}^{\infty} f(2\pi n) = \frac{1}{2\pi}\sum_{n=-\infty}^{\infty}\int_{-\infty}^{\infty} dt\ f(t)e^{int}, \qquad (17)$$

we find

$$\begin{aligned}
H(r', x^9) &= c + \frac{Q}{2R_9}\frac{1}{r'}\left[1 + \sum_{n=1}^{\infty} e^{\left(i\frac{x^9}{R_9} - \frac{r'}{R_9}\right)n} + \sum_{n=1}^{\infty} e^{\left(-i\frac{x^9}{R_9} - \frac{r'}{R_9}\right)n}\right] \qquad (18) \\
&= c + \frac{Q}{2R_9}\frac{1}{r'}\frac{\sinh(r'R_9^{-1})}{\cosh(r'R_9^{-1}) - \cos(x^9R_9^{-1})}. \qquad (19)
\end{aligned}$$

We note that the absolute convergences in both sides of the equation is needed for the use of the Poisson resummation formula (17). If the summation is divergent, we need to justify the formula by subtracting divergent pieces in the summation. This is the case in the following.

It is known that the H-monopole solution is obtained by smearing the NS5-brane along the x^9-direction. In (19), one observes that the relevant term of the H-monopole solution comes from the $n = 0$ sector. On the other hand, the $n \neq 0$ KK-modes break the isometry along the x^9-direction. Let us consider the $n \neq 0$ corrections from the T-dualized, namely, the KK-monopole side. They are obtained by interchanging x^9 and $\tilde{x}_9$. The result is

$$H(r', \tilde{x}_9) = c + \frac{Q'}{r'} \left[1 + \sum_{n \neq 0} \exp\left(in\frac{\tilde{x}_9}{\tilde{R}_9} - r' \left| \frac{n}{\tilde{R}_9} \right| \right) \right]. \tag{20}$$

Here $\tilde{R}_9$ is the dual radius of R_9. The expression (20) is nothing but the one for the geometry of KK-monopole with the string worldsheet instanton corrections [8].

We proceed the second T-duality transformation. The x^8- and x^9-directions are compactified to a torus T^2. Then the harmonic function becomes

$$H(\rho, \tilde{x}_8, \tilde{x}_9) = c + \sum_{m=-\infty}^{\infty} \sum_{n=-\infty}^{\infty} \frac{Q}{\rho^2 + (\tilde{x}_8 - 2\pi\tilde{R}_8 m)^2 + (\tilde{x}_9 - 2\pi\tilde{R}_9 n)^2}, \tag{21}$$

where $\rho^2 = (x^6)^2 + (x^7)^2$. Since the summation (21) diverges, we regularize the infinity and subtract the corresponding terms. It can be shown that the divergent pieces come from $n = m = 0$ parts [1]. Subtracting these terms and using the Poisson resummation formula (17), we find

$$H(\rho, \tilde{x}_8, \tilde{x}_9) = \frac{Q}{2\pi\tilde{R}_8\tilde{R}_9} \left[\log\frac{\mu}{\rho} + \sum_{n,m \neq (0,0)} e^{im\frac{\tilde{x}_8}{\tilde{R}_8}} e^{in\frac{\tilde{x}_9}{\tilde{R}_9}} K_0 \left(\rho\sqrt{\left(\frac{m}{\tilde{R}_8}\right)^2 + \left(\frac{n}{\tilde{R}_9}\right)^2} \right) \right], \tag{22}$$

where $K_0(x)$ is the modified Bessel function of the second kind and μ is a constant. If we consider the smearing by the limit $\tilde{R}_8, \tilde{R}_9 \to 0$ with $Q/2\pi\tilde{R}_8\tilde{R}_9$ fixed, the $n, m \neq 0$ terms in (22) vanish and H becomes the harmonic function for the exotic 5^2_2-brane of codimension two. The logarithmic behavior of the harmonic function implies that the 5^2_2-brane is not well-defined as a stand-alone object. On the other hand, if we take the limit $\tilde{R}_8 \to \infty$ with $Q/2\pi\tilde{R}_8$ and $\tilde{R}_9$ fixed, we find

$$H(\rho, \tilde{x}_9) = \frac{Q}{2\pi\tilde{R}_8\tilde{R}_9} \left[\log\frac{\mu}{\rho} + \sum_{n \neq 0} e^{in\frac{\tilde{x}_9}{\tilde{R}_9}} K_0 \left(\rho \left| \frac{n}{\tilde{R}_9} \right| \right) \right]. \tag{23}$$

It is obvious that the $n \neq 0$ terms break isometry along the $\tilde{x}_9$-direction. Two of the present authors studied the worldsheet instanton effects to the 5^2_2-brane geometry [12]. We have shown that the worldsheet instanton corrections to the exotic 5^2_2-brane geometry induce the $n \neq 0$ isometry-breaking terms in (23). Therefore it is natural to expect that the $(n, m) \neq (0, 0)$ contributions in (22) also come from worldsheet instantons. Indeed, the large-ρ expansion of the modified Bessel function shows the following relation:

$$e^{im\frac{\tilde{x}_8}{\tilde{R}_8} + in\frac{\tilde{x}_9}{\tilde{R}_9}} K_0 \left(\rho\sqrt{\left(\frac{m}{\tilde{R}_8}\right)^2 + \left(\frac{n}{\tilde{R}_9}\right)^2} \right) \sim \exp\left[-\rho\sqrt{\left(\frac{m}{\tilde{R}_8}\right)^2 + \left(\frac{n}{\tilde{R}_9}\right)^2} + i\left(m\frac{\tilde{x}_8}{\tilde{R}_8} + n\frac{\tilde{x}_9}{\tilde{R}_9} \right) \right]. \tag{24}$$

Here only the relevant terms are extracted. The characteristic exponential behavior suggests that the winding corrections in (22) are interpreted as the instanton effects. We will discuss this issue in the next section.

Journal of Physics: Conference Series **1416** (2019) 012017 doi:10.1088/1742-6596/1416/1/012017

4. Instantons in GLSM

As we have mentioned, it is crucial to formulate a GLSM to capture the instanton nature of the winding dependence. Since the five-brane backgrounds discussed above are expected to preserve fractions of 32 supersymmetries in type II superstrings, it is desirable that the GLSM is formulated in supersymmetric contexts. In [2], based on the two-dimensional $\mathcal{N} = (2,2)$ superfield languages, we construct GLSMs for a defect NS5-brane, a KK-vortex and a 5^2_2-brane. These GLSMs are organized into a GLSM which we call semi-doubled GLSM. We have confirmed that the semi-doubled GLSM is reduced in the IR to a semi-doubled NLSM that encompasses NLSMs for the defect NS5-brane, the KK-vortex and the 5^2_2-brane simultaneously.

It is known that the instanton analysis is performed in a specific parameter regime of the model [10]. In this regime and the supersymmetric vacuum, we obtain truncated models of GLSMs. For example, we find the following truncated model for the 5^2_2-brane [2]:

$$\begin{aligned}\mathcal{L}_{5^2_2} &= \frac{1}{2e^2}F_{12}^2 + \frac{1}{2\hat{e}^2}\hat{F}_{12}^2 + i\sqrt{2}\,\tilde{\vartheta}_8 F_{12} + i\sqrt{2}\,\tilde{\vartheta}_9 \hat{F}_{12} \\ &\quad + |D_m q|^2 + |D_m p|^2 + \frac{e^2}{2}\Big(|q|^2 - \sqrt{2}\zeta_8\Big)^2 + \frac{\hat{e}^2}{2}\Big(|p|^2 - \sqrt{2}\zeta_9\Big)^2, \end{aligned} \tag{25}$$

where F_{mn}, $\tilde{F}_{mn}$ and p, q are the gauge field strengths and the scalar fields in the $U(1)^2$ vector multiplets. $e, \hat{e}$, $\zeta_{8,9}$ and $\tilde{\vartheta}_{8,9}$ are the gauge, the Fayet-Iliopoulos and the topological parameters. The parameters $\tilde{\vartheta}_{8,9}$ are interpreted as the $\tilde{x}_{8,9}$ winding coordinates of the T^2-fiber in the IR. The Lagrangian (25) represents the $U(1)^2$ Abelian-Higgs model in the critical couplings. This is a natural extension of the $U(1)$ Abelian-Higgs model appeared in the truncated GLSM for the H-monopole [10]. In (25), one observes that when the gauge fields $(A_m, \hat{A}_m)$ take values in the non-trivial topological sectors, the isometries along the $\tilde{x}_{8,9}$ winding directions are broken. The Bogomol'nyi completion of the Lagrangian reads

$$\begin{aligned}\mathcal{L}_{5^2_2} &= \frac{1}{2e^2}\Big\{F_{12} \pm e^2(|q|^2 - \sqrt{2}\zeta_8)\Big\}^2 + \big|(D_1 \pm iD_2)q\big|^2 \\ &\quad + \frac{1}{2\hat{e}^2}\Big\{\hat{F}_{12} \pm \hat{e}^2(|p|^2 - \sqrt{2}\zeta_9)\Big\}^2 + \big|(D_1 \pm iD_2)p\big|^2 \\ &\quad \pm \sqrt{2}\,\zeta_8 F_{12} + i\sqrt{2}\,\tilde{\vartheta}_8 F_{12} \pm \sqrt{2}\,\zeta_9 \hat{F}_{12} + i\sqrt{2}\,\tilde{\vartheta}_9 \hat{F}_{12}\,. \end{aligned} \tag{26}$$

Thus the Euclidean action $S_{\rm E} = \frac{1}{2\pi}\int d^2x \mathcal{L}_{5^2_2}$ is bounded from below:

$$S_{\rm E} \;\geq\; \sqrt{2}\,\zeta_8\,|n_8| - i\sqrt{2}\,\tilde{\vartheta}_8\, n_8 + \sqrt{2}\,\zeta_9\,|n_9| - i\sqrt{2}\,\tilde{\vartheta}_9\, n_9\,. \tag{27}$$

Here the integers (n_8, n_9) are the topological numbers defined by the first Chern number:

$$n_8 = -\frac{1}{2\pi}\int d^2x\, F_{12}, \qquad n_9 = -\frac{1}{2\pi}\int d^2x\, \hat{F}_{12}. \tag{28}$$

The Euclidean action is most stringently bounded by

$$S_{\rm E} \;\geq\; \sqrt{2}\,|\vec{\zeta}|\,\sqrt{\vec{n}^2} - i\sqrt{2}\,(\vec{\vartheta}\cdot\vec{n}) \;=\; S_{\rm inst}\,. \tag{29}$$

Here we have defined $\vec{\zeta} = (\zeta_8, \zeta_9)$, $\vec{n} = (n_8, n_9)$ and $\vec{\vartheta} = (\tilde{\vartheta}_8, \tilde{\vartheta}_9)$. The inequality is saturated when $\vec{\zeta} \parallel \vec{n}$ and the following BPS equations are satisfied:

$$\begin{aligned} F_{12} &= \mp e^2\Big(|q|^2 - \sqrt{2}\zeta_8\Big), \quad (D_1 \pm iD_2)q = 0, \\ \hat{F}_{12} &= \mp \hat{e}^2\Big(|p|^2 - \sqrt{2}\zeta_9\Big), \quad (D_1 \pm iD_2)p = 0. \end{aligned} \tag{30}$$

This is a pair of the Abrikosov-Nielsen-Olesen (ANO) vortex equations and the solutions to these equations are nothing but the instantons in two dimensions.

The path-integral of the GLSM for the 5^2_2-brane around the classical solutions to (30) results in the broken isometries along the $\tilde{x}_{8,9}$-directions in the IR geometry. If the topological numbers (n_8, n_9) are identified with the winding-modes in T^2, the instanton corrections to the harmonic function should be proportional to the factor $e^{-S_{\rm inst}}$ where the instanton action is given in (29). This is indeed the case. After identifying appropriate parameters, we find that the instanton contributions $e^{-S_{\rm inst}}$ precisely match with the expression appeared in the DFT solution (24) [2]. The same is essentially true for the defect NS5-brane and the KK-vortex.

5. Conclusion

In this contribution, we discuss five-brane solutions in DFT. In particular, we have shown explicit five-brane solutions of codimension four. Solutions of other codimensions are found in our paper [1]. A remarkable fact about the solutions is that they depend on the string winding coordinates. The geometries corresponding to these solutions are interpreted as supergravity backgrounds corrected by string worldsheet instantons.

In order to understand the worldsheet instanton effects, it is useful to find a gauged linear sigma model that flows into a non-linear sigma model whose target space is a five-brane in IR. We have proposed a doubled GLSM that reproduces the defect NS5-brane, the KK-vortices and the 5^2_2-brane backgrounds appropriately. We call this the semi-doubled GLSM [2]. We have shown that the truncated model gives rise to the $U(1)^2$ Abelian-Higgs model whose BPS condition is the ANO vortex equations. This is nothing but the instantons in two dimensions. The instanton effects are interpreted as a generalization of disk instantons in the IR [1]. This effect precisely reproduces the nature of the DFT solutions.

One can proceed to the cases with more isometries, for example, the R-brane. The discussion is completely parallel. For example, the geometry of the 5^3_2-branes includes T^3-fibration over $\mathbb{R}$. Indeed, we can define a generalization of the disk map and we deduce that instanton action agrees with the exponential factor appeared in the solution. The same is true for the DFT wave solutions. However, this picture fails when we move to space-filling branes. Even more, it is not straightforward to construct a corresponding GLSM for the R-brane. It is interesting to find GLSMs for the R-brane and the space-filling branes and study the instanton effects to these branes. It is also interesting to study the winding solutions even in the exceptional field theory (EFT) which is a U-duality generalization of DFT.

Acknowledgments

The work of S.S. is supported by the Japan Society for the Promotion of Science (JSPS) KAKENHI Grant Number **JP17K14294**. The work of K.S. is supported by the Sasakawa Scientific Research Grant (No. 2019-2027) from the Japan Science Society.

References

[1] Kimura T, Sasaki S and Shiozawa K 2018 *JHEP* **1807** 001
[2] Kimura T, Sasaki S and Shiozawa K 2018 *JHEP* **1812** 095
[3] Hull C and Zwiebach B 2009 *JHEP* **0909** 099
[4] Obers N A and Pioline B 1999 *Phys. Rept.* **318** 133
[5] Berkeley J, Berman D S and Rudolph F J 2014 *JHEP* **1406** 006
[6] Berman D S and Rudolph F J 2015 *JHEP* **1505** 015
[7] Bakhmatov I, Kleinschmidt A and Musaev E T 2016 *JHEP* **1610** 076
[8] Harvey J A and Jensen S 2005 *JHEP* **0510** 028
[9] Witten E 1993 *Nucl. Phys.* B **403** 159
[10] Tong D 2002 *JHEP* **0207** 013
[11] Kimura T and Sasaki S 2013 *Nucl. Phys.* B **876** 493
[12] Kimura T and Sasaki S 2013 *JHEP* **1308** 126
[13] Hohm O, Hull C and Zwiebach B 2010 *JHEP* **1008** 008

Journal of Physics: Conference Series **1416** (2019) 012018 doi:10.1088/1742-6596/1416/1/012018

Open problems in the Kontsevich graph construction of Poisson bracket symmetries

Arthemy V Kiselev

Bernoulli Institute for Mathematics, Computer Science & Artificial Intelligence, University of Groningen, P.O. Box 407, 9700 AK Groningen, The Netherlands

E-mail: a.v.kiselev@rug.nl

Abstract. Poisson brackets admit infinitesimal symmetries which are encoded using oriented graphs; this construction is due to Kontsevich (1996). We formulate several open problems about combinatorial and topological properties of the graphs involved, about integrability and analytic properties of such symmetry flows (in particular, for known classes of Poisson brackets), and about cohomological, differential geometric, and quantum aspects of the theory.

1. Introduction. Whenever a (non)linear ODE system in a finite-dimensional model from celestial or rigid body mechanics, optimisation and control, optics and photonics, (bio)chemistry, or molecular biology is Poisson, $\frac{\mathrm{d}}{\mathrm{d}t}\boldsymbol{x} = \{\boldsymbol{x}, H\}_{\mathcal{P}}$, the problem of deforming this model into a continuous family (along a new parameter $\varepsilon \in \mathcal{I} \subseteq \mathbb{R}$) can be seen in two different ways. By modifying the Hamiltonian function H, one changes the available energy which is transformed into model's evolution by the Poisson bracket. On the other hand, by keeping the Hamiltonian unchanged but by shifting the Poisson bracket $\{\cdot,\cdot\}_{\mathcal{P}}$ into a family $\{\cdot,\cdot\}_{\mathcal{P}(\varepsilon)}$, one changes the algorithm how the given energy store is processed into dynamics and integrated over the time t into trajectories. A particular case of the latter is the infinitesimal shift $\mathcal{P} \mapsto \mathcal{P}+\varepsilon[\![\mathcal{P},\mathfrak{X}]\!]+\bar{o}(\varepsilon)$ of the Poisson bi-vector $\mathcal{P}$ (hence the bracket) by taking its own Schouten bracket with a globally defined (at all points $\boldsymbol{x} \in M^r$ of the model configuration space) vector field $\mathfrak{X}$. If so, the 'novelty' of the shifted Poisson structure is fictitious because this shift of $\mathcal{P}$ in fact amounts to an infinitesimal diffeomorphism of M^r along the integral trajectories of $\mathfrak{X}$. Integrating to finite ε from $\varepsilon_0 = 0$, one obtains a family of descriptions for the same Poisson model by using new variables $\mathbf{x}^{(}\varepsilon)$ instead of the old $\boldsymbol{x}$; neither the Hamiltonian H nor the bracket $\{\cdot,\cdot\}_{\mathcal{P}}$ would then change in earnest. Apart from all of the above, the scaling $\frac{\mathrm{d}}{\mathrm{d}\varepsilon}\mathcal{P} = \mathcal{P}$, which yields $\mathcal{P}(\varepsilon) = \exp(\varepsilon)\cdot\mathcal{P}\big|_{\varepsilon_0=0}$, accelerates or slows down the time pace (for $\varepsilon < 0$ and $\varepsilon > 0$, respectively).

We are interested in finding infinitesimal symmetries $\mathcal{Q}(\mathcal{P}) \in \mathrm{sym}\big([\![\mathcal{P},\mathcal{P}]\!] = 0\big)$ of the Jacobi identity $\frac{1}{2}[\![\mathcal{P},\mathcal{P}]\!] = 0$ for classical Poisson bi-vectors $\mathcal{P}$ such that the symmetries $\mathcal{Q}$ are (i) nonlinear in $\mathcal{P}$ (otherwise it would be the time rescaling), (ii) proper, $\mathcal{Q}(\mathcal{P}) \not\equiv 0$, in the sense that the shift of $\mathcal{P}$ to $\mathcal{P} + \varepsilon\mathcal{Q}(\mathcal{P}) + \bar{o}(\varepsilon)$ is not everywhere vanishing on M^r as soon as $\mathcal{P}$ is Poisson and $[\![\mathcal{P},\mathcal{P}]\!] = 0$, and (iii) universal w.r.t. all real (or complex) finite-dimensional affine Poisson manifolds $(M^{r<\infty}_{\text{aff}}, \mathcal{P})$. The Hamiltonian we keep unchanged.

It is standard that infinitesimal symmetries $\mathcal{Q}$ of Poisson brackets $\{\cdot,\cdot\}_{\mathcal{P}}$ are $\partial_{\mathcal{P}}$-cocycles w.r.t. the Poisson differential $\partial_{\mathcal{P}} = [\![\mathcal{P},\cdot]\!]$, whence every coboundary $\partial_{\mathcal{P}}(\mathfrak{X}) \in \ker\partial_{\mathcal{P}}$ is a cocycles for all vectors $\mathfrak{X}$ but maybe not *vice versa*. The (non)triviality of infinitesimal symmetries $\mathcal{Q}$ as the respective Poisson cocycles, i.e. the (non)solvability of the coboundary equation $\mathcal{Q}(\mathcal{P}) = [\![\mathcal{P},\mathcal{X}]\!] + \Diamond\big(\mathcal{P},[\![\mathcal{P},\mathcal{P}]\!]\big)$ for $\mathcal{X}$ and $\Diamond$ is a separate question (which can be addressed after such a symmetry $\mathcal{Q}$ is obtained).

A universal construction of nonlinear proper infinitesimal symmetries $\dot{\mathcal{P}} = \mathcal{Q}(\mathcal{P})$ by using oriented graphs, as well as by using the Lie algebra morphism $\mathrm{O\vec{r}}$ which takes suitable cocycles $\gamma \in \ker[\bullet\!\!-\!\!\bullet,\cdot]$ in the unoriented graph complex to Poisson cocycles $\mathcal{Q}(\mathcal{P}) = \mathrm{O\vec{r}}(\gamma)(\mathcal{P}) \in \ker[\![\mathcal{P},\cdot]\!]$, were designed by Kontsevich in [22]. (See [4] and references therein for a detailed analysis why the graph orientation morphism $\mathrm{O\vec{r}}$ actually works.) Examples of this construction for known graph cocycles $\gamma_{2\ell+1}$ on n vertices and $2n-2$ edges (namely, $n = 2\ell+2 = 4$, 6, and 8) have been given in [2, 3], [7], and [4], respectively. Practical calculation of graph cocycles is addressed in [8, 30]; the algorithms to verify the Poisson cocycle factorisation through the Jacobi identity are available from [9]. In the fundamental work [29], see also [27], Willwacher related the unoriented graph complex to generators of the Grothendieck–Teichmüller Lie algebra $\mathfrak{grt}$. In what follows, we expect some familiarity with the subject, e.g., on the basis of [22, 24] and [2, 4, 9, 19, 28] or the lectures [18] available on-line from the Steklov MI RAS; the notation is standard.

The aim of this paper is to bring together a set of open problems which relate to and interconnect various properties of the universal infinitesimal symmetries for Poisson brackets and of the unoriented graph cocycles that determine a class of such symmetries under the orientation morphism.

2. Initial datum: affine Poisson manifold $\big(M^r_{\text{aff}},\mathcal{P}\big)$.

The Kontsevich construction [22] works for every finite-dimensional affine Poisson manifold M^r_{aff}. Although the affine transition functions between local coordinate charts amount to the simplest possible linear operators and translations (consider the local coordinate 'angle' on the circle $\mathbb{S}^1 = \mathbb{R}/2\pi\mathbb{Z}$), the manifold M^r_{aff} can of course be topologically nontrivial (like $\mathbb{S}^1$). Moreover, the choice of an affine atlas on an even-dimensional manifold with a nondegenerate Poisson structure can prevent the attainability of Darboux coordinates. (Indeed, that would likely require a nonlinear change of variables.) On odd-dimensional Poisson manifolds, there cannot be any Darboux coordinates. This is important because the Kontsevich symmetries $\mathcal{Q}(\mathcal{P})$ are typically equal to zero at Poisson structures $\mathcal{P}$ with constant or very low polynomial degree coefficients (e.g., at the Kirillov–Kostant linear brackets or Sklyanin's brackets with quadratic coefficients), which is in contrast with the high-degree Nambu construction in §6 below (cf. [12]).

Problem 1. Describe ample classes of finite-dimensional *affine* Poisson manifolds $\big(M^r_{\text{aff}},\mathcal{P}\big)$.

Remark 1. The globalisation problem of extending the orgraph construction of a given Kontsevich symmetry $\mathcal{Q}(\mathcal{P}) \in \mathrm{sym}([\![\mathcal{P},\mathcal{P}]\!] = 0)$ to a *smooth* atlas requires extra structures and their correlation with the Poisson bracket (e.g., an affine connection ∇ on M^r and invariance of the bi-vector $\mathcal{P}$ with respect to it). The formula of extension $\dot{\mathcal{P}} = \widehat{\mathcal{Q}}(\nabla)(\mathcal{P})$ would typically involve the curvature terms (which vanish while the manifold is affine).

Problem 2. Find the smooth-atlas globalisation(s) of the Kontsevich symmetries $\mathcal{Q}_{1:\frac{6}{2}}(\mathcal{P}) = \mathrm{O\vec{r}}(\gamma_3)(\mathcal{P})$, $\mathcal{Q}_5(\mathcal{P}) = \mathrm{O\vec{r}}(\gamma_5)(\mathcal{P})$, and $\mathcal{Q}_7(\mathcal{P}) = \mathrm{O\vec{r}}(\gamma_7)(\mathcal{P})$, found in [2], [7], and [4] respectively.

3. Why are the graph cocycles so special ?

3.1. Are the graph cocycles truly graphs ?

In the papers [21] (see also [22] and [4, 8, 28] for a pedagogical review), Kontsevich introduced the graph complex – one of the many – with parity-even vertices, with a wedge ordering of parity-odd edges, and the differential $\mathrm{d} = [\bullet\!\!-\!\!\bullet,\cdot]$ produced by the graded commutator of graph insertions into vertices. This direction of research was furthered by Willwacher *et al.* [11, 14, 29]: in particular, in [30] a generating function counts the numbers of nonzero (i.e. not equal to minus itself) unoriented graphs with respect to their bi-grading by the vertex-edge numbers. Nontrivial unoriented graph cocycles can be built using these – partly relevant but largely irrelevant – graphs. Yet we wee that the cocycles, which are few compared with the total graph numbers, are clearly non-generic.

Problem 3. As soon as the vertex number n is large enough, are the combinatorial properties of graphs γ_a in a cocycle $\gamma = \sum_a c_a \cdot \gamma_a$ well defined, i.e. stable over the entire equivalence class $[\gamma] = [\gamma + \mathrm{d}(\lambda)]$ modulo arbitrary coboundaries from graphs λ on $n-1$ vertices? The properties to examine can be the distribution of diameters (over the set of graphs $\{\gamma_a\}$ forming the cocycle $\gamma = \sum_a c_a \cdot \gamma_a$) and distribution of vertex valencies.
• What is the large-n and asymptotic ($n \to \infty$) behaviour of the graph cohomology dimension (primarily, on or near the ray of vertex-edge bi-gradings $(n, 2n-2)$, which is relevant to the Poisson bracket symmetries); are the dimensions encoded by their own generating function?

Let us keep in mind that the realisation of a certain differential complex using unoriented graphs – visibly non-generic but boiled down to very low dimensional subspaces – might actually veil the true nature of cocycles if the chosen realisation is 'incidental'.

3.2. Wheels first, what next? We recall the established Lie algebra isomorphism [29] between the unoriented graph cohomology group of our interest and the Grothendieck–Teichmüller Lie algebra $\mathfrak{grt}$ introduced by Drinfeld. (On the other hand, the dgLa of these graph cocycles is mapped by the orientation homomorphism $\mathrm{O\vec{r}}(\cdot)(\mathcal{P})$ to the Lie algebra of Poisson bracket symmetries endowed with the Poisson differential $[\![\mathcal{P}, \cdot]\!]$.) Certain integral formulae relating graph cocycles to the $\mathfrak{grt}$ generators, known explicitly in low orders, are offered in [27, §6]. The composition of a cocycle in a given bi-grading, including the structure of the wheel cocycles related to the $\mathfrak{grt}$ Lie algebra, is in principle unknown.

Problem 4. Illustrate the work of Willwacher's isomorphism $\mathfrak{grt} \simeq H^0(\mathrm{Gra})$ at small vertex numbers $n = 4, 6, 8, 9, 10, \ldots$, where there are nontrivial cocycles in Gra.

3.3. Are there 'many more' symmetries of $\{\cdot,\cdot\}_{\mathcal{P}}$ than connected graph cocycles? Both the constructions of the vertex-blow-up differential $\mathrm{d} = [\bullet\!\!-\!\!\bullet, \cdot]$ and graph orientation morphism $\gamma = \sum_a c_a \cdot \gamma_a \longmapsto \dot{\mathcal{P}} = \mathrm{O\vec{r}}(\gamma)(\mathcal{P})$ are so robust that the connectedness assumption for the graphs γ_a without tadpoles is in fact excessive, although it is ofted adopted in the literature.[1] For instance, whenever $\gamma^1, \ldots, \gamma^k \in \ker \mathrm{d}$ are graph cocycles (not necessarily each on n^i vertices and exactly $\#E(\gamma^i) = 2n^i - 2$ edges) such that $\sum_i n^i = n$ and $\sum_i \#E(\gamma^i) = 2n-2$, then the formal sum $\gamma = \bigsqcup_i \gamma^i$ of disconnected graphs, where the disjoint union $\sqcup$ is linear over each sum $\gamma^i = \sum_a c_a^i \cdot \gamma_a^i$, itself is a graph cocycle, $\gamma \in \ker \mathrm{d}$, but now on n vertices and precisely $2n-2$ edges, so that the orientation morphism $\mathrm{O\vec{r}}$ yields a new universal symmetry of Poisson brackets. The differential-polynomial expression encoding that symmetry $\mathcal{Q}(\mathcal{P})$ is the skew-symmetrised (w.r.t. the bi-vector arguments) product of expressions each stemming from the respective cocycle γ^i.

Corollary 1. Every Kontsevich's symmetry $\mathcal{Q}(\mathcal{P}) = \mathrm{O\vec{r}}(\gamma)(\mathcal{P})$ can be multiplied by an arbitrary power of differential polynomial $\mathrm{O\vec{r}}(\omega)(\mathcal{P})$, where $\omega \in \ker \mathrm{d}$ is a graph cocycle on N vertices and exactly $2N$ edges.

The graph number Table 2(even•) in [30] (cf. [8, Tab. 2–3]) shows that cells along the ray $(\#V, \#E) = (N, 2N)$ are populated by nonzero graphs starting at $N \geqslant 7$.

Problem 5. Starting which $N' \geqslant 7$, are there nontrivial cocycles $\omega \in \ker \mathrm{d}$? And starting which $N'' \geqslant N'$, are these cocycles orientable (i.e. meeting no obstructions [9, App. A]) to nonzero Kontsevich's graphs built from wedges $\leftarrow\bullet\rightarrow$? Give explicit examples of new symmetries $\dot{\mathcal{P}} = \big(\mathrm{O\vec{r}}(\omega)(\mathcal{P})\big)^{m\neq 0} \cdot \mathrm{O\vec{r}}(\gamma)(\mathcal{P})$ for known cocycles γ and new such ω on the ray $(N, 2N)$.

In the above constructions, we meet $m \in \mathbb{Z}$ cocycle factors in the disjoint product of graphs.

4. What are the topological identities in the spaces of Leibniz graphs?
The Poisson cocycle condition, $[\![\mathcal{P}, \mathcal{Q}(\mathcal{P})]\!] \doteq 0$ to hold by force of $[\![\mathcal{P}, \mathcal{P}]\!] = 0$, is satisfied by

[1] The other traditional such (over)assumption is the strong connectedness, i.e. the absence of bottlenecks.

Kontsevich's symmetries $\mathcal{Q}(\mathcal{P}) = \vec{\text{Or}}(\gamma)(\mathcal{P})$ because the graph orientation morphism $\vec{\text{Or}}$ yields an explicit solution $\Diamond(\mathcal{P}, [\![\mathcal{P},\mathcal{P}]\!]) = \sum_i \vec{\text{Or}}(\gamma)(\mathcal{P},\ldots,\mathcal{P}_{i-1},[\![\mathcal{P},\mathcal{P}]\!]_i,\mathcal{P}_{i+1},\ldots,\mathcal{P})$ for the problem $[\![\mathcal{P},\mathcal{Q}(\mathcal{P})]\!] = \Diamond(\mathcal{P},[\![\mathcal{P},\mathcal{P}]\!])$ of factorisation via the Jacobi identity for Poisson structures $\mathcal{P}$ (see [22] and [4]). We observed in [2, 7] that these realisations by the Leibniz graphs, with the Jacobiator inside, are *not unique*: there exist more solutions, $\Diamond = \Diamond_1 \neq \Diamond_2$, etc. In other words, there are identities of the form $\Diamond_2 - \Diamond_1 \equiv 0$ in the spaces of Leibniz graphs: when all the Leibniz rules for the derivatives falling on the Jacobiator are worked out and each Jacobiator expands to the standard sum of three terms, the expansions of initially different Leibniz graphs overlap, having common Kontsevich's orgraphs. (The reverse of this was the core idea in our algorithm to generate an Ansatz with Leibniz graphs for the r.-h.s. of factorisation problems $[\![\mathcal{P},\mathcal{Q}(\mathcal{P})]\!] = \Diamond(\mathcal{P},[\![\mathcal{P},\mathcal{P}]\!])$, see [9, §1.2].)

Problem 6. Let us say that two Leibniz graphs are adjacent if their expansions have a Kontsevich orgraph in common. Now, describe the topological structure of the identities $\Diamond_2 - \Diamond_1 \equiv 0$ for known non-unique solutions of the factorisation problems. Otherwise speaking, describe the meta-graphs with Leibniz graphs as vertices: what is their genus, diameter, typical cycle length, etc. ?

- Investigate the saturation curve for the number of Leibniz graphs produced by the iterative algorithm (e.g., see [9, Table 1]): at which stage are the new Leibniz graphs typically leaves vs handles? What are the last new Leibniz graphs in this sense?

5. Are the evolution equations $\dot{\mathcal{P}} = \mathcal{Q}(\mathcal{P})$ integrable?

The graph orientation morphism $\vec{\text{Or}}$ yields the symmetry $\mathcal{Q}(\mathcal{P}) = \vec{\text{Or}}(\gamma)(\mathcal{P}) \in \text{sym}([\![\mathcal{P},\mathcal{P}]\!] = 0)$ for every graph cocycle $\gamma \in \ker[\bullet\!\!-\!\!\bullet, \cdot]$ on $n \geqslant 4$ vertices and $2n-2$ edges (see [22] and [4, 19]). The right-hand sides of evolutionary PDE systems $\frac{\mathrm{d}}{\mathrm{d}\varepsilon}\mathcal{P} = \mathcal{Q}(\mathcal{P})$ are nonlinear of polynomial degree n and, generally speaking, proper, i.e. $\mathcal{Q}(\mathcal{P}) \not\equiv 0$ if $\mathcal{P}$ is Poisson and γ is not a coboundary. Not originating from this graph cohomology, the scaling $\dot{\mathcal{P}} = \mathcal{P}$, which amounts to $\mathcal{P} \mapsto \exp(\varepsilon)\,\mathcal{P}$, is linear in $\mathcal{P}$ and proper.

Problem 7. Are there any *other* nonlinear proper infinitesimal symmetries $\mathcal{Q}(\mathcal{P}) \in \text{sym}([\![\mathcal{P},\mathcal{P}]\!] = 0)$ which are universal with respect to all (affine) Poisson manifolds of arbitrary finite dimension but which are not obtained from the graph cocycles by using Kontsevich's orientation morphism $\vec{\text{Or}}$?

The evolution $\dot{\mathcal{P}} = \vec{\text{Or}}(\gamma)(\mathcal{P})$ is formally integrable in terms of (un)oriented graphs: $\mathcal{P}(\varepsilon) = \exp\big(\varepsilon \vec{\text{Or}}(\gamma)(\mathcal{P}) \cdot \vec{\partial}/\partial\mathcal{P}\big)\big(\mathcal{P}\big|_{\varepsilon=0}\big)$.

Problem 8. Does the formal power series expansion $\mathcal{P}(\varepsilon)$ using graphs determine a globally well defined Poisson structure at all points of arbitrary affine Poisson manifold $(M^r_{\text{aff}}, \mathcal{P})$? In other words, do the power series converge (uniformly over M^r_{aff})?

The PDE systems $\dot{\mathcal{P}} = \mathcal{Q}(\mathcal{P})$ which we know from [3, 2, 4, 7] consist of $r(r-1)/2$ evolution equations with differential-polynomial right-hand sides.

Problem 9. What are the (Liouville, in particular super-)integrability properties of these nonlinear evolutionary PDE systems $\dot{\mathcal{P}} = \mathcal{Q}(\mathcal{P})$?

Conjecture 2. Natural mathematical structures themselves usually admit a structure of the same kind.

Problem 10. Does the evolution $\dot{\mathcal{P}} = \mathcal{Q}(\mathcal{P})$ of Poisson structures $\mathcal{P}$ obey the above meta-principle implying that this evolution itself ought to be Poisson, $\mathcal{Q}(\mathcal{P}) = \boldsymbol{\pi}\big(\boldsymbol{\chi}(\mathcal{P})\big)$, where $\boldsymbol{\pi}$ is a variational Poisson structure on the jet space of Poisson bi-vector components and $\boldsymbol{\chi}$ is a variational covector (possibly, the variational derivative $\delta\mathcal{H}/\delta\mathcal{P}$ of a well defined Hamiltonian functional, cf. [16, 26])?

Remark 2. Another realisation of this meta-principle could be that the graded Poisson structure $\pi = [\![\cdot,\cdot]\!]$ is the Schouten bracket and the "1-form" χ with respect to bi-vectors $\mathcal{P}$ is a $[(-1)+2]$-vector field $\mathcal{X}$, so that the Poisson dynamics is $\dot{\mathcal{P}} = [\![\mathcal{P},\mathcal{X}]\!]$, see §7 below.

6. Are the "generic" Poisson brackets on $\mathbb{R}^3$ preserved by the flows?

6.1. From the work of Nambu and other physicists in the 1970s (cf. [12, 15] and references therein) we recall a construction of an ample class of Poisson bi-vectors $da/\,dvol$ on $\mathbb{R}^3$: every such bracket contains two free functional parameters. Denote by x, y, z the Cartesian (or any other global) coordinates on $\mathbb{R}^3$. Fix a smooth function $a \in C^\infty(\mathbb{R}^3)$ and let there be another functional parameter $\varrho(x,y,z)$ referred to the chosen coordinate system. (In earnest, ϱ is a density with nontrivial behaviour under coordinate reparametrisations, see [3].) For any functions $f,g \in C^\infty(\mathbb{R}^3)$, put by definition $\{f,g\}_{a,\varrho}(x,y,z) = \varrho(x,y,z)\cdot \det\|D(a,f,g)/D(x,y,z)\|(x,y,z)$ using the Jacobian determinant in the right-hand side.

Lemma 3. The bracket $\{\cdot,\cdot\}_{a,\varrho}$ is Poisson, i.e. it is a bi-linear skew-symmetric bi-derivation satisfying the Jacobi identity.

Problem 11. Is the class of Poisson brackets $\{\cdot,\cdot\}_{a,\varrho}$ preserved by any of Kontsevich's infinitesimal symmetries $\dot{\mathcal{P}} = \mathcal{Q}(\mathcal{P})$? Specifically, for a given symmetry $\dot{P}^{ij} = Q^{ij}\big[P^{\alpha\beta}([a],\varrho)\big]$ of the Poisson bi-vector $\mathcal{P} = (P^{ij})$, does there exist a well defined evolutionary system $\dot{a} = A\big([a],[\varrho]\big)$, $\dot{\varrho} = R\big([a],[\varrho]\big)$ such that this vector field in the jet space with fibre variables a, ϱ is pushed forward by the tangent map τ_* to the symmetry $\mathcal{Q}\big[\mathcal{P}([a],\varrho)\big]$, here $\tau\colon (a,\varrho)(x,y,z) \mapsto \{\cdot,\cdot\}_{a,\varrho}\big|_{(x,y,z)}$ is the linear differential operator.

Note that the (non)existence of the lift $\dot{\mathcal{P}} \to \dot{a}, \dot{\varrho}$ can depend on a choice of the symmetry $\mathcal{Q}$.

6.2. Which Poisson brackets are *not* shifted? All symmetry problems are accompanied by the problem of finding invariant solutions.

Problem 12. Describe those two-parameter Poisson brackets $\{\cdot,\cdot\}_{a,\varrho}$ on $\mathbb{R}^3$ which stay invariant under a given Kontsevich's infinitesimal symmetry: $\mathcal{Q}\big(\{\cdot,\cdot\}_{a,\varrho}\big) \equiv 0$ at all $(x,y,z) \in \mathbb{R}^3$.

- More generally, describe (classes of) bi-vectors $\mathcal{P}$ which are $\mathcal{Q}$-invariant solutions of the Jacobi identity: $[\![\mathcal{P},\mathcal{P}]\!] = 0$ and $\mathcal{Q}(\mathcal{P}) = 0$ for a given $\mathcal{Q}$.

Problems 11–12 are typical in the geometry of differential equations (see [26] and [16, Lect. 3,7]): a symmetry is induced by the projection from a differential covering, and a PDE system is consistently constrained by its own symmetry.

7. Are the universal symmetries $\mathcal{Q}(\mathcal{P}) \in \ker[\![\mathcal{P},\cdot]\!]$ (non)trivial Poisson cocycles?

7.1. The $\partial_{\mathcal{P}}$-triviality of a given Poisson cocycle $\mathcal{Q}(\mathcal{P})$ by definition means the existence of a global vector field $\mathcal{X} \in \Gamma(TM^r)$ and, for the construction at the level of Kontsevich's orgraphs, of a sum $\Diamond$ of Leibniz graphs such that $\mathcal{Q}(\mathcal{P}) = [\![\mathcal{P},\mathcal{X}]\!] + \Diamond(\mathcal{P},[\![\mathcal{P},\mathcal{P}]\!])$. The term $\Diamond(\cdot,\cdot)$ is invisible if $\mathcal{P}$ is Poisson. It can be seen ([2, 22]) that for the Kontsevich tetrahedral cocycle $\gamma_3 \in \ker \mathrm{d}$ and symmetry $\dot{\mathcal{P}} = \mathrm{O\vec{r}}(\gamma_3)(\mathcal{P})$, there is no graph δ such that $\mathrm{O\vec{r}}(\delta)(\mathcal{P})$ would be the trivialising vector field $\mathcal{X}$. Nevertheless, in all the examples of Poisson bi-vectors $\mathcal{P}$ on affine manifolds M^r_{aff}, all the symmetries $\mathcal{Q}(\mathcal{P}) \in \mathrm{sym}([\![\mathcal{P},\mathcal{P}]\!] = 0)$ known so far are spontaneously trivial, i.e. they always appear to be $\partial_{\mathcal{P}}$-coboundaries w.r.t. seemingly *ad hoc* vector fields $\mathcal{X}$. (In particular, see [1] for illustration and [12, 25] for a store of Poisson brackets.)

Problem 13. Do there exist nonlinear, proper, and still $\partial_{\mathcal{P}}$-nontrivial infinitesimal symmetries $\mathcal{Q}(\mathcal{P}) \in \mathrm{sym}([\![\mathcal{P},\mathcal{P}]\!] = 0)$ of Poisson bi-vectors $\mathcal{P}$ on finite-dimensional affine Poisson manifolds M^r_{aff}? If the converse it true, what is then the mechanism of $\partial_{\mathcal{P}}$-triviality of all these Poisson cocycles (in particular, produced by the graph orientation morphism $\mathrm{O\vec{r}}$ from nontrivial graph cocycles $\gamma \in \ker \mathrm{d}$)?

Remark 3. Were the infinitesimal symmetries $\dot{\mathcal{P}} = \mathcal{Q}(\mathcal{P})$ generically *not* $\partial_{\mathcal{P}}$-exact, then near all or most of all Poisson systems in Nature (provided that their Poisson structures are sufficiently nonlinear) there would be many almost identical systems with the same parameter space and same energy – expressed in terms of these common parameters – but with a different finite-time behaviour. Do we observe this effect empirically? (One contrasts the mutations of specimen vs variability of profiles for a given specimen, e.g. shapes or colourings.)

Remark 4. Whenever the trivialising vector field $\mathcal{X}$ exists, it is not uniquely defined. Indeed, for any function $h \in C^\infty(M^r)$, one has $[\![\mathcal{P}, \mathcal{X}]\!] = [\![\mathcal{P}, \mathcal{X} + [\![\mathcal{P}, h]\!]]\!]$ because $\partial_{\mathcal{P}}^2 = 0$, so that the 1-vectors $\mathcal{X}$ themselves mark the $\partial_{\mathcal{P}}$-cohomology classes.

7.2. The frequent, generic, or immanent existence of global vector fields $\vec{X}(\mathcal{P})$ trivialising the symmetries $\mathcal{Q}(\mathcal{P}) = [\![\vec{X}, \mathcal{P}]\!]$ would mean that the evolution of coefficients of the Poisson bi-vector $\mathcal{P}$ at a point $\boldsymbol{a} \in M^r_{\text{aff}}$ – actually, to the same extent as the evolution $\dot{\Omega} = [\![\vec{X}, \Omega]\!]$ of *every* multi-vector at $\boldsymbol{a}$ – is just the reparametrisation of components of the tensor $\mathcal{P}$ (resp., Ω) under the infinitesimal shift along the integral trajectories of the vector field $\vec{X} \in \Gamma(TM^r_{\text{aff}})$. Specifically, the local description of this diffeomorphism at finite ε is the change of coordinates:

$$\mathbf{x}^{\varepsilon\neq0}_{\text{new}}(\text{point } \boldsymbol{a} \in M^r_{\text{aff}}) = \boldsymbol{x}^{\varepsilon=0}_{\text{old}}(\text{point } \boldsymbol{b} \in M^r_{\text{aff}} \mid \exp(\varepsilon\vec{X})(\boldsymbol{b}) = \boldsymbol{a}).$$

This standard lemma in analysis on manifolds is extended – from $\dot{Y} = [\vec{X}, Y]$ with the commutator of two vector fields to $\dot{\mathcal{P}} = [\![\vec{X}, \mathcal{P}]\!]$ and $\dot{\Omega} = [\![\vec{X}, \Omega]\!]$ with the Schouten bracket of 1-vector $\vec{X}$ and bi-vector $\mathcal{P}$ or any other multi-vector Ω – by the recursive definition of the Schouten bracket $[\![\cdot,\cdot]\!]$ as the graded Leibniz-rule extension of the commutator $[\cdot,\cdot]$ (see [25]).

Remark 5. Whenever the coefficients of Poisson bi-vector $\mathcal{P}$ are sufficiently nonlinear and if the differential order of the chosen universal symmetry $\mathcal{Q}$ does not yet trivialise $\mathcal{Q}(\mathcal{P})(\boldsymbol{a})$ at all points $\boldsymbol{a}$ of the manifold M^r_{aff} simultaneously, then the $\partial_{\mathcal{P}}$-trivialising vector field $\vec{X}(\mathcal{P})(\boldsymbol{a})$, if exists, would also be non-constant. Hence its trajectories along the charts of M^r_{aff} would also be nonlinear. This construction yields a *nonlinear* change of local coordinates $\mathbf{x}^{\varepsilon\neq0}_{\text{new}} \rightleftarrows \boldsymbol{x}^{\varepsilon=0}_{\text{old}}$ on an *affine* manifold!

Problem 14. For a given affine Poisson manifold $(M^r_{\text{aff}}, \mathcal{P})$, how many universal symmetries – taken from the countable set $\vec{\text{Or}}(\gamma \in \ker \text{d})(\mathcal{P})$ produced by the graph orientation morphism $\vec{\text{Or}}$ from the $\mathfrak{grt}$-related wheel cocycles and their iterated commutators [14, 29] – do not vanish identically as global sections from $\Gamma(\bigwedge^2 TM^r_{\text{aff}})$?

Problem 15. In the above setting, how ample is the set of trivialising vector fields $\vec{X} = \vec{X}(\gamma \in \ker \text{d}) \in \Gamma(TM^r_{\text{aff}})$ on a given manifold $(M^r_{\text{aff}}, \mathcal{P})$ with a fixed Poisson structure? Is the vector space of these fields $\vec{X}$ locally complete, to be sufficient for the approximation of arbitrary vector fields on M^r_{aff}?

In brief, are the exponents of the trivialising vector fields $\vec{X}(\gamma, \mathcal{P})$ enough to imitate a *smooth* structure on an *affine* Poisson manifold?

8. What is the quantisation of Poisson bracket symmetries?

8.1. The standard procedure of first quantisation in Poisson vs Quantum Mechanics is the replacement of dynamical functions from the algebra $C^\infty(M^r)$ on a Poisson (typically, symplectic) manifold $(M^r, \{\cdot,\cdot\}_{\mathcal{P}})$ by the associative noncommutative algebra of operators which act on a suitable complex Hilbert space. The correspondence between the classical Poisson bracket and commutator of operators is $\{\cdot,\cdot\}_{\mathcal{P}} \mapsto \frac{\boldsymbol{i}}{\hbar}[\cdot,\cdot]$; here $\hbar$ is the Planck constant and $\boldsymbol{i}^2 = -1$.

Problem 16. For a given infinitesimal symmetry $\mathcal{Q} \in \text{sym}\{\cdot,\cdot\}_{\mathcal{P}}$ which acts nontrivially on the Poisson structure $\{\cdot,\cdot\}_{\mathcal{P}}$ in a given classical model, so that $\mathcal{Q}(\{\cdot,\cdot\}_{\mathcal{P}}) \not\equiv 0$, is there always a well-defined post-quantisation response in either or both the Hilbert space of states or the operator algebra of observables?

8.2. The Kontsevich construction of deformation quantisation $\times \mapsto \star = \times + \hbar\,\{\cdot,\cdot\}_{\mathcal{P}} + \bar{o}(\hbar)$ on affine Poisson manifolds $(M^r_{\text{aff}}, \mathcal{P} \not\equiv 0)$ extends the usual product $\times$ in the algebra $C^\infty(M^r_{\text{aff}})$ to an associative noncommutative star-product $\star$ in the unital algebra $C^\infty(M^r_{\text{aff}})[[\hbar]]$ of formal power series in the deformation parameter $\hbar$ (see [23], also [5]); the construction is essentially indifferent to the choice of real or complex number field (cf. [10]). By adding power series tails to the arguments f, g of $\star$, i.e. by extending $f \mapsto T(f) = f + \hbar T_1(f) + \bar{o}(\hbar)$ and $g \mapsto T(g) = g + \hbar T_1(g) + \bar{o}(\hbar)$ using a sequence $T = \{T_n,\ n \geqslant 1\}$ of finite-order differential operators encoded by the Kontsevich graphs, one acts on the star-product by the gauge transformation $T\colon \star \mapsto \star'$; it is induced by the formula $f \star' g = T^{-1}\big(T(f) \star T(g)\big)$.

Problem 17. Describe and illustrate by examples the action of graph complex on the set of star-products (in particular, in the context of their gauge transformations $\star \rightleftarrows \star'$) whenever the Poisson bi-vector $\mathcal{P}$ in the leading deformation term $\hbar\mathcal{P}$ of $\star = \times + \hbar\,\{\cdot,\cdot\}_{\mathcal{P}} + \bar{o}(\hbar)$ is shifted by a symmetry $\mathcal{Q}(\mathcal{P}) = \text{O}\vec{\text{r}}(\gamma \in \ker[\bullet\!\!-\!\!\bullet, \cdot])(\mathcal{P})$ to $\hbar\mathcal{P} + \varepsilon\mathcal{Q}(\hbar\mathcal{P}) + \bar{o}(\varepsilon)$.

Example 1. The Kontsevich tetrahedral flow ([22, 24] and [2, 3]) perturbs the Poisson bracket by the terms starting at $\hbar^{4-1} = \hbar^3$, which shows up nontrivially at $\hbar^4$ in the expansion $\star$ mod $\bar{o}(\hbar^4)$, now available from [6].

Problem 18. How does the graph complex act – by the symmetries $\dot{\mathcal{P}} = \text{O}\vec{\text{r}}(\gamma)(\mathcal{P})$ – on the Ikeda–Izawa Poisson σ-model [13] in which the Feynman path integral expansions of correlation functions [10] encode the Kontsevich star-product $\star = \times + \hbar\,\{\cdot,\cdot\}_{\mathcal{P}} + \bar{o}(\hbar)$?

- Does the σ-model in [13] have symmetries not originating from the graphs $\gamma \in \ker \mathrm{d}$?

The presence of hidden (non)gauge degrees of freedom in the Kontsevich $\star$-products can be put in the further context of Drinfeld's associators.

Problem 19. Likewise, do the universal symmetries $\dot{\mathcal{P}} = \mathcal{Q}(\mathcal{P}) \in \text{sym}\big([\![\mathcal{P},\mathcal{P}]\!] = 0\big)$ persist in the course of the Batalin–Vilkovisky quantisation, $\big([\![\mathcal{P},\mathcal{P}]\!] = 0\big) \Longrightarrow \big(i\hbar\,\Delta\mathcal{P} + \tfrac{1}{2}[\![\mathcal{P},\mathcal{P}]\!] = 0\big)$, of Poisson bi-vectors $\mathcal{P}$?

Problem 20. Is there a well-defined – and universal w.r.t. the set $\mathcal{Q} \in \text{sym}([\![\mathcal{P},\mathcal{P}]\!] = 0)$ of graph-encoded symmetries of Poisson structures $\mathcal{P}$ – generalisation of the Kontsevich construction, originally on affine Poisson manifolds $\big(M^{r<\infty}_{\text{aff}}, \mathcal{P}\big)$, to the commutative non-associative calculus of cyclic words with their topological pair-of-pants multiplication $\mathbb{S}^1 \times \mathbb{S}^1 \to \mathbb{S}^1$ in the Kontsevich formal noncommutative (super)geometry (see [20] or [17] and references therein)?

Acknowledgements. The author is grateful to the Organizers of the 26th International Conference on Integrable Systems and Quantum Symmetries (ISQS26, 8–12 July 2019 at ČVUT Prague, Czech Republic) for warm atmosphere during the event. This research is supported by JBI RUG project 106552 and the IHÉS (in part, by the Nokia Fund). A part of this work was done while the author was visiting at IM NASU in Kiev, Ukraine. The author thanks M. Kontsevich, A. G. Nikitin, and R. Buring for helpful discussion.

References

[1] Bouisaghouane A 2017 The Kontsevich tetrahedral flow in 2*D*: a toy model *Preprint* arXiv:1702.06044 [math.DG]

[2] Bouisaghouane A, Buring R and Kiselev A 2017 The Kontsevich tetrahedral flow revisited *J. Geom. Phys.* **119**272–85 (*Preprint* arXiv:1608.01710 [q-alg])

[3] Bouisaghouane A and Kiselev A V 2017 Do the Kontsevich tetrahedral flows preserve or destroy the space of Poisson bi-vectors? *J. Phys.: Conf. Ser.* **804** Paper 012008 1–10 (*Preprint* arXiv:1609.06677 [q-alg])

[4] Buring R and Kiselev A V 2019 The orientation morphism: from graph cocycles to deformations of Poisson structures *J. Phys.: Conf. Ser.* **1194** Paper 012017 1–10 *Preprint* arXiv:1811.07878 [math.CO])

[5] Buring R and Kiselev A V 2019 Formality morphism as the mechanism of ⋆-product associativity: how it works *Collected works Inst. Math., Kyiv* **16**:1 *Symmetry & Integrability of Equations of Mathematical Physics* pp 22–43 (*Preprint* arXiv:1907.00639 [q-alg])

[6] Buring R and Kiselev A V 2019 The expansion ⋆ mod $\bar{o}(\hbar^4)$ and computer-assisted proof schemes in the Kontsevich deformation quantization *Experimental Math.* accepted 1–77 (*Preprint* arXiv:1702.00681 [math.CO])

[7] Buring R, Kiselev A V and Rutten N J 2018 Poisson brackets symmetry from the pentagon-wheel cocycle in the graph complex *PEPAN Letters* **49**(5) 924–8 (*Preprint* arXiv:1712.05259 [math-ph])

[8] Buring R, Kiselev A V and Rutten N J 2017 The heptagon-wheel cocycle in the Kontsevich graph complex *J. Nonlin. Math. Phys.* **24** Suppl. 1 157–73 (*Preprint* arXiv:1710.00658 [math.CO])

[9] Buring R, Kiselev A V and Rutten N J 2018 Infinitesimal deformations of Poisson bi-vectors using the Kontsevich graph calculus *J. Phys.: Conf. Ser.* **965** Paper 012010 1–12 (*Preprint* arXiv:1710.02405 [math.CO])

[10] Cattaneo A S and Felder G 2000 A path integral approach to the Kontsevich quantization formula *Comm. Math. Phys.* **212**:3 591–611 (*Preprint* arXiv:q-alg/9902090)

[11] Dolgushev V A, Rogers C L and Willwacher T H 2015 Kontsevich's graph complex, GRT, and the deformation complex of the sheaf of polyvector fields *Ann. Math.* **182**(3) 855–943 (*Preprint* arXiv:1211.4230 [math.KT])

[12] Grabowski J, Marmo G and Perelomov A M 1993 Poisson structures: towards a classification *Mod. Phys. Lett.* A **8**:18 1719–33

[13] Ikeda N 1994 Two-dimensional gravity and nonlinear gauge theory *Ann. Phys.* **235**:2 435–64

[14] Khoroshkin A, Willwacher T and Živković M 2017 Differentials on graph complexes *Adv. Math.* **307** 1184–214 (*Preprint* arXiv:1411.2369 [q-alg])

[15] Kiselev A V 2005 On associative Schlessinger–Stasheff algebras and Wronskian determinants *Fundam. Appl. Math.* **11**:1 159–80; *J. Math. Sci.* 2007 **141**:1 1016–30 (*Preprint* arXiv:math.RA/0410185)

[16] Kiselev A V 2012 The twelve lectures in the (non)commutative geometry of differential equations *Preprint* IHÉS/M/12/13 (Bures-sur-Yvette, France) 1–140

[17] Kiselev A V 2018 The calculus of multivectors on noncommutative jet spaces *J. Geom. Phys.* **130** 130–67 (*Preprint* arXiv:1210.0726 [math.DG])

[18] Kiselev A V 2019 *Deformations of Poisson Brackets: the Graph Complex and Graph Orientation Morphism* Lectures at the Independent University of Moscow (13–15 May 2019) url: `http://www.mathnet.ru/php/conference.phtml?option_lang=rus&eventID=25&confid=1591`

[19] Kiselev A V and Buring R 2019 The Kontsevich graph orientation morphism revisited *Preprint* `arXiv:1904.13293` [math.CO]

[20] Kontsevich M 1993 Formal (non)commutative symplectic geometry *The Gel'fand Mathematical Seminars 1990-1992* ed L Corwin, I Gelfand *et al* (Boston MA: Birkhäuser) pp 173–87

[21] Kontsevich M 1994 Feynman diagrams and low-dimensional topology, *Proc. First Europ. Congr. of Math.* **2** (Paris, 1992), Progr. Math. **120** (Basel: Birkhäuser) pp 97–121
Kontsevich M 1995 Homological algebra of mirror symmetry. *Proc. Intern. Congr. Math.* **1** (Zürich, 1994) (Basel: Birkhäuser) pp 120–39

[22] Kontsevich M 1997 Formality conjecture *Deformation Theory and Symplectic Geometry* (*Proc. Ascona 1996*) *Math. Phys. Stud.* vol 20 ed D Sternheimer, J Rawnsley *et al* (Dordrecht: Kluwer Acad. Publ.) pp 139–56

[23] Kontsevich M 2003 Deformation quantization of Poisson manifolds *Lett. Math. Phys.* **66**:3 157–216 (*Preprint* arXiv:q-alg/9709040)

[24] Kontsevich M 2017 Derived Grothendieck–Teichmüller group and graph complexes [after T. Willwacher] *Séminaire Bourbaki* (*69ème année, Janvier 2017*) **1126** 1–26

[25] Laurent–Gengoux C, Pichereau M and Vanhaecke P 2013 *Poisson Structures* (Grundlehren der Mathematischen Wissenschaften vol 347) (Berlin Heidelberg: Springer-Verlag)

[26] Olver P J 1993 *Applications of Lie Groups to Differential Equations* Grad. Texts in Math. **107** (2nd ed) (NY: Springer–Verlag)

[27] Rossi C A and Willwacher T 2014 P. Etingof's conjecture about Drinfeld associators *Preprint* arXiv:1404.2047 [q-alg]

[28] Rutten N J and Kiselev A V 2019 The defining properties of the Kontsevich unoriented graph complex *J. Phys.: Conf. Ser.* **1194** Paper 012095 1–10 (*Preprint* arXiv:1811.10638 [math.CO])

[29] Willwacher T 2015 M. Kontsevich's graph complex and the Grothendieck–Teichmüller Lie algebra *Invent. Math.* **200**(3) 671–760 (*Preprint* arXiv:1009.1654 [q-alg])

[30] Willwacher T and Živković M 2015 Multiple edges in M. Kontsevich's graph complexes and computations of the dimensions and Euler characteristics *Adv. Math.* **272** 553–78 (*Preprint* arXiv:1401.4974 [q-alg])

Journal of Physics: Conference Series **1416** (2019) 012019 doi:10.1088/1742-6596/1416/1/012019

Wreath products and multipartite quantum systems

Vladimir V Kornyak
Laboratory of Information Technologies, Joint Institute for Nuclear Research
141980 Dubna, Russia

E-mail: vkornyak@gmail.com

Abstract. A natural symmetry group of a multicomponent quantum system is a special combination of a symmetry group acting within a single component ("local group") and a group that permutes the components ("spatial symmetry group"). This combination is called the *wreath product*. Unitary representations of wreath products describe quantum evolutions of multipartite systems. It is known that any unitary representation of a finite group is contained in some permutation representation. We describe an algorithm for decomposing permutation representations of wreath products into irreducible components. This decomposition makes it possible to study the quantum behavior (entanglement, non-local correlations, etc.) of multipartite systems in invariant subspaces of the permutation Hilbert space.

1. Introduction

The Hilbert space of a multipartite quantum system is the tensor product of the Hilbert spaces of the components: $\widetilde{\mathcal{H}} = \bigotimes_{x=1}^{N} \mathcal{H}_x$. The states of the multipartite system that can be represented as a weighted sum of tensor products of the states of the components are called *separable*. States that are not separable are called *entangled*. The vast majority of the states of a typical multipartite system are entangled. The concept of entanglement is the basis of quantum informatics. Entanglement leads to such experimentally observable phenomena as nonlocal quantum correlations and quantum teleportation. Moreover, in recent years, the idea that physical space itself is not a fundamental entity, but emerges as an approximate phenomenological structure within "the Hilbert space of the Universe" as a result of some process of statistical selection has become increasingly popular. The idea of an emergent space is attractive, in particular, because of the possibility to reformulate the problem of reconciling quantum mechanics with gravity, where the main difficulties arise from the fact that quantum and space-time structures are considered to be equally fundamental. Typical approaches to identify geometric structures within a Hilbert space use the concept of entanglement (see, e.g., [1, 2]) and can be briefly described as follows. The quantum state in a large Hilbert space decomposes approximately into a tensor product of a large number of factors, which are interpreted as points (or bulks) of a geometric space. The distances between the points (metric) are determined by an *entanglement measure* — a typical example of which is *quantum mutual information*.

Let $X \cong \{1, \ldots, N\}$ be a space with a finite set of points. To reproduce the usual property of space, homogeneity, we will assume that the group of spatial symmetries $G = G(X)$ permutes transitively the components — indexed by the elements of X — of a multipartite system. In this case, the Hilbert space of the system can be written as

$$\widetilde{\mathcal{H}} = \mathcal{H}^{\otimes N}, \tag{1}$$

Journal of Physics: Conference Series **1416** (2019) 012019 doi:10.1088/1742-6596/1416/1/012019

where $\mathcal{H}$ is a representative of the G-orbit $\mathcal{H}_{xG}$. We will call $\mathcal{H}$ a *local Hilbert space.*

A quantum description can be made constructive if continuous groups of unitary evolutions are replaced in quantum formalism by unitary representations of finite groups. It is known that any linear (which is always unitary for a simple general reason) representation of a finite group is a subrepresentation of some permutation representation. In particular, the so-called *regular* representation, that is, the permutation representation of the action of a group on its own elements, contains all possible irreducible representations of a given group. Thus, we can embed any constructive quantum model into a suitable invariant subspace of some permutation representation [3, 4]. If we can decompose a permutation representation of a group into irreducible components, then we can construct any representation of the group. Here we describe an algorithm for decomposing Hilbert space (1) into invariant subspaces with respect to the natural symmetry group of a multipartite quantum system. In particular, the algorithm outputs a complete set of orthogonal projectors to irreducible invariant subspaces.

2. Irreducible invariant projectors of wreath product representation

Let $V \cong \{1, \ldots, M\}$ be a basis of the local Hilbert space $\mathcal{H}$ on which a group of *local symmetries* $F = F(V)$ acts. The sets X and V and the group F can be treated, respectively, as the *base*, the *typical fiber* and the *structure group* of a *fiber bundle.* A natural symmetry group that acts on the set V^X of the *bundle sections* and preserves the structure of the bundle is the *wreath product* of the groups F and G [5]. It is denoted as $\widetilde{W} = F \wr G$ and has the structure of a semidirect product: $\widetilde{W} \cong F^X \rtimes G$. The action of $\widetilde{W}$ on V^X is defined as $v(x)\left(f(x), g\right) = v\left(xg^{-1}\right) f\left(xg^{-1}\right)$, where $v \in V^X$, $f \in F^X$, $g \in G$. The *right-action* convention is used for all group actions.

In [6], we proposed an algorithm for decomposing representations of finite groups based on the construction of a *complete set of mutually orthogonal irreducible invariant projectors.* These projectors are special elements of the *centralizer algebra,* which is defined as the algebra of matrices that commute with all matrices of the representation. The dimension of the centralizer algebra is called the *rank of the representation.* The computer implementation of the algorithm in [6] proved to be very effective in problems with low ranks. In particular, the program coped with many high dimensional representations of simple groups and their "small" extensions (such representations typically have low ranks), presented in the ATLAS [7], in the computationally difficult case of characteristic zero [6].

The permutation representation $\widetilde{P}$ of the wreath product is a representation of $\widetilde{W}$ by $(0,1)$-matrices of the size $M^N \times M^N$ that have the form $\widetilde{P}(\widetilde{w})_{u,v} = \delta_{u\widetilde{w},v}$, where $\widetilde{w} \in \widetilde{W}$; $u, v \in V^X$; δ is the *Kronecker delta.* To obtain all irreducible components of the representation $\widetilde{P}$ in the Hilbert space (1) it is sufficient to assume that the base field of this space is some abelian extension of the field of rational numbers $\mathbb{Q}$ that is a splitting field for the local group F.

Unfortunately, to split the representation $\widetilde{P}$ we cannot apply the algorithm [6] because wreath products are far from simple groups and their representations have too high ranks. However, it is possible to express the irreducible invariant projectors for the wreath product representation in terms of the projectors for the local group representation.

Let $B_1, \ldots, B_L$ be the complete set of mutually orthogonal irreducible invariant projectors for the permutation representation of the local group F. Denote by $\overline{L}^X$ the set of all maps from X to $\overline{L}$, where $\overline{L} = \{1, \ldots, L\}$. The action of $g \in G$ on the map $\ell = [\ell_1, \ldots, \ell_N] \in \overline{L}^X$ is defined as $\ell g = [\ell_{1g}, \ldots, \ell_{Ng}]$. Then we can prove the following

Proposition. *The irreducible invariant projector for the wreath product representation $\widetilde{P}$ is*

$$\widetilde{B}_k = \sum_{\ell \in kG} B_{\ell_1} \otimes \cdots \otimes B_{\ell_N}\,, \tag{2}$$

where kG denotes the G-orbit of the map k on the set $\overline{L}^X$.

The easily verifiable completeness condition $\sum_{i=1}^{\widetilde{K}} \widetilde{B}_{k^{(i)}} = \mathbb{1}_{M^N}$ holds. Here $\widetilde{K}$ is the number of irreducible components of the wreath product representation, $k^{(i)}$ denotes some numbering of the orbits of G on $\overline{L}^X$, $\mathbb{1}_{M^N}$ is the identity matrix in the Hilbert space (1).

To compute the projectors (2), we wrote a program in C. The input data for the program are the generators of the spatial and local groups as well as the complete set of irreducible invariant projectors of the local group (obtained, e.g., by the program described in [6]).

3. Examples of calculations

We give here three examples of applying the program to wreath products constructed from the proper symmetry groups of the octahedron, icosahedron, and dodecahedron. The calculations were performed on a PC with 3.30GHz CPU and 16GB RAM. Irreducible representations are denoted below by their dimensions in bold, and permutation representations by dimensions in bold with underlining. The space X in all cases is the icosahedron, whereas the local basis V runs through all three Platonic solids. For the simplest example (V = *octahedron*), we give a more detailed (with inevitable gaps) output and only brief information in other cases.

3.1. Octahedron

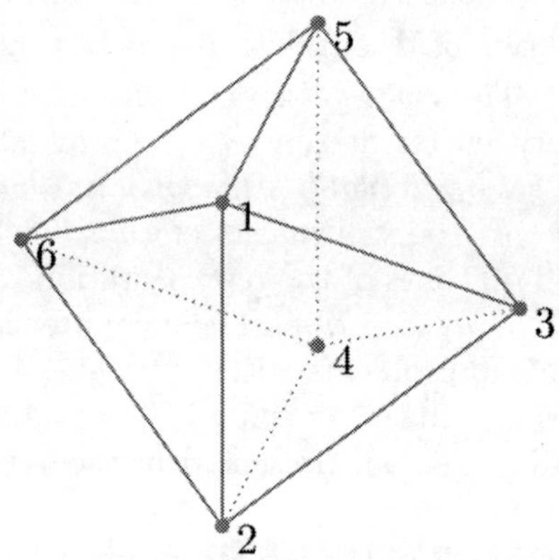

Figure 1. Octahedron.

- For the vertex numbering as in Figure 1, the proper symmetry group $G_O \cong S_4$ of the octahedron can be generated, e.g., by the two permutations $(1,3,5)(2,4,6)$ and $(1,2,4,5)$.
- The six-dimensional permutation representation of G_O, $\underline{\mathbf{6}}$, has rank 3 with the following basis of the centralizer algebra

$$A_1 = \mathbb{1}_6,\ A_2 = \begin{pmatrix} \mathbb{0}_3 & \mathbb{1}_3 \\ \mathbb{1}_3 & \mathbb{0}_3 \end{pmatrix},\ A_3 = \begin{pmatrix} Y & Y \\ Y & Y \end{pmatrix},\ \text{where } Y = \begin{pmatrix} 0 & 1 & 1 \\ 1 & 0 & 1 \\ 1 & 1 & 0 \end{pmatrix}. \quad (3)$$

- The decomposition into irreducible components has the form $\underline{\mathbf{6}} \cong \mathbf{1} \oplus \mathbf{2} \oplus \mathbf{3}$.
- The complete set of irreducible orthogonal projectors expressed in basis (3) is

$$B_1 = \frac{1}{6}(A_1 + A_2 + A_3),\ B_2 = \frac{1}{3}\left(A_1 + A_2 - \frac{1}{2}A_3\right),\ B_3 = \frac{1}{2}(A_1 - A_2). \quad (4)$$

Journal of Physics: Conference Series **1416** (2019) 012019 doi:10.1088/1742-6596/1416/1/012019

3.2. Icosahedron

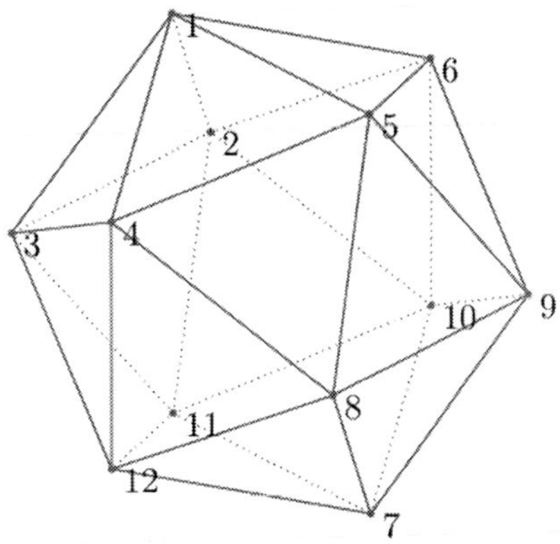

Figure 2. Icosahedron.

- The permutations $(1,7)(2,8)(3,12)(4,11)(5,10)(6,9)$ and $(2,3,4,5,6)(8,9,10,11,12)$ generate the proper symmetry group $G_{\mathrm{I}} \cong \mathsf{A}_5$ of the icosahedron.
- The four symmetric 12×12 matrices[1] $A_1 = \mathbb{1}_{12}$, A_2, A_3 and A_4 form the basis of the centralizer algebra of the permutation representation $\underline{\mathbf{12}}$ of the group G_{I}.
- The irreducible decomposition has the form $\underline{\mathbf{12}} \cong \mathbf{1} \oplus \mathbf{3} \oplus \mathbf{3}' \oplus \mathbf{5}$, where $\mathbf{3}$ and $\mathbf{3}'$ denote two nonequivalent three-dimensional irreducible representations of the group A_5.
- The complete set of irreducible orthogonal projectors is

$$B_{\mathbf{1}} = \frac{1}{12}\left(A_1 + A_2 + A_3 + A_4\right), \qquad B_{\mathbf{3}} = \frac{1}{4}\left(A_1 - A_2 + \frac{\sqrt{5}}{5}A_3 - \frac{\sqrt{5}}{5}A_4\right),$$

$$B_{\mathbf{3}'} = \frac{1}{4}\left(A_1 - A_2 - \frac{\sqrt{5}}{5}A_3 + \frac{\sqrt{5}}{5}A_4\right), \quad B_{\mathbf{5}} = \frac{5}{12}\left(A_1 + A_2 - \frac{1}{5}A_3 - \frac{1}{5}A_4\right).$$

3.3. Dodecahedron

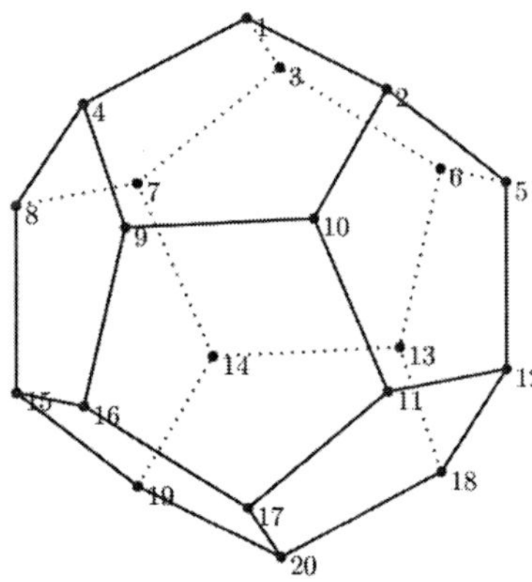

Figure 3. Dodecahedron.

[1] These matrices are too cumbersome to be given here, but can be easily computed with a simple procedure.

- The generators of the proper symmetry group $G_{\mathrm{D}} \cong \mathsf{A}_5$ of the dodecahedron are, e.g., $(2,3,4)(5,7,9)(6,8,10)(11,13,15)(12,14,16)(17,18,19)$ and $(1,2,5,6,3)(4,10,12,13,7)(8,9,11,18,14)(15,16,17,20,19)$.
- The rank of the permutation representation $\underline{\mathbf{20}}$ of the group G_{D} is equal to 8. The centralizer algebra basis for the representation $\underline{\mathbf{20}}$ consists of eight 20×20 matrices:

$$\overbrace{A_1 = \mathbb{1}_{20}, A_2, A_3, A_4, A_5, A_6}^{\text{symmetric matrices}}, A_7, A_8 = A_7^{\mathrm{T}}.$$

- The irreducible decomposition of $\underline{\mathbf{20}}$, which has the form $\underline{\mathbf{20}} \cong \mathbf{1} \oplus \mathbf{3} \oplus \mathbf{3}' \oplus \underbrace{\mathbf{4} \oplus \mathbf{4}} \oplus \mathbf{5}$, contains two copies of the four-dimensional irreducible representation of the group A_5.
- The complete set of irreducible orthogonal projectors is

$$B_{\mathbf{1}} = \frac{1}{20}\left(A_1 + A_2 + A_3 + A_4 + A_5 + A_6 + A_7 + A_8\right),$$

$$B_{\mathbf{3}} = \frac{3}{20}\left(A_1 - A_2 + \frac{\sqrt{5}}{3}A_3 - \frac{\sqrt{5}}{3}A_4 - \frac{A_5}{3} - \frac{A_6}{3} + \frac{A_7}{3} + \frac{A_8}{3}\right),$$

$$B_{\mathbf{3}'} = \frac{3}{20}\left(A_1 - A_2 - \frac{\sqrt{5}}{3}A_3 + \frac{\sqrt{5}}{3}A_4 - \frac{A_5}{3} - \frac{A_6}{3} + \frac{A_7}{3} + \frac{A_8}{3}\right),$$

$$B_{\mathbf{4}}^{(1)} = \frac{1}{5}\left(A_1 - \frac{A_3}{3} - \frac{A_4}{3} + \frac{2A_6}{3} - \frac{1 - \mathrm{i}\sqrt{11}}{6}A_7 - \frac{1 + \mathrm{i}\sqrt{11}}{6}A_8\right),$$

$$B_{\mathbf{4}}^{(2)} = \frac{1}{5}\left(A_1 - \frac{A_3}{3} - \frac{A_4}{3} + \frac{2A_6}{3} - \frac{1 + \mathrm{i}\sqrt{11}}{6}A_7 - \frac{1 - \mathrm{i}\sqrt{11}}{6}A_8\right),$$

$$B_{\mathbf{5}} = \frac{1}{4}\left(A_1 + A_2 + \frac{A_3}{3} + \frac{A_4}{3} - \frac{A_5}{3} - \frac{A_6}{3} - \frac{A_7}{3} - \frac{A_8}{3}\right).$$

3.4. *Computer outputs*

3.4.1. *Wreath product* $\mathsf{S}_4(\mathit{octahedron}) \wr \mathsf{A}_5(\mathit{icosahedron})$

`Representation dimension:` 2 176 782 336 `Rank:` 122 776
`Number of different suborbit lengths:` 46
`Wreath suborbit lengths:`
$1^{35}, 2^{249}, 3^{11}, 4^{258}, 5^{16}, 6^{403}, 8^{1442}, 16^{1890}, 24^{2418}, 32^{3943}, 48^{43}, 64^{5082}, 80^{51}, 96^{6482}, 128^{6629}, 256^{8858}, 384^{10735}, 512^{7237}, 768^{67}, 1024^{9901}, 1280^{58}, 1536^{\underline{12006}}, 2048^{5611}, 4096^{8209}, 6144^{9603}, 8192^{3093}, 12288^{39}, 16384^{4795}, 20480^{46}, 24576^{5558}, 32768^{1225}, 65536^{2171}, 98304^{2389}, 131072^{310}, 196608^{15}, 262144^{674}, 327680^{14}, 393216^{743}, 524288^{77}, 1048576^{146}, 1572864^{177}, 2097152^{18}, 4194304^{16}, 5242880^{2}, 6291456^{24}, \underline{16777216}^{7}$.
`Checksum =` 2176782336 `Maximum multiplicity =` 12006

Journal of Physics: Conference Series **1416** (2019) 012019 doi:10.1088/1742-6596/1416/1/012019

`Wreath invariant basis forms:`

$$\begin{aligned}
\widetilde{A}_1 &= A_1^{\otimes 12}\\
\widetilde{A}_2 &= A_1^{\otimes 5} \otimes A_2 \otimes A_1^{\otimes 2} \otimes A_2 \otimes A_1^{\otimes 3}\\
\widetilde{A}_3 &= A_1^{\otimes 4} \otimes A_2^{\otimes 2} \otimes A_1^{\otimes 2} \otimes A_2^{\otimes 2} \otimes A_1^{\otimes 2}\\
&\vdots\\
\widetilde{A}_{61387} &= A_2^{\otimes 3} \otimes A_1 \otimes A_3 \otimes A_1^{\otimes 2} \otimes A_3^{\otimes 2} \otimes A_2^{\otimes 2} \otimes A_3\\
&+ A_2 \otimes A_1 \otimes A_2^{\otimes 2} \otimes A_1 \otimes A_3 \otimes A_1 \otimes A_2 \otimes A_3^{\otimes 3} \otimes A_2\\
&+ A_1 \otimes A_2^{\otimes 2} \otimes A_3^{\otimes 3} \otimes A_2 \otimes A_1 \otimes A_3 \otimes A_1 \otimes A_2^{\otimes 2}\\
&+ A_1 \otimes A_3^{\otimes 2} \otimes A_2^{\otimes 2} \otimes A_3 \otimes A_2^{\otimes 2} \otimes A_1 \otimes A_3 \otimes A_1 \otimes A_2\\
\widetilde{A}_{61388} &= A_2 \otimes A_1 \otimes A_2^{\otimes 2} \otimes A_1 \otimes A_3 \otimes A_1 \otimes A_2^{\otimes 2} \otimes A_3^{\otimes 3}\\
&+ A_2^{\otimes 2} \otimes A_3 \otimes A_1 \otimes A_2 \otimes A_1^{\otimes 2} \otimes A_3^{\otimes 2} \otimes A_2^{\otimes 2} \otimes A_3\\
&+ A_1 \otimes A_3^{\otimes 2} \otimes A_2^{\otimes 2} \otimes A_3 \otimes A_2^{\otimes 2} \otimes A_1 \otimes A_2 \otimes A_1 \otimes A_3\\
&+ A_1 \otimes A_2 \otimes A_3^{\otimes 3} \otimes A_2^{\otimes 2} \otimes A_1 \otimes A_3 \otimes A_1 \otimes A_2^{\otimes 2}\\
\widetilde{A}_{61389} &= A_2 \otimes A_1 \otimes A_2^{\otimes 2} \otimes A_1 \otimes A_3 \otimes A_2^{\otimes 3} \otimes A_3^{\otimes 3}\\
&+ A_2^{\otimes 2} \otimes A_3 \otimes A_2^{\otimes 2} \otimes A_1^{\otimes 2} \otimes A_3^{\otimes 2} \otimes A_2^{\otimes 2} \otimes A_3\\
&+ A_1 \otimes A_3^{\otimes 2} \otimes A_2^{\otimes 2} \otimes A_3 \otimes A_2^{\otimes 2} \otimes A_1 \otimes A_2^{\otimes 2} \otimes A_3\\
&+ A_2^{\otimes 2} \otimes A_3^{\otimes 3} \otimes A_2^{\otimes 2} \otimes A_1 \otimes A_3 \otimes A_1 \otimes A_2^{\otimes 2}\\
&\vdots\\
\widetilde{A}_{122774} &= A_3^{\otimes 2} \otimes A_2 \otimes A_3^{\otimes 9} + A_3^{\otimes 3} \otimes A_2 \otimes A_3^{\otimes 8} + A_3^{\otimes 10} \otimes A_2 \otimes A_3 + A_3^{\otimes 11} \otimes A_2\\
\widetilde{A}_{122775} &= A_3^{\otimes 4} \otimes A_2 \otimes A_3^{\otimes 7} + A_3^{\otimes 5} \otimes A_2 \otimes A_3^{\otimes 6} + A_3^{\otimes 8} \otimes A_2 \otimes A_3^{\otimes 3}\\
&+ A_3^{\otimes 9} \otimes A_2 \otimes A_3^{\otimes 2}\\
\widetilde{A}_{122776} &= A_2 \otimes A_3^{\otimes 11} + A_3 \otimes A_2 \otimes A_3^{\otimes 10} + A_3^{\otimes 6} \otimes A_2 \otimes A_3^{\otimes 5} + A_3^{\otimes 7} \otimes A_2 \otimes A_3^{\otimes 4}
\end{aligned}$$

`Wreath product decomposition is multiplicity free`
`Number of irreducible components:` 122 776
`Number of different dimensions:` 134
`Irreducible dimensions:`
$1, 4^{6}, 6^{3}, 8^{6}, 9^{3}, 12^{15}, 16^{32}, 18^{7}, 20, 24^{70}, 32^{41}, 36^{86}, 45, 48^{191}, 54^{26}, 64^{84}, 72^{298}, 80^{4}, 81^{7}, 96^{412}, 108^{223}, 128^{114}, 144^{913}, 162^{54}, 180^{8}, 192^{704}, 216^{926}, 243^{4}, 256^{104}, 288^{1804}, 320^{7}, 324^{504}, 384^{772}, 405^{4}, 432^{2517}, 486^{99}, 512^{76}, 576^{2508}, 648^{1909}, 720^{17}, 729^{9}, 768^{705}, 864^{4303}, 972^{818}, 1024^{51}, 1152^{2562}, 1280^{3}, 1296^{4455}, 1458^{141}, 1536^{479}, 1620^{16}, 1728^{5322}, 1944^{2712}, 2048^{20}, 2187^{4}, 2304^{1935}, 2592^{6708}, 2880^{14}, 2916^{961}, 3072^{223}, 3456^{4575}, 3645^{7}, 3888^{5495}, 4096^{4}, 4374^{136}, 4608^{1004}, 5120, 5184^{6924}, 5832^{2754}, 6144^{59}, 6480^{18}, 6561^{9}, 6912^{2719}, 7776^{\underline{6966}}, 8192^{3}, 8748^{822}, 9216^{329}, 10368^{4760}, 11520^{10}, 11664^{4695}, 12288^{19}, 13122^{98}, 13824^{1011}, 14580^{13}, 15552^{5781}, 17496^{1999}, 18432^{83}, 19683^{3}, 20736^{2085}, 23328^{4826}, 25920^{16}, 26244^{511}, 27648^{260}, 31104^{2964}, 32805^{3}, 34992^{2775}, 36864^{5}, 39366^{55}, 41472^{534}, 46080, 46656^{3012}, 52488^{1023}, 55296^{15}, 58320^{19}, 59049^{5}, 62208^{877}, 69984^{2173}, 78732^{242}, 82944^{48}, 93312^{1038}, 103680^{4}, 104976^{1079}, 118098^{27}, 124416^{102}, 131220^{8}, 139968^{905}, 157464^{355}, 186624^{130}, 209952^{568}, 233280^{7}, 236196^{84}, 279936^{148}, 295245, 314928^{254}, 354294^{6}, 419904^{116}, 472392^{79}, 524880^{3}, 531441, 629856^{62}, 708588^{15}, 944784^{26}, 1180980, \underline{1417176}^{9}.$

`Checksum =` 2176782336 `Maximum number of equal dimensions =` 6966

```
Wreath irreducible projectors:
```

$$
\begin{aligned}
\widetilde{B}_1 =& B_1^{\otimes 12}\\
\widetilde{B}_2 =& B_1^{\otimes 3} \otimes B_2 \otimes B_1^{\otimes 6} \otimes B_2 \otimes B_1\\
\widetilde{B}_3 =& B_1^{\otimes 9} \otimes B_2 \otimes B_1^{\otimes 2} + B_1^{\otimes 4} \otimes B_2 \otimes B_1^{\otimes 7}\\
&\vdots\\
\widetilde{B}_{61387} =& B_2 \otimes B_3 \otimes B_1 \otimes B_2^{\otimes 2} \otimes B_1^{\otimes 3} \otimes B_3^{\otimes 2} \otimes B_2 \otimes B_3\\
&+ B_3 \otimes B_2 \otimes B_1 \otimes B_2 \otimes B_1 \otimes B_3 \otimes B_1 \otimes B_2 \otimes B_3 \otimes B_1 \otimes B_3 \otimes B_2\\
&+ B_1^{\otimes 2} \otimes B_3 \otimes B_2 \otimes B_3^{\otimes 2} \otimes B_2 \otimes B_3 \otimes B_1 \otimes B_2^{\otimes 2} \otimes B_1\\
&+ B_1 \otimes B_2^{\otimes 2} \otimes B_3 \otimes B_1 \otimes B_3^{\otimes 2} \otimes B_2 \otimes B_3 \otimes B_1 \otimes B_2 \otimes B_1\\
\widetilde{B}_{61388} =& B_2^{\otimes 2} \otimes B_3 \otimes B_1^{\otimes 2} \otimes B_3 \otimes B_1 \otimes B_2 \otimes B_3 \otimes B_2^{\otimes 2} \otimes B_3\\
&+ B_1 \otimes B_2 \otimes B_3 \otimes B_2^{\otimes 2} \otimes B_3 \otimes B_2^{\otimes 2} \otimes B_3 \otimes B_1^{\otimes 2} \otimes B_3\\
\widetilde{B}_{61389} =& B_1^{\otimes 2} \otimes B_2^{\otimes 3} \otimes B_1 \otimes B_2 \otimes B_3 \otimes B_2 \otimes B_1 \otimes B_3^{\otimes 2}\\
&+ B_2 \otimes B_1 \otimes B_2 \otimes B_3 \otimes B_2 \otimes B_1 \otimes B_3 \otimes B_2 \otimes B_1^{\otimes 2} \otimes B_2 \otimes B_3\\
&+ B_1 \otimes B_3 \otimes B_2 \otimes B_1^{\otimes 3} \otimes B_3 \otimes B_2^{\otimes 3} \otimes B_3 \otimes B_2\\
&+ B_3 \otimes B_2 \otimes B_3 \otimes B_2 \otimes B_1^{\otimes 2} \otimes B_2 \otimes B_1^{\otimes 2} \otimes B_2 \otimes B_3 \otimes B_2\\
&+ B_2 \otimes B_3^{\otimes 3} \otimes B_1 \otimes B_2 \otimes B_1^{\otimes 3} \otimes B_2^{\otimes 3}\\
&+ B_3 \otimes B_2^{\otimes 2} \otimes B_3 \otimes B_2^{\otimes 2} \otimes B_1 \otimes B_3 \otimes B_1^{\otimes 3} \otimes B_2\\
&\vdots\\
\widetilde{B}_{122774} =& B_3^{\otimes 2} \otimes B_2 \otimes B_3^{\otimes 9} + B_3^{\otimes 3} \otimes B_2 \otimes B_3^{\otimes 8} + B_3^{\otimes 10} \otimes B_2 \otimes B_3 + B_3^{\otimes 11} \otimes B_2\\
\widetilde{B}_{122775} =& B_3^{\otimes 2} \otimes B_2 \otimes B_3 \otimes B_2 \otimes B_3^{\otimes 7} + B_3^{\otimes 2} \otimes B_2 \otimes B_3^{\otimes 2} \otimes B_2 \otimes B_3^{\otimes 6}\\
&+ B_3^{\otimes 3} \otimes B_2 \otimes B_3^{\otimes 4} \otimes B_2 \otimes B_3^{\otimes 3} + B_3^{\otimes 5} \otimes B_2 \otimes B_3^{\otimes 4} \otimes B_2 \otimes B_3\\
&+ B_3^{\otimes 8} \otimes B_2 \otimes B_3^{\otimes 2} \otimes B_2 + B_3^{\otimes 9} \otimes B_2 \otimes B_3 \otimes B_2\\
\widetilde{B}_{122776} =& B_3^{\otimes 3} \otimes B_2^{\otimes 2} \otimes B_3^{\otimes 7} + B_2 \otimes B_3^{\otimes 4} \otimes B_2 \otimes B_3^{\otimes 6}\\
&+ B_3 \otimes B_2 \otimes B_3^{\otimes 3} \otimes B_2 \otimes B_3^{\otimes 6} + B_3^{\otimes 6} \otimes B_2 \otimes B_3 \otimes B_2 \otimes B_3^{\otimes 3}\\
&+ B_3^{\otimes 7} \otimes B_2^{\otimes 2} \otimes B_3^{\otimes 3} + B_3^{\otimes 9} \otimes B_2^{\otimes 2} \otimes B_3
\end{aligned}
$$

```
Time:  0.58 sec
Maximum number of tensor monomials:  531441
```

3.4.2. Wreath product A_5(*icosahedron*) ≀ A_5(*icosahedron*)

```
Representation dimension: 8 916 100 448 256      Rank: 3 875 157
Wreath product decomposition is multiplicity free
Number of irreducible components: 3 875 157
Number of different dimensions: 261
Time:  7.35 sec
Maximum number of tensor monomials:  16777216
```

3.4.3. Wreath product A_5(*dodecahedron*) ≀ A_5(*icosahedron*)

```
Representation dimension: 4 096 000 000 000 000      Rank: > 502 985 717
```

```
Wreath product decomposition has non-trivial multiplicities
Number of irreducible components: 502985717
Number of different dimensions: 1065
Time: 26 min 46.16 sec
Maximum number of tensor monomials: 2176782336
```

4. Remark on simulation of multiparticle quantum systems

Projection operators (2) obtained by our program are intended for calculations in models of multipartite quantum systems. These operators are matrices of huge dimensions, for example, about four quadrillions for the representation in section 3.4.3. Obviously, explicit computation of such matrices is impossible and actually not necessary. The presentation of projectors for wreath products in the form of tensor polynomials allows us to reduce the calculation of quantum correlations to a sequence of manipulations with small matrices of local projectors.

Consider, for example, the computation of a scalar product in an invariant subspace. Let $\left|1\right\rangle, \ldots, \left|M\right\rangle$ be an orthonormal basis in the local Hilbert space $\mathcal{H}$. Then the orthonormal basis in the Hilbert space $\widetilde{\mathcal{H}} = \mathcal{H}^{\otimes N}$ of the wreath product is formed by elements of the form $\left|m_1\right\rangle \otimes \cdots \otimes \left|m_N\right\rangle$, where $m = [m_1, \ldots, m_N] \in \overline{M}^N$, $M = |V|, N = |X|$, $\overline{M} = \{1, \ldots, M\}$ and $\overline{N} = \{1, \ldots, N\}$. General vectors in the wreath product Hilbert space can be written as follows:

$$\widetilde{\Phi} = \sum_{m \in V^X} \varphi_m \left|m_1\right\rangle \otimes \cdots \otimes \left|m_N\right\rangle \text{ and } \widetilde{\Psi} = \sum_{n \in V^X} \psi_n \left|n_1\right\rangle \otimes \cdots \otimes \left|n_N\right\rangle,$$

where φ_m and ψ_n are arbitrary scalars from the base field $\mathcal{F}$.[2] The scalar product of these vectors in the invariant subspace that is defined by projector (2) has the form

$$\left\langle \widetilde{\Phi} \middle| \widetilde{B}_k \middle| \widetilde{\Psi} \right\rangle = \sum_{m \in V^X} \sum_{n \in V^X} \sum_{\ell \in kG} \overline{\varphi}_m \psi_n A_{m,n,\ell},$$

where

$$A_{m,n,\ell} = \left\langle m_1\right| \otimes \cdots \otimes \left\langle m_N\right| B_{\ell_1} \otimes \cdots \otimes B_{\ell_N} \left|n_1\right\rangle \otimes \cdots \otimes \left|n_N\right\rangle \quad (5)$$

$$= \left\langle m_1\right| B_{\ell_1} \left|n_1\right\rangle \cdots \left\langle m_N\right| B_{\ell_N} \left|n_N\right\rangle \equiv \left(B_{\ell_1}\right)_{m_1 n_1} \cdots \left(B_{\ell_N}\right)_{m_N n_N}. \quad (6)$$

The last expression (6), which is simply the product of N scalars, is obtained by applying to (5) a general property of the tensor product — the identity $(A \otimes B)(C \otimes D) = (AC) \otimes (BD)$.

Acknowledgments

I thank Yu.A. Blinkov, V.P. Gerdt and N.N. Vassiliev for helpful discussions.

References

[1] Van Raamsdonk M 2010 Building up spacetime from quantum entanglement *Gen. Relativ. Grav.* **42**, 2323–9
[2] Cao C, Carroll S M and Michalakis S 2017 Space from Hilbert space: Recovering geometry from bulk entanglement *Phys. Rev. D* **95**, 024031
[3] Kornyak V V 2018 Modeling Quantum Behavior in the Framework of Permutation Groups *EPJ Web of Conferences* **173** 01007
[4] Kornyak V V 2018 Quantum models based on finite groups *J. Phys.: Conf. Series* **965** 012023
[5] Meldrum J D P 1995 *Wreath Products of Groups and Semigroups* (Longman/Wiley)
[6] Kornyak V V 2019 A new algorithm for irreducible decomposition of representations of finite groups *J. Phys.: Conf. Ser.* **1194** 012060
[7] Wilson R *et al. Atlas of finite group representations*

[2] We can restrict ourselves to natural φ_m and ψ_n, if we adhere to the ideology of the approach [3, 4].

Journal of Physics: Conference Series **1416** (2019) 012020 doi:10.1088/1742-6596/1416/1/012020

Soliton solutions of logarithmic wave equation and their application for polycrystalline metals

M Kraiev[1], K Domina[2], V Kraieva[3], and K G Zloshchastiev[4]

[1]Dniprovsky Metallurgical Plant, Mayakovsky Street, Dnipro 49064, Ukraine
[2]Z. I. Nekrasov Iron and Steel Institute of NAS of Ukraine, Academician Starodubov Square, Dnipro 49107, Ukraine
[3]Department of Physics, V. Lazaryan Dniprovsky National University of Railway Transport, Lazaryan Street, Dnipro 49010, Ukraine
[4]Institute of Systems Science, Durban University of Technology, Durban 4000, South Africa

E-mail: `k.g.zloschastiev@gmail.com`

Abstract. Wave equations with logarithmic nonlinearity are applied to Korteweg-type materials which can undergo liquid-solid or liquid-gas phase transitions. One of predictions of the theory is a periodical pattern for inhomogeneities of density, which can occur in the form of bubbles or cells. Such inhomogeneities are described by soliton and solitary wave solutions of the logarithmic wave equation in the vicinity of a liquid-solid phase transition. During the solidification process, these inhomogeneities become centers of nucleation of grains. Previous works were dealing with generic natural silicate materials in geophysics, such as magmas in volcanic conduits, where the (approximately) periodical flows and structures were observed. Here we report an experimental evidence of a large-scale periodicity in structure of grains in the structural steel S235/A570 Grade 36, copper C-Cu/C14200, stainless steel X10CrNiTi18-10/AISI 321, and aluminium-magnesium alloy 5083/5056.

1. Introduction

Wave equations with logarithmic nonlinearity find fruitful applications in different branches of physics - from nuclear physics and condensed-matter theory to particle physics, theory of physical vacuum and quantum gravity, to mention just a few literature landmarks [1, 2, 3, 4, 5, 6, 7, 8, 9, 10, 11, 12, 13, 14, 15, 16, 17]. Such a universality can be explained by the fact that one can derive the logarithmic nonlinearity for a large class of the strongly-interacting many-body systems in which interaction energies predominate kinetic ones, using simple statistical-mechanics arguments and Madelung hydrodynamical presentation [17]. Examples of such systems include the Korteweg-type materials which can undergo liquid-solid or liquid-gas phase transitions [8, 18, 19]. In these materials, capillarity and surface tension play a substantial role, which makes them useful for modeling various flows with non-negligible surface and interface effects [20, 21].

One of theory's predictions are large-scale periodical inhomogeneities of density and microhardness, bubbles, cells or grains, caused by existence of multiple soliton and Gaussian-shaped solitary wave solutions for an underlying logarithmic wave equation in the vicinity of a liquid-solid phase transition. Recently, a study of these solutions' application was done for generic natural silicate materials in geophysics [19], such as magmas in volcanic conduits, where the (approximately) periodical flows and structures were known to occur [18]. In this paper, we report an experimental evidence of above-mentioned periodicity in a grain structure of non-alloy structural steel S235/A570 Grade 36, copper C-Cu/C14200, stainless steel X10CrNiTi18-10/AISI 321, and aluminium-magnesium alloy 5083/5056.

2. Formalism

Let us use the fluid approximation of continuum mechanics and introduce the fluid wavefunction which encodes main properties of a flow. This function can be written in a Madelung form [22]

$$\Psi = \sqrt{\rho}\exp(iS), \tag{1}$$

where $\rho = \rho(\mathbf{x}, t)$ is a fluid density, and $S = S(\mathbf{x}, t)$ is a phase which is related to the fluid velocity $\mathbf{u} = \nabla S$ (we assume the irrotational flow for simplicity). This wavefunction should not be confused with particle wavefunctions in quantum mechanics, but rather it is a complex-valued function which stores macroscopic information about our fluid, such as the density and velocity of a fluid parcel [12, 22].

Nevertheless, some mathematical similarities between these categories of functions do exist and will be used in what follows. In particular, function Ψ obeys a normalization condition

$$\int_V |\Psi|^2 \mathrm{d}V = \int_V \rho \,\mathrm{d}V = M, \tag{2}$$

where M and V are the total mass and volume of the fluid. This poses restrictions upon fluid wave functions which are similar to a quantum-mechanical case: the set of all normalizable fluid wave functions must constitute a Hilbert space, such as L^2.

Let us assume that such a fluid is in an isothermal state for which the characteristic kinetic energies of its particles are smaller than interaction potentials energies between them. This can happens not only at low temperatures but also in high-density or effectively low-dimensional system, because it is the ratio of kinetic and potential energies which matters.

If this ratio is small then, following a line of reasoning of works [17, 19], one can find out that the equation, which controls the dynamics of a function Ψ, contains the logarithmic nonlinearity:

$$i\partial_t \Psi = \left[-\frac{\mathcal{D}}{2}\nabla^2 - b\ln\left(|\Psi|^2/\rho_0\right)\right]\Psi, \tag{3}$$

where b, ρ_0 and $\mathcal{D}$ are real-valued parameters. One can show that $b \sim T$ where T is an absolute temperature and notation "$\sim$" means "a linear function of". This equation must be supplemented with the above-mentioned normalization condition.

It should be noted that, for the same set of boundary conditions, eq. (3) allows multiple (eigen)solutions which correspond to excited states in the Hilbert space of the problem. This means that the Korteweg-type fluid can "choose" one of these solutions spontaneously and thus undergo spontaneous phase transitions. However, the ground state is a preferable one as it corresponds to a minimum of the wave-mechanical energy. This and other low-energy states will be discussed in the next section.

Furthermore, substituting eq. (1) into eq. (3), one can recover the hydrodynamic laws for mass and momentum conservation for a two-phase compressible inviscid fluid with internal capillarity which flow is irrotational and isothermal:

$$\partial_t \rho + \nabla\cdot(\rho\mathbf{u}) = 0, \tag{4}$$

$$\partial_t \mathbf{u} + \mathbf{u}\cdot\nabla\mathbf{u} - \frac{1}{\rho}\nabla\cdot\mathbb{T} = 0, \tag{5}$$

with the stress tensor $\mathbb{T}$ of the Korteweg form with capillarity effects [20], which is used to model fluid mixtures with phase changes and diffuse interfaces [21]:

$$\mathbb{T} = -\frac{\mathcal{D}}{4\rho}\nabla\rho\otimes\nabla\rho - \tilde{p}\mathbb{I}, \tag{6}$$

where $\mathbb{I}$ is the identity matrix, $\tilde{p} = p(\rho) - \frac{1}{4}\mathcal{D}\nabla^2\rho = -b\rho - \frac{1}{4}\mathcal{D}\nabla^2\rho$ is a capillary pressure, and $p(\rho) = -b\rho$ is a barotropic equation of state for the fluid pressure p [8, 18].

Thus, eq. (3) is as a matter of fact a short form of writing two equations of hydrodynamics. This makes it useful for studies, especially because a large amount information about logarithmic wave equations has been revealed in different branches of physics and mathematics.

3. Phase structure and solutions

Depending on whether the nonlinear coupling b is positive or negative, the Korteweg-type material can be in different phases. Since this coupling is related to temperature, these phases correspond to the material above and below a certain critical temperature.

Cellular phase. If the nonlinear coupling is positive, then the field-theoretical potential density, given by

$$\mathcal{V}(|\Psi|^2) = -b|\Psi|^2\left[\ln\left(|\Psi|^2/\rho_0\right) - 1\right] + \mathcal{V}_0, \tag{7}$$

has an upside-down Mexican-hat shape, with local degenerate maxima at $|\Psi| = |\Psi_e| \equiv \sqrt{\rho_0}$ [19]. In this case, one solitary wave solution of eq. (3) can be found analytically [1, 2]. It has the form of a Gaussian parcel

$$\Psi_g^{(+)}(\mathbf{x},t) = \pm\sqrt{\rho_g^{(+)}(\mathbf{x})}\exp\left(-i\omega_g^{(+)}t\right), \tag{8}$$

$$\rho_g^{(+)}(\mathbf{x}) = \tilde{\rho}\exp\left[-\frac{(\mathbf{x}-\mathbf{x}_0)^2}{\ell^2}\right], \tag{9}$$

$$\omega_g^{(+)} = b\left[\bar{d} - \ln\left(\frac{\tilde{\rho}}{\rho_0}\right)\right], \tag{10}$$

where $\bar{d}$ is the number of spatial dimensions of the fluid, and

$$\tilde{\rho} = M/\tilde{V},\ \tilde{V} = \pi^{\bar{d}/2}\ell^{\bar{d}},\ \ell = \sqrt{|\mathcal{D}/(2b)|} \tag{11}$$

are the density peak value, effective volume and Gaussian width, respectively. One can show that the solution $\Psi_g^{(+)}$ has a lowest eigenvalue of the frequency, $\omega^{(+)}$, which makes it an analogue of a ground state in wave mechanics.

Therefore, in this phase, our material tends to fragment into clusters of density inhomogeneities with a Gaussian profile, referred here as cells. If this phase is a melt of some metal, then these inhomogeneities are more likely to become centers of crystalline nucleation when solidification begins.

Foam phase. If the nonlinear coupling is negative, then the field-theoretical potential density (7) has a conventional Mexican-hat shape, with a local maximum at $|\Psi| = 0$ and local degenerate minima at $|\Psi| = |\Psi_e|$; the latter correspond to a state with the lowest eigenvalue of frequency $\omega_g^{(-)} = 0$. This suggests that the model contains multiple topological sectors, and the topologically nontrivial solitons exists which interpolate between local minima.

Unfortunately, no topological solutions of eq. (3) can be found analytically in this phase, but numerical study reveals their existence. In Cartesian coordinates, this solution is a product of $\bar{d}$ 1D kink solitons saturating the Bogomolny-Prasad-Sommerfield (BPS) bound,

$$\Psi_s^{(-)}(\mathbf{x},t) = \prod_{j=1}^{\bar{d}}\psi_j(x_j,t), \tag{12}$$

and each of these 1D kinks have a nonzero topological charge,

$$Q = \varrho_0^{-1/2} \left[\psi_j(x_j \to +\infty) - \psi(x_j \to -\infty) \right], \quad j = 1, .., \bar{d}, \tag{13}$$

where $\varrho_0 = \rho_0^{1/\bar{d}}$. Numerical plots of these solutions and their densities can be found in figure 2 of ref. [19]. Despite this solution is not the lowest-frequency one, a nonzero topological charge enhances its stability against decay into a ground state Ψ_e. According to a value of charge Q, all non-singular finite-energy solutions can be cast into four topological sectors [10]. For two of these sectors, topological charge does not vanish, which ensures stability of corresponding BPS solitons.

Furthermore, the mass density of the soliton, $|\Psi_s^{(-)}|^2$, grows from its center of mass outwards, cf. figure 2 of ref. [19], therefore this solution can be viewed as a bubble with a characteristic size ℓ. In a single-solution setup this solution would extend for the whole space, but in the real material the multi-soliton picture must occur: kinks and antikinks would alternate and match each other at distances of an order ℓ. Therefore, a Korteweg material tends to form a foam-like structure in this phase. This process facilitates the release of previously dissolved gas. It can result in either boiling during the liquid-gas transition or in the formation of cavities in the bulk of the material, such as pores and blowholes caused by decreased hydrogen solubility or carbon monoxide production during solidification. Therefore, the above-mentioned phases can be used for modeling of both the liquid-gas transitions and the liquid-solid ones, depending on whether the temperature of the foam phase is larger or smaller than the temperature of the cellular phase.

4. Large-scale structure of metals

The crystalline structure appears in metals during a solidification process. In pure metals, this structure is merely a crystal lattice, which can be of a simple cube, body-centered cube and face-centered cube type. In real metals, due to various impurities and inhomogeneities, a much more complicated pattern arises. Throughout the melt, multiple centers of nucleation occur. Crystals begin to grow from those centers until the reach each other in the bulk. This is where interfaces develop marking borders of single-crystal domains grains, each of about a micrometre scale. Finally, the non-crystal polycrystalline structure occurs as a macroscopic large-scale pattern of adjacent grains.

The question arises: how orderly is this large-scale pattern? Is the size and mass of each grain a totally random value? Apart from purely academic interest, such questions have an immediate practical importance. Each grain's microscopic hardness depends on its size and density distribution therein. Therefore, if one controls sizes of grains, at least on average, one can control the macroscopic properties of the whole material, such as hardness, plasticity and mass. This is of utmost importance for industrial applications.

The conventional statistical mechanics of the nucleation process is extremely complicated, due to an obvious complexity of this process and large number of effects and acting agents involved, both physical and chemical. In most cases, one ends up applying the "common-wisdom" arguments: since the impurities and hence nucleation centers are randomly distributed in the molt bulk, one would expect that distances between those centers are going to be random too. Therefore, one would expect that the distribution of sizes of grains would be close to uniform, hence no periodic pattern in the large-scale structure in the polycrystalline metals can form.

A theory described in previous sections gives a different answer to this question, which is also a more specific one. If the melt can be described by a Korteweg-type system, at least in the leading-order approximation, then, according to our model, its density has a tendency to form a periodic pattern of repeating solitonic solutions (gaussons or kinks, depending on a phase).

Journal of Physics: Conference Series **1416** (2019) 012020 doi:10.1088/1742-6596/1416/1/012020

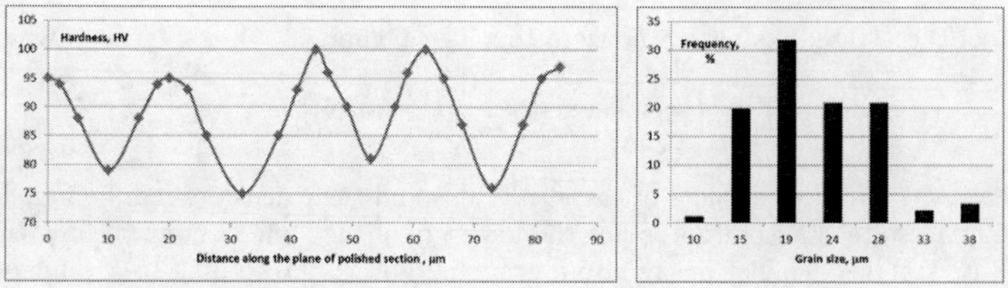

(*a*) (*b*)

Figure 1. Hardness profile (*a*) and distribution of grain sizes (*b*) for the structural steel S235/A570 Grade 36.

In other words, the density inhomogeneities emerge in a melt on a large scale which are not directly related to molecular bonding but they are a collective nonlinear phenomenon. During a solidification process, those points where density profile reaches extrema are more probable to become centers of nucleation. Therefore, an average size of a grain can not be a random value, but it must be close to a width ℓ of the solutions of a previous section, cf. formula (11).

5. Experiment

We consider a few most common polycrystalline metallic materials: non-alloy structural steel S235/A570 Grade 36, copper C-Cu/C14200, stainless steel X10CrNiTi18-10/AISI 321, and aluminium-magnesium alloy 5083/5056.

Microstructure of the steel S235/A570 consists of the ferrite and pearlite phases, with a 75 : 25 proportion, respectively. Ferrite grains have mainly an equiaxial structure, except a few stringer grains of an elongated shape. They have no particular spatial orientation, and their size varies around 20 μm. Sizes of the pearlitic microconstituent grains for this steel vary between 5 and 45 μm.

Microstructure of the copper specimen consists of approximately equiaxed grains of a size about 100 μm. As in the previous material, a few stringer grains of an elongated shape are present. Inside grains one can find deformation twins.

Microstructure of the stainless steel is austenitic and consists of polyhedral grains of a size about 100 μm. As in the copper specimen above, one can encounter deformation twins in this steel.

Microstructure of the aluminium alloy specimen consists of approximately equiaxed grains of a size about 30 μm. A small amount of secondary ferric and silicon phases can be found both inside grains and on their borders.

The microhardness measurements of microstructure components were performed for above-mentioned metals by using the Vickers hardness method. The load during the measurements was chosen according to a size of indentation in a specimen. The minimal load was chose to be either 0.025 or 0.05 N, depending on a grain's size. The points of load's application were in the following areas of a grain: grain's center, half-way to the grain's border, near the border, and on the border.

The profile of a microhardness throughout the structured steel's specimen is given in figure 1a. Since the hardness is related to a density of the material, its pattern immediately suggest a presence of a large-scale structure. The statistical analysis confirms that the average size of grains is not random: its distribution is far from being uniform but has a peak, cf. figure 1b. Other materials from those mentioned above demonstrate similar patterns to figure 1, and thus they are not listed here.

6. Conclusion

Wave equations with logarithmic nonlinearity are applied to Korteweg-type materials which can undergo liquid-solid or liquid-gas phase transitions. These materials have a two-phase structure where each phase is marked by a sign of the nonlinear coupling. Depending on whether the nonlinear coupling is positive or negative, solutions can have both trivial and non-trivial topology.

From a physical point of view, such solutions describe the inhomogeneities of density, which can occur in the form of bubbles or cells in the vicinity of liquid-solid phase transition. For instance, during the solidification process, these inhomogeneities become centers of nucleation of grains in the Korteweg-type material. One of theory's predictions is a large-scale periodical patterns, which is opposite to the randomness expectations based on plain statistical arguments. Due to numerous additional effects occurring during the solidification process, this periodicity gets distorted but not entirely destroyed.

Previous works were dealing with generic natural silicate materials in geophysics, such as magmas in volcanic conduits, where the (approximately) periodical flows and structures are known, see works [8, 18, 19] and references therein. We conducted experimental studies of structure of grains in structural steel S235/A570 Grade 36, copper C-Cu/C14200, stainless steel X10CrNiTi18-10/AISI 321, and aluminium-magnesium alloy 5083/5056. Based on results, we thus report an experimental evidence of the above-mentioned periodicity in these materials' structure. We show that an average size of grains in metals should be closely related to the average width of the soliton-type solutions ℓ.

Acknowledgments

K.G.Z.'s research is supported by Department of Higher Education and Training of South Africa and in part by National Research Foundation of South Africa.

References

[1] Rosen G 1968 *J. Math. Phys.* **9** 996
[2] Bialynicki-Birula I and Mycielski J 1976 *Ann. Phys. (N. Y.)* **100** 62
[3] Buljan H, Šiber A, Soljačić M, Schwartz T, Segev M and Christodoulides D N 2003 *Phys. Rev. E* **68** 036607
[4] Hefter E F 1985 *Phys. Rev. A* **32** 1201
[5] Yasue K 1978 *Ann. Phys. (N.Y.)* **114** 479
[6] Brasher J D 1991 *Int. J. Theor. Phys.* **30** 979
[7] Schuch D 1997 *Phys. Rev. A* **55** 935
[8] De Martino S, Falanga M, Godano C and Lauro G 2003 *Europhys. Lett.* **63** 472
[9] Zloshchastiev K G 2010 *Grav. Cosmol.* **16** 288
[10] Zloshchastiev K G 2011 *Acta Phys. Polon.* **42** 261
[11] Avdeenkov A V and Zloshchastiev K G 2011 *J. Phys. B: At. Mol. Opt. Phys.* **44** 195303
[12] Zloshchastiev K G 2012 *Eur. Phys. J. B* **85** 273
[13] Dzhunushaliev V and Zloshchastiev K G 2013 *Central Eur. J. Phys.* **11** 325
[14] Dzhunushaliev V, Makhmudov A and Zloshchastiev K G 2016 *Phys. Rev. D* **94** 096012
[15] Scott T C, Zhang X, Mann R B and Fee G J 2016 *Phys. Rev. D* **93** 084017
[16] Scott T C and Shertzer J 2018 *J. Phys. Commun.* **2** 075014
[17] Zloshchastiev K G 2018 *Z. Naturforsch. A* **73** 619
[18] Lauro G 2008 *Geophys. Astrophys. Fluid Dyn.* **102** 373
[19] Zloshchastiev K G 2018 *Europhys. Lett. (EPL)* **122** 39001
[20] Dunn J E and Serrin J B 1985 *Arch. Rat. Mech. Anal.* **88** 95
[21] Anderson D M, Mc Fadden G B and Wheeler A A 1998 *Annu. Rev. Fluid Mech.* **30** 139
[22] Rylov Yu A 1999 *J. Math. Phys.* **40** 256

Application of geometrical methods to study the systems of differential equations for quantum-mechanical problems

Nina Krylova[1,2], **Yanina Voynova**[3], **Vladimir Balan**[4]
[1]Belarusian State University, Minsk, Belarus
[2]Belarusian State Agrarian Technical University, Minsk, Belarus
[3]B.I.Stepanov Institute of Physics of the National Academy of Science of Belarus, Minsk, Belarus
[4]University Politehnica of Bucharest, Bucharest, Romania

E-mail: krylovang@bsu.by

Abstract. A geometrical method based on the structural stability theory is used to study systems of differential equations which arise in quantum-mechanical problems. We consider a 1/2-spin particle in external Coulomb field or in the presence of magnetic charge on the background of the de-Sitter space, and a free 3/2-spin particle in spherical coordinates of the flat space. It turns out that the first and the second Kosambi-Cartan-Chern invariants are nontrivial for the corresponding systems, while the 3-d, 4-th and 5-th invariants identically vanish. From physical point of view, the second invariant determines how rapidly the different branches of the solution diverge from or converge to the intersection points, while the most interesting are the singular points. The convergence (divergence) near the singular points $r = 0, \infty$ are shown to correlate with the behavior of solutions for quantum mechanical states (discrete and continuous spectra). The vanishing of the 3-d, 4-th and 5-th invariants geometrically implies the existence of a nonlinear connection on the tangent bundle, having zero torsion and curvature.

1. Introduction

In many situations of physical interest, the equations of motion of a dynamical system follow from a Lagrangian L via the Euler-Lagrange equations which may be written as:

$$\dot{y}^i(r) + 2Q^i(r, x, y) = 0, \tag{1}$$

where x^i are $r-$dependent coordinates along the paths, $y^i = dx^i/dr = \dot{x}^i$ and

$$Q^i = \frac{1}{4} g^{il} \left(\frac{\partial^2 L}{\partial x^k \partial y^l} y^k - \frac{\partial L}{\partial x^l} + \frac{\partial^2 L}{\partial y^l \partial r} \right), \quad g_{ij} = \frac{1}{2} \frac{\partial^2 L}{\partial y^i \partial y^j}. \tag{2}$$

The geometrical study of these systems of equations is based on the use of the KCC (Kosambi–Cartan–Chern) invariants. The general KCC-theory is developed in detail in numerous works, e.g., [1, 2, 3].

The first KCC-invariant ε^i, which can be interpreted as an external force, is given by

$$\varepsilon^i(r, x, y) = \frac{\partial Q^i}{\partial y^j} y^j - 2Q^i. \tag{3}$$

Journal of Physics: Conference Series **1416** (2019) 012021 doi:10.1088/1742-6596/1416/1/012021

The second KCC-invariant $P^i_{\ j}$ is defined by [1]:

$$P^i_{\ j} = 2\frac{\partial Q^i}{\partial x^j} + 2Q^s \frac{\partial^2 Q^i}{\partial y^j \partial y^s} - \frac{\partial^2 Q^i}{\partial y^j \partial x^s} y^s - \frac{\partial Q^i}{\partial y^s}\frac{\partial Q^s}{\partial y^j} - \frac{\partial^2 Q^i}{\partial y^j \partial r}, \tag{4}$$

and relates to the Jacobi stability of the dynamical system. There is an analogy between the equations of geodesic deviation expressed in terms of the Riemann curvature:

$$\frac{D^2\xi^i}{Ds^2} = R^i_{kjl}\frac{dx^k}{ds}\frac{dx^l}{ds}\xi^j = -K^i_j\xi^j$$

and in terms of the second KCC-invariant

$$\frac{D^2\xi^i}{Dr^2} = P^i_j\xi^j.$$

It is known that a pencil of geodesic curves emanating from some point r_0 converges (or diverges) if the real parts of all the eigenvalues of the 2-nd KCC-invariant $P^i_{\ j}$ are negative (or positive ones).

The third KCC-invariant

$$R^i_{jk} = \frac{1}{3}\left(\frac{\partial P^i_j}{\partial y^k} - \frac{\partial P^i_k}{\partial y^j}\right) \tag{5}$$

determines the torsion of the Berwald connection.

The fourth KCC-invariant is an extension of the Riemann–Christoffel tensor,

$$B^i_{jkl} = \frac{\partial R^i_{jk}}{\partial y^l}. \tag{6}$$

Finally, the fifth KCC-invariant, which extends the Douglas tensor, is provided by

$$D^i_{jkl} = \frac{\partial^3 Q^i}{\partial y^j \partial y^k \partial y^l}. \tag{7}$$

It is important to note that if the vector field Q^i is linear in the coordinates x^i and y^i, then the first and the second invariants are functions of the radial coordinate r, do not depend on x^i and y^i, and the remaining invariants identically vanish.

2. Spin 1/2 particle in the Coulomb field in Minkowski space

The Dirac equation for the particle in external Coulomb field gives the system [4]:

$$\begin{aligned} &\left(\frac{d}{dr} + \frac{\nu}{r}\right) f + \left(E + \frac{e}{r} + m\right) g = 0, \\ &\left(\frac{d}{dr} - \frac{\nu}{r}\right) g - \left(E + \frac{e}{r} - m\right) f = 0, \end{aligned} \tag{8}$$

where $\nu = j + 1/2$ and $j = 1/2, 3/2, \ldots$. In terms of the variable $z = 2\lambda r$, $\lambda = \sqrt{m^2 - E^2}$, and of the new functions f_1, g_1 given by the relations:

$$f = \sqrt{m+E}(f_1 + g_1)\phi(z), \quad g = \sqrt{m-E}(f_1 - g_1)\phi(z), \tag{9}$$

where $\mu = em/\lambda$, $\epsilon = eE/\lambda$, one gets the equations with separated variables:

$$\frac{d^2 f_1}{dz^2} + \left(\frac{2\epsilon+1}{2z} - \frac{1}{4} + \frac{1+4(\mu^2-\nu^2-\epsilon^2)}{4z^2}\right) f_1 = 0,$$
$$\frac{d^2 g_1}{dz^2} + \left(\frac{2\epsilon-1}{2z} - \frac{1}{4} + \frac{1+4(\mu^2-\nu^2-\epsilon^2)}{4z^2}\right) g_1 = 0. \tag{10}$$

Geometrically, the first and the second invariants (3–4) for the last system (10) are given by the formulas

$$\varepsilon^1 = \left(\tfrac{1}{4} - \tfrac{2\epsilon+1}{2z} - \tfrac{1+4(\mu^2-\nu^2-\epsilon^2)}{4z^2}\right) f_1,$$
$$\varepsilon^2 = \left(\tfrac{1}{4} - \tfrac{2\epsilon-1}{2z} - \tfrac{1+4(\mu^2-\nu^2-\epsilon^2)}{4z^2}\right) g_1,$$
$$(P^i{}_j) = \begin{pmatrix} -\frac{1}{4} + \frac{2\epsilon+1}{2z} + \frac{1+4(\mu^2-\nu^2-\epsilon^2)}{4z^2} & 0 \\ 0 & -\frac{1}{4} + \frac{2\epsilon-1}{2z} + \frac{1+4(\mu^2-\nu^2-\epsilon^2)}{4z^2} \end{pmatrix}.$$

We further study the behavior of the eigenvalues of $P^i{}_j$ near the singular points $z = 0$, $z = \infty$:

$$z \to 0 \quad \Rightarrow \lambda_{1,2} \to \tfrac{1+4(e^2-\nu^2)}{2z^2} \approx \tfrac{1-4\nu^2}{2z^2} < 0;$$
$$z \to \infty \quad \Rightarrow \lambda_{1,2} \to -\tfrac{1}{4} < 0.$$

This correlates with the behavior of solutions near the points $r = 0, \infty$ for quantum mechanical states (discrete and continuous spectra).

We can construct the Lagrangian in the form

$$L = g_{ij} y^i y^j + b_j(x, r) y^j. \tag{11}$$

Taking into account the linearity of the differential equations (10) in the coordinates, we will search the 1-form $b_j(x,r)$ as a linear combination of the coordinates, namely $b_j(x,r) = h_{ij}x^i$. Then, using the above definition (2), and substituting L from (11) ($L \rightsquigarrow Q^i$), one gets

$$Q^i = \frac{1}{4} g^{il}\left(h_{jl}y^j - h_{lj}y^l + \frac{\partial g_{jl}}{\partial r} y^j + \frac{\partial h_{jl}}{\partial r} x^j\right). \tag{12}$$

Assuming the symmetry $h_{jl} = h_{lj}$ and $g_{ij} = \delta_{ij}$, we find the semispray coefficients

$$Q^1 = \tfrac{1}{4}\tfrac{dh_{11}}{dz} = \tfrac{1}{2}\left(\tfrac{2\epsilon+1}{2z} - \tfrac{1}{4} + \tfrac{1+4(\mu^2-\nu^2-\epsilon^2)}{4z^2}\right),$$
$$Q^2 = \tfrac{1}{4}\tfrac{dh_{22}}{dz} = \tfrac{1}{2}\left(\tfrac{2\epsilon-1}{2z} - \tfrac{1}{4} + \tfrac{1+4(\mu^2-\nu^2-\epsilon^2)}{4z^2}\right).$$

By integrating the last relations, one gets the Lagragian:

$$L = \dot f_1^2 + \dot g_1^2 - \left(\frac{1}{2}z - 2\epsilon \ln z + \frac{1+4(\mu^2-\nu^2-\epsilon^2)}{2z}\right)\left(\dot f_1 f_1 + \dot g_1 g_1\right) + \ln z\left(\dot f_1 f_1 - \dot g_1 g_1\right).$$

A similar study can be done for the original system (8) in terms of $f(r)$ and $g(r)$. By differentiating the equations with respect to r, the system becomes

$$\tfrac{d^2 f}{dr^2} + \tfrac{\nu}{r}\tfrac{df}{dr} + \left(E + \tfrac{e}{r} + m\right)\tfrac{dg}{dr} - \tfrac{\nu f + eg}{r^2} = 0,$$
$$\tfrac{d^2 g}{dr^2} - \tfrac{\nu}{r}\tfrac{dg}{dr} - \left(E + \tfrac{e}{r} - m\right)\tfrac{df}{dr} + \tfrac{\nu g + ef}{r^2} = 0. \tag{13}$$

Now, the first and second invariants are

$$\begin{aligned}\varepsilon^1 &= -\tfrac{1}{2}\left(E+\tfrac{e}{r}+m\right)\tfrac{dg}{dr}-\tfrac{\nu}{2r}\tfrac{df}{dr}+\tfrac{e}{r^2}g+\tfrac{\nu}{r^2}f,\\ \varepsilon^2 &= \tfrac{1}{2}\left(E+\tfrac{e}{r}-m\right)\tfrac{df}{dr}+\tfrac{\nu}{2r}\tfrac{dg}{dr}-\tfrac{e}{r^2}f-\tfrac{\nu}{r^2}g,\end{aligned}$$

$$(P^i{}_j)=\begin{pmatrix}\frac{1}{4}\left(E+\frac{e}{r}\right)^2-\frac{m^2}{4}-\frac{\nu(\nu+2)}{4r^2} & -\frac{e}{2r^2}\\ \frac{e}{2r^2} & \frac{1}{4}\left(E+\frac{e}{r}\right)^2-\frac{m^2}{4}-\frac{\nu(\nu-2)}{4r^2}\end{pmatrix}.$$

The eigenvalues of $P^i{}_j$ are given by the expressions:

$$\begin{aligned}\lambda_1 &= \tfrac{-2\sqrt{\nu^2-e^2}-\nu^2}{4r^2}+\tfrac{1}{4}\left(E+\tfrac{e}{r}\right)^2-\tfrac{m^2}{4},\\ \lambda_2 &= \tfrac{2\sqrt{\nu^2-e^2}-\nu^2}{4r^2}+\tfrac{1}{4}\left(E+\tfrac{e}{r}\right)^2-\tfrac{m^2}{4}.\end{aligned}$$

Near the singularities of the system, the eigenvalues turn to

$$r\to 0\Rightarrow\begin{cases}\lambda_1\to\frac{1-\left(1+\sqrt{\nu^2-e^2}\right)^2}{4r^2}<0,\\ \lambda_2\to\frac{1-\left(1-\sqrt{\nu^2-e^2}\right)^2}{4r^2},\ \text{and } \lambda_2<0 \text{ for } \nu^2-e^2>4,\end{cases}$$

$$r\to\infty\Rightarrow\lambda_1=\lambda_2\to(E^2-m^2)/4.$$

The behavior of eigenvalues near the point $r=0$ is slightly different, comparing with the previous case of separated variables f_1, g_1. This may be explained as follows. Strictly speaking, direct differentiation of the first order system leads to the non-equivalent mathematical problem

$$\begin{aligned}&\hat{A}F=0\rightsquigarrow\tfrac{d}{dr}\hat{A}F=0,\\ &\hat{A}\tilde{F}=const\rightsquigarrow\tfrac{d}{dr}\hat{A}\tilde{F}=0,\quad \tilde{F}\neq F.\end{aligned}$$

The system for the radial functions $f(r),g(r)$ (8) may be written in the equivalent form as

$$\begin{aligned}&\tfrac{d^2f}{dr^2}-\left(\tfrac{\nu(\nu+1)}{r^2}-\left(E+\tfrac{e}{r}\right)^2+m^2\right)f-\tfrac{e}{r^2}g=0,\\ &\tfrac{d^2g}{dr^2}-\left(\tfrac{\nu(\nu-1)}{r^2}-\left(E+\tfrac{e}{r}\right)^2+m^2\right)g+\tfrac{e}{r^2}f=0.\end{aligned}\tag{14}$$

Now, the first and the second invariants are

$$\begin{aligned}\varepsilon^1 &= \left(\tfrac{\nu(\nu+1)}{r^2}-\left(E+\tfrac{e}{r}\right)^2+m^2\right)f+\tfrac{e}{r^2}g,\\ \varepsilon^2 &= \left(\tfrac{\nu(\nu-1)}{r^2}-\left(E+\tfrac{e}{r}\right)^2+m^2\right)g-\tfrac{e}{r^2}f,\end{aligned}$$

$$(P^i{}_j)=\begin{pmatrix}\left(E+\frac{e}{r}\right)^2-m^2-\frac{\nu(\nu+1)}{r^2} & -\frac{e}{r^2}\\ \frac{e}{r^2} & \left(E+\frac{e}{r}\right)^2-m^2-\frac{\nu(\nu-1)}{r^2}\end{pmatrix}.$$

The eigenvalues of $P^i{}_j$ are

$$\begin{aligned}\lambda_1 &= \tfrac{-\sqrt{\nu^2-e^2}-\nu^2}{r^2}+\left(E+\tfrac{e}{r}\right)^2-m^2,\\ \lambda_2 &= \tfrac{\sqrt{\nu^2-e^2}-\nu^2}{r^2}+\left(E+\tfrac{e}{r}\right)^2-m^2.\end{aligned}$$

Journal of Physics: Conference Series **1416** (2019) 012021 doi:10.1088/1742-6596/1416/1/012021

In this case, near the singular points $r = 0$ and ∞ the eigenvalues behave as

$$r \to 0 \Rightarrow \begin{cases} \lambda_1 \to \frac{1-\left(1+2\sqrt{\nu^2-e^2}\right)^2}{4r^2} < 0, \\ \lambda_2 \to \frac{1-\left(1-2\sqrt{\nu^2-e^2}\right)^2}{4r^2}, \text{ with } \lambda_2 < 0 \text{ for } \nu^2 - e^2 > 1, \end{cases}$$

$$r \to \infty, \Rightarrow \lambda_1 = \lambda_2 \to (E^2 - m^2).$$

So, for bound states, $(E^2 - m^2) < 0$. We conclude that the choice of the concrete representation for the system of equations slightly influences the behavior of the eigenvalues of the second invariant.

The corresponding Lagrangian is given by the expression:

$$L = \dot{f}^2 - \dot{g}^2 + 2\left(\frac{\nu^2 - e^2}{r} + (E^2 - m^2)\, r + 2eE \ln r\right)\left(\dot{f}f - \dot{g}g\right) + \frac{2\nu}{r}\left(\dot{f}f + \dot{g}g\right) + \frac{2e}{r}\left(g\dot{f} + f\dot{g}\right). \tag{15}$$

3. Spin 1/2 particle in the field of Abelian monopole on the background of de Sitter space-time

The study of Dirac equation in the field of Dirac monopole on the background of de Sitter space-time (in static coordinates) may be reduced to two separated equations [5]:

$$\frac{d^2 f}{dz^2} + \frac{3 + 4z\,(1-z)\,[4U_1(z) - 1]}{16z^2\,(1-z)^2}\, f(z) = 0,$$
$$\frac{d^2 g}{dz^2} + \frac{3 + 4z\,(1-z)\,[4U_2(z) - 1]}{16z^2\,(1-z)^2}\, g(z) = 0, \tag{16}$$

where the following notations are used:

$$U_1(z) = -\frac{1}{4}\left(M - \frac{i}{2}\right)^2 + \frac{\epsilon\,(\epsilon - i)}{4(1-z)} - \frac{\alpha(\alpha + 1t)}{4z},$$

$$U_2(z) = -\frac{1}{4}\left(M - \frac{i}{2}\right)^2 + \frac{\epsilon\,(\epsilon + i)}{4(1-z)} - \frac{\alpha(\alpha - 1)}{4z}.$$

We note the symmetry between the two equations:

$$f \rightsquigarrow g, \qquad \epsilon \rightsquigarrow -\epsilon, \qquad \alpha \rightsquigarrow -\alpha. \tag{17}$$

Now the first and second invariants are given by expressions:

$$\varepsilon^1 = \frac{4\epsilon(3-\epsilon)z - \left((3-2M)^2+4\right)z(z-1) - 4\alpha(\alpha+1)(z-1) - 3}{16(z-1)^2 z^2}\, f,$$
$$\varepsilon^2 = \frac{-4\epsilon(3+\epsilon)z - \left((3-2M)^2+4\right)z(z-1) - 4\alpha(\alpha-1)(z-1) - 3}{16(z-1)^2 z^2}\, g,$$

$$(P^i{}_j) = \begin{pmatrix} \frac{\left((3-2M)^2+4\right)z(z-1) + 4\epsilon(\epsilon-3)z + 4\alpha(\alpha+1)(z-1) + 3}{16(z-1)^2 z^2} & 0 \\ 0 & \frac{\left((3-2M)^2+4\right)z(z-1) + 4\epsilon(\epsilon+3)z + 4\alpha(\alpha-1)(z-1) + 3}{16(z-1)^2 z^2} \end{pmatrix}.$$

The behavior of the eigenvalues of the second invariant $P^i{}_j$ in the neighborhood of the singular points is as follows

$$z \to 0 \Rightarrow \lambda_{1,2} \to \frac{3-4\alpha(\alpha \pm 1)}{16z^2} < 0, \qquad \alpha = \sqrt{(j+1/2)^2 - k^2} > 1;$$

$$z \to 1 \Rightarrow \lambda_{1,2} \to \frac{3+4\epsilon(\epsilon \mp i)}{16(z-1)^2}, \quad \mathrm{Re}[\lambda_{1,2}] > 0.$$

The behavior at $z \to 1$ correlates with the structure of the solutions near the de Sitter horizon. For this model, the Lagrangian is determined by the expression

$$\begin{aligned} L &= \dot{f}^2 + \dot{g}^2 \\ &+ \tfrac{1}{8}\left[\tfrac{-3+4\epsilon(i-\epsilon)}{z-1} + \tfrac{-3+4\alpha(1+\alpha)}{z} + 2(-2+4\epsilon(i-\epsilon)+4\alpha(1+\alpha)+(i-2M)^2)\arctan[1-2z]\right] f\dot{f} \\ &+ \tfrac{1}{8}\left[\tfrac{-3-4\epsilon(i+\epsilon)}{z-1} + \tfrac{-3+4\alpha(\alpha-1)}{z} + 2(-2-4\epsilon(i+\epsilon)+4\alpha(\alpha-1)+(i-2M)^2)\arctan[1-2z]\right] g\dot{g}. \end{aligned}$$

4. Spherical solutions for the spin 3/2 particle

As has been shown in [6], the problem of finding of the spherically symmetric solutions of the wave equation for the spin 3/2 particle reduces to two systems, each of them consisting of three equations ($i, j \in \{1,2,3\}$)

$$\begin{aligned} &\frac{d^2F^i}{dr^2} + (\epsilon^2 - m^2)F^i - \frac{1}{r^2}T^i_jF^j = 0, \\ &\frac{d^2G^i}{dr^2} + (\epsilon^2 - m^2)G^i - \frac{1}{r^2}T'^i_jG^j = 0, \end{aligned} \tag{18}$$

where the mixing matrices T and T' have the form

$$T^i{}_j = \begin{pmatrix} b^2 & \sqrt{2}b & b \\ \sqrt{2}b & a^2+a+2 & \sqrt{2}(a+1) \\ b & \sqrt{2}(a+1) & b^2+2 \end{pmatrix}, \tag{19}$$

$$T'^i{}_j = \begin{pmatrix} b^2 & \sqrt{2}b & -b \\ \sqrt{2}b & a^2-a+2 & \sqrt{2}(a-1) \\ -b & \sqrt{2}(a-1) & b^2+2 \end{pmatrix}; \tag{20}$$

the parameters a and b depend on the total angular momentum

$$a = j + 1/2, \quad b = \sqrt{(j-1/2)(j+3/2)}.$$

The second invariants in explicit form are

$$(P_F)^i{}_j = \begin{pmatrix} -\frac{b^2}{r^2} + \epsilon^2 - m^2 & -\frac{\sqrt{2}b}{r^2} & -\frac{b}{r^2} \\ -\frac{\sqrt{2}b}{r^2} & -\frac{a^2+a+2}{r^2} + \epsilon^2 - m^2 & -\frac{\sqrt{2}(a+1)}{r^2} \\ -\frac{b}{r^2} & -\frac{\sqrt{2}(a+1)}{r^2} & -\frac{b^2+2}{r^2} + \epsilon^2 - m^2 \end{pmatrix}, \tag{21}$$

$$(P_G)^i{}_j = \begin{pmatrix} -\frac{b^2}{r^2} + \epsilon^2 - m^2 & -\frac{\sqrt{2}b}{r^2} & \frac{b}{r^2} \\ -\frac{\sqrt{2}b}{r^2} & -\frac{a^2-a+2}{r^2} + \epsilon^2 - m^2 & -\frac{\sqrt{2}(a-1)}{r^2} \\ \frac{b}{r^2} & -\frac{\sqrt{2}(a-1)}{r^2} & -\frac{b^2+2}{r^2} + \epsilon^2 - m^2 \end{pmatrix}. \tag{22}$$

The eigenvalues λ_F^i and λ_G^i for the invariants P_F and P_G are given by the formulas:

$$\lambda_F^1 = \lambda_F^2 = \epsilon^2 - m^2 - \frac{(2j-1)(2j+1)}{4r^2}, \quad \lambda_F^3 = \epsilon^2 - m^2 - \frac{(2j+3)(2j+5)}{4r^2};$$
$$\lambda_G^1 = \lambda_G^2 = \epsilon^2 - m^2 - \frac{(2j+1)(2j+3)}{4r^2}, \quad \lambda_G^3 = \epsilon^2 - m^2 - \frac{(2j-1)(2j-3)}{4r^2}. \tag{23}$$

Near the singular point, they become simpler

$$r \to 0 \Rightarrow \lambda_i \to -\infty < 0, \qquad r \to \infty \Rightarrow \lambda_i \to \epsilon^2 - m^2 > 0,$$

which agree with the structure of the solution of continuous spectrum.

5. Conclusions

We have considered a 1/2-spin particle in Coulomb field and in the presence of magnetic charge on the background of the de-Sitter space, and a free 3/2-spin particle in spherical coordinates of the flat space. It turns out that the first and the second Kosambi-Cartan-Chern invariants are nontrivial for these systems, while the 3-d, 4-th and 5-th invariants identically vanish. The first invariant determines the vector field on the configuration space of the differential system, and is interpreted as an external field potential. From physical point of view, the second invariant determines how rapidly the different branches of the solutions diverge from or converge to the intersection point, while the singular points are most relevant. The convergence (divergence) near the singular points $r = 0, \infty$ correlates with the behavior of solutions for quantum mechanical states (discrete and continuous spectra).

It is shown that the choice of the concrete representation for the system of equations slightly influences on behavior of the eigenvalues of the second invariant.

The vanishing of the 3-d, 4-th and 5-th invariants means that, in geometrical terms, there exists a nonlinear connection on the tangent bundle, with zero torsion and curvature.

6. Acknowledgments

The present work was developed under the auspices of the Project No F19M-032 from Belarusian Republican Foundation for Fundamental Research and the Project BRFFR-RA No F18RA-015, within the cooperation framework between Romanian Academy and Belarusian Republican Foundation for Fundamental Research.

7. References

[1] Atanasiu G, Balan V, Brinzei N and Rahula M 2010 *Differential geometry of the second order and applications: Miron-Atanasiu theory* (Moscow: URSS).
[2] Antonelli P L, Bevilacqua L and Rutz S F 2003 *Nonlinear Analysis* **4** 743
[3] Antonelli P L and Bucataru I 2001 *An.St. Univ. "Al.I.Cuza" Iasi, Mat. N.S.* **47** 405
[4] Balan V, Manukyan A M, Ovsiyuk E M, Redkov V M and Veko O V 2015 *Applied Sciences* **17** 9
[5] Veko O V, Kazmerchuk K V, Ovsiyuk E M, Kisel V V and Redkov V M 2014 *J. Nonlinear Phenomena in Complex Systems* **17, No 4** 461
[6] Ivashkevich A V, Ovsiyuk E M, Kisel V V and Redkov V M 2019 *Reports of NASB* **63, No 3** 282

Thermodynamics of AdS_5 black holes: holographic QCD and Stückelberg model

Eugenio Megías[1,2]

[1]Departamento de Física Atómica, Molecular y Nuclear and Instituto Carlos I de Física Teórica y Computacional, Universidad de Granada, Avenida de Fuente Nueva s/n, E-18071 Granada, Spain

[2]Departamento de Física Teórica, Universidad del País Vasco UPV/EHU, Apartado 644, E-48080 Bilbao, Spain

E-mail: emegias@ugr.es

Abstract. We explore the thermodynamics of AdS_5 black holes in two models: i) an improved holographic QCD model with a simple dilaton potential, and ii) the Stückelberg model in 5D. In the former case, by applying techniques of singular perturbation theory, we obtain a resummation of the naive expansion at high temperatures, providing a good fit to the lattice data for the trace anomaly. In the latter, we find a solution of the equations of motion by considering an expansion in the conformal dimension of the current associated to the gauge field.

1. Introduction

The gauge/gravity duality is a powerful tool to study the properties of gauge theories, and in particular of QCD, in their strongly coupled regime. Using this duality, the thermodynamics of a field theory can be obtained from the classical computation of the thermodynamics of black holes in the gravity dual. The entropy of a black hole can be computed from the famous Bekenstein-Hawking entropy formula [1, 2, 3, 4, 5]. In conformal AdS_5 the entropy scales like $S_{\text{Black Hole}} \propto r_h^3 \propto T^3$; however, in order to have a reliable extension of this duality to $SU(N_c)$ Yang-Mills theory, the first task is to control the breaking of conformal invariance. Conformal transformations are broken by quantum corrections due to the necessary regularization of the UV divergences, yielding the so-called *trace anomaly* [6, 7]

$$\langle T^\mu_\mu \rangle = -\varepsilon + 3p = -\frac{\beta(g)}{2g^3}\langle (F^a_{\mu\nu})^2 \rangle \,, \tag{1}$$

where $T^{\mu\nu}$ is the energy-momentum tensor, ε is the energy density, p is the pressure and β is the beta function. We will explore the application of techniques of singular perturbation theory to get a better control of the high temperature expansion of the thermodynamic quantities.

On the other hand, theories with a massive gauge field in the bulk of AdS_5 have been proposed as holographic duals of QCD with dynamical gauge fields. The relation of the Stückelberg field in holography to the axial anomaly has been first pointed out in Ref. [8]. Recent applications of the Stückelberg mechanism include the computation of anomalous effective actions [9], anomalous transport and their renormalization properties [10, 11], AdS/QCD [12] and beyond the Standard Model scenarios [13]. In the last part of this work we will study the thermodynamics of the Stückelberg model in 5D as well as the consistency of the entropy formula in this model.

2. The holographic QCD model

The bottom-up approach is based on the building of a gravity dual of QCD, including the main properties of this theory. We introduce in this section the model and study the solutions.

2.1. The model

One of the most successful holographic models of QCD within the bottom-up scenario is the 5D Einstein-dilaton model, with the Euclidean action [2]

$$S = \frac{1}{2\kappa^2} \int d\rho \, d^4x \sqrt{G} \Big(-R + G^{\mu\nu} \partial_\mu \Phi \partial_\nu \Phi + 2V(\Phi) \Big) + S_{\mathrm{GH}} \,, \tag{2}$$

where $\kappa^2 = 1/(2M^3)$ is the 5D Newton constant with M the Planck mass, and Φ is a scalar field to be identified with the Yang-Mills coupling through $g^2 = e^{\gamma\Phi}$. The boundary term S_{GH} is the usual Gibbons-Hawking contribution built up from the extrinsic curvature. The introduction of a scalar field breaks conformal invariance, and the form of the scalar potential $V(\Phi)$ is usually phenomenologically adjusted to describe some observables of QCD. We can introduce the superpotential, whose relation with the scalar potential is $V(\Phi) = \frac{1}{2} W'(\Phi)^2 - \frac{2}{3} W(\Phi)^2$. In this work we will consider the simple form of the superpotential

$$W(\Phi) = \ell^{-1}(3 + v_0 e^{\gamma\Phi}) \,, \tag{3}$$

where ℓ is the radius of the asymptotically AdS_5 background. The combination $\gamma^2 v_0/2 \equiv b_0$ can be taken from the one-loop β-function of a pure SU(N_c) gauge theory, i.e.

$$\beta(g) := \mu \frac{dg}{d\mu} = -b_0 g^3 = -\frac{11\,\mathrm{N_c}}{3(4\pi)^2} g^3 \,. \tag{4}$$

2.2. Equations of motion

We are interested in the study of finite temperature solutions of the Einstein-scalar gravity model corresponding to a black hole. In Fefferman-Graham coordinates the metric is

$$ds^2 = G_{\mu\nu} dx^\mu dx^\nu = \frac{\ell^2}{4\rho^2} d\rho^2 + \frac{\ell^2}{\rho} g_{\tau\tau}(\rho) d\tau^2 + \frac{\ell^2}{\rho} g_{xx}(\rho) d\vec{x}^2 \,. \tag{5}$$

The asymptotic expansions for the metric and the dilaton field in holographic QCD models have been studied in details in these coordinates [14]. This metric has a regular horizon at $\rho = \rho_h$, i.e. $g_{\tau\tau}(\rho_h) = g'_{\tau\tau}(\rho_h) = 0$, and the temperature and entropy density of the black hole are given by

$$T = \frac{1}{2\pi} \sqrt{2\rho_h g''_{\tau\tau}(\rho_h)} \,, \qquad s = \frac{2\pi}{\kappa^2} \left(\frac{\ell^2}{\rho_h} g_{xx}(\rho_h) \right)^{3/2} \,. \tag{6}$$

Using this metric, the equation for the scalar field, which is a second order differential equation, can be decoupled by an additional differentiation. Then it can be converted in autonomous by making the change of variables, $(\rho, \Phi) \to (\Phi, u(\Phi))$, where $u(\Phi) \equiv \rho \Phi'(\rho)$, leading to [1]

$$u''(\Phi) = \frac{\left(u'(\Phi)\right)^2}{u(\Phi)} - \frac{3\ell^2 V'(\Phi)}{4u(\Phi)^2} u'(\Phi) + \frac{\ell^4 \left(V'(\Phi)\right)^2}{16 u(\Phi)^3} + \frac{\ell^2 \left(8V(\Phi) + 3V''(\Phi)\right)}{12 u(\Phi)} \,. \tag{7}$$

At zero temperature, when $g_{xx} = g_{\tau\tau}$, the solution corresponds to a domain wall configuration, $u_0(\Phi) = \frac{\ell}{2} W'(\Phi) = \frac{\gamma v_0}{2} e^{\gamma\Phi}$, and it is given by

$$e^{\gamma\Phi(\rho)} = \frac{2}{\gamma^2 v_0 \log(\rho_0/\rho)} \,, \qquad g_{xx}(\Phi) = g_{\tau\tau}(\Phi) = g_{(0)} \cdot e^{-2\Phi/(3\gamma)} \,. \tag{8}$$

[1] We refer the reader to Ref. [15] for the remaining field equations involving $g_{xx}(\Phi)$ and $g_{\tau\tau}(\Phi)$

Using $g^2 = e^{\gamma\Phi}$ and Eq. (4), it turns out that the radial coordinate $\rho^{-1/2}$ can be interpreted as the renormalization group (RG) scale, and $\rho_0^{-1/2} \equiv \Lambda_{\rm QCD}$ is an integration constant corresponding to the location of the Landau pole of QCD. The other integration constant $g_{(0)}$ is arbitrary.

2.3. Asymptotically AdS black hole solutions from resummation

We now consider a black hole solution specified by the horizon data (ρ_h, Φ_h). Local analysis of the equations of motion shows that, in order to have a regular horizon at $\Phi = \Phi_h$, it is necessary that $\lim_{\Phi\to\Phi_h}\left(u(\Phi)g'_{xx}(\Phi) - g_{xx}(\Phi)\right) = 0$, which fixes the first term of the power series solution

$$u(\Phi) = \sum_{n=0}^{\infty} a_n(\Phi_h)(\Phi_h - \Phi)^{n+1/2}\,, \tag{9}$$

and similar expressions for $g_{xx}(\Phi)$ and $g_{\tau\tau}(\Phi)$. We will study the analytical solution in the regime $\nu \equiv \gamma v_0 e^{\gamma\Phi_h} \ll 1$ for any $\gamma > 0$ by using a boundary-layer analysis. In this regime the leading part of each coefficient of the series (9) behaves as $a_n(\Phi_h) \propto \nu^{-n+1/2}$. This suggests to consider an expansion of the form $u^{\rm in}(\Phi) = \sum_{n=1}^{\infty}\nu^n U_n((\Phi_h - \Phi)/\nu)$ which is valid in the region of boundary-layer near the horizon, where the inner variable, $\Psi \equiv (\Phi_h - \Phi)/\nu$, is $O(1)$. The first order of this expansion reads $u^{\rm in}(\Phi) = \frac{\nu}{2}\sqrt{1 - e^{-4\Psi}} + O(\nu^2)$. For the regime outside the boundary layer we tried a solution of the form $u^{\rm out}(\Phi) = \sum_{n=1}^{\infty}\nu^n u_n(\Phi_h - \Phi)$, and the corresponding substitution in Eq. (7) leads to $u^{\rm out}(\Phi) = \frac{\nu}{2}e^{\gamma(\Phi-\Phi_h)}$ to all orders in ν, so that the outer solution has the same form as the zero temperature solution. Thus the asymptotic matching produces a uniform approximation valid for $(-\infty, \Phi_h)$ given by

$$u^{\rm unif}(\Phi) = \frac{v_0\gamma}{2}e^{\gamma\Phi} + u^{\rm in}(\Phi) - \frac{v_0\gamma}{2}e^{\gamma\Phi_h} + O(\nu^2)\,, \tag{10}$$

so that $|u(\Phi) - u^{\rm unif}(\Phi)| = O(\nu^2)$. The last term in Eq. (10) is the common limit of the inner and outer approximation in the matching region, $\nu \ll \Phi_h - \Phi \ll 1$.

From the knowledge of the lowest orders of $u^{\rm in}(\Phi)$, it is possible to get in closed form the lowest orders in ν for the inner solutions of the metric, $g_{ij}^{\rm in}(\Phi) = \sum_n \nu^n g_{(n)ij}(\Psi)$, while the outer solution reduces to the zero temperature result given by Eq. (8). Finally, the asymptotic matching of the outer solution with the inner expansion determines the horizon quantities

$$g_{xx}(\Phi_h) = 2g_{(0)}e^{-2\Phi_h/(3\gamma)}\left(1 - \frac{\nu\log 2}{3\gamma} + O(\nu^2)\right)\,, \tag{11}$$

$$g'_{\tau\tau}(\Phi_h) = -\frac{8g_{(0)}e^{-2\Phi_h/(3\gamma)}}{\nu}\left(1 + \frac{\nu}{12\gamma}(4 + 3\gamma^2 - 4\log 2) + O(\nu^2)\right)\,. \tag{12}$$

Note that the second derivative in Eq. (6) is written as $g''_{\tau\tau}(\rho_h) = \rho_h^{-2}\lim_{\Phi\to\Phi_h}\left(u(\Phi)u'(\Phi)g'_{\tau\tau}(\Phi)\right)$, so that these quantities are directly related to the temperature and entropy of the black hole.

3. Thermodynamics of holographic QCD

Using the previous results, we can apply the holographic prescription [16] for the derivation of the trace anomaly of the Yang-Mills theory (see e.g. Ref. [2] for full details on the procedure).

3.1. Trace Anomaly

The perturbative running of the coupling constant in Yang-Mills theory induces a deformation in the action which may be written as

$$\delta S = \delta\left(\frac{1}{2g^2}\right)\int d^4x\,{\rm Tr}\,F^2 = \int d^4x\left(\frac{\gamma}{2}e^{-\gamma\Phi}\delta\Phi\right)\left(-{\rm Tr}\,F^2\right)\,. \tag{13}$$

This gives rise to the trace anomaly of the stress tensor

$$T^\mu_\mu = -\frac{\beta(g)}{g^3}\,\mathrm{Tr}\,F^2 = b_0\,\mathrm{Tr}\,F^2\,. \tag{14}$$

To apply the holographic prescription we can assume that the boundary value of the combination $\frac{\gamma}{2b_0}e^{-\gamma\Phi(\rho)}\delta\Phi(\rho)$ plays the role of a source that couples to the dual operator $\mathcal{O} = -b_0\,\mathrm{Tr}\,F^2$. The one-point function $\langle\mathcal{O}\rangle$ is therefore proportional to the gluon condensate.

With the zero temperature solution $\Phi_0(\rho)$ given by Eq. (8) and $\gamma^2 v_0/2 = b_0$, one can prove that a change in the scale $\delta\rho_0$, induces a change in the scalar field $\delta\Phi$ of the form $\frac{\gamma}{2b_0}e^{-\gamma\Phi_0}\delta\Phi = -\frac{\delta\rho_0}{2\rho_0}$. Then, according with Eq. (13), the derivative of the free energy density w with respect to $\log\rho_0$ must be proportional to the gluon condensate, leading to

$$2\rho_0\frac{\partial w}{\partial\rho_0} = b_0\langle\mathrm{Tr}\,F^2\rangle \qquad \text{where} \qquad S = \frac{V_3}{T}\,w\,. \tag{15}$$

3.2. First law of thermodynamics

We then assume that the free energy density is a function of the source ρ_0 and the temperature T, whose variation is $dw = -s\,dT - \langle\mathcal{O}\rangle d\rho_0/(2\rho_0)$. This assumption requires the condition for integrability $2\rho_0\partial s/\partial\rho_0 = \partial\langle\mathcal{O}\rangle/\partial T$, which has been checked in Ref. [15] by using holographic renormalization techniques. Since there is no conserved charge, the Euler identity adopts the form $w = \varepsilon - Ts$. Then, it follows that the first law of thermodynamics is

$$d\varepsilon = T\,ds - \langle\mathcal{O}\rangle\frac{d\rho_0}{2\rho_0}\,. \tag{16}$$

At this point one can observe that there is a similarity to the thermodynamic identities of elastic bodies, where the work done by an applied stress τ_{ij} is given by $dW = -V_3\tau_{ij}\,d\eta_{ij}$, being η_{ij} the strain deformation and V_3 the volume, see e.g. Refs. [17, 18]. This means that the source $\sim\log\sqrt{\rho_0}$ plays the role of the deformation, while the condensate $\langle\mathcal{O}\rangle$ is the analogous of the stress tensor. There is no need to interpret the source as any kind of chemical potential [19, 20], and this analogy with the thermodynamics of deformation seems more suitable.

3.3. Equation of state when $\nu \ll 1$

Using dimensional analysis one finds that a compact way of writing the pressure is

$$p(T,\rho_0) = -w = aT^4 + aT^4\,\mathcal{N}(T^2\rho_0)\,, \tag{17}$$

where a is a constant and $\mathcal{N}(x)$ is a function to be determined. We may now compute the pressure from the information about the horizon encoded in Eqs. (11) and (12), which determine the temperature and entropy density when $T\sqrt{\rho_h} \gg 1$ and $\Phi_h \to -\infty$. When using the relation $\rho_h \simeq 2\rho_0\exp\left(-\frac{2}{\gamma^2 v_0}e^{-\gamma\Phi_h}\right)$, one obtains for the entropy density [2]

$$\frac{s}{T^3} = T^{-3}\frac{\partial p}{\partial T} = 4a + 4a\,\mathcal{N}(T^2\rho_0) + 2a\,T^2\rho_0\,\mathcal{N}'(T^2\rho_0)\,, \tag{18}$$

with $a = \ell^3\pi^4/(2\kappa^2)$ and $\mathcal{N}(x) = -\frac{3}{\mathcal{W}\left(3v_0^{-1}(g_{(0)}^{-1}\pi^2 x)^{3\gamma^2/2}\right)} + O(\mathcal{W}^{-2})$, where $\mathcal{W}(z)$ denotes the principal branch of the Lambert $\mathcal{W}$-function. The trace of the thermal stress tensor can be obtained from the relation, $\varepsilon + p = Ts = T\partial_T p$, as

$$\varepsilon - 3p = -b_0\langle\mathrm{Tr}\,F^2\rangle = \frac{\ell^3\pi^4\gamma^2T^4}{2\kappa^2}\mathcal{N}(T^2\rho_0)^2 + \ldots\,. \tag{19}$$

[2] If one chooses the 5D Newton constant κ^2 to reproduce the Stefan-Boltzmann limit of the entropy density in gluodynamics at high temperatures, $s_{\mathrm{gluons}} \to (\mathrm{N}_c^2 - 1)\frac{4\pi^2}{45}T^3$, then one has $\frac{\ell^3}{\kappa^2} = (\mathrm{N}_c^2 - 1)\frac{2}{45\pi^2}$

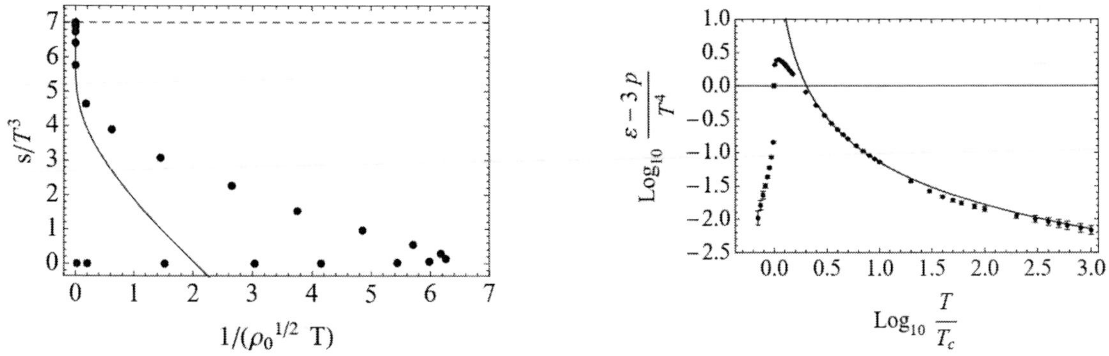

Figure 1. Left panel: s/T^3 as a function of $(\rho_0^{1/2}T)^{-1}$. We display as dots the result from a numerical computation of the equations of motion, while the solid (blue) line corresponds to the analytical result of Eq. (18). The horizontal dashed lines correspond to the Stefan-Boltzmann limit. We have used $\mathrm{N}_c = 3$ and $\gamma = \sqrt{2/3}$. Right panel: $(\varepsilon - 3p)/T^4$ as a function of T. The dots correspond to the lattice data for gluodynamics from [21] with $\mathrm{N}_c = 3$. We show as a solid (blue) line the analytical result from Eq. (19) with the values of the parameters given in Sec. 3.4.

The asymptotic behavior $\mathcal{W}(z) \sim \log z$ as $z \to \infty$ is consistent with the logarithmic behavior of the trace anomaly at high temperature. Note that in the opposite limit, $\mathcal{W}(z) \approx z$ as $z \to 0$, the model seems to predict the existence of power corrections in temperature. Although these corrections are desirable, as we will see below they don't play any phenomenological role within this model. However, it is remarkable that the resummation performed above leads to such a result. Finally, we display in the left panel of Fig. 1 the entropy density (normalized to T^3) as a function of the inverse of temperature. Note that this is a multivalued function with two regimes corresponding to big (upper branch) and small (lower branch) black holes.

3.4. Phenomenological consequences for QCD

We can now compare the results of the model by considering γ and ρ_0 as free parameters, with the lattice data of the equation of state of gluodynamics in SU(3). The best fit to the lattice data of the trace anomaly of Ref. [21] for temperatures $3T_c \leq T \leq 1000T_c$ leads to $\gamma = 2.375(19)$ and $\rho_0 T_c^2/g_{(0)} = 0.0555(10)$, with $\chi^2/\mathrm{dof} = 0.088$. The result for the trace anomaly is displayed in the right panel of Fig. 1. We conclude that the model leads to an accurate description of the lattice data in the whole regime $3 \leq T/T_c$. For temperatures closer to T_c, it has been discussed in the literature that a good description of the lattice data requires the existence of power corrections in T^2, see e.g. Refs. [22, 23]. However, from this analysis it seems to be that the model does not lead to power corrections in this regime. In either case, we cannot exclude that power corrections could become relevant after considering the effects of RG flows connecting two fixed points, a study already outlined in Ref. [15], or in a sophisticated version of the model.

4. The Stückelberg model

In the last part of this work we will study the thermodynamics of the Stückelberg model in 5D. This model is an extension of the Einstein-Maxwell theory including a massive gauge field.

4.1. The model

The Stückelberg model in 5D is defined by the Euclidean action

$$S = \int dz d^4x \sqrt{G} \left[\frac{1}{2\kappa^2} \left(R + \frac{12}{\ell^2} \right) - \frac{1}{4g_5^2} F_{\mu\nu} F^{\mu\nu} - \frac{1}{2g_5^2} G^{\mu\nu} (\partial_\mu \Phi - m A_\mu)(\partial_\nu \Phi - m A_\nu) \right] + S_{\mathrm{GH}} \,, \quad (20)$$

where A_μ is a U(1) gauge field with mass m. The mass of the gauge field is related to the conformal dimension $\tilde{\Delta}$ of the U(1) current operator J^μ dual to the gauge potential as $m^2\ell^2 = (\Delta-3)(\Delta-1)$ with $3 \leq \Delta < 4$, where for convenience we have defined $\Delta := 3 - \tilde{\Delta}$. Let us consider for the moment the metric in conformal coordinates

$$ds^2 = \frac{\ell^2}{z^2}\left[dz^2 + g_{ij}dx^i dx^j\right] , \tag{21}$$

which is asymptotically AdS$_5$ near the boundary, $z \to 0$. The behavior of the time component of the gauge field in this regime is

$$A_\tau(z) = A_{(0)\tau}\left(\frac{z}{\ell}\right)^{3-\Delta} + \mathcal{O}_A^\tau\left(\frac{z}{\ell}\right)^{\Delta-1} + \cdots , \tag{22}$$

where $A_{(0)\tau}$ corresponds to a deformation of the conformal field theory (CFT), with energy scale $\Lambda = 1/\ell$, i.e.

$$\int d^4x\, A_{(0)\tau}\mathcal{J} = \int d^4x\, \Lambda^{3-\Delta}A_{(0)\tau}\mathcal{O}_A^\tau . \tag{23}$$

In this expression $\mathcal{O}_A^\tau \equiv \mathcal{J}\Lambda^{\Delta-3}$ is a condensate of dimension Δ, while $A_{(0)\tau}$ plays the role of a chemical potential.

4.2. Black hole solution

The equations of motion of the model follow from the variation of the action with respect to scalar and gauge fields, and the metric. Since we are interested in black hole solutions, we will consider the following ansatz

$$ds^2 = \frac{r^2 f(r)}{\ell^2}d\tau^2 + \frac{\ell^2}{r^2 g(r)}dr^2 + \frac{r^2}{\ell^2}d\vec{x}^2 , \qquad A(r) = j(r)d\tau , \qquad \Phi(r) = \text{const} , \tag{24}$$

where we have used $r = \ell^2/z$, and seek for perturbative solutions to first order in $\Delta - 3$. The leading order solution, $\Delta = 3$, corresponds to the well-known result in the Einstein-Maxwell theory in 5D, which reads

$$f_0(r) = g_0(r) = \left(1 - \frac{r_h^2}{r^2}\right)\left(1 + \frac{r_h^2}{r^2} - \frac{q^2}{r_h^2 r^4}\right) , \qquad j_0(r) = A_{(0)\tau}\left(1 - \frac{r_h^2}{r^2}\right) , \tag{25}$$

where the charge associated to the U(1) symmetry is $q = \sqrt{\frac{2}{3}}\frac{A_{(0)\tau}\ell r_h^2 \kappa}{g_5}$, and the temperature of the black hole is $T_0 = \frac{r_h}{\pi\ell^2}\left(1 - \frac{q^2}{2r_h^6}\right)$. We use the parametrization

$$f(r) = f_0(r)\left(1 + F(r)\right) , \qquad g(r) = f_0(r)\left(1 + F(r) + G(r)\right) , \qquad j(r) = j_0(r) + J(r) , \tag{26}$$

where F, G, J are corrections of $O(\Delta - 3)$. By plugging these functions into the equations of motion, one gets two first order differential equations for F and G, and one second order differential equation for J. Then there are four integration constants that can be fixed from: i) asymptotically AdS$_5$ of the solution, i.e. $f(r) \to 1$ as $r \to \infty$; ii) regularity of $F(r)$ at the horizon; iii) vanishing of $A_\tau(r)$ at the horizon, or equivalently $J(r_h) = 0$; and finally iV) the behavior of A_τ near the boundary must be $A_\tau(r) \sim A_{(0)\tau}\left(\frac{z}{\ell}\right)^{\Delta-3}$, cf. Eq. (22). These conditions fix the constants and determine the temperature as a function of r_h and $A_{(0)\tau}$, so that

$$T(r_h, A_{(0)\tau}) = T_0\left(1 + \frac{\delta T}{T_0}\right) \quad \text{where} \quad \frac{\delta T}{T_0} = F(r_h) + \frac{1}{2}G(r_h) . \tag{27}$$

The explicit expressions for $f(r)$, $g(r)$ and $j(r)$ are too complicated to be shown here, but we will provide below the explicit results for the thermodynamical quantities which simplify a lot.

Journal of Physics: Conference Series **1416** (2019) 012022 doi:10.1088/1742-6596/1416/1/012022

4.3. Renormalization

In order to get the renormalized action, we need to add appropriate counterterms to cancel the divergences of the action at the boundary. In Fefferman-Graham coordinates the metric becomes as in Eq. (5), while the gauge field is $A = A_\tau(\rho)d\tau$ with $\rho = z^2$. From $g(r)$ one can obtain the asymptotic series giving r in terms of z. The expansion near the boundary turns out to be

$$r = \frac{\ell^2}{z} + a_1 z + a_3 z^3 + a_5 z^5 + b_5 z^5 \log z + \cdots , \qquad z \to 0 , \tag{28}$$

with the lowest order coefficient given by $a_1 = -q^2 m^2/(8r_h^4)$. This series expansion yields explicit expressions for the metric components $g_{xx}(\rho)$ and $g_{\tau\tau}(\rho)$ near the boundary. After a straightforward computation, one can write the bulk contribution of the action as a total derivative, and then identify the counterterm needed to cancel the divergences. The result is

$$S_{\rm ct} = \frac{1}{2\kappa^2}\int d^4x\sqrt{h}\,\frac{6}{\ell} + \frac{\ell}{4g_5^2}\int d^4x\sqrt{h}\,h^{ij}(\partial_i\Phi - mA_i)(\partial_j\Phi - mA_j) , \tag{29}$$

where h is the induced metric at the boundary. The renormalized action is then related to the thermodynamics of the system as

$$S_{\rm ren} = S_{\rm reg} + S_{\rm ct} = \frac{V_3}{T}\, w(r_h, q) , \tag{30}$$

where w is the free energy. In the last part of this section we will obtain explicit expressions for some of the thermodynamical variables.

4.4. Thermodynamical variables

These variables can be obtained from the thermodynamic potential Eq. (30) by using the standard thermodynamical relations. At finite temperature and chemical potential, these read

$$p = -w , \qquad \varepsilon = -T^2 \frac{\partial}{\partial T}\left(\frac{w}{T}\right)\bigg|_\mu , \qquad s = -\frac{\partial w}{\partial T}\bigg|_\mu , \qquad n = -\frac{\partial w}{\partial \mu}\bigg|_T , \tag{31}$$

where n and μ are the charge density and chemical potential, respectively. These quantities fulfill the Euler identity $\varepsilon + p = Ts + \mu n$. Other quantity of interest is the speed of sound, which is defined as a variation at fixed entropy per particle, i.e.

$$c_s^2 = \left(\frac{\partial p}{\partial \varepsilon}\right)_{s/n} = \left(\frac{\partial p}{\partial \varepsilon}\right)_n + \frac{n}{\varepsilon + p}\left(\frac{\partial p}{\partial n}\right)_\varepsilon . \tag{32}$$

Using these formulas and the free energy computed as explained in Sec. 4.3, one obtains that in the high temperature limit, $|A_{(0)\tau}| \ll T$, the trace anomaly and speed of sound write

$$\varepsilon - 3p = (\Delta - 3)\frac{2\pi^2\ell}{g_5^2} A_{(0)\tau}^2 T^2 + \cdots , \qquad c_s^2 = \frac{1}{3} - \frac{(\Delta - 3)A_{(0)\tau}^2}{3\xi^2 T^2} + \cdots , \tag{33}$$

where $\xi = (\sqrt{3}\pi g_5 \ell)/(2\kappa)$, and the dots stand for higher powers of $1/T$. Note that $\varepsilon - 3p \geq 0$ and $c_s^2 \leq \frac{1}{3}$, so that the terms $\propto (\Delta - 3)$ correspond to non-conformal corrections induced by the deformation of the CFT, cf. Sec. 4.1. Finally, one can check the validity of the Bekenstein-Hawking entropy formula in this model. Using Eqs. (31) and (33) one finds $s = 2\pi r_h^3/(\kappa^2 \ell^3)$. A trivial comparison with the area of the event horizon, $\mathcal{A}$, computed with the metric Eq. (24) leads to $s = 2\pi\mathcal{A}/\kappa^2$, confirming the validity of the entropy relation. This means that the usual laws of black hole thermodynamics are fulfilled also when $\Delta > 3$, at least up to order $O(\Delta - 3)$.

5. Conclusions

In this work we have studied the thermodynamics of a holographic model for QCD by applying techniques of singular perturbation theory. We have seen that with these techniques the leading asymptotics of the trace anomaly $\sim (\log T)^{-2}$ is resummed to give a dependence controlled by the Lambert $\mathcal{W}$-function. This reproduces the expected logarithmic suppression at high temperature, and it produces an accurate description of the lattice data in the regime $T \geq 3T_c$, even with a simple model having two parameters.

In the second part of this work we have studied the thermodynamical properties of the Stückelberg model in 5D. We have obtained analytical results by performing an expansion in the conformal dimension of the current operator, and found that the expressions are governed by pieces that break conformal invariance. The Bekenstein-Hawking entropy formula turns out to be still valid, at least at first order in the expansion.

Acknowledgments

Part of this work is based on Ref. [15], co-authored with Manuel Valle. I would like to thank him for collaboration and enlightening discussions. This research has been supported by Spanish MINEICO and European FEDER funds (Grant No. FIS2017-85053-C2-1-P), Plan Nacional de Altas Energías Spanish MINEICO (Grant No. FPA2015-64041-C2-1-P), Junta de Andalucía (Grant No. FQM-225), Basque Government (Grant No. IT979-16), and Consejería de Conocimiento, Investigación y Universidad of the Junta de Andalucía and European Regional Development Fund (ERDF) (Grant No. SOMM17/6105/UGR), as well as by Spanish MINEICO Ramón y Cajal Program (Grant No. RYC-2016-20678), and by Universidad del País Vasco UPV/EHU, Bilbao, Spain, through a Visiting Professor appointment.

References

[1] Gubser S S and Nellore A 2008 *Phys. Rev.* **D78** 086007 (*Preprint* 0804.0434)
[2] Gursoy U, Kiritsis E, Mazzanti L and Nitti F 2009 *JHEP* **05** 033 (*Preprint* 0812.0792)
[3] Alanen J, Kajantie K and Suur-Uski V 2009 *Phys. Rev.* **D80** 126008 (*Preprint* 0911.2114)
[4] Megias E, Pirner H J and Veschgini K 2011 *Phys. Rev.* **D83** 056003 (*Preprint* 1009.2953)
[5] Li D, He S, Huang M and Yan Q S 2011 *JHEP* **09** 041 (*Preprint* 1103.5389)
[6] Collins J C, Duncan A and Joglekar S D 1977 *Phys. Rev.* **D16** 438–449
[7] Landsman N P and van Weert C G 1987 *Phys. Rept.* **145** 141
[8] Klebanov I R, Ouyang P and Witten E 2002 *Phys. Rev.* **D65** 105007 (*Preprint* hep-th/0202056)
[9] Coriano C, Irges N and Morelli S 2007 *JHEP* **07** 008 (*Preprint* hep-ph/0701010)
[10] Gursoy U and Jansen A 2014 *JHEP* **10** 092 (*Preprint* 1407.3282)
[11] Jimenez-Alba A, Landsteiner K and Melgar L 2014 *Phys. Rev.* **D90** 126004 (*Preprint* 1407.8162)
[12] Casero R, Kiritsis E and Paredes A 2007 *Nucl. Phys.* **B787** 98–134 (*Preprint* hep-th/0702155)
[13] Kiritsis E 2005 *Phys. Rept.* **421** 105–190 [Erratum: Phys. Rept.429,121(2006)] (*Preprint* hep-th/0310001)
[14] Papadimitriou I 2011 *JHEP* **08** 119 (*Preprint* 1106.4826)
[15] Megias E and Valle M 2018 *Fortsch. Phys.* **66** 1800035 (*Preprint* 1707.04747)
[16] Klebanov I R and Witten E 1999 *Nucl. Phys.* **B556** 89–114 (*Preprint* hep-th/9905104)
[17] Landau L D and Lifshitz E M 1986 *Theory of Elasticity* (Butterworth-Heinemann; 3 edition)
[18] Wallace D C 1998 *Thermodynamics of Crystals* (Dover Publications)
[19] Buchel A and Liu J T 2003 *JHEP* **11** 031 (*Preprint* hep-th/0305064)
[20] Lu H, Pope C N and Wen Q 2015 *JHEP* **03** 165 (*Preprint* 1408.1514)
[21] Borsanyi S, Endrodi G, Fodor Z, Katz S D and Szabo K K 2012 *JHEP* **07** 056 (*Preprint* 1204.6184)
[22] Pisarski R D 2007 *Prog. Theor. Phys. Suppl.* **168** 276–284 (*Preprint* hep-ph/0612191)
[23] Megias E, Ruiz Arriola E and Salcedo L L 2009 *Phys. Rev.* **D80** 056005 (*Preprint* 0903.1060)

Journal of Physics: Conference Series **1416** (2019) 012023 doi:10.1088/1742-6596/1416/1/012023

Vaisman Algebroid and Doubled Structure of Gauge Symmetry in Double Field Theory

Haruka Mori[1], **Shin Sasaki**[2], **Kenta Shiozawa**[3]

Department of Physics, Kitasato University, Sagamihara 252-0373, Japan

E-mail: [1]h.mori@sci-kitasato-u.ac.jp
[2]shin-s@kitasato-u.ac.jp
[3]k.shiozawa@sci.kitasato-u.ac.jp

Abstract. The Vaisman algebroid is a kind of algebroid structure. It is defined by an extension of the Courant algebroid, and phisically related to the gauge symmetry in Double Field Theory (DFT), which is an effective theory of string theory. DFT has T-duality as a manifest symmetry. In this study, we focus on the "doubled structure" in the Vaisman algebroid. It is already well known that some kind of Lie algebras are obtained by the Drinfel'd double of Lie bialgebras. The Courant algebroid is obtaind by the Drinfel'd double of Lie bialgebroids. We find that the Vaisman algebroid can be obtained by an analogue of the "Drinfel'd double" of Lie algebroids. We discuss the algebraic origin of the strong constraint in DFT.

1. Introduction

T-duality is a duality in superstring theory. It is important to understand T-duality to study the nature of space-time. From the viewpoint of geometry, it has been studied in the framework of Hitchin's generalized geometry [1]. The natural algebraic structure which appears generalized geometry is the Courant bracket. Mathematically, it is known that a Courant bracket defines a Courant algebroid (for example [2]).

Double Field Theory (DFT) is a kind of effective theory of superstring theory [3]. DFT exhibits T-duality manifestly and it has other various symmetries. For example, DFT has a gauge symmetry. It is a T-duality covariantized symmetry which originates from the diffeomorphism and the $U(1)$ gauge symmetry of the NSNS B-field. The bracket which describes the gauge symmetry in DFT is called the C-bracket. Since the C-bracket does not satisfy the Jacobi identity, the C-bracket defines not a Lie algebra but a *metric algebroid*. This structure is first presented in [4]. We call this structure Vaisman algebroid. A Vaisman algebroid can be defined as an extension of a Courant algebroid. This is because a Vaisman algebroid violates some of the axioms that a Courant algebroid should satisfy. The Vaisman algebroid is known as the pre-DFT algebroid in [5].

In this proceeding, we show that the Vaisman algebroid has some kind of a doubled structure. In other words, we show that the Vaisman algebroid is constructed by a pair of Lie algebroids. This doubled structure is closely related with the Drinfel'd double. It is known that a Courant algebroid can be obtained by the Drinfel'd double of a Lie bialgebroid [6]. Our work is the generalization of this result. The contents of this proceeding is based on the first half of our paper [7].

Journal of Physics: Conference Series **1416** (2019) 012023 doi:10.1088/1742-6596/1416/1/012023

The DFT geometry is called "doubled geometry". Mathematically, the doubled geometry is related to the para-Hermitian geometry [8] and Born geometry [9]. It is known that the C-bracket in DFT appears as a Vaisman bracket on the para-Hermitian geometry. In the second half of our paper [7], we constructed the double structure of the Vaisman algebroid on the para-Hermitian geometry and we also discussed the algebraic origin of the strong constraint which is the physical condition in DFT.

2. Lie bialgebra

First, we introduce a Lie algebra. We define a Lie algebra $\mathfrak{g}$ on a field K as a set of a vector space V and a Lie bracket $[\cdot,\cdot]$. The Lie bracket $[\cdot,\cdot]: V \times V \to V$ satisfies the Jacobi identity. We can define the dual Lie bracket $[\cdot,\cdot]_*$ and the dual Lie algebra $\mathfrak{g}^*$ by considering the dual vector space V^*. The inner product $\langle\cdot,\cdot\rangle$ is defined between $\mathfrak{g}$ and $\mathfrak{g}^*$ naturally. $\langle\cdot,\cdot\rangle$ take the value on K.

Here, we can introduce a p-th exterior product of $\mathfrak{g}$: $\wedge^p\mathfrak{g} = \mathfrak{g} \wedge \mathfrak{g} \wedge \cdots \wedge \mathfrak{g}$. As an analogue of an exterior derivative d in the cotangent bundle on a manifold, we can define the exterior derivative $\mathrm{d} : \wedge^p\mathfrak{g}^* \to \wedge^{p+1}\mathfrak{g}^*$, which satisfies $\mathrm{d}^2 = 0$. Similarly, we can define an exterior derivative $\mathrm{d}^* : \wedge^p\mathfrak{g} \to \wedge^{p+1}\mathfrak{g}$. The Lie bracket can be extended to the case where the argument is $\wedge^p\mathfrak{g}$. This bracket is called the Schouten-Nijenhuis bracket and satisfies the following properties.

(i) $[a,b]_{\mathrm{S}} = -(-)^{(p-1)(q-1)}[b,a]_{\mathrm{S}}$.

(ii) $[a, b\wedge c]_{\mathrm{S}} = [a,b]_{\mathrm{S}} \wedge c + (-)^{(p-1)q} b \wedge [a,c]_{\mathrm{S}}$.

(iii) $(-)^{(p-1)(r-1)}[a,[b,c]_{\mathrm{S}}]_{\mathrm{S}} + (-)^{(q-1)(r-1)}[b,[c,a]_{\mathrm{S}}]_{\mathrm{S}} + (-)^{(r-1)(q-1)}[c,[a,b]_{\mathrm{S}}]_{\mathrm{S}} = 0$.

(iv) The bracket of an element $\wedge^p\mathfrak{g}$ and an element in $\wedge^0\mathfrak{g} = K$ is 0.

Where $a \in \wedge^p\mathfrak{g}$, $b \in \wedge^q\mathfrak{g}, c \in \wedge^r\mathfrak{g}$. The Schouten-Nijenhuis bracket defines a Gerstenhaber algebra.

Next, we examine the algebraic structure between $\mathfrak{g}$ and $\mathfrak{g}^*$. If we consider the Lie bracket $[\cdot,\cdot]$ as a bilinear map $\mu : \wedge^2\mathfrak{g} \mapsto \mathfrak{g}$, we can define the co-bracket $\delta : \mathfrak{g} \mapsto \wedge^2\mathfrak{g}$. Similarly, a dual Lie bracket can be defined by $\mu^* : \wedge^2\mathfrak{g}^* \mapsto \mathfrak{g}^*$. This μ^* is an adjoint of δ which means the relation $\langle x, \mu_*\xi\rangle = \langle \mu_*^* x, \xi\rangle$ holds. Then, μ_*^* corresponds to the co-bracket. The co-bracket satisfies the co-Jacobi identity. This is equivalent to the Jacobi identity for the dual Lie bracket $[\cdot,\cdot]_*$. Even more, the co-bracket satisfies the following 1-cocycle condition:

$$\delta([x,y]) = \mathrm{ad}_x^{(2)}\delta(y) - \mathrm{ad}_y^{(2)}\delta(x), \quad x,y \in \mathfrak{g}. \tag{1}$$

Then, the set $(\mathfrak{g}, \mu, \delta)$ is called Lie bialgrbra. The same Lie bialgebra can be defined as $(\mathfrak{g}^*, \mu^*, \delta^*)$.

3. From Lie algebra to Lie algebroid

A Lie algebroid is a structure defined as a generalization of Lie algebras, This is defined by a vector bundle E on the manifold M, an anchor map $\rho : E \to TM$ and a Lie algebroid bracket $[\cdot,\cdot]_E : \Gamma(E) \times \Gamma(E) \to \Gamma(E)$ satisfying the Jacobi identity. The following two expressions are required for ρ and $[\cdot,\cdot]_E$:

$$[X, fY]_E = (\rho(X)\cdot f)Y + f[X,Y]_E, \tag{2}$$

$$\rho([X,Y]_E) = [\rho(X), \rho(Y)]. \tag{3}$$

Here, $f \in C^\infty(M)$, and $[\cdot,\cdot]$ is the Lie bracket on $\Gamma(TM)$. (2) is the Leibniz rule of the Lie algebroid bracket. (3) is the homomorphism of ρ with respect to $[\cdot,\cdot]_E$. Considering dual vector bundle E^* for E, a dual Lie algebroid $(E^*, \rho_*, [\cdot,\cdot]_{E^*})$ can also be defined on the same manifold M. The inner product $\langle\cdot,\cdot\rangle$ can be defined naturally between E and E^*.

We can define the multi-vectors $\Gamma(\wedge^\bullet E)$ and the multi-forms $\Gamma(\wedge^\bullet E^*)$. We define the exterior derivative d for $\Gamma(\wedge^\bullet E^*)$. The exterior derivative d is defined as follows [10]:

$$\begin{aligned} \mathrm{d}\xi(X_1,\ldots,X_{p+1}) =& \sum_{i=1}^{p+1}(-)^{i+1}\rho(X_i)\cdot(\xi(X_1,\ldots,\check{X}_i,\ldots,X_{p+1})) \\ &+\sum_{i<j}(-)^{i+j}\xi([X_i,X_j]_E,X_1,\ldots,\check{X}_i,\ldots,\check{X}_j,\ldots,X_{p+1}), \end{aligned} \tag{4}$$

where $\xi \in \Gamma(\wedge^p E^*), X_i \in \Gamma(E)$. The notation $\check{X}_i$ stands for that the term is omitted in the expression. Similarly, a Lie derivative $\mathcal{L}_X : \Gamma(\wedge^p E^*) \to \Gamma(\wedge^p E^*)$ by $X \in \Gamma(E)$ is defined by

$$\mathcal{L}_X(\xi)(Y_1,\ldots,Y_p) = \rho(X)\cdot(\xi(Y_1,\ldots,Y_p)) - \sum_{i=1}^{p}\xi\left(Y_1,\ldots,[X,Y_i]_E,\ldots,Y_p\right). \tag{5}$$

where $Y_1,\ldots,Y_p \in \Gamma(E)$, $\xi \in \Gamma(\wedge^p E^*)$. The interior product $\iota_X : \Gamma(\wedge^p E^*) \to \Gamma(\wedge^{p-1}E^*)$ by $X \in \Gamma(E)$ is defined by

$$\iota_X(\xi(Y_1,\ldots,Y_{p-1})) = \xi(X,Y_1,\ldots,Y_{p-1}), \tag{6}$$

where $Y_1,\ldots,Y_{p-1} \in \Gamma(E), \xi \in \Gamma(\wedge^p E^*)$. We also define the exterior derivative d_*, Lie derivative and the interior product ι_* for $\Gamma(\wedge^\bullet E)$.

For $X \in \Gamma(\wedge^{p+1}E), Y \in \Gamma(\wedge^{q+1}E)$, the Schouten-Nijenhuis bracket $[\cdot,\cdot]_\mathrm{S}$ is defined as an extension of the Lie algebroid bracket to multi vectors (or forms). $[\cdot,\cdot]_\mathrm{S}$ has the following properties.

(i) $[X,Y]_\mathrm{S} = -(-)^{pq}[Y,X]_\mathrm{S}$.
(ii) $[X,f]_\mathrm{S} = \rho(X)\cdot f$ for $X \in \Gamma(E)$.
(iii) For $X \in \Gamma(\wedge^{p+1}E)$, the bracket $[X,\cdot]_\mathrm{S}$ acts on $\Gamma(\wedge^q E)$ as a degree-p derivation.

If the derivation condition

$$\mathrm{d}_*[X,Y]_\mathrm{S} = [\mathrm{d}_*X,Y]_\mathrm{S} + [X,\mathrm{d}_*Y]_\mathrm{S} \tag{7}$$

is satisfied between these dual Lie algebroids, this Lie algebroid pair becomes a Lie bialgebroid [10]. A Lie bialgebroid is a generalization of Lie bialgebras. The derivation condition corresponds to the 1-cocycle condition (1) of Lie bialgebra. It is known that we obtain a Courant algebroid structure by the Drinfel'd double of a Lie bialgebroid [6].

4. Doubled structure of Vaisman algebroid

A Vaisman algebroid is defined by a vector bundle $\mathcal{V}$ on a manifold M, an anchor map $\rho : \mathcal{V} \to TM$, a non-degenerate symmetric bilinear form $(\cdot,\cdot)$ and a Vaisman bracket $[\cdot,\cdot]_\mathrm{V} : \Gamma(\mathcal{V})\times\Gamma(\mathcal{V}) \to \Gamma(\mathcal{V})$. The Vaisman algebroid $(\mathcal{V},\rho,(\cdot,\cdot),[\cdot,\cdot]_\mathrm{V})$ satisfies the following two axioms.

Axiom V1. $[e_1,fe_2]_\mathrm{V} = f[e_1,e_2]_\mathrm{V} + (\rho(e_1)\cdot f)e_2 - (e_1,e_2)\mathcal{D}f$.

Axiom V2. $\rho(e_1)\cdot(e_2,e_3) = ([e_1,e_2]_\mathrm{V} + \mathcal{D}(e_1,e_2),e_3) + (e_2,[e_1,e_3]_\mathrm{V} + \mathcal{D}(e_1,e_3))$.

Here $e_1,e_2,e_3 \in \Gamma(\mathcal{V})$ and $\mathcal{D}$ is a differential operator over $f \in C^\infty(M)$.

In our paper, we have shown that a Vaisman algebroid can be obtained by a double of a pair of Lie algebroids. In the following, only the brief flow of proofs is introduced. The details can be found in [7].

First, we show that the Axioms V1 and V2 are satisfied only by the properties of a Lie algebroid. In the following, we consider $\mathcal{V} = E \oplus E^*$ and $X \in \Gamma(E), \xi \in \Gamma(E^*)$. The Vaisman bracket is defined by

$$\begin{aligned}[e_1, e_2]_{\mathrm{V}} = [X_1, X_2]_E + \mathcal{L}_{\xi_1} X_2 - \mathcal{L}_{\xi_2} X_1 - \mathrm{d}_*(e_1, e_2)_- \\ + [\xi_1, \xi_2]_{E^*} + \mathcal{L}_{X_1}\xi_2 - \mathcal{L}_{X_2}\xi_1 + \mathrm{d}(e_1, e_2)_-,\end{aligned} \tag{8}$$

where

$$(e_1, e_2)_\pm = \frac{1}{2}\Big(\langle \xi_1, X_2\rangle \pm \langle \xi_2, X_1\rangle\Big). \tag{9}$$

We employ $(\cdot,\cdot)_+$ as the non-degenerate symmetric bilinear form $(\cdot,\cdot)$.

Axiom V1 is the Leibniz rule for the Vaisman bracket. If we calculate the left hand side of Axiom V1, we obtain

$$[e_1, fe_2]_{\mathrm{V}} = [X_1, fX_2]_{\mathrm{V}} + [X_1, f\xi_2]_{\mathrm{V}} + [\xi_1, fX_2]_{\mathrm{V}} + [\xi_1, f\xi_2]_{\mathrm{V}}, \tag{10}$$

Terms in the right hand side of (10) becomes

$$\begin{aligned}[X_1, f\xi_2]_{\mathrm{V}} &= f[X_1, \xi_2]_{\mathrm{V}} + (\rho(X_1)\cdot f)\xi_2 - \frac{1}{2}\mathcal{D}g\langle \xi_2, X_1\rangle, \\ [\xi_1, fX_2]_{\mathrm{V}} &= f[\xi_1, X_2]_{\mathrm{V}} + (\rho_*(\xi_1)\cdot f)X_2 - \frac{1}{2}\mathcal{D}f\langle \xi_1, X_2\rangle, \\ [X_1, fX_2]_{\mathrm{V}} &= f[X_1, X_2]_{\mathrm{V}} + (\rho(X_1)\cdot f)X_2, \\ [\xi_1, f\xi_2]_{\mathrm{V}} &= f[\xi_1, \xi_2]_{\mathrm{V}} + (\rho_*(\xi_1)\cdot f)\xi_2.\end{aligned} \tag{11}$$

If we add all the terms in the right hand side of (10) by using (11), we obtain

$$[e_1, fe_2]_{\mathrm{V}} = f[e_1, e_2]_{\mathrm{V}} + (\rho(e_1)\cdot f)e_2 - (e_1, e_2)_+\mathcal{D}f. \tag{12}$$

Therefore, Axiom V1 is satisfied.

Axiom V2 is the compatibility condition between $(\cdot,\cdot)$ and $\mathcal{D}$. Due to the property of $(\cdot,\cdot)$, the left hand side of Axiom V2 is written as follows:

$$\rho_{\mathrm{V}}(e)\cdot(e_1, e_2)_+ = ([e, e_1]_{\mathrm{V}}, e_2)_+ + (e_1, [e, e_2]_{\mathrm{V}})_+ + \frac{1}{2}\rho_{\mathrm{V}}(e_1)\cdot(e, e_2)_+ + \frac{1}{2}\rho_{\mathrm{V}}(e_2)\cdot(e, e_1)_+. \tag{13}$$

If we use the following relation

$$\begin{aligned}\frac{1}{2}\rho_{\mathrm{V}}(e_1)\cdot(e, e_2)_+ &= (\mathcal{D}(e, e_2), e_1)_+, \\ \frac{1}{2}\rho_{\mathrm{V}}(e_2)\cdot(e, e_1)_+ &= (\mathcal{D}(e, e_1), e_2)_+,\end{aligned}$$

then, we obtain the following result:

$$\rho(e_1)\cdot(e_2, e_3)_+ = ([e_1, e_2]_{\mathrm{V}} + \mathcal{D}(e_1, e_2)_+, e_3)_+ + (e_2, [e_1, e_3]_{\mathrm{V}} + \mathcal{D}(e_1, e_3)_+)_+ \tag{14}$$

Therefore Axiom V2 holds.

The structures of the algebroid are different from those in a Courant algebroid which appears in the generalized geometry. This is because the three axioms that Courant algebroids should

Journal of Physics: Conference Series **1416** (2019) 012023 doi:10.1088/1742-6596/1416/1/012023

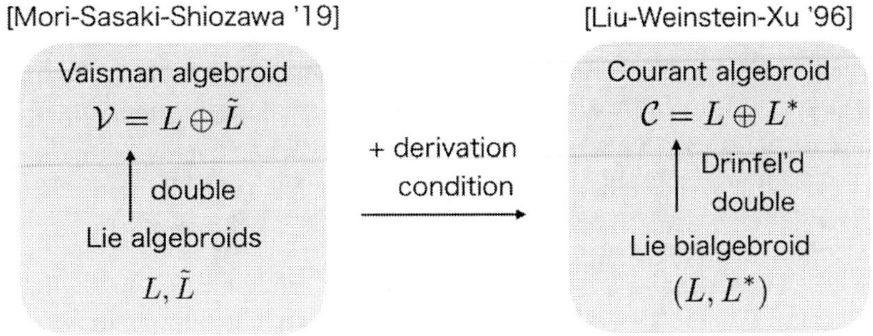

Figure 1. The relationship between [6] and [7]. We can obtain different structures with or without the derivation condition.

satisfy are not included in this algebroid (Axiom C1, C2, C4 in our paper [7]). Axiom C1 is a deformed Jacobi identity. The Jacobiator is calculated as follows:

$$[[e_1, e_2]_V, e_3]_V + \text{c.p.} = \mathcal{D}T(e_1, e_2, e_3) - (J_1 + J_2 + \text{c.p.}). \tag{15}$$

where

$$T(e_1, e_2, e_3) = \frac{1}{3}\Big(([e_1, e_2]_V, e_3)_+ + \text{c.p.}\Big), \tag{16}$$

$$J_1 = \iota_{X_3}\Big(\mathrm{d}[\xi_1, \xi_2]_V - \mathcal{L}_{\xi_1}\mathrm{d}\xi_2 + \mathcal{L}_{\xi_2}\mathrm{d}\xi_1\Big) + \iota_{\xi_3}\Big(\mathrm{d}_*[X_1, X_2]_V - \mathcal{L}_{X_1}\mathrm{d}_*X_2 + \mathcal{L}_{X_2}\mathrm{d}_*X_1\Big), \tag{17}$$

$$J_2 = \Big(\mathcal{L}_{\mathrm{d}_*(e_1,e_2)_-}\xi_3 + [\mathrm{d}(e_1, e_2)_-, \xi_3]_V\Big) - \Big(\mathcal{L}_{\mathrm{d}(e_1,e_2)_-}X_3 + [\mathrm{d}_*(e_1, e_2)_-, X_3]_V\Big). \tag{18}$$

Since the term $(J_1 + J_2 + \text{c.p.})$ remains, Axiom C1 does not hold under the current conditions. Here, c.p. is the cyclic permutation.

Axiom C2 means the homomorphism of ρ with respect to $[\cdot, \cdot]_V$, this axiom corresponds to the relation (3). By subtracting the right hand side from the left hand side of (3) for the Vaisman bracket $[\cdot, \cdot]_V$, we obtain

$$\begin{aligned} &\rho([e_1, e_2]_V) \cdot f - [\rho(e_1), \rho(e_2)]f \\ &= -\langle \xi_1, (\mathcal{L}_{\mathrm{d}f}X_2 - [X_2, \mathrm{d}_*f]_E)\rangle + \langle \xi_2, (\mathcal{L}_{\mathrm{d}f}X_1 - [X_1, \mathrm{d}_*f]_E)\rangle \\ &\quad + \frac{1}{2}\Big(\rho\rho_*^* + \rho_*\rho^*\Big)\mathrm{d}_0(\langle \xi_1, X_2\rangle - \langle \xi_2, X_1\rangle)f, \end{aligned} \tag{19}$$

where d_0 is the usual exterior derivative for $\Gamma(T^*M)$. We have used the relation, $\mathrm{d} = \rho^*\mathrm{d}_0, \mathrm{d}_* = \rho_*^*\mathrm{d}_0$. Axiom C2 holds when this right hand side in (19) vanishes. However, the result is generically non-zero.

If we calculate the left hand side of Axiom C4, we obtain

$$(\mathcal{D}f, \mathcal{D}g)_+ = \frac{1}{2}\Big(\rho_*\rho^* + \rho\rho_*^*\Big)(\mathrm{d}_0 f)g. \tag{20}$$

Axiom C4 holds when the right hand side vanished. However, the result is generically non-zero.

If a derivation condition (7) is imposed between the dual Lie algebroids and the condition of Lie bialgebroid holds, the above Axioms C1, C2, and C4 are satisfied. Therefore, imporsing the derivation condition to the Vaisman algebroid constructed in this way gives a Courant algebroid of [6] (see figure 1). Conversely, we can say that a Vaisman algebroid can be composed of two Lie algebroids, not of Lie bialgebroids. According to the DFT context, this derivation condition corresponds to the strong constraint [7].

5. Conclusion and discussions

In this proceeding, we consider the Vaisman algebroid as the gauge symmetry algebra in DFT. We have shown that the Vaisman algebroid can be obtained by the double of Lie algebroids. From the physical viewpoint, the double structure of the gauge symmetry in DFT originates from the symmetry between the Kaluza-Klein modes and winding modes of strings. Mathematically, it was understood as an analogue of the Drinfel'd double for Lie algebroids. In our paper, we discuss the Vaisman algebroid based on a doubled structure which appears naturally on a para-Hermitian manifold. Based on $\Gamma(E)$ and $\Gamma(E^*)$ in section 2 in this proceeding, a concrete construction of the exterior algebras in DFT was also performed in our paper.

The algebroid structures discussed in this proceeding provide the local structure in the para-Hermitian manifold. The problem of finding the corresponding groupoid structure describing the global nature of the symmetry is called *coquecigrue problem* (first appearance in [11]). Examining the groupoid structure corresponding for the Vaisman algebroid in DFT is an interesting problem [12].

Acknowledgments

The work of S.S. is supported by the Japan Society for the Promotion of Science (JSPS) KAKENHI Grant Number JP17K14294. The work of K.S. is supported by the Sasakawa Scientific Research Grant (No. 2019-2027) from the Japan Science Society.

References

[1] Hitchin N 2003 Generalized Calabi-Yau manifolds *Quart. J. Math.* **54** 281 (*Preprint* math/0209099)
[2] Courant T J 1990 Dirac Manifolds *Trans. Amer. Math. Soc.* **319** 631
[3] Hull C and Zwiebach B 2009 Double Field Theory *J. High Energy Phys.* JHEP09(2009)099 (*Preprint* 0904.4664)
[4] Vaisman I 2012 On the geometry of double field theory *J. Math. Phys.* **53** 033509 (*Preprint* 1203.0836)
[5] Chatzistavrakidis A, Jonke L, Khoo F S and Szabo R J 2018 Double Field Theory and Membrane Sigma-Models *J. High Energy Phys.* JHEP07(2018)015 (*Preprint* 1802.07003)
[6] Liu Z J, Weinstein A and Xu P 1997 Manin Triples for Lie Bialgebroids *J. Differential Geom.* **45** 547 (*Preprint* dg-ga/9508013)
[7] Mori H, Sasaki S and Shiozawa K 2019 Doubled Aspects of Vaisman Algebroid and Gauge Symmetry in Double Field Theory (*Preprint* 1901.04777)
[8] Svoboda D 2018 Algebroid Structures on Para-Hermitian Manifolds *J. Math. Phys.* **59** 122302 (*Preprint* 1802.08180)
[9] Freidel L, Rudolph F J and Svoboda D 2019 A Unique Connection for Born Geometry *Commun. Math. Phys.* (*Preprint* 1806.05992)
[10] Mackenzie K C H and Xu P 1994 Lie bialgebroids and Poisson groupoids *Duke Math. J.* **73** 415
[11] Loday J L 1993 Une version non commutative des algèbres de Lie : les algèbres de Leibniz *Les rencontres physiciens-mathématiciens de Strasbourg -RCP25* **44** 25
[12] Mori H, Sasaki S and Shiozawa K, Work in progress

Discrete analysis on non-cubic lattices

Maryna Nesterenko[†], **Severin Pošta**[‡]

[†] Institute of Mathematics of NAS of Ukraine, 3, Tereshchenkivs'ka str, Kyiv, 01004, Ukraine,
[‡] Department of Mathematics, Faculty of Nuclear Sciences and Physical Engineering, Czech Technical University in Prague, Trojanova 13, CZ-120 00 Prague, Czech Republic

E-mail: maryna@imath.kiev.ua, severin.posta@fjfi.cvut.cz

Abstract. The paper proposes practical and computation methods for discrete analysis of functions defined on the weight lattices or model sets of semisimple Lie groups. They are entirely group theoretical, being based on finite groups and their duals. Numerical computations approximate target functions uniformly on the entire infinite space. Several examples are considered.

1. Introduction

Discrete Fourier transform is one of the common tools for the spectral analysis of signals and for the general analysis of functions. It is widely used in various branches of science and technology, e.g. for mp3 and jpg coding, in statistics, etc. There are many other discrete transformations (Hartley, cosine, wavelet, Lapped, finite Legendre, etc.), but there are a few methods for processing the rapidly increasing amount of 3D digital data gathered today and methods that take into account the non-cubic atomic structures of the solid bodies. Moreover, the existence of the materials with the structures that are shift non-invariant, Faraday wave experiments, potential field of a physical quasicrystal and other physical phenomena require even more sophisticated approaches to the Fourier analysis.

In this paper we consider analogs of multidimensional Fourier transforms for non-cubic translation-invariant lattices and model sets (quasicrystals).

First we explore families of special functions depending on the weight lattices of simple Lie algebras, which are not only cubic and point out their most valuable properties:

- The functions are defined for each compact semisimple Lie group G. There are several infinite families of functions per G. The number of continuous variables on which the functions depend, is equal to the rank of G.
- Orbit functions have well defined symmetries with respect to the affine Weyl group of G and these functions (within each family) are orthogonal when integrated over the fundamental region of F of the maximal torus T of G.
- The functions can be sampled on the lattice fragment $F_M = F \cap L/M$ of the weight lattice L refined by $M \in N$. There is a finite subset Λ_M of such 'digital' functions that are pairwise orthogonal when summed up over the points of F_M.
- Each function is an eigenfunction of the Laplace operator appropriate for G and their eigenvalues are known explicitly. Therefore orbit functions can be applied to the solution

of the corresponding Neumann and Dirichlet boundary-value problems on the fundamental domains of the Weyl groups.

- Each function can be transformed to the polynomial form in several ways.

Then we consider the problem of discrete methods for dealing with functions that are intrinsically almost periodic, but not actually periodic (the functions that depend on quasicrystals). Usually such functions are approximated as periodic ones with the loose of some information, but it is not entirely satisfactory.

The main idea (due to H. Bohr) is to build the analysis around the theory of quasicrystals using the local hulls and dynamical systems. The n-dimensional Fourier analysis for such functions was set in [5] and some examples of functions based on the standard Fibonacci quasicrystal were considered.

The key features of the method:

- retains the essential feature of the almost periodicity;
- uses all the spectral components of the natural Fourier module;
- creates approximations that are valid uniformly on the entire space of quasicrystal;
- involves methods of fast discrete Fourier analysis, namely use finite groups;
- takes advantage of internal symmetry.

2. Functions on non-cubic lattices

The main task of this section is to give all the tools for analog of the discrete Fourier transformation not with the complete system of exponents on cubic lattices, but with orbit functions on non-cubic lattices generated by fundamental region of a semisimple group.

There are seven compact semisimple Lie groups of rank 3 $SU(2) \times SU(2) \times SU(2)$, $SU(3) \times SU(2)$, $O(5) \times SU(2)$, $G(2) \times SU(2)$, $SU(4)$, $O(7)$ and $Sp(6)$ with the respective Lie algebras $A_1 \times A_1 \times A_1$, $A_2 \times A_1$, $C_2 \times A_1$, $G_2 \times A_1$, A_3, B_3 and C_3.

Let R^n is the real Euclidean space spanned by the simple roots α of a simple Lie algebra and ω-basis is the basis of fundamental weights.

In the weight lattice P, we define the cone of dominant weights P^+ and its subset of strictly dominant weights P^{++}: $P = Z\omega_1 + Z\omega_2 + \cdots + Z\omega_n \supset P^+ = Z^{\geq 0}\omega_1 + \cdots + Z^{\geq 0}\omega_n \supset P^{++} = Z^{>0}\omega_1 + \cdots + Z^{>0}\omega_n$.

The Weyl group W is the finite group generated by the reflections $r_{\alpha_i}x = x - \frac{2\langle x, \alpha_i\rangle}{\langle \alpha_i, \alpha_i\rangle}\alpha_i$, $x \in R^n$.

For any simple Lie algebra there is a unique highest root $\xi = m_1\alpha_1 + m_2\alpha_2 + \cdots + m_n\alpha_n$ and a reflection through it's middle is $r_0 x = x - \frac{2\langle x, \xi\rangle}{\langle \xi, \xi\rangle}\xi + \frac{2\xi}{\langle \xi, \xi\rangle}$.

The group generated by $n+1$ reflections $r_0, r_1, \ldots, r_n$, is the affine Weyl group W^{aff} and the even subgroup of it is W_e.

The fundamental region $F \subset R^n$ for any W^{aff} is the convex hull of the vertices $F(G) = \{0, \frac{\breve{\omega}_1}{m_1}, \frac{\breve{\omega}_2}{m_2}, \ldots, \frac{\breve{\omega}_n}{m_n}\}$ (for the group W_e we take a doubled fundamental region). All possible three-dimensional fundamental regions are presented on Figures 1–6, note, that only one of them ($A_1 \times A_1 \times A_1$) is rectangular.

We define basic orbit functions C-, S- and E-functions, specified by a given point $\lambda \in Z^n$ and a chosen semisimple Lie group G with the corresponding Weyl group W, its even subgroup W^e and fundamental region F:

$C_\lambda(x) = \sum_{\mu \in W_\lambda} e^{2\pi i\langle \mu, x\rangle}$, where $x \in R^n$, $\lambda \in P^+$;

$S_\lambda(x) = \sum_{\mu \in W_\lambda} (-1)^{l(\mu)} e^{2\pi i\langle \mu, x\rangle}$, where $x \in R^n$, $\lambda \in P^{++}$ and $l(\mu)$ is the number of reflections from λ to μ;

$E_\lambda(x) = \sum_{\mu \in W_{e\lambda}} e^{2\pi i\langle \mu, x\rangle}$, where $x \in R^n$, $\lambda \in P^+$.

Journal of Physics: Conference Series **1416** (2019) 012024 doi:10.1088/1742-6596/1416/1/012024

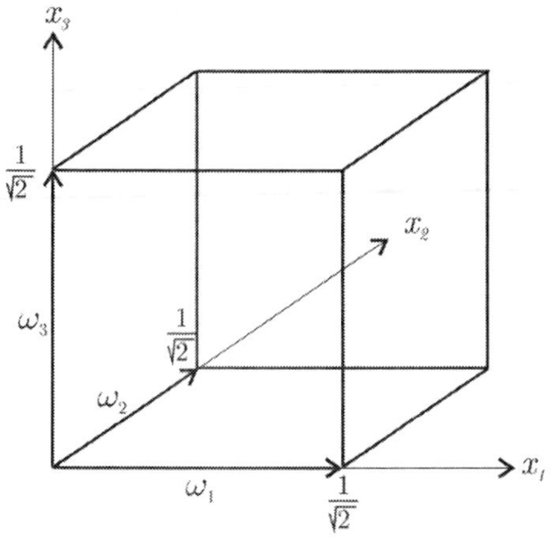

Figure 1. The fundamental region of the Lie algebra $A_1 \times A_1 \times A_1$.

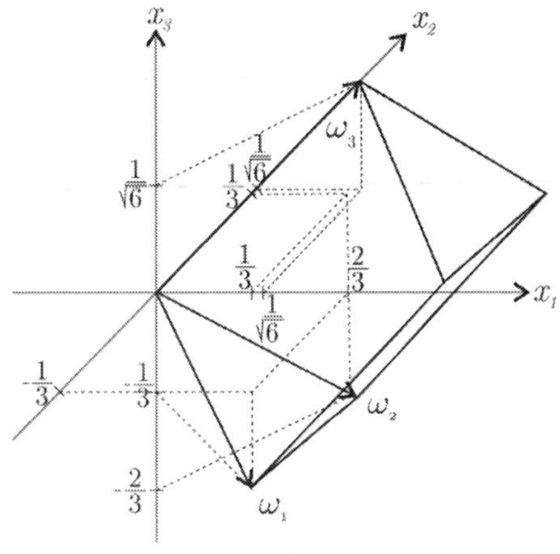

Figure 2. The fundamental region of the Lie algebra $A_2 \times A_1$.

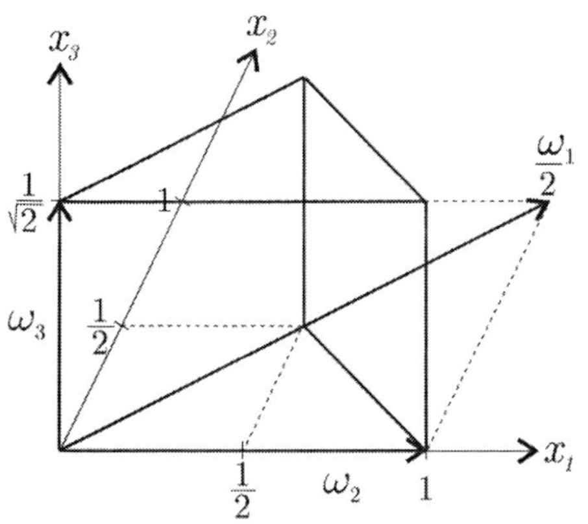

Figure 3. The fundamental region of the Lie algebra $C_2 \times A_1$.

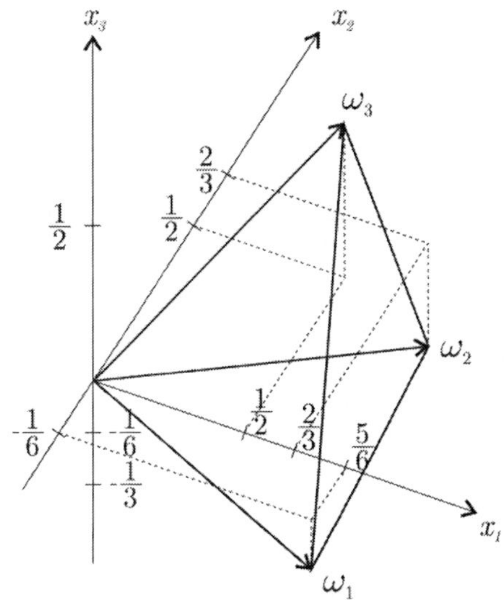

Figure 4. The fundamental region of the Lie algebra A_3.

For any complex square integrable functions $\phi(x)$ and $\psi(x)$, we define a continuous scalar product $\langle \phi(x), \psi(x) \rangle := \int_F \phi(x)\overline{\psi(x)} \mathrm{d}x$. The integration is taken with respect to the Euclidean measure, the bar means complex conjugation, and $x \in F$.

$\langle C_\lambda(x), C_{\lambda'}(x) \rangle = |W_\lambda| \cdot |F| \cdot \delta_{\lambda\lambda'}$, $\langle S_\lambda(x), S_{\lambda'}(x) \rangle = |W| \cdot |F| \cdot \delta_{\lambda\lambda'}$, here $|W|$ is the order of Weyl group, $|W_\lambda|$ is the size of Weyl group orbit, and $|F|$ is the volume of fundamental region. The similar relation is true for E-functions as well, see the details [1].

Therefore, each family of orbit functions forms an orthogonal basis in the Hilbert space of squared integrable functions $\mathcal{L}^2(F)$. Hence functions given on F can be expanded in terms of linear combinations of C-, S- or E-functions.

Each continuous function on the fundamental region with continuous derivatives can be expanded as the sum of C-, S- or E-functions.

Let $f(x)$ be a function defined on F (or F_e for E-functions), then it may be written that

Journal of Physics: Conference Series **1416** (2019) 012024 doi:10.1088/1742-6596/1416/1/012024

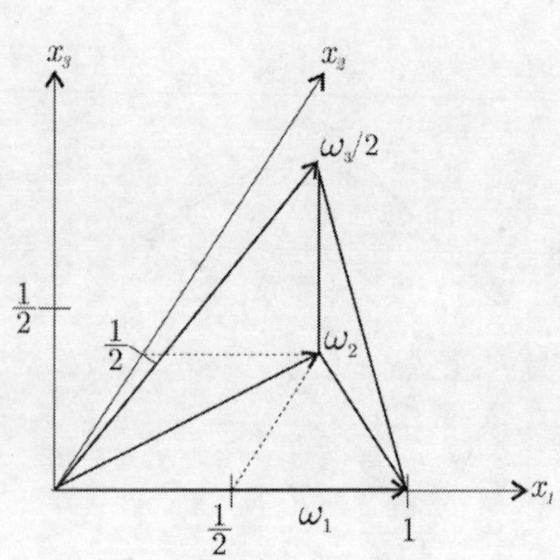

Figure 5. The fundamental region of the Lie algebra B_3.

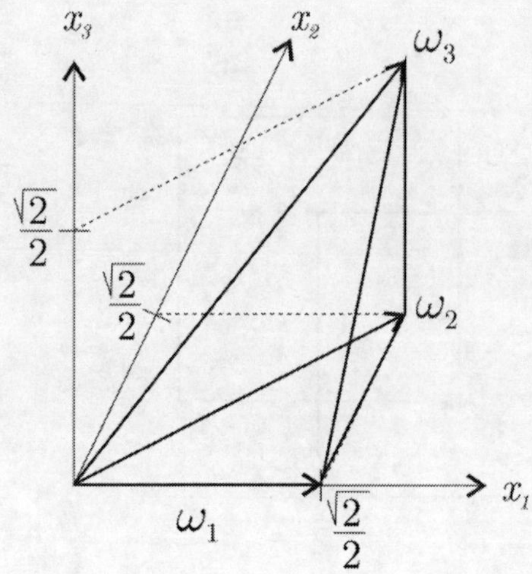

Figure 6. The fundamental region of the Lie algebra C_3.

$f(x) = \sum_{\lambda \in P^+} c_\lambda C_\lambda(x), \qquad c_\lambda = |W_\lambda|^{-1}|F|^{-1}\langle f(x), C_\lambda(x)\rangle;$
$f(x) = \sum_{\lambda \in P^{++}} c_\lambda S_\lambda(x), \qquad c_\lambda = |W|^{-1}|F|^{-1}\langle f(x), S_\lambda(x)\rangle;$
$f(x) = \sum_{\lambda \in P_e} c_\lambda E_\lambda(x), \qquad c_\lambda = |W_{e\ \lambda}|^{-1}|F_e|^{-1}\langle f(x), E_\lambda(x)\rangle.$

The respective discrete transform can be used for the continuous interpolation of values of a function $f(x)$ between its given values on some grid. There is a uniform way to introduce a grid for any dimension and any semisimple Lie algebra.

The grid $F_M \subset F$ depends on a natural number M and consists of the points
$F_M = \left\{ \frac{s_1}{M}\breve{\omega}_1 + \cdots + \frac{s_n}{M}\breve{\omega}_n \mid s_1, \ldots, s_n \in Z^{\geq 0},\ \sum_{i=1}^{n} s_i m_i \leq M \right\}.$

Then we have the following discrete transforms for the function $f(x)$:

$f(x) = \sum_{\lambda \in \Lambda_M} b_\lambda C_\lambda(x), \quad x \in F_M, \qquad b_\lambda = \frac{\langle f, C_\lambda \rangle_M}{\langle C_\lambda, C_\lambda \rangle_M};$
$f(x) = \sum_{\lambda \in \Lambda_M} b_\lambda S_\lambda(x), \quad x \in F_M, \qquad b_\lambda = \frac{\langle f, S_\lambda \rangle_M}{\langle S_\lambda, S_\lambda \rangle_M};$
$f(x) = \sum_{\lambda \in \Lambda_M} b_\lambda E_\lambda(x), \quad x \in F_{e\ M}, \qquad b_\lambda = \frac{\langle f, E_\lambda \rangle_M}{\langle E_\lambda, E_\lambda \rangle_M}.$

Here $\langle \cdot, \cdot \rangle_M$ denotes the discrete scalar product, see [1].Once the coefficients b_λ are calculated, the function $f(x)$ smoothly interpolates the values of $f(x_i)$, $i = 1, 2, \ldots, |F_M|$ and using the symmetry (C-, E-functions) or antisymmetry (S-functions) we get the approximation on the whole Euclidian space R^n.

Consider the orbit functions in the orthogonal bases (the orthogonal bases for these algebras are well known, see e.g. [2]).

The Laplace operator in the orthogonal coordinates gives the same eigenvalues on every exponential function summand of an orbit function with eigenvalue $-4\pi\langle\lambda, \lambda\rangle$. Therefore, the functions $C_\lambda(x)$, $E_\lambda(x)$ and $S_\lambda(x)$ are eigenfunctions of the Laplace operator:

$$\Delta \begin{pmatrix} C_\lambda(x) \\ E_\lambda(x) \\ S_\lambda(x) \end{pmatrix} = -4\pi^2\langle\lambda, \lambda\rangle \begin{pmatrix} C_\lambda(x) \\ E_\lambda(x) \\ S_\lambda(x) \end{pmatrix}.$$

Thereby, a Laplace operator for each Lie group is given in a different set of coordinates. In Table 1 the explicit forms of the 3d Laplacians are given in the ω-basis for all semisimple Lie algebras.

On the boundary of F, the C-functions have a vanishing normal derivative, while S-functions reach zero at the boundary, what makes it possible to solve the following boundary value

Journal of Physics: Conference Series **1416** (2019) 012024 doi:10.1088/1742-6596/1416/1/012024

Table 1. Three-dimensional Laplacians of semi-simple Lie algebras.

Lie algebra	Laplace operator
$A_1 \times A_1 \times A_1$	$\Delta = \partial^2_{x_1} + \partial^2_{x_2} + \partial^2_{x_3}$
$A_2 \times A_1$	$\Delta = \partial^2_{x_1} - \partial_{x_1}\partial_{x_2} + \partial^2_{x_2} + \partial^2_{x_3}$
$C_2 \times A_1$	$\Delta = 2\partial^2_{x_1} - 2\partial_{x_1}\partial_{x_2} + \partial^2_{x_2} + \partial^2_{x_3}$
A_3	$\Delta = \partial^2_{x_1} - \partial_{x_1}\partial_{x_2} + \partial^2_{x_2} - \partial_{x_2}\partial_{x_3} + \partial^2_{x_3}$
B_3	$\Delta = \partial^2_{x_1} - \partial_{x_1}\partial_{x_2} + \partial^2_{x_2} - 2\partial_{x_2}\partial_{x_3} + 2\partial^2_{x_3}$
C_3	$\Delta = 2\partial^2_{x_1} - 2\partial_{x_1}\partial_{x_2} + 2\partial^2_{x_2} - 2\partial_{x_2}\partial_{x_3} + \partial^2_{x_3}$

problems.

C-function is a solution of the Neumann boundary value problem on n-dimensional simplex F
$\Delta f(x) = \Lambda f(x)$, $\frac{\partial f(x)}{\partial \nu} = 0$ for $x \in \partial F$.

S-function is a solution of the Dirichlet boundary value problem on n-dimensional simplex F
$\Delta f(x) = \Lambda f(x)$, $\qquad f(x) = 0$ for $x \in \partial F$.

Let the continuous variable x is given relative to the ω-basis and Δ denotes the Laplace operator, where the differentiation ∂_{x_i} is made with respect to the direction given by ω_i. Then $\Delta = \sum_{i,j=1}^{n} \frac{C_{ij}}{\langle\alpha_i,\alpha_i\rangle} \partial_{x_i}\partial_{x_j}$, where C is the Cartan matrix.

It is known in Lie theory that the matrix of scalar products of the simple roots is positive definite, moreover our definition makes matrix $\frac{C_{ij}}{\langle\alpha_i,\alpha_i\rangle}$ symmetric, hence it can be diagonalized and the Laplace operator could be transformed to the sum of second derivatives by an appropriate change of variables.

Note, that the discretization procedure works well for the both Neumann and Dirichlet boundary value problems.

One more feature that can be exploit in discrete analysis by means of orbit functions is the relation to orthogonal polynomials. Polynomials can be obtained by the exponential substitution, that is rather straightforward [3, 4], namely $X_j := e^{2\pi i x_j}$, $x_j \in R$, $j = 1, 2, \ldots, n$. Two other possibilities are the polynomials obtained by trigonometric substitution and polynomials obtained by the recursive method, see [7].

3. Discrete methods for almost periodic functions

Our key objective is to describe a method for discrete computational Fourier analysis of a function defined on aperiodic atomic arrangement with highly structured long-range order (quasicrystal).

Let R^d is a real Euclidean space of finite dimension d and $D_r(R^d)$ are all point sets $\Lambda \subset R^d$ for which the distance $|x - y| \geq r > 0\ \forall x, y \in \Lambda$.

The *local topology* on $D_r(R^d)$ is introduced as follows. Two sets Λ_1 and Λ_2 of $D_r(R^d)$ are 'close' if, for some large R and some small ϵ, one has $\Lambda_1 \cap B_R \subset \Lambda_2 + B_\epsilon$, and $\Lambda_2 \cap B_R \subset \Lambda_1 + B_\epsilon$.

For $\Lambda \in D_r(R^d)$ the *local hull* of Λ is $X(\Lambda) = \overline{\{t + \Lambda \,:\, t \in R^d\}} \subset D_r(R^d)$, i.e. take all translates of Λ and take their closure in the local topology.

The translation action of R^d on Λ lifts to a translation action on $X(\Lambda)$. The local hull $X(\Lambda)$ is compact and the R^d-action on it is continuous.

Consider a function $F : X(\Lambda) \longrightarrow \mathbb{C}$. We can define from it a function $f : R^d \longrightarrow \mathbb{C}$ by $f(t) = F(t + \Lambda)$.

If F is continuous then we note that for all $t_1, t_2 \in R^d$, $t_1 + \Lambda$ and $t_2 + \Lambda$ are close $\Longrightarrow$ $F(t_1 + \Lambda)$ and$F(t_2 + \Lambda)$ are close $\Longrightarrow$ $f(t_1)$ and $f(t_2)$ are close. Thus continuity of F implies continuity of f.

A function $f : R^d \longrightarrow \mathbb{C}$ is called *local with respect to a set* $\Lambda \in D_r(R^d)$, or Λ-*local*, if for all $\epsilon' > 0$ there exist R and ϵ so that whenever $t_1, t_2 \in R^d$ satisfy that $t_1 + \Lambda$ and $t_2 + \Lambda$ are (R, ϵ)-close then $|f(t_1) - f(t_2)| < \epsilon'$.

Using locality, we can go in the opposite direction.

Let $\Lambda \in D_r(R^d)$ and let $f : R^d \to \mathbb{C}$ be local with respect to Λ. Define $F : \{t + \Lambda \ : \ t \in R^d\} \longrightarrow \mathbb{C}$ (so that F is a function on a part of $D_r(R^d)$) by $F(t + \Lambda) = f(t)$. Then F is continuous on $\{t + \Lambda \ : \ t \in R^d\}$ w.r.t the local topology.

For each local function f with respect to Λ there is a unique continuous function F on the local hull, whose restriction to the orbit of Λ is f. Every continuous function on the local hull of Λ arises in this way. Thus a locality with respect to Λ and the existence and continuity of an extension function on $X(\Lambda)$ amount to the same thing.

In the situation that $X(\Lambda)$ is equipped with an R^d-invariant probability measure μ the action of R^d on $X(\Lambda)$ leads to unitary action T of R^d on $L^2(X(\Lambda), \mu)$. Namely for all $F \in L^2(X(\Lambda), \mu)$ and for all $t \in R^d$, $T_t F$ is the function defined by $T_t F(\Gamma) = F(-t + \Gamma)$ and with $\langle F \mid G\rangle := \int\limits_{X(\Lambda)} F\overline{G}\, d\mu$ we have $\langle T_t F \mid T_t G\rangle = \langle F \mid G\rangle$.

$$\begin{array}{ccccc} R^d & \longleftarrow(\|) & R^d \times R^d & (\perp)\longrightarrow & R^d \\ & & \cup & & \\ L & \longleftarrow(1-1) & \widetilde{L} & (\text{dense image})\longrightarrow & L' \end{array}$$

Here $\widetilde{L}$ is a lattice in $R^d \times R^d$ which is oriented so that the projections into R^d are $1-1$ and dense respectively.

The left-hand R^d is *physical space* (in which Λ is going to lie).

The right-hand R^d is *internal space* (is used to control the projection of the lattice $\widetilde{L}$ into physical space).

L denotes the image of $\widetilde{L}$ under projection into physical space. Since this projection is one-one, L and $\widetilde{L}$ similar groups, so L is a free Abelian group of rank $2d$, i.e. it has a Z-basis of $2d$ elements. However, it necessarily has accumulation points, and the typical situation is that L is dense in physical space.

Choose a subset Ω in internal space. This *window* is assumed to be compact, equal to the closure of its interior, and to have boundary of measure 0. Using it we define $\Lambda = \Lambda(\Omega) = \{x : \tilde{x} \in \widetilde{L},\, x' \in \Omega\}$.

Sets of the form $t + \Lambda(\Omega)$, $t \in R^d$, are called *cut and project sets* or *model sets*. In particular, for each $(x, y) \in R^d \times R^d$ we define $\Lambda_{(x,y)} = x + \Lambda(-y + \Omega)$.

If $(x, y) = (x', y')\text{mod}\widetilde{L}$, then $x + \Lambda(-y + \Omega) = x' + \Lambda(-y' + \Omega)$. Thus these model sets are parameterized by the torus of dimension $2d$ $(R^d \times R^d)/\widetilde{L} =: T$.

Denote $(x, y)_L$ the congruence class $(x, y)\text{mod}\widetilde{L}$.

There is a natural measure, the Haar measure, θ_T on T. It is invariant under the R^d-action.

Suppose f is a local function with respect to the model set Λ. From the local function f we have its extension $F \in L^2(X(\Lambda), \mu)$ which is continuous. Then we obtain $\widetilde{F} \in L^2(T, \theta)$, where $\widetilde{F}((t, 0)_L) = \widetilde{F}(\beta(t + \Lambda)) = F(t + \Lambda) = f(t)$, and we can write

$\widetilde{F}(\cdot) = \sum\limits_{k \in L^\circ} a_k,\, e^{2\pi i \langle \tilde{k} | \cdot \rangle}$, where L° is the dual of L.

The *Fourier-Bohr expansion* of the local function f is

$f(t) = F(t + \Lambda) = \widetilde{F}((t, 0)_L) = \sum\limits_{k \in L^\circ} a_k\, e^{2\pi i \langle \tilde{k} | (t,0) \rangle} = \sum\limits_{k \in L^\circ} a_k\, e^{2\pi i \langle k | t \rangle}$.

Unfortunately, we don't have total control over $\widetilde{F}$. We know it only on $(R^d, 0)_L$. To compute a_k out of f alone, we use the analog of Birkhoff ergodic theorem (transformed from Keller 1998) for all continuous functions $\widetilde{G}$ on T, $\int\limits_T \widetilde{G}(x)\, d\theta_T(x) = \lim_{R \to \infty} \frac{1}{\text{vol} B_R} \int\limits_{B_R} \widetilde{G}((t, 0)_L) dt$.

Journal of Physics: Conference Series **1416** (2019) 012024 doi:10.1088/1742-6596/1416/1/012024

Thus $a_k = \lim_{R\to\infty} \frac{1}{\mathrm{vol}B_R} \int\limits_{B_R} e^{-2\pi i \langle \tilde{k} | (t,0) \rangle} \widetilde{F}((t,0)_L)\, dt = \lim_{R\to\infty} \frac{1}{\mathrm{vol}B_R} \int\limits_{B_R} e^{-2\pi i \langle k|t \rangle} f(t)\, dt.$

Here we use $F((t,0)_L) = f(t)$ and $\tilde{k} = (k,k')$, so $\langle \tilde{k} \mid (t,0) \rangle = \langle k,t \rangle + \langle k',0 \rangle = \langle k,t \rangle$.

For discretization we begin with the cut and project scheme with torus T and note the natural extension of the mapping $(\cdot)'$ to the rational span of the module L:

$$\begin{array}{ccccc} QL & \longleftarrow(1-1) & Q\widetilde{L} & \longrightarrow & QL' \\ x & \leftarrow & \tilde{x} = (x,x') & \rightarrow & x' \end{array}$$

We assume that R^{2d} is supplied with the standard dot product (denoted $\langle \cdot | \cdot \rangle$), and then define the dual lattice: $\widetilde{L}^\circ = \{ Y \in R^d \times R^d \ :\ \langle Y \mid \tilde{x} \rangle \in Z \text{ for all } \tilde{x} \in \widetilde{L} \}$.

Then $\widetilde{L}^\circ$ is a Z-module of the same rank as $\widetilde{L}$, namely $2d$. There is a cut and project scheme of which $\widetilde{L}^\circ$ is the lattice:

$$\begin{array}{ccccc} R^d & \longleftarrow(||) & R^d \times R^d & (\perp)\longrightarrow & R^d \\ & & \cup & & \\ L^\circ & \longleftarrow(1-1) & \widetilde{L}^\circ & (\text{dense image})\longrightarrow & (L^\circ)' \ . \end{array}$$

The data points in R^d at which computations of our functions will be made come by projection into physical space of a suitable set of coset representatives of $\widetilde{L_N}$ modulo $\widetilde{L}$. The corresponding frequencies (wave vectors) k are chosen from the dual lattice L°. The choice of values of k at which we should evaluate the Fourier coefficients a_k come by selecting suitable representatives of $\widetilde{L}^\circ$ modulo $\widetilde{L_N}^\circ$. The key point is the duality

$\langle \cdot,\cdot \rangle : \widetilde{L}^\circ / \widetilde{L}_N^\circ \times \widetilde{L_N}/\widetilde{L} \longrightarrow \frac{1}{N} Z/Z.$

The main effort required is the creation of the data points. Once this is done, the same set of data points and Fourier frequencies will work for any almost periodic function arising from the same cut and project scheme.

The approximating functions are not just local approximations, they are also *global* approximations, in the sense that they provide finite Fourier series that approximate f throughout its entire domain (namely R^d).

Below on Figures 7 and 8 we show two examples of approximation, which use the one-dimensional Fibonacci model set. The functions to be approximated are: a distance function ($f(t)$ = the distance of t to the nearest point of Λ) and a step function ($f(x) = 1$, if x is in a long interval, and $f(x) = -1$, if x is in a short interval).

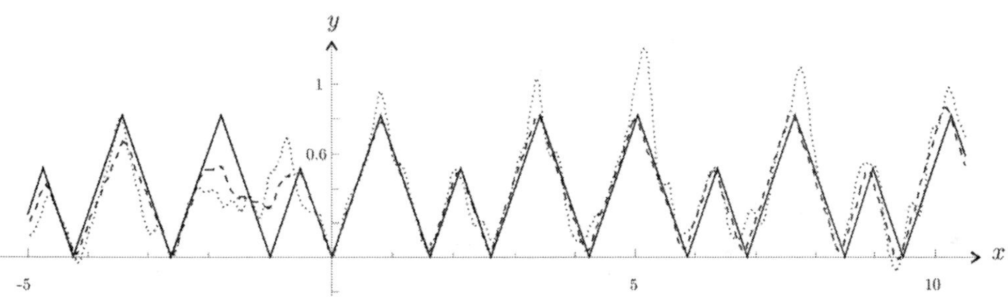

Figure 7. The distance function and approximants are drawn for the data sampled on 121 points

4. Final Comments

Concerning orbit functions, we insisted throughout this paper that the underlying group is semisimple. The extension to the finite Coxeter groups that are not Weyl groups of a simple Lie algebra is straightforward. Many of the properties of orbit functions extend to these cases. Only their orthogonality, continuous or discrete, has not been shown so far.

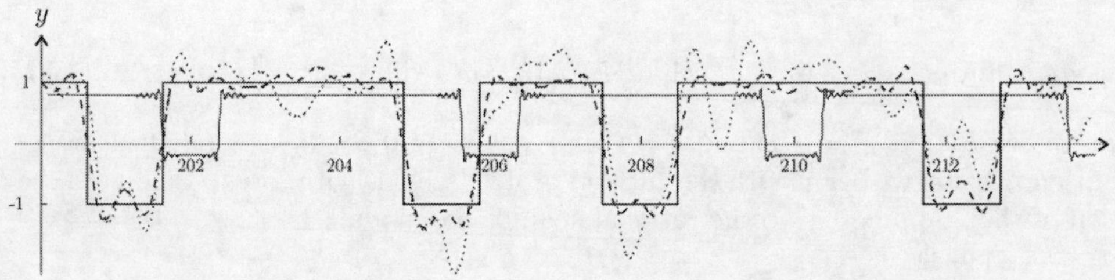

Figure 8. Here we adduce the step function and approximants are drawn on the interval [200, 215] for 49 data points.

Let $\Lambda_1 \in R^{d_1}$ and $\Lambda_2 \in R^{d_2}$ are two cut-and-project sets (quasicrystalls), then their cartesian product $\Lambda = \Lambda_1 \times \Lambda_2$ is non-periodic as well. Therefore one-dimensional Fibonacci quasicrystalls can be used for the construction of higher-dimensional ones. Unfortunately, to apply the proposed approximation we have to construct all cut-and-project ingredients, torus parametrization and discretization from the every beginning. As the initial lattice for the cut-and-project method the direct sum of the initial lattices of the quasicrystalls Λ_1 and Λ_2 is to be taken.

Finally, the way in which finite groups and their duals are used for orbit functions and model sets makes the Fourier approximations amenable to the technique of the fast Fourier transform.

Acknowledgments
SP acknowledges the support of SGS18/189/OHK4/3T/14, project of the Czech Technical University in Prague. MN is grateful for the hospitality extended to her at the Department of mathematics, FNSPE, Czech Technical University in Prague.

References

[1] Hrivnák J and Patera J 2009 *J. Phys. A: Math. Theor.* **42**, 385208
[2] Humphreys J E 1972 *Introduction to Lie algebras and representation theory*, New York, Springer
[3] Klimyk A and Patera J 2006 *SIGMA* **2**, 006
[4] Klimyk A and Patera J 2007 *SIGMA* **3**, 023
[5] Moody R V, Nesterenko M and Patera J 2008 *Acta Crystallographica A* **A64** 654
[6] Moody R V 1997 *The mathematics of long-range order, ed. R.V. Moody*, NATO ASI, Series **C489**, Kluwer, Dordrecht 239–268
[7] Nesterenko M, Patera J, Szajewska M and Tereszkiewicz A 2010 *J. Phys. A: Math. and Theor.* **43**(49):495207

Journal of Physics: Conference Series **1416** (2019) 012025 doi:10.1088/1742-6596/1416/1/012025

Hom-Lie structures on 3-dimensional skew symmetric algebras

Elvice Ongong'a[1,2], Johan Richter[3] and Sergei Silvestrov[1]

[1] Division of Applied Mathematics, Mälardalen University, Box 833, Västerås 72123, Sweden
[2] School of Mathematics, University of Nairobi, Box 30197-00100, Nairobi, Kenya
[3] Deparment of Mathematics and Natural Sciences, Blekinge Institute of Technology, SE-371 79 Karlskrona, Sweden

E-mail: elvice.omondi.ongonga@mdh.se, johan.richter@bth.se, sergei.silvestrov@mdh.se

Abstract. We describe the dimension of the space of possible linear endomorphisms that turn skew-symmetric three-dimensional algebras into Hom-Lie algebras. We find a correspondence between the rank of a matrix containing the structure constants of the bilinear product and the dimension of the space of Hom-Lie structures. Examples from classical complex Lie algebras are given to demonstrate this correspondence.

1 Introduction

Hom-Lie algebras were first introduced by Hartwig, Larsson and Silvestrov in [1] by studying some examples of deformed Lie algebras which arise from twisted discretizations of vector fields. Hom-Lie algebras are therefore generalisations of Lie algebras by having an additional twist α, a linear endomorphism. The space of such linear endomorphisms that turn a skew symmetric algebra into a Hom-Lie algebra forms a vector subspace, known as Hom-Lie structures. In [3], it is proved that every 3-dimensional skew-symmetric algebra can be turned into a Hom-Lie algebra. The authors in [2] give results where the dimension of the space of such linear endomorphisms is 6.

This work gives results for the remaining dimensions of the Hom-Lie structures in 3-dimension case. The first section gives preliminaries on constructing polynomial equations from the Hom-Jacobi identity which are linear in structure constants of the linear map and and gives specific equations for the dimension 3 case. The second section of this work proceeds to give results showing correspondence between the dimension of the space of possible linear endomorphisms and the rank of a 3 by 3 matrix of all structure constants of the bilinear map, denoted by C. The last section demonstrates the results obtained using examples from classical Lie algebras. The Hom-Lie structures for Simple Lie algebra $\mathfrak{sl}(2,\mathbb{C})$ have been studied previously, in for example, [7] and [8]. Hom-Lie structures for multiplicative Heisenberg Lie algebras have also been studied in [6].

2 Preliminaries

All algebras in this article are over an algebraically closed field, $\mathbb{K}$, of characteristic 0. We begin by giving the definition of a Hom-Lie algebra.

Journal of Physics: Conference Series **1416** (2019) 012025 doi:10.1088/1742-6596/1416/1/012025

Definition 2.1. *A Hom-Lie algebra is a triple $(V, [\cdot,\cdot], \alpha)$ consisting of a linear space V, bilinear map $[\cdot,\cdot] : V \times V \longrightarrow V$ and a linear space homomorphism $\alpha : V \to V$ satisfying*

$$[x, y] = -[y, x] \qquad (skew - symmetry) \tag{1}$$

$$[\alpha(x), [y, z]] + [\alpha(y), [z, x]] + [\alpha(z), [x, y]] = 0 \qquad (Hom - Jacobi\ identity) \tag{2}$$

for all $x, y, z \in V$.

Let the structure constants associated to both the bilinear product and linear map be given by $\{C_{ij}^k\}_{i<j}$ and $\{a_{it}\}$ respectively. We have the following equations involving the structure constants in general for n-dimensional case:

$$[e_i, e_j] = \sum_{s=1}^{n} C_{ij}^s e_s \text{ and } \alpha(e_i) = \sum_{t=1}^{n} a_{it} e_t \tag{3}$$

Replacing equations in (3) in the Hom-Jacobi identity and writing the equations as linear in a_{it}, we have the following;

$$\sum_{t=1}^{n} \left[a_{it} \left(\sum_{s=1}^{n} C_{jk}^s C_{ts}^r \right) + a_{jt} \left(\sum_{s=1}^{n} C_{ki}^s C_{ts}^r \right) + a_{kt} \left(\sum_{s=1}^{n} C_{ij}^s C_{ts}^r \right) \right] = 0 \tag{4}$$

with $1 \le i < j < k \le n,\ r = 1, 2, \ldots, n$.

We denote the matrix of linear transformation with coefficients involving structure constants of the bilinear map by M and the column matrix involving the structure constants of linear map α by a_α. In n-dimensional case the matrix M has n^2 columns and $\binom{n}{3} n$ rows. The elements of M are the coefficients associated to a_{it}, a_{jt}, a_{kt} given in (4). A general construction of M is presented in [2].

Thus (4) can be written as $Ma_\alpha = 0$ and the matrix M represents a linear transformation $L : \mathbb{K}^9 \to \mathbb{K}^3$. In order to realize a Hom-Lie algebra, it is then required that $a_\alpha \in \ker L$. We have that $6 \le \dim \ker L \le 9$. Hom-Lie structures, denoted by HomLie(μ), is the space of all linear endomorphisms that satisfy the Hom-Jacobi identity for some skew symmetric algebra H. That is the vector subspace

$$HomLie(\mu) = \{\alpha \in End\ V \mid \circlearrowleft_{x,y,z} [\alpha(x), [y, z]] = 0\}$$

where $\circlearrowleft_{x,y,z} = [\alpha(x), [y, z]] + [\alpha(y), [z, x]] + [\alpha(z), [x, y]]$.

This study concerns describing the dimensions of the space of such Hom-Lie structures in 3-dimensional case. The matrix a_α is given as $a_\alpha = \begin{pmatrix} a_{11} & a_{21} & a_{31} & a_{12} & a_{22} & a_{32} & a_{13} & a_{23} & a_{33} \end{pmatrix}^t$ and matrix M associated to the bilinear product is given as

$$M = \begin{pmatrix} M_{1,1} & M_{1,2} & M_{1,3} & M_{1,4} & M_{1,5} & M_{1,6} & M_{1,7} & M_{1,8} & M_{1,9} \\ M_{2,1} & M_{2,2} & M_{2,3} & M_{2,4} & M_{2,5} & M_{2,6} & M_{2,7} & M_{2,8} & M_{2,9} \\ M_{3,1} & M_{3,2} & M_{3,3} & M_{3,4} & M_{3,5} & M_{3,6} & M_{3,7} & M_{3,8} & M_{3,9} \end{pmatrix} \tag{5}$$

where

$M_{r,1} = (C_{23}^2 C_{12}^r + C_{23}^3 C_{13}^r)$, $M_{r,2} = (-C_{13}^2 C_{12}^r - C_{13}^3 C_{13}^r)$, $M_{r,3} = (C_{12}^2 C_{12}^r + C_{12}^3 C_{13}^r)$

$M_{r,4} = (C_{23}^3 C_{23}^r - C_{23}^1 C_{12}^r)$, $M_{r,5} = (C_{13}^1 C_{12}^r - C_{13}^3 C_{23}^r)$, $M_{r,6} = (C_{12}^3 C_{23}^r - C_{12}^1 C_{12}^r)$

$M_{r,7} = (-C_{23}^1 C_{13}^r - C_{23}^2 C_{23}^r)$, $M_{r,8} = (C_{13}^1 C_{13}^r + C_{13}^2 C_{23}^r)$, $M_{r,9} = (-C_{12}^1 C_{13}^r - C_{12}^2 C_{23}^r)$

$r = 1, 2, 3$.

We denote by C the matrix of structure constants $\{C_{ij}^k\}_{i<j}$ for $i, j, k = 1, 2, 3$ given as

$$C = \begin{pmatrix} C_{12}^1 & C_{12}^2 & C_{12}^3 \\ C_{13}^1 & C_{13}^2 & C_{13}^3 \\ C_{23}^1 & C_{23}^2 & C_{23}^3 \end{pmatrix}.$$

3 Hom-Lie structure dimension

In this section, we show that the dimension of $\mathrm{HomLie}(\mu)$ can be described by the matrix C. We show such correspondence between the ranks of matrix C and M. In [2] the authors prove the following proposition.

Proposition 3.1. *Let H be a 3-dimensional skew symmetric algebra with structure constants $\{C_{ij}^k\}_{i<j}$ and C be the matrix of structure constants. The dimension of the space of possible endomorphisms, $HomLie(\mu)$, attains minimum dimension 6, if and only if $\det C$ is non-zero.*

From proposition 3.1, we see that if C is not of full rank then $\dim HL(\mu) \geq 7$. For $\dim HL(\mu) = 7$ we require that $\operatorname{rank} M = 2$. Let us denote by $(m, n)_i$ any 2 by 2 sub-determinant of M involving columns m and n with $1 \leq m < n \leq 9$ and $i = 1, 2, 3$. We proceed to list all 108 such 2 by 2 sub-determinants of M.

For $i = 1, 2, 3$

$(1,2)_i = A_i \cdot D_3 \quad (1,3)_i = A_i \cdot B_3 \quad (2,3)_i = A_i \cdot A_3 \quad (4,5)_i = B_i \cdot D_2 \quad (4,6)_i = B_i \cdot B_2$

$(5,6)_i = A_i \cdot A_2 \quad (7,8)_i = D_i \cdot D_1 \quad (7,9)_i = D_i \cdot B_1 \quad (8,9)_i = D_i \cdot A_1$

$(1,4)_i = C_{23}^1 C_{23}^3 A_i + C_{23}^2 C_{23}^3 B_i + (C_{23}^3)^2 D_i$

$(1,5)_i = -C_{13}^1 C_{23}^3 A_i - C_{23}^2 C_{13}^3 B_i - C_{13}^3 C_{23}^3 D_i$

$(1,6)_i = C_{12}^1 C_{23}^3 A_i + C_{23}^2 C_{12}^3 B_i + C_{12}^3 C_{23}^3 D_i$

$(1,7)_i = -C_{23}^1 C_{23}^2 A_i - (C_{23}^2)^2 B_i - C_{13}^2 C_{23}^3 D_i$

$(1,8)_i = C_{13}^1 C_{23}^2 A_i + C_{13}^2 C_{23}^2 B_i + C_{13}^2 C_{23}^3 D_i$

$(1,9)_i = -C_{12}^1 C_{23}^2 A_i - C_{12}^2 C_{23}^2 B_i - C_{12}^2 C_{23}^3 D_i$

$(2,4)_i = -C_{23}^1 C_{13}^3 A_i - C_{13}^2 C_{23}^3 B_i - C_{13}^3 C_{23}^3 D_i$

$(2,5)_i = C_{13}^1 C_{13}^3 A_i + C_{13}^2 C_{13}^3 B_i + (C_{13}^3)^2 D_i$

$(2,6)_i = -C_{12}^1 C_{13}^3 A_i - C_{13}^2 C_{12}^3 B_i - C_{12}^3 C_{13}^3 D_i$

$(2,7)_i = C_{23}^1 C_{13}^2 A_i + C_{13}^2 C_{23}^2 B_i + C_{23}^2 C_{13}^3 D_i$

$(2,8)_i = -C_{13}^1 C_{13}^2 A_i - (C_{13}^2)^2 B_i - C_{13}^2 C_{13}^3 D_i$

$(2,9)_i = C_{12}^1 C_{13}^2 A_i + C_{12}^2 C_{13}^2 B_i + C_{12}^2 C_{13}^3 D_i$

$(3,4)_i = C_{23}^1 C_{12}^3 A_i + C_{12}^2 C_{23}^3 B_i + C_{12}^3 C_{23}^3 D_i$

$(3,5)_i = -C_{13}^1 C_{12}^3 A_i - C_{12}^2 C_{13}^3 B_i - C_{12}^3 C_{13}^3 D_i$

$(3,6)_i = C_{12}^1 C_{12}^3 A_i + C_{12}^2 C_{12}^3 B_i + (C_{12}^3)^2 D_i$

$(3,7)_i = -C_{23}^1 C_{12}^2 A_i - C_{12}^2 C_{23}^2 B_i + C_{23}^2 C_{12}^3 D_i$

$(3,8)_i = C_{13}^1 C_{12}^2 A_i + C_{12}^2 C_{13}^2 B_i + C_{13}^2 C_{12}^3 D_i$

$(3,9)_i = -C_{12}^1 C_{12}^2 A_i - (C_{12}^2)^2 B_i - C_{12}^2 C_{12}^3 D_i$

$(4,7)_i = (C_{23}^1)^2 A_i + C_{23}^1 C_{23}^2 B_i + C_{23}^1 C_{23}^3 D_i$

$(4,8)_i = -C_{13}^1 C_{23}^1 A_i - C_{23}^1 C_{13}^2 B_i - C_{13}^1 C_{23}^3 D_i$

$(4,9)_i = C_{12}^1 C_{23}^1 A_i + C_{23}^1 C_{12}^2 B_i + C_{12}^1 C_{23}^3 D_i$

$(5,7)_i = -C_{13}^1 C_{23}^1 A_i - C_{13}^1 C_{23}^2 B_i - C_{23}^1 C_{13}^3 D_i$

$(5,8)_i = (C_{13}^1)^2 A_i + C_{13}^1 C_{13}^2 B_i + C_{13}^1 C_{13}^3 D_i$

$(5,9)_i = -C_{12}^1 C_{13}^1 A_i - C_{13}^1 C_{12}^2 B_i - C_{12}^1 C_{13}^3 D_i$

$(6,7)_i = C_{12}^1 C_{23}^1 A_i + C_{12}^1 C_{23}^2 B_i + C_{23}^1 C_{12}^3 D_i$

$(6,8)_i = -C_{12}^1 C_{13}^1 A_i - C_{12}^1 C_{13}^2 B_i - C_{13}^1 C_{12}^3 D_i$

$(6,9)_i = (C_{12}^1)^2 A_i + C_{12}^1 C_{12}^2 B_i + C_{12}^1 C_{12}^3 D_i$

where

$$A_1 = \begin{vmatrix} C_{12}^1 & C_{12}^2 \\ C_{13}^1 & C_{13}^2 \end{vmatrix}, \quad A_2 = \begin{vmatrix} C_{12}^1 & C_{12}^3 \\ C_{13}^1 & C_{13}^3 \end{vmatrix} \quad A_3 = \begin{vmatrix} C_{12}^2 & C_{12}^3 \\ C_{13}^2 & C_{13}^3 \end{vmatrix}$$

Journal of Physics: Conference Series **1416** (2019) 012025 doi:10.1088/1742-6596/1416/1/012025

$$B_1 = \begin{vmatrix} C_{12}^1 & C_{12}^2 \\ C_{23}^1 & C_{23}^2 \end{vmatrix} \quad B_2 = \begin{vmatrix} C_{12}^1 & C_{12}^3 \\ C_{23}^1 & C_{23}^3 \end{vmatrix} \quad B_3 = \begin{vmatrix} C_{12}^2 & C_{12}^3 \\ C_{23}^2 & C_{23}^3 \end{vmatrix}$$

$$D_1 = \begin{vmatrix} C_{13}^1 & C_{13}^2 \\ C_{23}^1 & C_{23}^2 \end{vmatrix}, \quad D_2 = \begin{vmatrix} C_{13}^1 & C_{13}^3 \\ C_{23}^1 & C_{23}^3 \end{vmatrix} \quad D_3 = \begin{vmatrix} C_{13}^2 & C_{13}^3 \\ C_{23}^2 & C_{23}^3 \end{vmatrix}$$

Thus from the computations, we observe that all 2 by 2 minors of M are either factored into 2 by 2 minors of C or can be written as sums involving such minors of C. We give the following proposition.

Proposition 3.2. *Let H be a 3-dimensional skew symmetric algebra with structure constants $\{C_{ij}^k\}_{i<j}$ and C be the matrix of structure constants. If rank C is 2 then the dimension of the space of possible endomorphisms, HomLie(μ), is 7.*

Proof. From previous result in proposition 3.1, M is of full rank if and only if $\det C = 0$. This implies that if rank C is 2 then M is not of full rank. We show that if rank C is 2 then there exists some non-zero 2 by 2 minor of M. With rank $C = 2$ then it means that one of the rows can be written as a linear combination of the other two rows. That is Row $i = \alpha$ Row $j + \beta$ Row k for rows i, j, k and $\alpha, \beta \in \mathbb{K}$. We consider all the possible cases.

Case 1: $\alpha, \beta \neq 0$.

We begin with the case where the first two rows are linearly independent.

If Row 3 $= \alpha$ Row 1 $+\beta$ Row 2 then

$$D_i = -\alpha A_i \quad \text{and} \quad B_i = \beta A_i \tag{6}$$

At least one of the 2 by 2 minors is non-zero since C is of rank 2. Thus since A_i is a common factor for both B_i and D_i for $i = 1, 2, 3$, it is enough to consider only A_i. So at least one A_i must be non-zero. We consider all such possible cases.

(a) $A_i \neq 0 \implies (2,3)_1 = A_1 \cdot A_3 \neq 0$.

(b) $A_1 \neq 0, A_2 \neq 0, A_3 = 0 \implies (5,6)_1 = -B_1 \cdot A_2 \neq 0$ since $B_1 \neq 0$.

(c) $A_1 \neq 0, A_2 = 0, A_3 \neq 0 \implies (1,3)_3 = A_3 \cdot B_3 \neq 0$ since $B_3 \neq 0$.

(d) $A_1 = 0, A_2 \neq 0, A_3 \neq 0 \implies (2,3)_3 = A_3 \cdot A_3 \neq 0$.

(e) $A_1 = 0, A_2 = 0, A_3 \neq 0 \implies (2,3)_3 = A_3 \cdot A_3 \neq 0$.

(f) $A_1 = 0, A_2 \neq 0, A_3 = 0 \implies (5,6)_2 = A_2 \cdot A_2 \neq 0$.

(g) $A_1 \neq 0, A_2 = 0, A_3 = 0 \implies (8,9)_1 = D_1 \cdot A_1 \neq 0$ since $D_1 \neq 0$.

If Row 2 $= \alpha$ Row 1 $+\beta$ Row 3 then

$$D_i = \alpha B_i \quad \text{and} \quad A_i = \beta B_i \implies D_i = \alpha' A_i \quad \text{and} \quad B_i = \beta' A_i$$

which is the same case as in (6) since $\alpha', \beta' \neq 0$.

Similarly, if Row 1 $= \alpha$ Row 2 $+\beta$ Row 3 then

$$A_i = -\beta D_i \quad \text{and} \quad B_i = \alpha D_i \implies D_i = \beta' A_i \quad \text{and} \quad B_i = \alpha' A_i$$

which is the same case shown from (6) since $\beta', \alpha' \neq 0$.

Case 2: Either $\alpha = 0$ or $\beta = 0$ but not both.

Journal of Physics: Conference Series **1416** (2019) 012025 doi:10.1088/1742-6596/1416/1/012025

In this case, it implies we check the cases where one of the rows can be written as a scalar product of another row.

Row 2 = k Row 1

We have

$$D_i = kB_i \quad \text{and} \quad A_i = 0 \tag{7}$$

So at least one B_i must be non-zero. We consider all such possible cases.

(a) $B_i \neq 0 \implies (4,6)_1 = B_1 \cdot B_2 \neq 0$.
(b) $B_1 \neq 0, B_2 \neq 0, B_3 = 0 \implies (4,6)_1 = B_1 \cdot B_2 \neq 0$.
(c) $B_1 \neq 0, B_2 = 0, B_3 \neq 0 \implies (7,8)_1 = D_1 \cdot D_1 \neq 0$ since $D_1 \neq 0$.
(d) $B_1 = 0, B_2 \neq 0, B_3 \neq 0 \implies (4,5)_2 = -B_2 \cdot D_2 \neq 0$ since $D_2 \neq 0$.
(e) $B_1 = 0, B_2 = 0, B_3 \neq 0 \implies$ Either $(3,4)_3$ or $(1,6)_3$ is non-zero.
(f) $B_1 = 0, B_2 \neq 0, B_3 = 0 \implies (4,5)_2 = -B_2 \cdot D_2 \neq 0$ since $D_2 \neq 0$.
(g) $B_1 \neq 0, B_2 = 0, B_3 = 0 \implies (7,8)_1 = D_1 \cdot D_1 \neq 0$ since $D_1 \neq 0$.

Row 3 = k Row 1

We have

$$D_i = kA_i \quad \text{and} \quad B_i = 0 \tag{8}$$

So at least one A_i must be non-zero. We consider all such possible cases.

(a) $A_i \neq 0 \implies (2,3)_1 = A_1 \cdot A_3 \neq 0$.
(b) $A_1 \neq 0, A_2 \neq 0, A_3 = 0 \implies (8,9)_1 = D_1 \cdot A_1 \neq 0$ since $D_1 \neq 0$.
(c) $A_1 \neq 0, A_2 = 0, A_3 \neq 0 \implies (8,9)_1 = D_1 \cdot A_1 \neq 0$ since $D_1 \neq 0$.
(d) $A_1 = 0, A_2 \neq 0, A_3 \neq 0 \implies (2,3)_2 = A_2 \cdot A_3 \neq 0$.
(e) $A_1 = 0, A_2 = 0, A_3 \neq 0 \implies (2,3)_3 = A_3 \cdot A_3 \neq 0$.
(f) $A_1 = 0, A_2 \neq 0, A_3 = 0 \implies$ Either $(3,5)_2$ or $(2,6)_2$ is non-zero.
(g) $A_1 \neq 0, A_2 = 0, A_3 = 0 \implies (8,9)_1 = D_1 \cdot A_1 \neq 0$ since $D_1 \neq 0$.

Row 3 = k Row 2

We have

$$B_i = kA_i \quad \text{and} \quad D_i = 0 \tag{9}$$

So at least one A_i must be non-zero. We consider all such possible cases.

(a) $A_i \neq 0 \implies (2,3)_1 = A_1 \cdot A_3 \neq 0$.
(b) $A_1 \neq 0, A_2 \neq 0, A_3 = 0 \implies (5,6)_1 = -B_1 \cdot A_2 \neq 0$ since $B_1 \neq 0$.
(c) $A_1 \neq 0, A_2 = 0, A_3 \neq 0 \implies (1,3)_1 = -A_1 \cdot B_3 \neq 0$ since $B_3 \neq 0$.
(d) $A_1 = 0, A_2 \neq 0, A_3 \neq 0 \implies (2,3)_3 = A_3 \cdot A_3 \neq 0$.
(e) $A_1 = 0, A_2 = 0, A_3 \neq 0 \implies (2,3)_3 = A_3 \cdot A_3 \neq 0$.
(f) $A_1 = 0, A_2 \neq 0, A_3 = 0 \implies (5,6)_2 = -B_2 \cdot A_2 \neq 0$ since $B_2 \neq 0$.
(g) $A_1 \neq 0, A_2 = 0, A_3 = 0 \implies$ Either $(3,8)_1$ or $(2,9)_1$ is non-zero.

Case 3: $\alpha, \beta = 0$

In this case, it implies we check the cases where one of the rows is zero while the other two are linearly independent.

Row 3 = 0

We have

$$B_i = D_i = 0 \text{ and } A_i \neq 0 \tag{10}$$

So at least one A_i must be non-zero. We consider all such possible cases.

(a) $A_i \neq 0 \implies (2,3)_1 = A_1 \cdot A_3 \neq 0$.
(b) $A_1 \neq 0, A_2 \neq 0, A_3 = 0 \implies$ Either $(6,9)_2$ or $(5,8)_2$ is non-zero.
(c) $A_1 \neq 0, A_2 = 0, A_3 \neq 0 \implies (2,3)_1 = A_3 \cdot A_3 \neq 0$.
(d) $A_1 = 0, A_2 \neq 0, A_3 \neq 0 \implies (2,3)_3 = A_3 \cdot A_3 \neq 0$.
(e) $A_1 = 0, A_2 = 0, A_3 \neq 0 \implies (2,3)_3 = A_3 \cdot A_3 \neq 0$.
(f) $A_1 = 0, A_2 \neq 0, A_3 = 0 \implies$ Either $(6,9)_2$ or $(5,8)_2$ is non-zero.
(g) $A_1 \neq 0, A_2 = 0, A_3 = 0 \implies$ Either $(3,8)_1$ or $(2,9)_1$ is non-zero.

Row 2 = 0

We have

$$A_i = D_i = 0 \text{ and } B_i \neq 0 \tag{11}$$

So at least one B_i must be non-zero. We consider all such possible cases.

(a) $B_i \neq 0 \implies (4,6)_1 = B_1 \cdot B_2 \neq 0$.
(b) $B_1 \neq 0, B_2 \neq 0, B_3 = 0 \implies (4,6)_1 = B_1 \cdot B_2 \neq 0$.
(c) $B_1 \neq 0, B_2 = 0, B_3 \neq 0 \implies$ Either $(3,9)_3$ or $(1,7)_3$ is non-zero.
(d) $B_1 = 0, B_2 \neq 0, B_3 \neq 0 \implies (4,6)_3 = B_3 \cdot B_2 \neq 0$.
(e) $B_1 = 0, B_2 = 0, B_3 \neq 0 \implies$ Either $(3,9)_3$ or $(1,7)_3$ is non-zero.
(f) $B_1 = 0, B_2 \neq 0, B_3 = 0 \implies (4,6)_2 = B_2 \cdot B_2 \neq 0$.
(g) $B_1 \neq 0, B_2 = 0, B_3 = 0 \implies$ Either $(3,9)_1$ or $(1,7)_1$ is non-zero.

Row 1 = 0

We have

$$A_i = B_i = 0 \text{ and } D_i \neq 0 \tag{12}$$

So at least one D_i must be non-zero. We consider all such possible cases.

(a) $D_i \neq 0 \implies (7,8)_1 = D_1 \cdot D_1 \neq 0$.
(b) $D_1 \neq 0, D_2 \neq 0, D_3 = 0 \implies (7,8)_2 = D_2 \cdot D_1 \neq 0$.
(c) $D_1 \neq 0, D_2 = 0, D_3 \neq 0 \implies (7,8)_3 = D_3 \cdot D_1 \neq 0$.
(d) $D_1 = 0, D_2 \neq 0, D_3 \neq 0 \implies$ Either $(4,8)_2$ or $(5,7)_2$ is non-zero.
(e) $D_1 = 0, D_2 = 0, D_3 \neq 0 \implies$ Either $(1,8)_3$ or $(2,6)_3$ is non-zero.
(f) $D_1 = 0, D_2 \neq 0, D_3 = 0 \implies$ Either $(4,8)_2$ or $(5,7)_2$ is non-zero.
(g) $D_1 \neq 0, D_2 = 0, D_3 = 0 \implies (7,8)_1 = D_1 \cdot D_1 \neq 0$.

Thus we have shown that there always exists a non-zero 2×2 sub-determinant of M whenever rank C is 2. This implies that dimension of HomLie(μ) is 7. □

From the list of all 2×2 minors of M given, it follows that if rank C is 1 then all such minors are zero and hence the dimension of HomLie(μ) > 7. So far, we have seen there is a correspondence between the rank of C and M, for rank $C = 2$ and 3. However, if rank $C = 1$ this dimension can either be 8 or 9. It is obvious that if rank C is 0 then M is of rank 0 and consequently dim HomLie(μ) $= 9$. We give possible cases for rank $C = 1$ with dim HomLie(μ) $= 9$.

$$\mathfrak{C}_1 = \begin{pmatrix} C_{23}^3 & C_{12}^2 & -\dfrac{C_{12}^2\, C_{23}^3}{C_{13}^1} \\ C_{13}^1 & \dfrac{C_{12}^2\, C_{13}^1}{C_{23}^3} & -C_{12}^2 \\ -\dfrac{C_{13}^1\, C_{23}^3}{C_{12}^2} & -C_{13}^1 & C_{23}^3 \end{pmatrix}, C_{12}^2, C_{13}^1, C_{23}^3 \neq 0, \quad \mathfrak{C}_2 = \begin{pmatrix} 0 & 0 & 0 \\ 0 & 0 & 0 \\ C_{23}^1 & 0 & 0 \end{pmatrix}, C_{23}^1 \neq 0$$

$$\mathfrak{C}_3 = \begin{pmatrix} C_{23}^3 & 0 & C_{12}^3 \\ 0 & 0 & 0 \\ \dfrac{(C_{23}^3)^2}{C_{12}^3} & 0 & C_{23}^3 \end{pmatrix}, C_{12}^3 \neq 0, \qquad \mathfrak{C}_4 = \begin{pmatrix} 0 & 0 & 0 \\ C_{13}^1 & C_{13}^2 & 0 \\ \dfrac{-(C_{13}^1)^2}{C_{13}^2} & -C_{13}^1 & 0 \end{pmatrix}, C_{13}^2 \neq 0$$

$$\mathfrak{C}_5 = \begin{pmatrix} 0 & C_{12}^2 & C_{12}^3 \\ 0 & \dfrac{(C_{12}^2)^2}{C_{12}^3} & -C_{12}^2 \\ 0 & 0 & 0 \end{pmatrix}, C_{12}^3 \neq 0$$

We now give some cases for rank $C = 1$ with dim HomLie$(\mu) = 8$.

$$\mathfrak{C}_1' = \begin{pmatrix} C_{12}^1 & 0 & 0 \\ 0 & 0 & 0 \\ 0 & 0 & 0 \end{pmatrix}, C_{12}^1 \neq 0, \quad \mathfrak{C}_2' = \begin{pmatrix} 0 & C_{12}^2 & 0 \\ 0 & 0 & 0 \\ 0 & 0 & 0 \end{pmatrix}, C_{12}^2 \neq 0, \quad \mathfrak{C}_3' = \begin{pmatrix} 0 & 0 & 0 \\ C_{13}^1 & 0 & 0 \\ 0 & 0 & 0 \end{pmatrix}, C_{13}^1 \neq 0$$

$$\mathfrak{C}_4' = \begin{pmatrix} 0 & 0 & 0 \\ 0 & 0 & C_{13}^3 \\ 0 & 0 & 0 \end{pmatrix}, C_{13}^3 \neq 0, \quad \mathfrak{C}_5' = \begin{pmatrix} 0 & 0 & 0 \\ 0 & 0 & 0 \\ 0 & C_{23}^2 & 0 \end{pmatrix}, C_{23}^2 \neq 0, \quad \mathfrak{C}_6' = \begin{pmatrix} 0 & 0 & 0 \\ 0 & 0 & 0 \\ 0 & 0 & C_{23}^3 \end{pmatrix}, C_{23}^3 \neq 0$$

Proposition 3.3. *If two skew-symmetric algebras (L, μ) and (L', μ') are isomorphic, then HomLie(μ) is isomorphic to HomLie(μ').*

Proof. Let $\phi : L \to L'$ be the isomorphism of the algebras. Let $\alpha \in$ HomLie(μ) and $\beta \in$ HomLie(μ'). The isomorphism between the Hom-Lie structures $\varphi : HomLie(\mu) \to HomLie(\mu')$ is defined by $\beta = \varphi(\alpha) := \phi \circ \alpha \circ \phi^{-1}$. Given $x, y, z \in L$, then there exists $x', y', z' \in L'$ as images of x, y and z respectively under the isomorphism ϕ. That is, $\phi(x) = x', \phi(y) = y'$ and $\phi(z) = z'$. $\beta = \phi \circ \alpha \circ \phi^{-1}$ implies $\beta(x') = \phi(\alpha(x)), \beta(y') = \phi(\alpha(y))$ and $\beta(z') = \phi(\alpha(z))$. We show that if $\alpha \in$ HomLie(μ) then $\beta \in$ HomLie(μ').

$$\begin{aligned} &\mu'(\beta(x'), \mu'(y', z')) + \mu'(\beta(y'), \mu(z', x')) + \mu'(\beta(z'), \mu'(x', y')) \\ &= \mu'(\phi(\alpha(x)), \mu'(\phi(y), \phi(z)) + \mu'(\phi(\alpha(y)), \mu(\phi(z), \phi(x)) + \mu'(\phi(\alpha(z)), \mu'(\phi(x), \phi(y)) \\ &= \phi[\mu((\alpha(x)), \mu(y, z)) + \mu((\alpha(y)), \mu(z, x)) + \mu((\alpha(z)), \mu(x, y))] \\ &= 0 \end{aligned}$$

□

Thus using proposition (3.3) we can describe dimension of all skew-symmetric algebras including Lie algebras by using representatives of such algebras.

4 3-dimensional Lie algebras

In this section, we use our results to deduce dim HomLie(μ) of classical complex Lie algebras. Such non-isomorphic representatives of all complex 3-dimensional Lie algebras can be seen in [4] and [5]. Let $\{e_1, e_2, e_3\}$ be the basis of the 3-dimensional Lie algebras.

Journal of Physics: Conference Series **1416** (2019) 012025 doi:10.1088/1742-6596/1416/1/012025

Lie algebra	C	dim HomLie(μ)
$[e_1,e_2]=e_3,\ [e_1,e_3]=0,\ [e_2,e_3]=0$ (Heisenberg Lie algebra)	$\begin{pmatrix}0&0&1\\0&0&0\\0&0&0\end{pmatrix}$	9 (Lie algebra in $\mathfrak{C}_3$)
$[e_1,e_2]=e_1,\ [e_1,e_3]=0,\ [e_2,e_3]=0$	$\begin{pmatrix}1&0&0\\0&0&0\\0&0&0\end{pmatrix}$	8 (Lie algebra in $\mathfrak{C}'_1$)
$[e_1,e_2]=e_2,\ [e_1,e_3]=\lambda\, e_3,\ [e_2,e_3]=0,\ \lambda\in\mathbb{C}^*$	$\begin{pmatrix}0&1&0\\0&0&\lambda\\0&0&0\end{pmatrix}$	7
$[e_1,e_2]=e_2,\ [e_1,e_3]=0,\ [e_2,e_3]=e_2+e_3$	$\begin{pmatrix}0&1&0\\0&0&0\\0&1&1\end{pmatrix}$	7
$[e_1,e_2]=-2e_1,\ [e_1,e_3]=e_2,\ [e_2,e_3]=-2e_3$ (Simple Lie algebra)	$\begin{pmatrix}-2&0&0\\0&1&0\\0&0&-2\end{pmatrix}$	6 ($\det C=4\neq 0$)

Table 1: Lie algebras and dimension of Hom-Lie structures

In [7] (proposition 7.1) the matrix of linear endomorphism α for all Hom-Lie algebras of $\mathfrak{sl}(2)$ type is in the form $\begin{pmatrix}a_{11}&a_{12}&a_{13}\\2a_{13}&a_{22}&a_{31}\\2a_{12}&a_{32}&a_{22}\end{pmatrix}$, with 6 parameters $a_{11},a_{12},a_{13},a_{13},a_{22},a_{32}\in\mathbb{K}$. Thus the space of HomLie(μ) for this case is of dimension 6.

Acknowledgments

Elvice Ongong'a is grateful to the International Science Program, Uppsala University for the financial support within the network of the Eastern Africa Universities Mathematics Programme (EAUMP) and to the research environment in Mathematics and Applied Mathematics (MAM), the Division of Applied Mathematics of the School of Education, Culture and Communication at Mälardalen University for hospitality and creating excellent conditions for research, research education and cooperation.

References

[1] Hartwig J T, Larsson D and Silvestrov S D 2006 Deformations of Lie algebras using $\sigma-$derivations *J. Algebra* **295** 314-361
[2] Ongong'a E, Richter J, and Silvestrov S 2019 Classification of 3-dimensional Hom-Lie algebras *J. Phys.: Conf Ser.* **1194** 012084
[3] Remm, E and Goze M 2017 On the algebraic variety of Hom-Lie algebras arXiv:1706.02484v1 [math.RA]
[4] Erdmann, K and Wildon, M J 2006 *Introduction to Lie algebras,* Springer Science & Business Media.
[5] Jacobson, N 1972 *Lie algebras,* Courier Corporation.
[6] Nejib, S 2018 *Cohomology of Heisenberg Hom-Lie algebras,* arXiv:1701.07090v4 [math.RA]
[7] Makhlouf A and Silvestrov S 2008 Hom-algebra structures *J. Gen. Lie Theory Appl.* **2** 51-64
[8] Makhlouf A and Zusmanovich P 2018 Hom-Lie structures on Kac-Moody algebras. Journal of Algebra **515**, 278-297

Journal of Physics: Conference Series **1416** (2019) 012026 doi:10.1088/1742-6596/1416/1/012026

Scalar particle with Darwin-Cox structure in external Coulomb field

E M Ovsiyuk[1], A D Koral'kov[1], A V Chichurin[2], Ya A Voynova[3] and V M Red'kov[3]

[1] Mozyr State Pedagogical University, Studencheskaya st. 28, Mozyr 247760, Belarus
[2] The John Paul II Catholic University of Lublin, Lublin, Poland 20-708
[3] B.I. Stepanov Institute of Physics of NAS of Belarus, Independence ave. 68-2, 220072 Minsk

E-mail: e.ovsiyuk@mail.ru , artemkoralkov@gmail.com , achichurin@gmail.com , voinovayanina@mail.ru , v.redkov@ifanbel.bas-net.by

Abstract. Generalized Klein-Fock-Gordon equation for a scalar particle with Darwin-Cox structure, which takes into account distribution of electric charge of the particle inside a finite spherical region is studied in external Coulomb field. Corresponding radial equation has two irregular singular points, $r = 0$ of the rank 3, $r = \infty$ of the rank 2, and four regular singular points. In the case of minimal angular momentum, $l = 0$, the structure of singularities becomes simpler: the points $r = 0, r = \infty$ are both of the rank 2, and four regular points remain the same. There are constructed formally exact Frobenius type solutions of the derived equations, convergence of relevant power series, with 8-term and 7-term recurrent relations respectively, is studied. As analytical quantization rule is taken so-called transcendence conditions. It provides us with 4-th order algebraic equation with respect to energy values, which has four sets of roots. Only one set of roots, $0 < E_{l,k} < 1$, depending on angular momentum $l = 0, 1, 2, ...$ and main quantum number $n = 0, 1, 2, ...$ may be interpreted as corresponding to some bound states of the particle in the Coulomb field. In the same manner, a generalized nonrelativistic Schrödinger equation for such a particle was studied, the final results are similar.

1. Introduction, Cox equation

In the frames of the theory of generalized relativistic wave equations, a special model for a spin zero particle was proposed by W. Cox [1]. Cox constructed the wave equation for a scalar particle with a larger set of tensor functions than it exists in the Proca's approach. He used the set of a scalar, 4-vector, antisymmetric and (irreducible) symmetric tensor, thus starting with the 20-component wave function. In particular, the Cox electromagnetic structure in presence of external electric field may be associated with the known C.G. Darwin interaction term in nonrelativistic Schrödinger equation, this interaction is related to non-point-like distribution of the electric charge in the finite volume inside the particle see.

In the present paper we start with Cox covariant system (let $D_\alpha = i\nabla_\alpha - \frac{e}{\hbar c}A_\alpha$):

$$\lambda_1 D^\beta \Phi_\beta - \frac{Mc}{\hbar}\Phi = 0, \quad \lambda_1^* D_\beta \Phi + \lambda_2 D^\alpha \Phi_{[\alpha\beta]} - \lambda_3 D^\alpha \Phi_{(\alpha\beta)} - \frac{Mc}{\hbar}\Phi_\beta = 0,$$
$$\lambda_2^*(D_\alpha\Phi_\beta - D_\beta\Phi_\alpha) - \frac{Mc}{\hbar}\Phi_{[\alpha\beta]} = 0, \quad \lambda_3^*(D_\alpha\Phi_\beta + D_\beta\Phi_\alpha - \frac{1}{2}g_{\alpha\beta}D^\rho\Phi_\rho) - \frac{Mc}{\hbar}\Phi_{(\alpha\beta)} = 0, \qquad (1)$$

Journal of Physics: Conference Series **1416** (2019) 012026 doi:10.1088/1742-6596/1416/1/012026

auxiliary numerical parameters $\lambda_1, \lambda_2, \lambda_3$ subject to the constraints

$$\lambda_2\lambda_2^* - \lambda_3\lambda_3^* = 0\,, \qquad \lambda_1\lambda_1^* - \frac{3}{2}\lambda_3\lambda_3^* = 1. \tag{2}$$

With the help of the third and the fourth equations in (1), we can exclude tensor components, $\Phi_{(\alpha\beta)}, \Phi_{[\alpha\beta]}$, so deriving a generalized Proca system

$$D^\beta\Phi_\beta - \frac{Mc}{\hbar}\Phi = 0, \quad D_\beta\Phi - \lambda(F_{\beta\alpha} + i\frac{\hbar c}{e}R_{\beta\alpha})\Phi^\alpha - \frac{Mc}{\hbar}\Phi_\beta = 0, \tag{3}$$

note that λ is an imaginary number

$$\lambda = i\frac{\hbar^2}{Mc}\frac{e}{\hbar c}(2\lambda_3\lambda_3^*)\,. \tag{4}$$

In (3) we can see a non-minimal interaction term of the Cox's scalar particle with external electromagnetic field and with geometric background through the Ricci tensor.

2. Setting the problem

In flat space, equations can be presented as

$$(mc/\hbar)(\delta_\alpha^\beta + \frac{\lambda}{mc/\hbar}F_\alpha^{\ \beta})\Phi_\beta = D_\alpha\Phi\,, \quad D^\alpha\Phi_\alpha = \frac{mc}{\hbar}\Phi\,, \tag{5}$$

where $D_\alpha = i\nabla_\alpha + (e/\hbar c)A_\alpha$ (note that parameter e is positive). Shortly, they read (Λ stands for 4×4 matrix)

$$\frac{mc}{\hbar}\Lambda_\alpha^{\ \beta}\Phi_\beta = D_\alpha\Phi\,, \quad D^\alpha\Phi_\alpha = \frac{mc}{\hbar}\Phi\,, \tag{6}$$

whence it follows a generalized Klein-Fock-Gordon equation for scalar function $\Phi(x)$:

$$[D_\rho(\Lambda^{-1})^{\rho\alpha}(x)D_\alpha\Phi \; - \frac{m^2c^2}{\hbar^2}]\Phi = 0\,. \tag{7}$$

Explicit form of eq. (7) in space with metric $g_{\alpha\beta}(x)$ is as follows

$$\left[(\frac{i}{\sqrt{-g}}\frac{\partial}{\partial x^\rho}\sqrt{-g} + \frac{e}{c\hbar}A_\rho)(\Lambda^{-1}(x))^{\rho\alpha}(i\frac{\partial}{\partial x^\alpha} + \frac{e}{c\hbar}A_\alpha) \; - \frac{m^2c^2}{\hbar^2}\right]\Phi = 0\,.$$

3. Separating the variables

We consider the particle in external Coulomb field, using spherical coordinates (r, θ, ϕ):

$$A_0 = \frac{e}{r}, \quad e > 0, \quad F_{r0} = -\frac{e}{r^2}.$$

The relevant matrix $\Lambda_\alpha^{\ \beta}$ has the structure

$$\Lambda = (\Lambda_\alpha^{\ \beta}) = \left| \begin{array}{cccc} 1 & \frac{\mu}{r^2} & 0 & 0 \\ \frac{\mu}{r^2} & 1 & 0 & 0 \\ 0 & 0 & 1 & 0 \\ 0 & 0 & 0 & 1 \end{array} \right|, \quad \mu = -\frac{\lambda e}{mc/\hbar}\,;$$

its inverse matrix is given by expression (let $\Lambda^{-1} = (K_\beta{}^\rho)$)

$$(K_\beta{}^\rho) = \begin{vmatrix} \frac{r^4}{r^4-\mu^2} & -\frac{\mu r^2}{r^4-\mu^2} & 0 & 0 \\ -\frac{\mu r^2}{r^4-\mu^2} & \frac{r^4}{r^4-\mu^2} & 0 & 0 \\ 0 & 0 & 1 & 0 \\ 0 & 0 & 0 & 1 \end{vmatrix}, (K^{\beta\rho}) = \begin{vmatrix} \frac{r^4}{r^4-\mu^2} & -\frac{\mu r^2}{r^4-\mu^2} & 0 & 0 \\ +\frac{\mu r^2}{r^4-\mu^2} & -\frac{r^4}{r^4-\mu^2} & 0 & 0 \\ 0 & 0 & -1/r^2 & 0 \\ 0 & 0 & 0 & -1/r^2\sin^2\theta \end{vmatrix}.$$

Explicitly, the wave equation looks as (let $\alpha = e^2/\hbar c$)

$$\Big\{ \Big(i\partial_0 + \frac{\alpha}{r}\Big)^2 \frac{r^4}{r^4-\mu^2} - \Big(i\partial_0 + \frac{\alpha}{r}\Big) \frac{\mu r^2}{r^4-\mu^2} i\partial_r$$
$$+ \frac{i}{r^2}\partial_r r^2 \frac{\mu r^2}{r^4-\mu^2}\Big(i\partial_0 + \frac{\alpha}{r}\Big) - \frac{i}{r^2}\partial_r r^2 \frac{r^4}{r^4-\mu^2} i\partial_r$$
$$+\frac{1}{r^2}\Big(\frac{1}{\sin\theta}\partial_\theta \sin\theta\partial_\theta + \frac{1}{\sin^2\theta}\partial_\phi\partial_\phi\Big) - \frac{m^2c^2}{\hbar^2}\Big\}\Phi = 0. \tag{8}$$

After separating the variables with the help of the usual substitution

$$\Phi(x) = e^{-i\epsilon x^0}\, Y_{lm}(\theta,\phi)\, R(r), \qquad \epsilon = \frac{E}{c\hbar}, \quad [\epsilon] = \frac{1}{\text{meter}}$$

we derive the radial equation

$$\frac{d^2R}{dr^2} + \Big[\frac{2}{r} + \frac{4\gamma^2}{r(r^4+\gamma^2)}\Big]\frac{dR}{dr} + \Big[\epsilon^2 - M^2 + \frac{2\alpha\epsilon}{r} + \frac{\alpha^2 - l(l+1)}{r^2}$$
$$+\frac{4\gamma\epsilon}{r^3} + \frac{3\alpha\gamma - \gamma^2M^2}{r^4} - \frac{l(l+1)\gamma^2}{r^6} - \frac{4(\gamma\epsilon r + \gamma\alpha)}{r^4+\gamma^2}\Big] R = 0\,, \tag{9}$$

where $i\mu = \gamma$ is real-valued. Here we have four regular points and two irregular ones:

$$r^4 + \gamma^2 = (r - e^{+i\pi/4}\sqrt{\gamma})(r + e^{+i\pi/4}\sqrt{\gamma})(r - e^{-i\pi/4}\sqrt{\gamma})(r + e^{-i\pi/4}\sqrt{\gamma})\,,$$

$$r = 0,\ \text{Rank} = 3\,, \qquad r = \infty,\ \text{Rank} = 2\,;$$

note identities

$$\sigma \equiv (-\gamma^2)^{1/4}, \quad \sigma^2 = \sqrt{-\gamma^2} = i\gamma, \quad \sigma^4 = -\gamma^2,$$
$$\frac{1}{r^4+\gamma^2} = \frac{1}{4\sigma^3}\Big(\frac{1}{r-\sigma} - \frac{1}{r+\sigma} + \frac{i}{r-i\sigma} - \frac{i}{r+i\sigma}\Big).$$

Behavior of solutions near four regular points is trivial:

$$r \to \pm\sigma, \quad R \sim (r \mp \sigma)^\rho, \quad \rho = 0, 2\,; \tag{10}$$
$$r \to \pm i\sigma, \quad R \sim (r \mp -i\sigma)^\rho, \quad \rho = 0, 2\,. \tag{11}$$

Substitution for local Frobenius solutions of Eq. (9) in vicinity of the point $r = 0$ is as follows

$$l = 0, \quad R(r) = r^C e^{Ar} e^{B/r} f(r)\,; \quad l = 1, 2, ..., \ R(r) = r^C e^{Ar} e^{B/r} e^{D/r^2} f(r)\,. \tag{12}$$

4. States with minimal orbital momentum

Let us consider the case of $l = 0$, the relevant radial equation becomes simpler

$$\frac{d^2R}{dr^2} + \left(\frac{6}{r} - \frac{4r^3}{r^4+\gamma^2}\right)\frac{dR}{dr} \tag{13}$$

$$+\left(\epsilon^2 - M^2 + \frac{2\alpha\epsilon}{r} + \frac{\alpha^2}{r^2} + \frac{4\gamma\epsilon}{r^3} + \frac{3\alpha\gamma - \gamma^2M^2}{r^4} - \frac{4\gamma(\epsilon r+\alpha)}{r^4+\gamma^2}\right)R = 0.$$

Here we have four regular points, and two irregular points, $r = 0$ and $r = \infty$, both of Rank 2. The main equation may be presented in more symmetrical form

$$\frac{d^2R}{dr^2} + \left(\frac{6}{r} - \frac{1}{r-\sigma} - \frac{1}{r-i\sigma} - \frac{1}{r+\sigma} - \frac{1}{r+i\sigma}\right)\frac{dR}{dr}$$

$$+\left[\epsilon^2 - M^2 + \frac{2\,\alpha\,\epsilon}{r} + \frac{\alpha^2}{r^2} + \frac{4\,\gamma\epsilon}{r^3} + \frac{3\,\alpha\,\gamma - \gamma^2M^2}{r^4}\right.$$

$$\left.-\frac{\gamma\,(\epsilon\,\sigma+\alpha)}{(r-\sigma)\,\sigma^3} + \frac{\gamma\,(\epsilon\,\sigma+i\alpha)}{(r+i\sigma)\,\sigma^3} + \frac{\gamma\,(-\epsilon\,\sigma+\alpha)}{(r+\sigma)\,\sigma^3} - \frac{\gamma\,(-\epsilon\,\sigma+i\alpha)}{(r-i\sigma)\,\sigma^3}\right]R = 0\,.$$

Solutions are constructed with the use of substitution $R(r) = r^C e^{Ar} e^{B/r}\, f(r)$; further we get four sets of parameters and an equation for function $f(r)$:

$$A = \pm\sqrt{M^2-\epsilon^2}\,, \quad B = \delta\sqrt{\gamma(\gamma M^2 - 3\,\alpha)}\,, \quad C = -2 + \delta\frac{2\gamma\epsilon}{\sqrt{\gamma(\gamma M^2-3\,\alpha)}}\,, \tag{14}$$

where $\delta = \pm 1$. To describe bound states, we should use negative values for A: $A = -\sqrt{M^2-\epsilon^2}$. As for B, first let us consider the possibility of imaginary values for B:

$$e^{B/r} = e^{\pm iR/r}, \quad R > 0\,, \quad \gamma \in (0, \frac{3\alpha}{M^2}). \tag{15}$$

For this case, we have divergent and oscillating behavior for solutions near the point $r = 0$:

$$r \to 0, \qquad R \sim \frac{1}{r^2} e^{\mp i\frac{2\gamma\epsilon}{R}\ln r} e^{\pm iR/r}\,; \tag{16}$$

hardly this behavior may be appropriate for bound states, By this reason, in the following we will assume that B is real-valued, this results in

$$B = -\sqrt{\gamma(\gamma M^2 - 3\,\alpha)}\,, \quad \gamma < 0\,. \tag{17}$$

Taking in mind these restrictions for parameters, we get the following equation for $f(r)$ (its short form is written down):

$$f'' + \left(a + \frac{a_1}{r} + \frac{a_2}{r^2} - \frac{1}{r+\sigma} - \frac{1}{r-\sigma} - \frac{1}{r+i\sigma} - \frac{1}{r-i\sigma}\right) f'$$

$$+\left(\frac{b_1}{r} + \frac{b_2}{r^2} + \frac{\beta_1}{r+\sigma} + \frac{\beta_2}{r-\sigma} + \frac{\beta_3}{r+i\sigma} + \frac{\beta_4}{r-i\sigma}\right) f = 0\,. \tag{18}$$

Solutions are searched as power series: $f(r) = \sum_{n=0}^{\infty} c_n r^n$. Then we derive 7-term recurrent relations for coefficients:

$$k = 5,\, 6,\, 7, ..., \quad [a(k-5) + (b_1 + \beta_1 + \beta_2 + \beta_3 + \beta_4)]c_{k-5}$$

$$+[(k-4)(k-5)+(a_1-4)(k-4)+\{(-\beta_1+\beta_2-i\beta_3+i\beta_4)\sigma+b_2\}]c_{k-4}$$
$$+[a_2\,(k-3)+\sigma^2(\beta_1+\beta_2-\beta_3-\beta_4)]\,c_{k-3}+(-\beta_1+\beta_2+i\beta_3-i\beta_4)\sigma^3 c_{k-2}$$
$$+[-\sigma^4 a(k-1)-b_1\sigma^4]c_{k-1}+[-\sigma^4 k(k-1)-\sigma^4 a_1 k-b_2\,\sigma^4]c_k-\sigma^4 a_2(k+1)c_{k+1}=0. \quad (19)$$

In accordance with Poincaré-Peronne method, we should divide the last relationship by $k^2 c_{k-5}$:

$$\frac{1}{k^2}\,[a\,(k-5)+(b_1+\beta_1+\beta_2+\beta_3+\beta_4)]$$
$$+\frac{1}{k^2}[(k-4)(k-5)+(a_1-4)(k-4)+\{(-\beta_1+\beta_2-i\beta_3+i\beta_4)\sigma+b_2\}]\frac{c_{k-4}}{c_{k-5}}$$
$$+\frac{1}{k^2}[a_2\,(k-3)+\sigma^2(\beta_1+\beta_2-\beta_3-\beta_4)]\,\frac{c_{k-3}}{c_{k-4}}\frac{c_{k-4}}{c_{k-5}}$$
$$+\frac{1}{k^2}(-\beta_1+\beta_2+i\beta_3-i\beta_4)\sigma^3\,\frac{c_{k-2}}{c_{k-3}}\frac{c_{k-3}}{c_{k-4}}\frac{c_{k-4}}{c_{k-5}}$$
$$+\frac{1}{k^2}[-\sigma^4 a\,(k-1)-b_1\,\sigma^4]\,\frac{c_{k-1}}{c_{k-2}}\frac{c_{k-2}}{c_{k-3}}\frac{c_{k-3}}{c_{k-4}}\frac{c_{k-4}}{c_{k-5}}$$
$$+\frac{1}{k^2}[-\sigma^4\,k(k-1)-\sigma^4 a_1 k-b_2\sigma^4]\frac{c_k}{c_{k-1}}\frac{c_{k-1}}{c_{k-2}}\frac{c_{k-2}}{c_{k-3}}\frac{c_{k-3}}{c_{k-4}}\frac{c_{k-4}}{c_{k-5}}$$
$$-\frac{1}{k^2}\sigma^4 a_2(k+1)\frac{c_{k+1}}{c_k}\frac{c_k}{c_{k-1}}\frac{c_{k-1}}{c_{k-2}}\frac{c_{k-2}}{c_{k-3}}\frac{c_{k-3}}{c_{k-4}}\frac{c_{k-4}}{c_{k-5}}=0\,,$$

and then tend k to infinity $k\longrightarrow\infty$. This results in an algebraic equation which determines possible convergence radii $R_{conv}=\frac{1}{|r|}$:

$$\lim_{k\to\infty}\frac{c_{k-4}}{c_{k-5}}=r,\quad r-\sigma^4 r^5=0,\quad R_{conv}=\frac{1}{|\sigma|}=|\sqrt{\gamma}|,\infty\,.$$

The problem is how to quantize the energy values. To this end, we will apply the constraint which separates from all Frobenius solutions the so-called transcendental solutions. To clarify this matter, we are to consider the above recurrent formula (19)

$$P_{k-5}c_{k-5}+P_{k-4}c_{k-4}+P_{k-3}c_{k-3}+P_{k-2}c_{k-2}+P_{k-1}c_{k-1}+P_k c_k+P_{k+1}c_{k+1}=0,$$

and impose a requirement to have vanished the coefficient at c_{k-5} (assuming $k\geq 5$):

$$P_{k-5}=a(k-5)+(b_1+\beta_1+\beta_2+\beta_3+\beta_4)=0\,. \quad (20)$$

If additionally, we assume the vanishing the coefficients

$$c_{k-4}=0,\quad c_{k-3}=0,\quad c_{k-2}=0,\quad c_{k-1}=0,\quad c_k=0, \quad (21)$$

then due to recurrent formula all remaining coefficients $c_{k+1},...$ will vanish identically, and the series will becomes polynomials. Evidently, the last (polynomial) method is worked only if the system of algebraic equations (20) and (21) has joint and physically admissible roots. However, numerical study shows that here it is not possible.

Journal of Physics: Conference Series **1416** (2019) 012026 doi:10.1088/1742-6596/1416/1/012026

5. Energy quantization for states with $l = 0$

Before numerical study, it is convenient to transform all relations to dimensionless quantities:

$$Mr = x\,, \qquad \frac{\epsilon}{M} = \frac{E'}{c\hbar}\frac{\hbar}{mc} = \frac{E'}{mc^2} = E, \quad \gamma M^2 = \Gamma\,, \tag{22}$$

and the radial equation (for states with $l = 0$) takes the form

$$\frac{d^2R}{dx^2} + \left(\frac{6}{x} - \frac{4x^3}{x^4+\Gamma^2}\right)\frac{dR}{dx} + \left(E^2 - 1 + \frac{2\alpha E}{x} + \frac{\alpha^2}{x^2} + \frac{4\Gamma E}{x^3} + \frac{3\alpha\Gamma - \Gamma^2}{x^4} - \frac{4\Gamma(Ex+\alpha)}{x^4+\Gamma^2}\right)R = 0\,.$$

Substitution for Frobenius solutions is $R(x) = x^C e^{Ax} e^{B/x} f(x)$, where

$$A = -\sqrt{1-E^2}\,,\;\; B = -\sqrt{-\Gamma(3\alpha-\Gamma)}\,,\;\; C = \frac{-2\Gamma E}{\sqrt{-\Gamma(3\alpha-\Gamma)}} - 2\,,\; \Gamma < 0\,,$$

the differential equation for $f(r)$ is

$$f'' + \left(a + \frac{a_1}{x} + \frac{a_2}{x^2} - \frac{1}{x+\Sigma} - \frac{1}{x-\Sigma} - \frac{1}{x+i\Sigma} - \frac{1}{x-i\Sigma}\right)f'$$
$$+ \left(\frac{b_1}{x} + \frac{b_2}{x^2} + \frac{\beta_1}{x+\Sigma} + \frac{\beta_2}{x-\Sigma} + \frac{\beta_3}{x+i\Sigma} + \frac{\beta_4}{x-i\Sigma}\right)f = 0\,. \tag{23}$$

Now we are to specify the above transcendency condition $k \geq 5,\ P_{k-5} = 0$. In explicit form this constraint reads

$$2\,A\,k - 8A + 2\,\alpha\,E + 2\,AC = 0\,,$$

whence we obtain a 4-th order algebraic equation with respect to the variable E:

$$\frac{4\,\Gamma^2 E^4}{\Gamma\,(-3\,\alpha+\Gamma)} - \frac{4\,(k-6)\,\Gamma E^3}{\sqrt{\Gamma\,(-3\,\alpha+\Gamma)}}$$
$$+ \left[(k-6)^2 + \alpha^2 - \frac{4\,\Gamma^2}{\Gamma\,(-3\,\alpha+\Gamma)}\right]E^2 + \frac{4\,(k-6)\,\Gamma\,E}{\sqrt{\Gamma\,(-3\,\alpha+\Gamma)}} - (k-6)^2 = 0\,.$$

Let $\Gamma = -0.001$, and $k = 5, 6, ..., 30$; then the needed roots are (only one series is physically interpretable as corresponding to energies of bound states: $0 < E^1_{l=0,k} < 1$, also see Fig. 1))

Table 1. Energy values at $l = 0$.

E = 0.9999213823	E = 0.9998474908	E = 0.9999867506
E = 0.9999954435	E = 0.9999977197	E = 0.9999986352
E = 0.9999990925	E = 0.9999993533	E = 0.9999995159
E = 0.9999996241	E = 0.9999996997	E = 0.9999997545
E = 0.9999997957	E = 0.9999998272	E = 0.9999998520
E = 0.9999998718	E = 0.9999998879	E = 0.9999999012
E = 0.9999999122	E = 0.9999999215	E = 0.9999999293
E = 0.9999999361	E = 0.9999999419	E = 0.9999999470

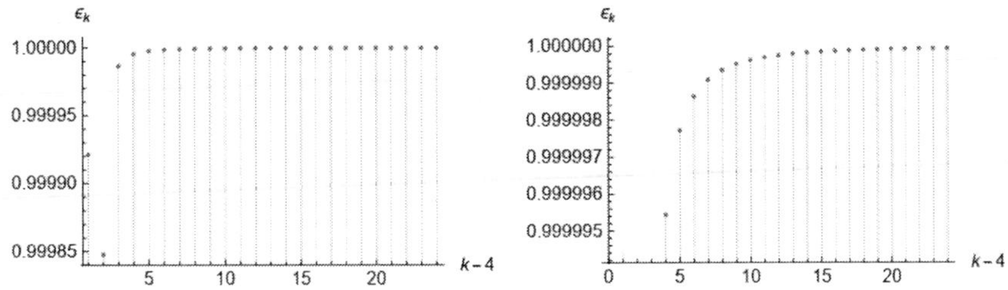

Figure 1. Energy levels, $l = 0, \Gamma = -10^{-3}$

6. Energy quantization for states with $l = 1, 2, ...$

Analogous study of the radial equation for states with higher values of l gives similar results. We have 8-term recurrent relations for coefficients of series. In this case, the relevant transcendency condition $P_{k-6} = 0$ takes on the form of quadratic equation with respect to the variable E:

$$(2k - 13)\sqrt{1 - E^2} = 2\alpha E + \frac{\sqrt{1 - E^2}(3\alpha + \gamma)}{\sqrt{l(l+1)}}. \tag{24}$$

At the given parameters $\alpha = \frac{1}{137}$, $\Gamma = -10^{-3}$, $l = 1$, $k = 6, ..., 30$ we numerically get the following values for energy levels:

Table 2. Energy values at $l = 1$.

E = 0.9998965381495	E = 0.9998902389493	E = 0.9999880428546
E =0.9999957123741	E = 0.9999978161294	E = 0.9999986801326
E = 0.9999991169800	E = 0.9999993680399	E =0.9999995254716
E = 0.9999996306431	E = 0.9999997043640	E = 0.9999997580299
E = 0.9999997983067	E = 0.9999998293044	E = 0.9999998536688
E = 0.9999998731660	E = 0.9999998890111	E = 0.9999999020622
E = 0.9999999129398	E = 0.9999999221009	E =0.9999999298885
E = 0.9999999365642	E = 0.9999999423299	E = 0.9999999473438

which are illustrated by Fig. 2.

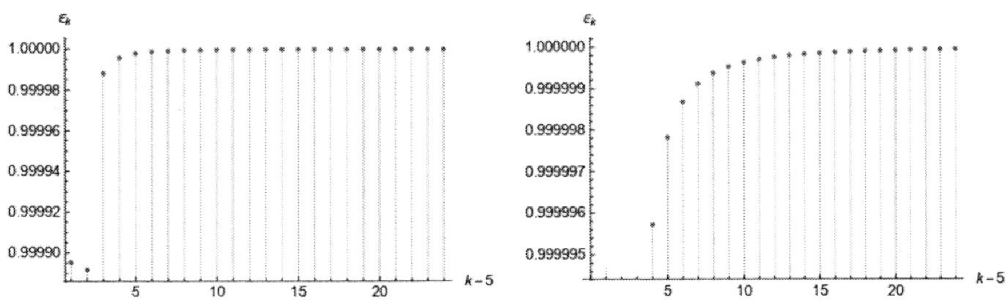

Figure 2. Energy levels, $l = 1, \Gamma = -10^{-3}$

Similarly, for parameters $\alpha = \frac{1}{137}$, $\Gamma = -10^{-3}$, $l = 2$, $k = 6, ..., 30$ we get the following energy levels (also see Fig. 3)

Table 3. Energy values at $l = 2$.

E= 0.9998952529767	E= 0.9998916171901	E= 0.9999880927290
E= 0.9999957230969	E= 0.9999978200294	E= 0.9999986819655
E= 0.9999991179832	E= 0.9999993686474	E =0.9999995258669
E= 0.9999996309146	E= 0.9999997045584	E= 0.9999997581739
E= 0.9999997984163	E= 0.9999998293897	E= 0.9999998537365
E= 0.9999998732207	E= 0.9999998890558	E= 0.9999999020993

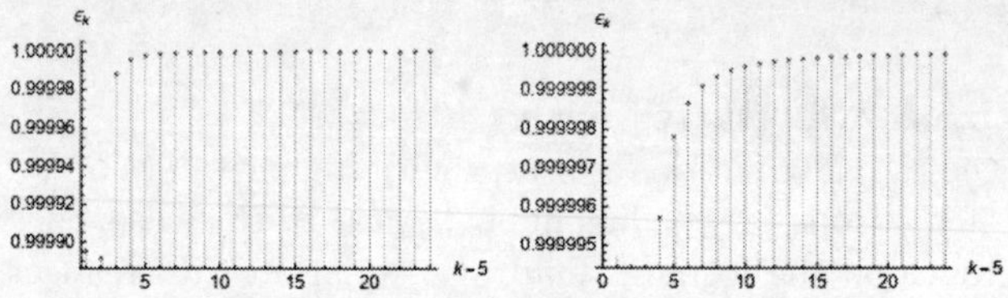

Figure 3. Energy levels, $l = 2$, $\Gamma = -10^{-3}$

Acknowledgments

The present work was developed within the framework of the project BRFFR A18-RA-015, set within the framework of the agreement between the National Academy of Sciences of Belarus and the Romanian Academy, and of the project BRFFR F18Y-009.

Special thanks to EU Project MOST, Grant R-f26C-50842, for financial support of attending the 26-th Internatinal Conference on Integrable Systems and Quantum Symmetries, July 8–12, 2019, Prague.

References

[1] Cox W 1982 *J. Phys. Math. Gen.* **15** 627
[2] Veko O V, Kazmerchuk K V, Kisel V V, Ovsiyuk E M and Red'kov V M 2014 *Nonlinear Phenomena in Complex Systems* **17** 464
[3] Ovsiyuk E M 2015 *Ukr. J. Phys.* **60** 485
[4] Kazmerchuk K V and Ovsiyuk E M 2015 *Ukr. J. Phys.* **60** 389
[5] Veko O V 2016 *Nonlinear Phenomena in Complex Systems* **19** 50
[6] Veko O V, Ovsiyuk E M and Red'kov V M 2017 *Proceedings of the National Academy of Sciences of Belarus. Ser. fiz.-mat.* **2** 71
[7] Veko O V, Voynova Ya A, Ovsiyuk E M and Red'kov V M 2017 *Proceedings of the National Academy of Sciences of Belarus. Ser. fiz.-mat.* **4** 55
[8] Chichurin A V, Ovsiyuk E M and Red'kov V M 2018 *Computers and Mathematics with Applications* **75** 1550-65
[9] Ronveaux A 1995 *Heun's differential equation* (Oxford: Oxford University Press)
[10] Slavyanov S Yu and Lay W 2000 *Special functions. A unified theory based on singularities* (Oxford: Oxford University Press)
[11] Kisel V V, Ovsiyuk E M, Veko O V, Voynova Ya A, Balan V and Red'kov V M 2018 *Elementary Particles with Internal Structure in External Fields. Vol I. General Theory; Vol II. Physical Problems* (New York: Nova Science Publishers Inc.)

Journal of Physics: Conference Series **1416** (2019) 012027 doi:10.1088/1742-6596/1416/1/012027

Symmetries in the projection evolution model

Pędrak Aleksandra
National Centre for Nuclear Research, Warsaw, Poland
E-mail: aleksandra.pedrak@ncbj.gov.pl

Góźdź Andrzej
Faculty of Mathematics, Physics and Computer Science,
University of Maria Curie–Skłodowska, Lublin, Poland
E-mail: andrzej.gozdz@umcs.lublin.pl

Góźdź Marek
Faculty of Mathematics, Physics and Computer Science,
University of Maria Curie–Skłodowska, Lublin, Poland
E-mail: mgozdz@kft.umcs.lublin.pl

Abstract.
Some introductory results concerning symmetries and conservation laws within the Projection Evolution (PEv) formalism are presented. It is shown that the PEv formalism conserves the quantum probability distributions after unitary transformations of the evolution operators represented by the orthogonal resolution of unity. Some conditions for transformations selecting the states with equal transition probabilities are found. A relation between a given symmetry represented by a Lie group and PEv conservation of the eigenvalues of the Casimir operator is derived.

1. Introduction

Until the end of the XIX century, in classical physics, time played the role of a parameter which orders events. In the next century, the relativistic theory demonstrated the necessity of treating time on the same footing as the space variables. The development of similar ideas in quantum mechanics was stopped for many years by the Pauli theorem [1, 2, 3]. This theorem states that it is not possible to form a Hermitian operator representing the position in time. The conclusion from this principle is that time cannot be a dynamical variable but only a parameter, similarly as it was in the classical mechanics. A detailed look at the theorem shows, however, that some of its assumptions can be weakened [4, 5]. Thanks to this, the construction of the time operator is possible. This paves the way to work on problem of time as a variable which plays equivalent role to the space variables.

The main goal of the projection evolution model is to rebuild the idea of the evolution and to change the role of time. The basic assumption in the projection evolution model is that the state of the Universe, which is described by the four-dimensional spacetime, can spontaneously change, according to some probability distribution. The probability depends on the state at the previous

Journal of Physics: Conference Series **1416** (2019) 012027 doi:10.1088/1742-6596/1416/1/012027

step of the evolution and the resulting state. Thus the evolution means the transformation of the state from one step of the evolution to another driven by operators which fulfill appropriate conditions. This excludes the unitary evolution where time is a parameter, as it is in the standard quantum mechanics. This type of evolution should be regarded as an approximation of the PEv only.

This approach to the problem of time in quantum physics gives rise to the projection evolution model [6]–[14]. This model allows to describe many problems and physical phenomena in a natural way. It allows to discuss the position in time operator [6], the interference in time [7, 9], particle oscillations [10], delayed-choice experiments [11] and others.

The idea of symmetries is a very strong, well developed and widely used tool in physics. In the projection evolution model, due to its specificity it is still an open problem. On the one hand, because in PEv the time is a quantum observable, any symmetry analysis has to be performed in the full spacetime. On the other hand, in this model the evolution is driven not only by the unitary transformations, which requires to discover new rules for symmetries and their relations to some conservation laws.

This leads to two separate types of problems in the symmetry analysis. One is to consider symmetry as an invariance in a single step of the evolution. In other words, we have to analyze the invariance of the states and required operators in the four dimensional spacetime. This analysis should give results analogical to the symmetry analysis in the standard quantum mechanics. The second problem is to consider symmetry along the evolution path, i.e., the chain of states which is formed by the subsequent steps in the evolution process. This is a new kind of problem because it requires to find invariance properties of every state in the evolution path. Such an analysis requires new theoretical methods which have to be developed. Because of the current lack of such methods, the symmetry analysis which is introduced in this paper does not form a systematical review. It is rather a piecemeal set of ideas how to approach this problem.

In the first section of the paper the main idea of the projection evolution model is introduced. It contains the description of the main theoretical assumptions of the model and introduces the mathematical tools used later. In the second section some elements of the symmetry analysis along the evolution path of the system are introduced. The last section contains a few remarks and open questions which appeared during the work on this problem and which can be the starting point for further research.

2. The projection evolution

The main postulate in the projection evolution (PEv) is the evolution principle. The evolution principle states that the evolution of a quantum system is a random process caused by the changes of the Universe. Changes of the Universe are the primary process. Every state of the Universe and any of its subsystems is, in this model, a function in the spacetime, which contains the full history including the past, the present, and the future. This happens, however, only until the next step of the evolution. The next step of the evolution modifies the state in the full spacetime, creating new history of the system. The change of the state can even affect the past, probably on very short time intervals only. It divests the time of the role of a parameter for the ordering of events. The problem of the broken causality can be observed even in the case of a semiclassical treatment of time [8]. In this model, the causality ordering relation is not necessarily determined by the direction of time. This property is a completely new idea and carry many new consequences.

The random process of spontaneous changes of the Universe is parametrized by the parameter τ. The parameter τ should be understood as an ordering parameter, thus in the set τ belongs to, a linear order relation must be defined. The ordering in the set of τ has a physical meaning of the quantum causality. τ is not understood as an additional dimension of the spacetime.

As the results of an evolution process one obtains the chain of states numbered by τ. At every evolution step τ_n, for the system under consideration there is a set of available states $\rho(\tau_n;\nu_k^{(n)})$. The operator $\rho(\tau_n;\nu_k^{(n)})$ denotes the density operator which represents the state of the system. The first parameter τ_n denotes the evolution step and the second parameter $\nu_k^{(n)}$ distinguishes between different states available at the step τ_n. The realization of such a step-by-step evolution is called the evolution path,

$$\rho(\tau_1;\nu_{k_1}^{(1)}) \to \rho(\tau_2;\nu_{k_2}^{(2)}) \to \ldots \to \rho(\tau_n;\nu_{k_n}^{(n)}) \to \ldots \tag{1}$$

The labels $\nu_k^{(n)}$ can be understood as quantum numbers which are attached to quantum observables. We write $\nu^{(n)} \in Q(\tau_n)$, where $Q(\tau_n)$ is a family of sets of quantum numbers defining available states for the evolution step τ_n. The set of available states for a given τ_n is denoted by:

$$S(\tau_n) = \left\{\rho(\tau_n;\nu_k^{(n)}) : \nu_k^{(n)} \in Q(\tau_n)\right\}. \tag{2}$$

The transitions among the states are described by the evolution operator $\mathbb{F}(\tau;\nu,\rho)$:

$$\mathbb{F}(\tau;\nu;\cdot) : \mathcal{T}_1^+(\mathcal{K}) \to \mathcal{T}^+(\mathcal{K}), \tag{3}$$

where $\mathcal{K}$ represents the state space of our physical system, $\mathcal{T}_1^+(\mathcal{K})$ is the space of operators with trace equal to one and $\mathcal{T}^+(\mathcal{K})$ is the space of operators with positive and finite trace.

The operator $\mathbb{F}(\tau;\nu,\rho)$ is Hermitian and positively defined:

$$\mathbb{F}(\tau;\nu,\rho)^\dagger = \mathbb{F}(\tau;\nu,\rho), \qquad \mathbb{F}(\tau;\nu,\rho) \geq 0. \tag{4}$$

For every fixed step of the evolution, the sum over available states of traces of $\mathbb{F}(\tau;\nu,\rho)$ is finite:

$$\sum_\nu \mathrm{Tr}[\mathbb{F}(\tau;\nu,\rho)] < \infty. \tag{5}$$

During the evolution, the next state is obtained from the previous one by using the operator $\mathbb{F}(\tau;\nu,\rho)$:

$$\rho(\tau_n;\nu_k^{(n)}) = \frac{\mathbb{F}\left(\tau_n;\nu_k^{(n)},\rho(\tau_{n-1};\nu_l^{(n-1)})\right)}{\mathrm{Tr}\left[\mathbb{F}\left(\tau_n;\nu_k^{(n)},\rho(\tau_{n-1};\nu_l^{(n-1)})\right)\right]}. \tag{6}$$

It is chosen randomly from the set of available states with the probability distribution for the chooser given by

$$\mathrm{pev}(\tau_n;\nu_l^{(n-1)} \to \nu_k^{(n)}) \equiv \mathrm{Prob}\left(\rho(\tau_{n-1};\nu_l^{(n-1)}) \to \rho(\tau_n;\nu_k^{(n)})\right). \tag{7}$$

The function (7) denotes the probability of the transition from the previous state $\rho(\tau_{n-1};\nu_l^{(n-1)})$, defined by the quantum numbers $\nu_l^{(n-1)} = \left(\nu_{l_1}^{(n-1)},\nu_{l_2}^{(n-1)},\ldots,\nu_{l_f}^{(n-1)}\right)$, to the next state $\rho(\tau_n;\nu_k^{(n)})$ defined by the quantum numbers $\nu_k^{(n)} = \left(\nu_{k_1}^{(n)},\nu_{k_2}^{(n)},\ldots,\nu_{k_f}^{(n)}\right)$.

The most important realization of the evolution operators $\mathbb{F}(\tau;\nu,\rho)$ can be constructed from the Krauss–like operators [15]:

$$\mathbb{F}(\tau;\nu,\rho) = \sum_a \mathbb{E}(\tau;\nu,a)\ \rho\ \mathbb{E}(\tau;\nu,a)^\dagger, \tag{8}$$

where the summation over a dependents on the quantum numbers ν. This form of the operators $\mathbb{F}(\tau;\nu,\rho)$ is important for a few reasons. Firstly, if for a given step of the evolution the $\mathbb{E}(\tau;\nu)$ operators form an orthogonal resolution of unity,

$$\mathbb{E}(\tau_n;\nu_k^{(n)}) \equiv \mathbb{E}(\tau_n;\nu_k^{(n)}, a=1) = \text{projection operators}, \tag{9}$$

$$\mathbb{E}(\tau_n;\nu_k^{(n)})\ \mathbb{E}(\tau;\nu_{k'}^{(n)}) = \delta_{kk'}\mathbb{E}(\tau;\nu_k^{(n)}), \tag{10}$$

$$\sum_k \mathbb{E}(\tau_n;\nu_k^{(n)}) = 1, \tag{11}$$

then the transition probability (7) is given in the following form:

$$\mathrm{pev}(\tau_n;\nu_l^{(n-1)} \to \nu_k^{(n)}) = \mathrm{Tr}\left[\mathbb{E}(\tau_n;\nu_k^{(n)})\rho(\tau_{n-1};\nu_l^{(n-1)})\right]. \tag{12}$$

In general, the method of finding the transition probability distribution from one mixed state, described by the density operator $\rho(\tau_n;\nu_k^{(n)})$, to another mixed state at the next step of the evolution is still an open problem. The above formula is valid for a special case in which the transition probability distribution is known. In the remainder of the paper, unless noted, the operator $\mathbb{F}(\tau;\nu,\rho)$ will be implemented by operators $\mathbb{E}(\tau;\nu)$ in the form (8).

The second important property of the decomposition (8) is, that this kind of the evolution operators can be defined by means of the so-called evolution generators. The projection evolution generators $\hat{\mathbb{W}}$ for a given τ are defined as a family of Hermitian operators whose spectral decompositions (plus possible constraints) give the orthogonal resolution of unity. They serve as the evolution operators:

$$\hat{\mathbb{W}}(\tau_n) = \sum_k w_k\ \mathbb{E}(\tau_n;\nu_k^{(n)}). \tag{13}$$

By using the appropriate generator one can reproduce the standard quantum mechanical evolution as a limiting case. For example, the generators:

$$\hat{\mathbb{W}}(\tau) = \hat{p}_0 - \hat{\mathcal{H}}(\tau) \qquad \text{(Schrödinger)}, \tag{14}$$

$$\hat{\mathbb{W}}(\tau) = \hat{p}_\mu \hat{p}_\nu \qquad \text{(free Klein-Gordon)}, \tag{15}$$

$$\hat{\mathbb{W}}(\tau) = \gamma^\mu \hat{p}_\mu \qquad \text{(free Dirac)}, \tag{16}$$

result in the Schrödinger, the free Klein-Gordon, and the free Dirac evolution, respectively. This fact shows that the projection evolution model reproduces the results from the standard quantum mechanics.

3. Symmetries

The symmetry of an object is a property which relates the invariance of this object with respect to an operation. In this paper we use a very general definition of symmetry. Let X be a set of objects and relation $\sim$ be an equivalence relation defined on X. The operation $S: X \to X$ is a symmetry of $x \in X$ if and only if it does not change the relation $\sim$,

$$x \sim S(x). \tag{17}$$

The set of operations S forms a group.

Depending on the physical requirements, the relation $\sim$ can be defined differently. For example, it can be defined as an invariance of a special functional, as it is done in the case of the action functional in the classical mechanics, the relativity relations among equivalent observers.

Because of the generality of this definition, it contains a wide spectrum of applications to physical problems.

In this paper we discuss the problem of symmetry analysis along the evolution path only. In the following we consider the evolution operators $\mathbb{E}(\tau;\nu)$ which are either a combination of the unitary operators or they form an orthogonal resolution of unity.

The problem is to find either the observable, or the expectation value of an observable, or the property of the physical state, or a similar object which is invariant at every step of the evolution along the evolution path. In other words, transformation from one evolution step to another does not change this object. The main motivation for studying of this problem is the unknown relation between the symmetries and the conservation laws.

We start from the definition of transformations of the operator $\mathbb{F}(\tau;\nu,\rho)$. Because $\mathbb{F}(\tau;\nu,\rho)$ transforms the quantum state ρ into another quantum state (not normalized), the resulting image of $\mathbb{F}(\tau;\nu,\rho)$ has to be compatible with the transformation of arguments.

To explain this, let us consider the problem of the rotation of some vector function. Let $f: R^3 \to R^3$. The values of the rotated function f' on the rotated argument x' should be equal to the rotation of the value of the original function on the original argument, $f'(x') = \hat{\mathcal{R}}f(x)$. In an analogical way the transformation of $\mathbb{F}(\tau;\nu,\rho)$ is defined.

Let $\mathbb{F}(\tau_n;\nu,\rho)$ be the evolution operator defined in (3). Let G be a group with two realizations $S_1(g): \mathcal{T}_1^+(\mathcal{K}) \to \mathcal{T}_1^+(\mathcal{K})$ and $S(g): \mathcal{T}^+(\mathcal{K}) \to \mathcal{T}^+(\mathcal{K})$. The transformation of the evolution operator $\mathbb{F}$ is defined as:

$$\mathbb{F}'(\tau_n;\nu,\rho') = S(g)\ \mathbb{F}(\tau_n;\nu,\rho)\ S(g^{-1}), \tag{18}$$

i.e., the resultant evolution operator for the transformed state is equal to the appropriately chosen transformation of the evolution operator for the original state. This idea can be expressed in a more convenient form:

$$\mathbb{F}'(\tau_n;\nu,\rho) = S(g)\ \mathbb{F}(\tau_n;\nu,S_1(g^{-1})\rho S_1(g))\ S(g^{-1}). \tag{19}$$

The group G provides the same physical interpretation for the operators $S(g)$ and $S_1(g)$ in the above definition.

In the same way one can define the transformation (19) for the evolution generated by the operators $\mathbb{E}(\tau_n;\nu_k^{(n)})$. For shortness we use the notation $S(g) = S_1(g) = g$. For the n-th step of the evolution, equation (19) has the following form:

$$\left[\mathbb{E}(\tau_n;\nu_k^{(n)})\rho(\tau_{n-1};\nu_l^{(n-1)})\mathbb{E}(\tau_n;\nu_k^{(n)})^\dagger\right]' = g\mathbb{E}(\tau_n;\nu_k^{(n)})g^{-1}\ \rho(\tau_{n-1};\nu_l^{(n-1)})\ g\mathbb{E}(\tau_n;\nu_k^{(n)})^\dagger g^{-1}. \tag{20}$$

The above equation shows that the transformation (19) for the operator $\mathbb{E}(\tau_n,\nu_k^{(n)})$ is equivalent to the transformation by the group element g,

$$\left[\mathbb{E}(\tau_n;\nu_k^{(n)})\right]' = g\mathbb{E}(\tau_n;\nu_k^{(n)})g^{-1}. \tag{21}$$

Definition (18) allows to conserve the probability distributions for unitary equivalent images of quantum mechanics. Let us assume that the transition probability from the evolution step τ_{n-1} to τ_n is given by

$$\mathrm{Prob}\,[\rho(\tau_{n-1};\nu_l^{(n-1)}) \to \rho(\tau_n;\nu_k^{(n)})] = \mathrm{Tr}[\mathbb{F}(\tau_n;\nu_k^{(n)},\rho(\tau_{n-1};\nu_l^{(n-1)}))]. \tag{22}$$

Similarly, the transition probability of the transformed by the group G state $\rho'(\tau_{n-1};\nu_l^{(n-1)})$ and, at the same time, the transformed evolution operator $\mathbb{F}'$, is given by

$$\mathrm{Prob}\,[\rho'(\tau_{n-1};\nu_l^{(n-1)}) \to \rho'(\tau_n;\nu_k^{(n)})] = \mathrm{Tr}[\mathbb{F}'(\tau_n;\nu_k^{(n)},\rho'(\tau_{n-1};\nu_l^{(n-1)}))]. \tag{23}$$

Journal of Physics: Conference Series **1416** (2019) 012027 doi:10.1088/1742-6596/1416/1/012027

It follows from equation (18) that both probabilities are equal:

$$\text{Prob}\,[\rho(\tau_{n-1};\nu_l^{(n-1)}) \to \rho(\tau_n;\nu_k^{(n)})] = \text{Prob}\,[\rho'(\tau_{n-1};\nu_l^{(n-1)}) \to \rho'(\tau_n;\nu_k^{(n)})]. \tag{24}$$

The consequence of such a symmetry for every step of the evolution is the fact that g does not change the structure of the possible evolution paths, which means that the probability distribution of the potential evolution paths is conserved. In this case, the operations $g \in \mathrm{G}$ can also be interpreted as operations which transform among equivalent descriptions of a given model (covariance), or as the transformation between equivalent observers (a kind of "relativity").

As the second kind of symmetry let us consider invariance of the transition probabilites when the final states are transformed by the group G. In this case, the symmetry group G is responsible for creating the sets of states which are chosen by the evolution in a completely random way. This specific conservation of transition probability can be defined as:

$$\text{Prob}\,[\rho(\tau_{n-1};\nu_l^{(n-1)}) \to \rho(\tau_n;\nu_k^{(n)})] = \text{Prob}\,[\rho(\tau_{n-1};\nu_l^{(n-1)}) \to \rho'(\tau_n;\nu_k^{(n)})]. \tag{25}$$

The transition probabilities among the transformed states are:

$$\text{Prob}\left[\rho(\tau_{n-1};\nu_l^{(n-1)}) \to \rho'(\tau_n;\nu_k^{(n)})\right] = \text{Tr}\left[g\mathbb{E}(\tau_n;\nu_k^{(n)})g^{-1}\,\rho(\tau_{n-1};\nu_l^{(n-1)})\; g\mathbb{E}(\tau_n;\nu_k^{(n)})^{\dagger}g^{-1}\right]. \tag{26}$$

It is easy to see that in two special cases,

$$[g, \mathbb{E}(\tau_n, \nu^{(n)})] = 0 \quad \text{or} \quad [g, \rho(\tau_{n-1}, \nu^{(n-1)})] = 0, \tag{27}$$

this probability is the same as for the original states.

The next important problem is the relation between symmetries and conservation laws. The correct formulation of this problem in the projection evolution model is problematic. In the most intuitive way one can say that we are looking for conditions under which the expectation value of given observable A is conserved within the PEv approach:

$$\langle A\rangle_{\rho(\tau_0;\nu_k^{(0)})} = \langle A\rangle_{\rho(\tau_1;\nu_{k'}^{(1)})} = \ldots = \langle A\rangle_{\rho(\tau_n;\nu_{k''}^{(n)})} = \ldots \tag{28}$$

The required conditions may involve special relations between the evolution operators, density operators and the observable operator. In the case when the evolution is described by the operators $\mathbb{E}(\tau_n;\nu_k^{(n)})$, the conservation of the expectation value $\langle A\rangle$ has the form:

$$\text{Tr}[A\,\rho(\tau_{n-1};\nu_l^{(n-1)})] = \frac{\text{Tr}[A\mathbb{E}(\tau_n;\nu_k^{(n)})\,\rho(\tau_{n-1};\nu_l^{(n-1)})\,\mathbb{E}(\tau_n;\nu_k^{(n)})^{\dagger}]}{\text{Tr}[\mathbb{E}(\tau_n;\nu_k^{(n)})\,\rho(\tau_{n-1};\nu_l^{(n-1)})\,\mathbb{E}(\tau_n;\nu_k^{(n)})^{\dagger}]} \tag{29}$$

for a series of subsequent n.

Assume that the operator $\mathbb{E}(\tau_n;\nu_k^{(n)})$ is unitary. Under this assumption the condition (29) takes the following form:

$$\text{Tr}\left[A\rho(\tau_n;\nu_k^{(n)})\right] = \frac{\text{Tr}[A\mathbb{E}(\tau_n;\nu_k^{(n)})\,\rho(\tau_{n-1};\nu_l^{(n-1)})\,\mathbb{E}(\tau_n;\nu_k^{(n)})^{\dagger}]}{\text{Tr}[\mathbb{E}(\tau_n;\nu_k^{(n)})\,\rho(\tau_{n-1};\nu_l^{(n-1)})\,\mathbb{E}(\tau_n;\nu_k^{(n)})^{\dagger}]} \tag{30}$$

$$= \text{Tr}[\mathbb{E}(\tau_n;\nu_k^{(n)})^{\dagger}\,A\,\mathbb{E}(\tau_n;\nu_k^{(n)})\,\rho(\tau_{n-1};\nu_l^{(n-1)})]. \tag{31}$$

In this case the expectation value of the observable A is conserved if the operator A commutes with the evolution operator, i.e., $[A, \mathbb{E}(\tau_n;\nu_k^{(n)})] = 0$. This fact has its counterpart in the standard

Journal of Physics: Conference Series **1416** (2019) 012027 doi:10.1088/1742-6596/1416/1/012027

quantum mechanics – if the Hamiltonian commutes with the operator A, its expectation value is conserved during the unitary evolution generated by this Hamiltonian.

The requirement of the commutation of the observable and the evolution operator is a rather strong condition. In some cases, it is possible to weaken this condition considering an approximate invariance of $\langle A\rangle_\rho$ for the unitary evolution. Assume that the observable A and the evolution operator do not commute, $[A, \mathbb{F}(\tau_n;\nu_k^{(n)})] \neq 0$, and let the evolution operator be unitary in the form:

$$\mathbb{F}(\tau_n;\nu_k^{(n)}) = \exp(-i\sigma_n\hat{S}) = \mathbb{I} + (e^{-i\sigma_n} - 1)\hat{S}, \tag{32}$$

where $\hat{S}$ is a projection operator and σ_n is a number which depends on the evolution step τ_n. The expectation value of A in the step n is in this case given by

$$\begin{aligned}\mathrm{Tr}[A\rho(\tau_n;\nu_k^{(n)})] = \;& \mathrm{Tr}[A\rho(\tau_{n-1};\nu_l^{(n-1)})] + (e^{i\sigma_n} - 1)\mathrm{Tr}[\hat{S}A\rho(\tau_{n-1};\nu_l^{(n-1)})] \\ &+(e^{-i\sigma_n} - 1)\mathrm{Tr}[A\hat{S}\rho(\tau_{n-1};\nu_l^{(n-1)})] \\ &+2(1-\cos\sigma_n)\mathrm{Tr}[A\hat{S}\rho(\tau_{n-1};\nu_l^{(n-1)})\hat{S}].\end{aligned} \tag{33}$$

The expectation value of A for the step n is equal to the sum of the expectation value of A at the previous evolution step and the part which depend on σ_n. Approximate invariance can be achieved by taking σ_n to minimize the part which depends on this parameter.

The idea of the unitary evolution can be generalized in different ways. For example, it is possible to consider the state at the evolution step n as a result of a unitary evolution of the previous state through a few different channels. In this case, the state for the evolution step n is a linear combination of the products of different unitary evolutions of the previous state,

$$\rho(\tau_n;\nu_k^{(n)}) = \frac{\sum_{m=1}^{N} U_m(\tau_n)\rho(\tau_{n-1};\nu_l^{(n-1)})U_m(\tau_n)^\dagger}{\mathrm{Tr}[\sum_{m=1}^{N} U_m(\tau_n)\rho(\tau_{n-1};\nu_l^{(n-1)})U_m(\tau_n)^\dagger]}, \tag{34}$$

where $U_m^\dagger = U_m^{-1}$ for $m = 1,\ldots,N$.

The second example, which leads to a unitary evolution, is when the evolution generator $\hat{\mathbb{W}}(\tau_n)$ evolves unitarily $\hat{\mathbb{W}}(\tau_n) = U(\tau_n)\hat{\mathbb{W}}_0 U(\tau_n)^\dagger$. Both examples open a new problem in the analysis of symmetries in the PEv model and they require further investigation.

The next interesting case of relations between the conservation laws and symmetries is the invariance of both the evolution operator and the quantum observable.

Let G be a group, $\mathcal{C}$ be a Casimir operator of the group G, and $|\kappa\Gamma a\rangle$ be a vector of an irreducible representation. Assume that the operators $\mathbb{F}(\tau_n)$ commute with the group $[G, \mathbb{F}(\tau_n)] = 0$. Then the Casimir operator and the density operator can be described in a basis of irreducible representation in the following way:

$$\mathcal{C} = \sum_\Gamma C_\Gamma \sum_{\kappa,a} |\kappa\Gamma a\rangle\langle\kappa\Gamma a| \quad \text{and} \quad \mathbb{F}(\tau_n,\nu_k^{(n)}) = \sum_{\kappa a} |\kappa\Gamma a\rangle\langle\kappa\Gamma a|, \tag{35}$$

where Γ depends on ν_n. Then

$$\mathrm{Tr}[\mathcal{C}\rho(\tau_n;\nu_k^{(n)})] = \frac{\mathrm{Tr}[\mathcal{C}\mathbb{F}(\tau_n;\nu_k^{(n)})\rho(\tau_{n-1};\nu_l^{(n-1)})\mathbb{F}(\tau_n;\nu_k^{(n)})]}{\mathrm{Tr}[\mathbb{F}(\tau_n;\nu_k^{(n)})\rho(\tau_{n-1};\nu_l^{(n-1)})\mathbb{F}(\tau_n;\nu_k^{(n)})]} = C_\Gamma. \tag{36}$$

It follows that if the evolution operators are invariant with respect to the group G, the expectation value of the Casimir operator $\mathcal{C}$ of the group G is conserved during the evolution.

This fact has its analogy in the standard quantum mechanics. Let us assume that the Hamiltonian H is invariant with respect to a group G. The eigenvectors of H belong to the invariant subspaces spanned by the bases of the irreducible representations of the group G. In this case the expectation value of the Casimir operator is conserved during the unitary evolution generated by this Hamiltonian.

4. Summary

In this preliminary report a few special cases of symmetries and possible conservation laws have been considered. Because the projection evolution allows for different types of evolution, we have focused on the most typical cases.

The first one is the unitary evolution, where the evolution parameter τ enumerates subsequent evolution steps. The unitary evolution operators $\mathbb{E}(\tau;\nu) = U(\tau)$ act in the full spacetime, contrary to the standard unitary evolution operator $\exp(-iHt/\hbar)$ which acts in the three dimensional space only, with time t playing the role of the events ordering parameter.

The second, very probably the most interesting case is the projection evolution generated by the evolution operators $\mathbb{E}(\tau;\nu)$ which furnish the orthogonal resolution of unity. The standard, Schrödinger type unitary evolution is a special case of the projection evolution generated by the appropriate orthogonal resolution of unity with the wave functions sharply localized in time.

As the first result in this paper we have found the basic conditions which have to be fulfilled to satisfy the covariance properties of the projection evolution with respect to a given group of transformations.

When the evolution operator is a four-dimensional unitary evolution, the results are similar to those which can be obtained in the standard Schrödinger type evolution. In this case the commutation of an observable A and the evolution operators is a sufficient condition for the conservation of the expectation value $\langle A \rangle_\rho$ between the steps of the projection evolution.

The last case shows a simple relation between the given symmetry group and the conservation of appropriate Casimir operators while the system evolves according to a specific form of the evolution operators.

It is obvious that the symmetries and conservation laws are the most basic features of any quantum formalism. The PEv approach also requires the full analysis of these problems. They will be the subject of further investigation.

References

[1] Pauli W 1926 *Handbuch der Physik* vol 23, ed H Geiger, K Scheel (Springer Berlin) pp 1-278
[2] Pauli W 1926 *Handbuch der Physik* vol 24, ed H Geiger, K Scheel (Springer Berlin) pp 83-272
[3] Pauli W 1958 *Handbuch der Physik* vol 5, ed Fludge (Springer Berlin) pp 1-168
[4] Galapon E A 2002 *Proc. R. Soc. Lond.* **A458** 451
[5] Galapon E A 2009 *Time in Quantum Mechanics- Vol.2, Lect. Notes Phys.* 789, ed Muga G, Ruschhaupt A and del Campo A (Springer Berlin Heidelberg) pp 25-63
[6] Dębicki M, Góźdź A and Stefańska K 2007 *Int. J. Mod. Phys. E* **16** 616
[7] Góźdź A, Dębicki M and Stefańska K 2008 *Phys. At. Nucl.* **71** 892
[8] Zych M, Costa F, Pikovski I, Brukner Č 2019 *Nature Communications* **10** 3772
[9] Góźdź A and Dębicki M 2007 *Phys. At. Nucl.* **70** 529
[10] Góźdź M and Góźdź A 2014 *Phys. Scr.* **89** 054010
[11] Góźdź A and Stefańska K 2008 *J. Phys.: Conf. Ser.* **104** 012007
[12] Góźdź A and Stefańska K 2008 *Int.J. Mod. Phys. E* **17** 217
[13] Góźdź A and Góźdź M 2017 *Phys. At. Nucl.* **80** 373
[14] Góźdź A, Rybak K, Pędrak A and Góźdź M 2015 *Acta Phys. Pol. B Proc. Suppl.* **8** 591
[15] Krauss K 1983 *States, Effects and Operations: Fundamental Notions of Quantum Theory* ed Bohm A, Dollard J D and Wootters W H (Springer-Verlag Berlin Hedelberg) pp 1-154

Journal of Physics: Conference Series **1416** (2019) 012028 doi:10.1088/1742-6596/1416/1/012028

Blocks in the category of finite-dimensional representations of principal W-algebra for $Q(2)$

Elena Poletaeva

School of Mathematical and Statistical Sciences, University of Texas Rio Grande Valley, Edinburg, TX 78539, USA

E-mail: elena.poletaeva@utrgv.edu

Abstract. We describe the blocks in the category of finite-dimensional representations of the principal finite W-algebra for the Lie superalgebra $Q(2)$.

Introduction

In the classical case when $\mathfrak{g}$ is a complex semi-simple Lie algebra and e is a nilpotent element in $\mathfrak{g}$, a *finite W-algebra* for $\mathfrak{g}$ is a quantization of the Poisson structure on the Slodowy slice (a transversal slice to the orbit of e in the adjoint representation). The general definition of a finite W-algebra was given by A. Premet in [16]. For a Lie superalgebra $\mathfrak{g} = \mathfrak{g}_{\bar{0}} \oplus \mathfrak{g}_{\bar{1}}$ with a reductive even part $\mathfrak{g}_{\bar{0}}$, the finite W-algebra is associated with an even nilpotent element $e \in \mathfrak{g}_{\bar{0}}$. It is denoted by W.

In the case when $\mathfrak{g} = \mathfrak{gl}(m|n)$ and e is the even principal nilpotent, J. Brown, J. Brundan and S. Goodwin classified irreducible representations of W and explored the connection with the category $\mathcal{O}$ for $\mathfrak{g}$ using coinvariants functor [1, 2].

In [13] we considered W associated with an even nilpotent element $\varphi \in \mathfrak{g}_{\bar{0}}^* \subset \mathfrak{g}^*$ in the coadjoint representation (this means that for the algebraic reductive group $G_{\bar{0}}$ of $\mathfrak{g}_{\bar{0}}$, the closure of the $G_{\bar{0}}$-orbit of φ in $\mathfrak{g}_{\bar{0}}^*$ contains zero). We proved that if φ is the *principal* nilpotent, i.e. the dimension of the even part of the annihilator of φ in $\mathfrak{g}$ is minimal, and $\mathfrak{g}$ is isomorphic to $\mathfrak{sl}(m|n)$, $\mathfrak{osp}(2|2n)$ or $Q(n)$, then every simple W-module is finite-dimensional. In [15] we classified simple W-modules for $\mathfrak{g} = Q(n)$ associated with the principal nilpotent coadjoint orbits (Theorem 4.6). The technique we used is completely different from one used in [1] due to the lack of triangular decomposition of W in our case. Instead, we described the restriction of simple $U(\mathfrak{h})$-modules to W and proved that any simple W-module occurs as a constituent of this restriction.

We consider the category $W-$mod of finite-dimensional W-modules. The natural problem is to describe the *blocks* in this category. In this work we make a step in this direction and describe the blocks in the case when $\mathfrak{g} = Q(2)$ (Theorem 8). Our results should have applications to the study of primitive ideals of $U(\mathfrak{g})$ in the sense of I. Losev (see [8, 9, 10]). In the super case the theory of the primitive ideals is even more complicated (see [3]). We also intend to apply these results to classify simple modules for super-Yangians of type Q.

All results are joint work with V. Serganova.

Journal of Physics: Conference Series **1416** (2019) 012028 doi:10.1088/1742-6596/1416/1/012028

1. Finite W-algebra for $Q(n)$

In this paper we consider the Lie superalgebra $\mathfrak{g} = Q(n)$ defined as follows (see [6]). Equip $\mathbb{C}^{n|n}$ with the odd operator ζ such that $\zeta^2 = -\mathrm{Id}$. Then $Q(n)$ is the centralizer of ζ in the Lie superalgebra $\mathfrak{gl}(n|n)$. It is easy to see that $Q(n)$ consists of matrices of the form $\left(\begin{array}{c|c} A & B \\ \hline B & A \end{array}\right)$, where A, B are $n \times n$ matrices. We fix the Cartan subalgebra $\mathfrak{h} \subset \mathfrak{g}$ to be the set of matrices with diagonal A and B. By n^+ (respectively, $\mathfrak{n}^-$) we denote the nilpotent subalgebras consisting of matrices with strictly upper triangular (respectively, low triangular) A and B. The Lie superalgebra $\mathfrak{g}$ has the triangular decomposition $\mathfrak{g} = \mathfrak{n}^- \oplus \mathfrak{h} \oplus \mathfrak{n}^+$ and we set $\mathfrak{b} = \mathfrak{n}^+ \oplus \mathfrak{h}$.

Denote by W the finite W-algebra associated with a principal even nilpotent element φ in the coadjoint representation of $Q(n)$. Let us recall the definition (see [16]). Let $\{e_{i,j}, f_{i,j} \,|\, i, j = 1, \dots, n\}$ denote the basis consisting of elementary even and odd matrices:

$$e_{i,j} = \left(\begin{array}{c|c} E_{ij} & 0 \\ \hline 0 & E_{ij} \end{array}\right), \quad f_{i,j} = \left(\begin{array}{c|c} 0 & E_{ij} \\ \hline E_{ij} & 0 \end{array}\right).$$

Choose $\varphi \in \mathfrak{g}^*$ such that

$$\varphi(f_{i,j}) = 0, \quad \varphi(e_{i,j}) = \delta_{i,j+1}.$$

Let I_φ be the left ideal in $U(\mathfrak{g})$ generated by $x - \varphi(x)$ for all $x \in \mathfrak{n}^-$. Let $\pi : U(\mathfrak{g}) \to U(\mathfrak{g})/I_\varphi$ be the natural projection. Then

$$W = \{\pi(y) \in U(\mathfrak{g})/I_\varphi \mid \mathrm{ad}(x)y \in I_\varphi \text{ for all } x \in \mathfrak{n}^-\}.$$

Using identification of $U(\mathfrak{g})/I_\varphi$ with the Whittaker module $U(\mathfrak{g}) \otimes_{U(\mathfrak{n})} \mathbb{C}_\varphi \simeq U(\mathfrak{b}) \otimes \mathbb{C}$ we can consider W as a subalgebra of $U(\mathfrak{b})$. The natural projection $\vartheta : U(\mathfrak{b}) \to U(\mathfrak{h})$ with the kernel $\mathfrak{n}^+U(\mathfrak{b})$ is called the *Harish-Chandra homomorphism.* It is proven in [13] that the restriction of ϑ to W is injective.

Set

$$\xi_i := (-1)^{i+1} f_{i,i}, \; x_i := \xi_i^2 = e_{i,i},$$

then

$$U(\mathfrak{h}) \simeq \mathbb{C}[\xi_1, \dots, \xi_n]/(\xi_i\xi_j + \xi_j\xi_i)_{i<j\leq n}.$$

We will identify W with $\vartheta(W)$ and use the generators in W introduced in [14] (Corollary 5.15):

$$u_k(0) := [\sum_{1\leq i_1<i_2<\dots<i_k\leq n} (x_{i_1} + (-1)^{k+1}\xi_{i_1}) \cdots (x_{i_{k-1}} - \xi_{i_{k-1}})(x_{i_k} + \xi_{i_k})]_{even},$$

$$u_k(1) := [\sum_{1\leq i_1<i_2<\dots<i_k\leq n} (x_{i_1} + (-1)^{k+1}\xi_{i_1}) \cdots (x_{i_{k-1}} - \xi_{i_{k-1}})(x_{i_k} + \xi_{i_k})]_{odd}.$$

For convenience we assume $u_k(0) = u_k(1) = 0$ for $k > n$.

2. Simple modules over associative superalgebras

We work in the category of vector superspaces over $\mathbb{C}$. We denote the *parity* of a homogeneous vector v of a superspace by $\bar{v} \in \mathbb{Z}_2$. All tensor products are over $\mathbb{C}$ unless specified otherwise.

Let $\mathcal{A}$ be a superalgebra. By an $\mathcal{A}$-module M we mean a $\mathbb{Z}_2$-graded left $\mathcal{A}$-module. A submodule of M is a $\mathbb{Z}_2$-graded submodule. By Π we denote the functor of parity switch $\Pi(M) = M \otimes \mathbb{C}^{0|1}$. For a module M over an associate superalgebra $\mathcal{A}$, ΠM has the same underlying vector space but with the opposite $\mathbb{Z}$-grading. The new action of $a \in \mathcal{A}$ on $m \in \Pi M$ is given in terms of the old action by $a \cdot m := (-1)^{\bar{a}} am$.

Recall that if M is a simple finite-dimensional $\mathcal{A}$-module over some associative superalgebra $\mathcal{A}$, then by Schur's Lemma $\mathrm{End}_{\mathcal{A}}(M)$ is either one-dimensional, or two-dimensional and has basis $\{\mathrm{Id}_M, \epsilon_M\}$, where ϵ_M is a (unique up to a sign) odd involution on M: $\epsilon_M^2 = \mathrm{Id}_M$. Note that ϵ_M provides an $\mathcal{A}$ isomorphism $M \longrightarrow \Pi(M)$. We say that M is an *irreducible of* M-*type* in the former case and an *irreducible of* Q-*type* in the latter (see [7, 4]).

Let $\mathcal{A}$ and $\mathcal{B}$ be two superalgebras. The tensor product $\mathcal{A} \otimes \mathcal{B}$ is again a superalgebra, where multiplication is given by

$$(a_1 \otimes b_1)(a_2 \otimes b_2) = (-1)^{\bar{b}_1 \bar{a}_2} a_1 a_2 \otimes b_1 b_2$$

for $a_i \in \mathcal{A}, b_i \in \mathcal{B}$. Let M and N be two modules over associative superalgebras $\mathcal{A}$ and $\mathcal{B}$. Then $M \otimes N$ is naturally a module over $\mathcal{A} \otimes \mathcal{B}$ where

$$(a \otimes b)(m \otimes n) = (-1)^{\bar{b}\bar{m}} am \otimes bn,$$

where $a \in \mathcal{A}, b \in \mathcal{B}$ and $m \in M, n \in N$. If M and N are two simple finite-dimensional modules over associative superalgebras $\mathcal{A}$ and $\mathcal{B}$, then the module $M \otimes N$ might be not simple. In fact, if M and N are both of M-type, then $M \otimes N$ is simple of M-type. If one of these modules is of M-type, and the other is of Q-type, then $M \otimes N$ is simple of Q-type. However, if M and N are both of Q-type, then $M \otimes N$ is not simple. Let ϵ_M and ϵ_N be odd involutions of M and N, respectively. Then the map $\epsilon_M \otimes \epsilon_N$ defined by

$$(\epsilon_M \otimes \epsilon_N)(m \otimes n) = (-1)^{\bar{m}} \epsilon_M(m) \otimes \epsilon_N(n)$$

is an even $\mathcal{A} \otimes \mathcal{B}$-automorphism of $M \otimes N$, and its square is $-\mathrm{Id}_{M \otimes N}$. In this case $M \otimes N$ decomposes into a direct sum of two $\mathcal{A} \otimes \mathcal{B}$-submodules, which are formed by the $\pm\mathbf{i}$-eigenspaces of $\epsilon_M \otimes \epsilon_N$. We can choose either submodule and denote it by $M \boxtimes N$. Then

$$M \otimes N \simeq M \boxtimes N \oplus \Pi(M \boxtimes N).$$

Both submodules are simple and of M-type.

3. Irreducible representations of W

3.1. *Representations of* $U(\mathfrak{h})$

Let $\mathfrak{s} = (s_1, \ldots, s_n) \in \mathbb{C}^n$. We call $\mathfrak{s}$ *regular* if $s_i \neq 0$ for all $i \leq n$ and *typical* if $s_i + s_j \neq 0$ for all $i \neq j \leq n$.

It follows from the representation theory of Clifford algebras that all irreducible representations of $U(\mathfrak{h})$ up to change of parity can be parameterized by $\mathfrak{s} \in \mathbb{C}^n$. Indeed, let M be an irreducible representation of $U(\mathfrak{h})$. By Schur's lemma every x_i acts on M as a scalar operator $s_i\mathrm{Id}$. Let $I_\mathfrak{s}$ denote the ideal in $U(\mathfrak{h})$ generated by $x_i - s_i$, then the quotient algebra $U(\mathfrak{h})/I_\mathfrak{s}$ is isomorphic to the Clifford superalgebra $C_\mathfrak{s}$ (we consider Clifford algebras as superalgebras with the natural $\mathbb{Z}_2$-grading) associated with the quadratic form:

$$B_\mathfrak{s}(\xi_i, \xi_j) = \delta_{ij} s_i.$$

Then M is a simple $C_\mathfrak{s}$-module.

The radical $R_\mathfrak{s}$ of $C_\mathfrak{s}$ is generated by the kernel of the form $B_\mathfrak{s}$. Let $m(\mathfrak{s})$ be the number of non-zero coordinates of $\mathfrak{s}$, then $C_\mathfrak{s}/R_\mathfrak{s}$ is isomorphic to the matrix superalgebra $M(2^{\frac{m}{2}-1}|2^{\frac{m}{2}-1})$ for even m and to the superalgebra $M(2^{\frac{m-1}{2}}) \otimes \mathbb{C}[\epsilon]/(\epsilon^2 - 1)$ for odd m.

Therefore $C_\mathfrak{s}$ has one (up to isomorphism) simple $\mathbb{Z}_2$-graded module $V(\mathfrak{s})$ of type Q for odd $m(\mathfrak{s})$, and two simple modules $V(\mathfrak{s})$ and $\Pi V(\mathfrak{s})$ of type M for even $m(\mathfrak{s})$ (see [11]). In the case when $\mathfrak{s}$ is regular, the form $B_\mathfrak{s}$ is non-degenerate and the dimension of $V(\mathfrak{s})$ equals 2^k, where $k = \lceil n/2 \rceil$. In general, $\dim V(\mathfrak{s}) = 2^{\lceil m(\mathfrak{s})/2 \rceil}$.

Journal of Physics: Conference Series **1416** (2019) 012028 doi:10.1088/1742-6596/1416/1/012028

Let $i+j=n$. We have the natural embedding of the Lie superalgebras $Q(i)\oplus Q(j)\hookrightarrow Q(n)$. If $\mathfrak{h}_r$ denotes the Cartan subalgebra of $Q(r)$, the above embedding induces the isomorphism

$$U(\mathfrak{h})\simeq U(\mathfrak{h}_i)\otimes U(\mathfrak{h}_j).$$

It induces an isomorphism of $U(\mathfrak{h})$-modules

$$V(\mathfrak{s})\simeq V(s_1,\dots,s_i)\boxtimes V(s_{i+1},\dots,s_n).$$

3.2. Restriction from $U(\mathfrak{h})$ to W

We denote by the same symbol $V(\mathbf{s})$ the restriction to W of the $U(\mathfrak{h})$-module $V(\mathfrak{s})$. We proved the following two statements in [15].

Proposition 1 ([15], Proposition 4.1). Let S be a simple W-module. Then S is a simple constituent of $V(\mathfrak{s})$ for some $\mathfrak{s}\in\mathbb{C}^n$.

Theorem 2 ([15], Theorem 4.2). If $\mathfrak{s}$ is typical, then $V(\mathfrak{s})$ is a simple W-module.

3.3. Simple W-modules for $n=2$

Let $n=2$. Let $s_1\neq -s_2$ and assume that $s_1,s_2\neq 0$. Then by Theorem 2, $V(\mathfrak{s})$ is simple as W-module. First, we describe the action of $U(\mathfrak{h})$ in $V(s_1,s_2)\simeq V(s_1)\boxtimes V(s_2)$. Note that

$$U(\mathfrak{h})\simeq U(\mathfrak{h}_1)\otimes U(\mathfrak{h}_1),$$

where $\mathfrak{h}_1$ is the Cartan subalgebra of $Q(1)$. Clearly, $U(\mathfrak{h}_1)\cong\mathbb{C}[\xi]$. Let $x=\xi^2$. For some suitable bases in $V(s_1)$ and $V(s_2)$, namely, $V(s_1)=<v_1\mid v_2>$, $V(s_2)=<w_1\mid w_2>$, where $\bar{v}_1=\bar{w}_1=0$ and $\bar{v}_2=\bar{w}_2=1$, the action of $U(\mathfrak{h}_1)$ in $V(s_i)$ is given by

$$\xi\mapsto\begin{pmatrix}0 & \sqrt{s_i}\\ \sqrt{s_i} & 0\end{pmatrix},\quad x\mapsto\begin{pmatrix}s_i & 0\\ 0 & s_i\end{pmatrix}\qquad \text{for } i=1,2.$$

We identify the elements ξ_i, x_i of $U(\mathfrak{h})$ is follows:

$$\xi_1\leftrightarrow\xi\otimes 1,\quad \xi_2\leftrightarrow 1\otimes\xi,\quad x_1\leftrightarrow x\otimes 1,\quad x_2\leftrightarrow 1\otimes x.$$

Then $V(s_1)\otimes V(s_2)=V(s_1,s_2)\oplus\Pi V(s_1,s_2)$, where

$$V(s_1,s_2)=<v_1\otimes w_1+\mathbf{i}v_2\otimes w_2\mid v_2\otimes w_1+\mathbf{i}v_1\otimes w_2>,\tag{1}$$

$$\Pi V(s_1,s_2)=<-v_1\otimes w_1+\mathbf{i}v_2\otimes w_2\mid v_2\otimes w_1-\mathbf{i}v_1\otimes w_2>.\tag{2}$$

Hence the action of $U(\mathfrak{h})$ in $V(s_1,s_2)$ is given by the following formulas in basis (1):

$$\xi_1\mapsto\begin{pmatrix}0 & \sqrt{s_1}\\ \sqrt{s_1} & 0\end{pmatrix},\quad \xi_2\mapsto\begin{pmatrix}0 & \sqrt{s_2}\mathbf{i}\\ -\sqrt{s_2}\mathbf{i} & 0\end{pmatrix}.$$

Note that W is generated by ϕ_0,ϕ_1,z_0 and z_1, where

$$\phi_0:=u_1(1)=\xi_1+\xi_2,\ \phi_1:=-u_2(1)=x_2\xi_1-x_1\xi_2,\ z_0=u_1(0)=x_1+x_2,\ z_1=u_2(0)=x_1x_2-\xi_1\xi_2.$$

Then we obtain the following formulas for the action of the generators of W:

$$\phi_0\mapsto\begin{pmatrix}0 & \sqrt{s_1}+\sqrt{s_1}\mathbf{i}\\ \sqrt{s_1}-\sqrt{s_2}\mathbf{i} & 0\end{pmatrix},\quad \phi_1\mapsto\sqrt{s_1s_2}\begin{pmatrix}0 & \sqrt{s_2}-\sqrt{s_1}\mathbf{i}\\ \sqrt{s_2}+\sqrt{s_1}\mathbf{i} & 0\end{pmatrix},\tag{3}$$

$$z_0 \mapsto (s_1+s_2)\mathrm{Id}, \quad z_1 \mapsto \begin{pmatrix} s_1s_2+\sqrt{s_1s_2}\mathbf{i} & 0 \\ 0 & s_1s_2-\sqrt{s_1s_2}\mathbf{i} \end{pmatrix}. \tag{4}$$

Note that formulas (3) and (4) hold when $s_1 \neq -s_2$.
Assume that $s_1 = -s_2$. If $s_1, s_2 = 0$ then $V(\mathfrak{s})$ is isomorphic to $\mathbb{C} \oplus \Pi\mathbb{C}$, where $\mathbb{C}$ is the trivial module. If $s_1 \neq 0$, we choose $\sqrt{s_1}, \sqrt{s_2}$ so that $\sqrt{s_2} = \sqrt{s_1}\mathbf{i}$. Note that the choice of sign controls the choice of the parity of $V(\mathfrak{s})$. The following exact sequence easily follows from (3) and (4):

$$0 \to \Pi\Gamma_{-s_1^2+s_1} \to V(\mathfrak{s}) \to \Gamma_{-s_1^2-s_1} \to 0, \tag{5}$$

where Γ_t is the simple module of dimension $(1|0)$ on which ϕ_0, ϕ_1 and z_0 act by zero and z_1 acts by the scalar t. The sequence splits only in the case $s_1 = 0$, when $\Gamma_0 \simeq \mathbb{C}$ is trivial. Thus, using Proposition 1, Theorem 2 and (5) we obtain
Lemma 3. If $n = 2$, then every simple W-module is isomorphic to one of the following
(1) $V(s_1, s_2)$ or $\Pi V(s_1, s_2)$ for $s_1 \neq -s_2$, $s_1, s_2 \neq 0$;
(2) $V(s, 0)$ if $s \neq 0$;
(3) Γ_t or $\Pi\Gamma_t$.

3.4. Invariance under permutations

Theorem 4 ([15], Theorem 4.4). Let $\mathfrak{s}' = \sigma(\mathfrak{s})$ for some permutation of coordinates.
(1) If $\mathfrak{s}$ is typical, then $V(\mathfrak{s})$ is isomorphic to $V(\mathfrak{s}')$ as a W-module.
(2) If $\mathfrak{s}$ is arbitrary, then $[V(\mathfrak{s})] = [V(\mathfrak{s}')]$ or $[\Pi V(\mathfrak{s}')]$, where $[X]$ denotes the class of X in the Grothendieck group.
Proof. We will prove the statement for $n = 2$. Assume first that $s_2 \neq -s_1$. In this case $V(s_1, s_2)$ is a $(1|1)$-dimensional simple W-module.

Let

$$D = \begin{pmatrix} \sqrt{s_2}+\sqrt{s_1}\mathbf{i} & 0 \\ 0 & \sqrt{s_1}+\sqrt{s_2}\mathbf{i} \end{pmatrix}.$$

Then by direct computation we have

$$D\phi_0 D^{-1} = \begin{pmatrix} 0 & \sqrt{s_2}+\sqrt{s_1}\mathbf{i} \\ \sqrt{s_2}-\sqrt{s_1}\mathbf{i} & 0 \end{pmatrix} \text{ and } D\phi_1 D^{-1} = \sqrt{s_1s_2}\begin{pmatrix} 0 & \sqrt{s_1}-\sqrt{s_2}\mathbf{i} \\ \sqrt{s_1}+\sqrt{s_2}\mathbf{i} & 0 \end{pmatrix}.$$

Therefore D defines an isomorphism between $V(s_1, s_2)$ and $V(s_2, s_1)$.

Now consider the case $s_1 = -s_2$. Then the structure of $V(s_1, -s_1)$ is given by the exact sequence (5). Let $V(\mathfrak{s}') = V(-s_1, s_1)$, then analogously we have the exact sequence

$$0 \to \Pi\Gamma_{-s_1^2-s_1} \to V(\mathfrak{s}') \to \Gamma_{-s_1^2+s_1} \to 0. \tag{6}$$

The statement (2) now follows directly from comparison of (5) and (6). The proof for an arbitrary n see in [15].

□

4. Central characters

The center of $U(\mathfrak{g})$ for $\mathfrak{g} = Q(n)$ is described in [18]. The center of $U(\mathfrak{h})$ coincides with $\mathbb{C}[x_1, \ldots, x_n]$ and the image of the center of $U(\mathfrak{g})$ under the Harish-Chandra homomorphism is generated by the polynomials $p_k = x_1^{2k+1} + \ldots + x_n^{2k+1}$ for all $k \in \mathbb{N}$. These polynomials are called Q-symmetric polynomials.

In [13] we proved that the center Z of W coincides with the image of the center of $U(\mathfrak{g})$ and hence can be also identified with the ring of Q-symmetric polynomials.

Every $\mathfrak{s}$ defines the central character $\chi_{\mathfrak{s}} : Z \to \mathbb{C}$. Furthermore, it follows from the description of simple W-modules in [15] (Theorem 4.6) that every simple W-module admits central character $\chi_{\mathfrak{s}}$ for some $\mathfrak{s}$. For every $\mathfrak{s} = (s_1, \dots, s_n)$ we define the *core* $c(\mathfrak{s}) = (s_{i_1}, \dots, s_{i_m})$ as a subsequence obtained from $\mathfrak{s}$ by removing all $s_j = 0$ and all pairs (s_i, s_j) such that $s_i + s_j = 0$. Up to a permutation this result does not depend on the order of removing. Thus, the core is well defined up to permutation. We call m the length of the core. The notion of core is very useful for describing the blocks in the category of finite-dimensional $Q(n)$-modules, see [12, 17].

Example 5. Let $\mathfrak{s} = (1, 0, 3, -1, -1)$, then $c(\mathfrak{s}) = (3, -1)$.

The following is a reformulation of the central character description in [18].

Lemma 6. Let $\mathfrak{s}, \mathfrak{s}' \in \mathbb{C}^n$. Then $\chi_{\mathfrak{s}} = \chi_{\mathfrak{s}'}$ if and only if $\mathfrak{s}$ and $\mathfrak{s}'$ have the same core (up to permutation).

It follows from Lemma 6 that the core depends only on the central character $\chi_{\mathfrak{s}}$, we denote it $c(\chi)$.

5. The category of finite-dimensional W-modules and blocks

Let $W-$mod be the *category of finite-dimensional W-modules.* A W-module M has *generalized central character* χ, if for any $z \in Z$ and $m \in M$, there exists $n \in \mathbb{Z}_{\geq 0}$ such that $(z - \chi(z))^n \cdot m = 0$. Let $W^\chi -$ mod be the full *subcategory of modules admitting generalized central character* χ. The category $W -$ mod is the direct sum of the subcategories $W^\chi -$ mod, as χ ranges over the central characters.

The *blocks* in the category $W -$ mod are equivalence class of linked objects. Each block lies in a single $W^\chi -$ mod, however, different blocks can belong to the same $W^\chi -$ mod, see [5].

5.1. Blocks in the category of finite-dimensional W-modules for Q(2)

Lemma 7. Let $n = 2$. A simple W-module S belongs to $W^\chi -$ mod if and only if one of the following three cases takes place:

(1) $S \simeq V(s_1, s_2)$ for $s_1 \neq s_2, s_1, s_2 \neq 0$ and $c(\chi) = (s_1, s_2)$,
(2) $S \simeq V(s, 0)$ for $s \neq 0$ and $c(\chi) = (s)$,
(3) $S \simeq \Gamma_t$ or $\Pi\Gamma_t$ and $\chi = 0$.

Proof. We have to compute the central character of the simple W-module. For a Q-symmetric polynomial $p_k = x_1^{2k+1} + x_2^{2k+1}$ we have that

$$p_k(S) = \begin{cases} s_1^{2k+1} + s_2^{2k+1} & \text{if } (1) \\ s^{2k+1} & \text{if } (2) \\ 0 & \text{if } (3) \end{cases}$$

Since p_k generate the center of W the statement follows.

□

Theorem 8. (1) Each simple W-module $V(s_1, s_2)$ for $s_1 \neq s_2, s_1, s_2 \neq 0$ forms a block in W^χ-mod, where $c(\chi) = (s_1, s_2)$.
(2) Each simple W-module $V(s, 0)$ for $s \neq 0$ forms a block in W^χ-mod, where $c(\chi) = (s)$.
(3) The blocks in the subcategory W^χ-mod, where $\chi = 0$, are described as follows. Let $a \in \mathbb{C}$. Define

$$a_n = a - n^2 + n\sqrt{1 - 4a} \text{ for } n = 0, \pm 1, \pm 2, \dots \tag{7}$$

Then Γ_a lies in the block formed by Γ_{a_n} if n is even and $\Pi\Gamma_{a_n}$, if n is odd. $\Pi\Gamma_a$ lies in the block formed by $\Pi\Gamma_{a_n}$ if n is even and Γ_{a_n}, if n is odd.

Journal of Physics: Conference Series **1416** (2019) 012028 doi:10.1088/1742-6596/1416/1/012028

Proof. Statements (1) and (2) follow from Lemma 6 and Lemma 7. To prove (3), first we will show that Γ_a is linked to $\Pi\Gamma_b$ if and only if

$$b = a - 1 \pm \sqrt{1-4a}. \tag{8}$$

Recall that W is generated by ϕ_0, ϕ_1, z_0 and z_1, where

$$\phi_0 = \xi_1 + \xi_2,\ \ \phi_1 = x_2\xi_1 - x_1\xi_2,\ \ z_0 = x_1 + x_2,\ \ z_1 = x_1x_2 - \xi_1\xi_2.$$

We have

$$[z_1, \phi_1] = 2z_1\phi_0 + 2\phi_1, \tag{9}$$

$$[z_1, \phi_0] = -2\phi_1. \tag{10}$$

Suppose that Γ_a is linked to $\Pi\Gamma_b$. The generators ϕ_0 and z_1 act in the vector superspace $\Gamma_a \oplus \Pi\Gamma_b$ as follows:

$$\phi_0 = \begin{pmatrix} 0 & 1 \\ 0 & 0 \end{pmatrix},\quad z_1 = \begin{pmatrix} a & 0 \\ 0 & b \end{pmatrix}.$$

Then from (10)

$$\phi_1 = \begin{pmatrix} 0 & \frac{b-a}{2} \\ 0 & 0 \end{pmatrix} \text{ and hence } [z_1, \phi_1] = \begin{pmatrix} 0 & \frac{(a-b)(b-a)}{2} \\ 0 & 0 \end{pmatrix}.$$

On the other hand, from (9)

$$[z_1, \phi_1] = 2\begin{pmatrix} a & 0 \\ 0 & b \end{pmatrix}\begin{pmatrix} 0 & 0 \\ 0 & 1 \end{pmatrix} + 2\begin{pmatrix} 0 & \frac{b-a}{2} \\ 0 & 0 \end{pmatrix}.$$

Hence

$$\frac{(a-b)(b-a)}{2} = 2a + b - a.$$

Then

$$b^2 + (2-2a)b + (a^2 + 2a) = 0. \tag{11}$$

This implies (8).

Conversely, if $a \neq 0$, set $s_1 = \frac{1-\sqrt{1-4a}}{2}$ and consider $V(-s_1, s_1)$ $(s_1 \neq 0)$. Recall that we have the non-split exact sequence (6):

$$0 \to \Pi\Gamma_{-s_1^2-s_1} \to V(-s_1, s_1) \to \Gamma_{-s_1^2+s_1} \to 0,$$

which becomes

$$0 \to \Pi\Gamma_b \to V(-s_1, s_1) \to \Gamma_a \to 0, \tag{12}$$

with $b = a - 1 + \sqrt{1-4a}$.

If we set $s_1 = \frac{1+\sqrt{1-4a}}{2}$, we obtain (12) with $b = a - 1 - \sqrt{1-4a}$.

If $a = 0$, set $s_1 = 1$ in (12). Then

$$0 \to \Pi\Gamma_{-2} \to V(-1, 1) \to \Gamma_0 \to 0.$$

Let V be a $(1|1)$-dimensional module on which z_0, z_1 and ϕ_1 act by zero and ϕ_0 acts by $\begin{pmatrix} 0 & 1 \\ 0 & 0 \end{pmatrix}$. Then

$$0 \to \Gamma_0 \to V \to \Pi\Gamma_0 \to 0.$$

Finally, the sum of the roots of equation (11) is $2a-2$. This gives the relation

$$a_{n-1}+a_{n+1}=2a_n-2 \quad (a_n=a). \tag{13}$$

Then (8) and (13) imply (7).

□

Example 9.
(1) $a=0$, then $a_n=n(1-n)$ and Γ_0 lies in the block

$$\dots,\Gamma_{-30},\Pi\Gamma_{-20},\Gamma_{-12},\Pi\Gamma_{-6},\Gamma_{-2},,\Pi\Gamma_0,\Gamma_0,\Pi\Gamma_{-2},\Gamma_{-6},\Pi\Gamma_{-12},\Gamma_{-20},\Pi\Gamma_{-30},\dots$$

(2) $a=\frac{1}{4}$, then $a_n=\frac{1}{4}-n^2$ and $\Gamma_{\frac{1}{4}}$ lies in the block

$$\Gamma_{\frac{1}{4}},\Pi\Gamma_{-\frac{3}{4}},\Gamma_{-\frac{15}{4}},\dots$$

(3) $a=1$, then $a_n=1-n^2+n\sqrt{-3}$ and Γ_1 lies in the block

$$\dots,\Pi\Gamma_{-3\sqrt{-3}-8},\Gamma_{-2\sqrt{-3}-3},\Pi\Gamma_{-\sqrt{-3}},\Gamma_1,\Pi\Gamma_{\sqrt{-3}},\Gamma_{2\sqrt{-3}-3},\Pi\Gamma_{3\sqrt{-3}-8},\dots$$

6. Acknowledgments

This work was supported by a grant from the Simons Foundation (#354874, Elena Poletaeva). I thank the organizers of the XXVIth Colloquium on Integrable Systems and Quantum Symmetries (ISQS-26) Czech Technical University, Prague, Czech Republic, July 8-12, 2019, for the very interesting conference and hospitality. I am grateful to V. Serganova for very helpful discussions.

References

[1] Brown J, Brundan J and Goodwin S 2013 *Algebra Number Theory* **7** 1849-82
[2] Brundan J and Goodwin S 2019 *Adv. Math.* **347** 273-339
[3] Coulembier K and Musson I 2018 *Tohoku Math. J. (2)* **70** 225-66
[4] Cheng S-J and Wang W 2012 *Dualities and Representations of Lie Superalgeras* Graduate Studies in Mathematics vol 144 (Providence, RI: American Mathematical Society) p 94
[5] Humphreys J 2008 *Representations of Semisimple Lie Algebras in the BGG Category* $\mathcal{O}$ Graduate Studies in Mathematics vol 94 (Providence, RI: American Mathematical Society) pp 30-32
[6] Kac V G 1977 *Adv. Math.* **26** 8-96
[7] Kac V G 1978 *Differential geometrical methods in mathematical physics, II (Proc. Conf., Univ. Bonn, Bonn, 1977) Lecture Notes in Math.* vol 676 (Berlin: Springer) pp 597-626
[8] Losev I 2010 Finite W-algebras *Proc. Intern. Congr. of Mathematicians* vol 3 (New Delhi: Hindustan Book Agency) 1281–1307 (*Preprint* arXiv:1003.5811 (v1) [math-RT])
[9] Losev I 2010 *J. Amer. Math. Soc.* **23** 35-59
[10] Losev I 2011 *Duke Math. J.* **159** 99-143
[11] Meinrenken E 2013 *Clifford Algebras and Lie Theory* A Series of Modern Surveys in Mathematics vol 58 (Heidelberg: Springer) pp 23-48
[12] Penkov I 1986 *Funktional. Anal. i Prilozhen.* **20** 37-45
[13] Poletaeva E and Serganova V 2016 On Kostant's theorem for the Lie superalgebra $Q(n)$ *Adv. Math.* **300** 320-59 (*Preprint* arXiv:1403.3866 (v1) [math-RT]).
[14] Poletaeva E and Serganova V 2017 On the finite W-algebra for the Lie superalgebra $Q(n)$ in the non-regular case *J. Math. Phys.* **58** 111701 (*Preprint* arXiv:1705.10200 (v1) (math-RT])
[15] Poletaeva E and Serganova V 2019 Representations of principal W-algebra for the superalgebra $Q(n)$ *Preprint* arXiv:1903.05272 (v1) [math-RT]
[16] Premet A 2002 *Adv. Math.* **170** 1-55
[17] Serganova V 2014 *Proc. Intern. Congr. of Mathematicians-Seoul 2014* vol 1 (Seoul: Kyung Moon Sa) pp 603-32
[18] Sergeev A 1983 *Lett. Math. Phys.* **7** 177-79

Journal of Physics: Conference Series **1416** (2019) 012029 doi:10.1088/1742-6596/1416/1/012029

Integrable surfaces induced by generalized Landau-Lifshitz equation with self-consistent potential

Zh. Sagidullayeva, G. Nugmanova and R. Myrzakulov

L.N. Gumilyov Eurasian National University, Nur-Sultan, Kazakhstan

E-mail: sagidullayeva.zh@gmail.com

Abstract. In this paper, we present integrable surfaces associated with generalized Landau-Lifshitz equation with self-consistent potential. We obtained the first and second fundamental forms. The first fundamental form allow us to calculate the curvature and metric properties of a surface, particularly, length and area of related space. The second fundamental form determines the external geometry of the surface in the vicinity of this point. Together they permit to define extrinsic invariants of the surface and its principal curvatures. The results can be used to describe spin waves in magnets and ferromagnets.

1. Introduction

The growing interest in the physics of magnetic phenomena is due to the difference in the interaction between the local magnetic moments of ions. They can diverse in their nature and have a wide range of energy scales; also, due to the characteristics of the crystal structure, they differ greatly between magnetic ions neighboring in different directions. The anisotropy property can lead to the formation of ferro- or antiferromagnetic states, as well as unusual magnetic structures spin waves or magnons.

In this work, we consider generalized Landau-Lifshitz equation (GLLE) with self-consistent potential. GLLE is a well-known equation describing magnetization waves in magnets, or simply, spin waves [1]-[7]. We present integrable surfaces associated with GLLE with self-consistent potential. The first and the second fundamental forms were obtained. The first fundamental form is used to calculate the curvature and metric properties of a surface, particularly, length and area of related space. The second fundamental form determines the external geometry of the surface in the vicinity of this point. Together they permit to define extrinsic invariants of the surface and its principal curvatures. The results can be used to describe spin waves in magnets and ferromagnets.

2. Landau-Lifshitz equation with self-consistent potential

The generalized Landau-Lifshitz equation with self-consistent potential is one of the generalizations of Landau-Lifshitz equation. It reads as

$$\mathbf{S}_t + \frac{1}{2}\mathbf{S} \wedge \mathbf{S}_{xx} + \frac{2}{a}\mathbf{S} \wedge \mathbf{W} = 0, \tag{1}$$

Journal of Physics: Conference Series **1416** (2019) 012029 doi:10.1088/1742-6596/1416/1/012029

$$\mathbf{W}_x + 2a\mathbf{S}\wedge\mathbf{W} \;=\; 0, \tag{2}$$

where $\wedge$ denotes a vector product, $\mathbf{S} = (S_1, S_2, S_3)$, $\mathbf{W} = (W_1, W_2, W_3)$ are vectors with lengths $\mathbf{S}^2 = S_1^2 + S_2^2 + S_3^2 = 1$, $\mathbf{W}^2 = W_1^2 + W_2^2 + W_3^2 = b$ and a, b are const. Also it would be important to note that in case of $W = 0$ system of equations (1)-(2) reduced to Classical Heisenberg Model. In most cases, it is convenient to work with the matrix form of this equation, which has the form

$$iS_t + \frac{1}{2}[S, S_{xx}] + \frac{1}{a}[S, W] \;=\; 0, \tag{3}$$
$$iW_x + a[S, W] \;=\; 0, \tag{4}$$

where $a = const$, $S = \Sigma_{j=1}^3 S_j(x, y, t)\sigma_j$ is a matrix analogue of the spin vector, W - potential with the matrix form $W = \Sigma_{j=1}^3 W_j(x, y, t)\sigma_j$,

$$S(x,t) = \begin{pmatrix} S_3 & S^- \\ S^+ & -S_3 \end{pmatrix}, \quad W(x,t) = \begin{pmatrix} W_3 & W^- \\ W^+ & -W_3 \end{pmatrix}, \tag{5}$$

and σ_j are Pauli matrices. The GLLE with self-consistent potential is gauge equivalent to Schrödinger-Maxwell-Bloch equation and corresponding investigations were made in works [8]-[11].

In mathematics, in the theory of integrable systems, a Lax pair is a pair of time-dependent matrices or operators that satisfy a corresponding differential equation are called the Lax equation. Lax pairs were introduced by Peter Lax to discuss solitons in continuous media [12]. The inverse scattering transform makes use of the Lax equations to solve such systems.

The LLE equation with self-consistent potential is integrable and its Lax representation can be written in the form

$$\Phi_x \;=\; U\Phi, \tag{6}$$
$$\Phi_t \;=\; V\Phi, \tag{7}$$

where the matrix operators U and V have the form

$$U \;=\; -i\lambda S, \tag{8}$$
$$V \;=\; \lambda^2 V_2 + \lambda V_1 + \left(\frac{i}{\lambda + a} - \frac{i}{a}\right) W. \tag{9}$$

Here

$$V_2 \;=\; -2iS, \quad V_1 = SS_x. \tag{10}$$

The compatibility condition of equations (6)-(7) give us

$$U_t - V_x + [U, V] = 0, \tag{11}$$

which is the zero curvature condition. Substituting U and V operators (8)-(10) in equation (11) we obtain the GLLE with self-consistent potential in matrix form (3)-(4).

Journal of Physics: Conference Series **1416** (2019) 012029 doi:10.1088/1742-6596/1416/1/012029

3. The fundamental forms for the GLLE with self-consistent potential

In general, the first and second fundamental forms (FF) are

$$I = g_{ij}dx^i dx^j, \tag{12}$$

$$II = b_{ij}dx^i dx^j, \tag{13}$$

here g_{ij}, b_{ij} are matrices

$$g_{ij} = \begin{pmatrix} E & F \\ F & G \end{pmatrix}, \tag{14}$$

$$b_{ij} = \begin{pmatrix} L & M \\ M & N \end{pmatrix}. \tag{15}$$

The FF can be presented through position and normal vectors:

$$I = \mathbf{dr} \cdot \mathbf{dr} = \mathbf{r}_x^2 dx^2 + 2\mathbf{r}_x \mathbf{r}_t dxdt + \mathbf{r}_t^2 dt^2, \tag{16}$$

or

$$I = Edx^2 + 2Fdxdt + Gdt^2 \tag{17}$$

and

$$II = -\mathbf{dn} \cdot \mathbf{dr} = (\mathbf{n} \cdot \mathbf{r}_{xx})\, dx^2 + 2\,(\mathbf{n} \cdot \mathbf{r}_{xt})\, dxdt + (\mathbf{n} \cdot \mathbf{r}_{tt})\, dt^2 \tag{18}$$

or

$$II = Ldx^2 + 2Mdxdt + Ndt^2. \tag{19}$$

The position vector is

$$\mathbf{r} = (r_1, r_2, r_3) \tag{20}$$

and normal to the surface is

$$\mathbf{n} = (n_1, n_2, n_3), \quad \mathbf{n}^2 = 1. \tag{21}$$

Using Sym-Tafel formula [13]

$$r = \Phi^{-1}\Phi_\lambda, \tag{22}$$

we can obtain this vectors in matrix forms

$$r_x = \Phi^{-1}U_\lambda\Phi, \quad r_t = \Phi^{-1}V_\lambda\Phi. \tag{23}$$

Relations between derivations of vector and matrix form of r with respect to x and t:

$$\mathbf{r}_x^2 = \frac{1}{2}tr\left(r_x^2\right), \tag{24}$$

$$\mathbf{r}_t^2 = \frac{1}{2}tr\left(r_t^2\right), \tag{25}$$

$$\mathbf{r}_x\mathbf{r}_t = \frac{1}{2}tr\left(r_x r_t\right). \tag{26}$$

Now, we obtain the necessary quantities

$$r_x^2 = \Phi^{-1}U_\lambda^2\Phi, \tag{27}$$

$$r_t^2 = \Phi^{-1}V_\lambda^2\Phi, \tag{28}$$

$$r_x r_t = \Phi^{-1}U_\lambda V_\lambda\Phi. \tag{29}$$

Journal of Physics: Conference Series **1416** (2019) 012029 doi:10.1088/1742-6596/1416/1/012029

The first fundamental form for the GLLE with self-consistent potential can be obtained by collecting and putting into (17) and necessary coefficients

$$\begin{aligned} I &= dx^2 + \left(8\lambda - \frac{1}{(\lambda+a)^2}Tr\{S,W\}\right)dxdt + \left(16\lambda^2 - \frac{1}{2}Tr((SS_x)^2)\right)dt^2 \\ &+ \left(\frac{2\lambda}{(\lambda+a)^2}Tr\{S,W\} + \frac{i\lambda}{2(\lambda+a)^2}Tr\{SS_x,W\} + \frac{1}{2(\lambda+a)^2}Tr(W^2)\right)dt^2, \end{aligned} \tag{30}$$

where $\{S,W\} = SW + WS$ - anticommutator. When $\lambda = \lambda_0 = 0$ then we can rewrite the first fundamental form as

$$\begin{aligned} I &= dx^2 - \frac{1}{a^2}Tr(SW)dxdt + \\ &+ \frac{1}{2}\left(Tr((SS_x)^2) - \frac{i}{a^2}Tr\{SS_x,W\} - \frac{4}{a^2}Tr(W^2)\right)dt^2, \end{aligned} \tag{31}$$

To obtain the second fundamental form, let us find the following

$$r_{xx} = \Phi^{-1}[U_\lambda, U]\Phi, \tag{32}$$

$$r_{xt} = \Phi^{-1}[U_\lambda, V]\Phi, \tag{33}$$

$$r_{tt} = \Phi^{-1}[V_\lambda, V]\Phi. \tag{34}$$

And a normal to surface can be calculated by

$$\mathbf{n} = \frac{\mathbf{r}_x \wedge \mathbf{r}_t}{|\mathbf{r}_x \wedge \mathbf{r}_t|}, \tag{35}$$

or

$$n = \frac{\Phi^{-1}[U_\lambda, V_\lambda]\Phi}{\sqrt{\frac{1}{2}tr\left([U_\lambda, V_\lambda]^2\right)}}. \tag{36}$$

Traces are determined as

$$tr(n \cdot r_{xx}) = \frac{tr([U_\lambda, V_\lambda][U_\lambda, U])}{\sqrt{\frac{1}{2}tr\left([U_\lambda, V_\lambda]^2\right)}}, \tag{37}$$

$$tr(n \cdot r_{xt}) = \frac{tr([U_\lambda, V_\lambda][U_\lambda, V])}{\sqrt{\frac{1}{2}tr\left([U_\lambda, V_\lambda]^2\right)}}, \tag{38}$$

$$tr(n \cdot r_{tt}) = \frac{tr([U_\lambda, V_\lambda][V_\lambda, V])}{\sqrt{\frac{1}{2}tr\left([U_\lambda, V_\lambda]^2\right)}}. \tag{39}$$

Here we came to interesting results normal vector to surface is equal to zero, cause trace of commutator $Tr[U_\lambda, V_\lambda] = 0$. Consequently, construction of the second fundamental form for considered model by this approach is impossible. It means that GLLE with self-consistent potential associated only with trivial soliton surface which proved by the existence of the first fundamental form only. For some equations soliton surfaces does not the same as integrable surface.

Journal of Physics: Conference Series **1416** (2019) 012029 doi:10.1088/1742-6596/1416/1/012029

4. Integrable surfaces induced by GLLE with self-consistent potential

In this section we want to present integrable surfase induced by GLLE with self-consistent potential. In a purpose of this lets make the following transformation $\mathbf{r}_x \equiv \mathbf{S}$. Then we can rewrite $\mathbf{r}_t$ as

$$\mathbf{r}_t = \frac{i}{2}\mathbf{r}_x \wedge \mathbf{r}_{xx} + \frac{1}{a^2}\mathbf{W}. \tag{40}$$

In this case, the first fundamental form denotes as

$$\begin{aligned} I &= dx^2 + \frac{2}{a^2}\mathbf{S}\mathbf{W}dxdt + \\ &+ \left(\frac{1}{4}(\mathbf{S}\wedge\mathbf{S}\mathbf{S}_x)^2 - \frac{1}{a^2}(\mathbf{S}\wedge\mathbf{S}_x)\mathbf{W} + \frac{1}{a^4}\mathbf{W}^2\right)dt^2, \end{aligned} \tag{41}$$

Now using (35) we can define normal to surface

$$\mathbf{n} = \frac{\frac{1}{2}\mathbf{r}_{xx} - \frac{1}{a^3}\mathbf{W}_x}{|\frac{1}{2}\mathbf{r}_{xx} - \frac{1}{a^3}\mathbf{W}_x|}, \tag{42}$$

or

$$\mathbf{n} = \frac{\mathbf{S}_x - \frac{1}{a^3}\mathbf{W}_x}{|\mathbf{S}_x - \frac{1}{a^3}\mathbf{W}_x|}. \tag{43}$$

Then determine coefficients of the 2nd fundamental form

$$L = \mathbf{n}\cdot\mathbf{r}_{xx}, \tag{44}$$

$$M = \mathbf{n}\cdot\left(\frac{1}{2}(\mathbf{r}_x\wedge\mathbf{r}_{xx})_x + \frac{1}{a^2}\mathbf{W}_x\right), \tag{45}$$

$$N = \mathbf{n}\cdot\left(\frac{1}{2}(\mathbf{r}_x\wedge\mathbf{r}_{xx})_t\right), \tag{46}$$

or in term of $\mathbf{S}$ will look like

$$L = \frac{(\mathbf{S}_x - \frac{1}{a^3}\mathbf{W}_x)\cdot\mathbf{S}_x}{|\mathbf{S}_x - \frac{1}{a^3}\mathbf{W}_x|}, \tag{47}$$

$$M = \frac{\mathbf{S}_x - \frac{1}{a^3}\mathbf{W}_x}{|\mathbf{S}_x - \frac{1}{a^3}\mathbf{W}_x|}\cdot\left(\frac{1}{a^2}\mathbf{W}_x - \frac{1}{2}(\mathbf{S}\wedge\mathbf{S}_x)_x\right), \tag{48}$$

$$N = -\frac{1}{2}\frac{\mathbf{S}_x - \frac{1}{a^3}\mathbf{W}_x}{|\mathbf{S}_x - \frac{1}{a^3}\mathbf{W}_x|}\cdot(\mathbf{S}\wedge\mathbf{S}_x)_t. \tag{49}$$

The second fundamental form can be written as

$$\begin{aligned} II &= \frac{(\mathbf{S}_x - \frac{1}{a^3}\mathbf{W}_x)\cdot\mathbf{S}_x}{|\mathbf{S}_x - \frac{1}{a^3}\mathbf{W}_x|}dx^2 + \\ &2\,\frac{\mathbf{S}_x - \frac{1}{a^3}\mathbf{W}_x}{|\mathbf{S}_x - \frac{1}{a^3}\mathbf{W}_x|}\cdot\left(\frac{1}{a^2}\mathbf{W}_x - \frac{1}{2}(\mathbf{S}\wedge\mathbf{S}_x)_x\right)dxdt + \\ &- \frac{1}{2}\frac{\mathbf{S}_x - \frac{1}{a^3}\mathbf{W}_x}{|\mathbf{S}_x - \frac{1}{a^3}\mathbf{W}_x|}\cdot(\mathbf{S}\wedge\mathbf{S}_x)_t dt^2. \end{aligned} \tag{50}$$

5. Area of surface for the GLLE with self-consistent potential

Surface's area is given in the form

$$S = \int \sqrt{g}\,dxdt = \int\int |\mathbf{r}_x \wedge \mathbf{r}_t|\,dxdt. \tag{51}$$

Then our required surface area S will look like

$$S = \int \left| \frac{i}{a^3}\mathbf{W} - \frac{1}{2}\mathbf{r}_x \right| dt, \tag{52}$$

or

$$S = \int \left| \frac{i}{a^3}\mathbf{W} - \frac{1}{2}\mathbf{S} \right| dt. \tag{53}$$

The Gaussian curvature of the surface is given by

$$K = \frac{LN - M^2}{EG - F^2}. \tag{54}$$

Which is for our model will be

$$K = -\frac{\left[\frac{1}{2}(\mathbf{S}_x - \frac{1}{a^3}\mathbf{W}_x)\cdot\mathbf{S}_x\right]\left[\mathbf{S}_x - \frac{1}{a^3}\mathbf{W}_x\cdot(\mathbf{S}\wedge\mathbf{S}_x)_t\right] + \left[\mathbf{S}_x - \frac{1}{a^3}\mathbf{W}_x\cdot\left(\frac{1}{a^2}\mathbf{W}_x - \frac{1}{2}(\mathbf{S}\wedge\mathbf{S}_x)_x\right)\right]^2}{\left[|\mathbf{S}_x - \frac{1}{a^3}\mathbf{W}_x|^2\right]\left[\frac{1}{4}(\mathbf{S}\wedge\mathbf{S}\mathbf{S}_x)^2 - \frac{1}{a^2}(\mathbf{S}\wedge\mathbf{S}_x)\mathbf{W} + \frac{1}{a^4}\mathbf{W}^2 - \left(\frac{1}{a^2}\mathbf{S}\mathbf{W}\right)^2\right]}, \tag{55}$$

and finally we found Gaussian curvature in terms of $\mathbf{S}$.

6. CONCLUSIONS

In this paper, using the Sym-Tafel formula, we obtained the first fundamental form of GLLE with self-consistent potential. Proved that for considered equation soliton surface does not equal to integrable surface induced by them. Then to construct integrable surface induced by GLLE with self-consistent potential we found the first and the second fundamental forms,area of surface and a Gaussian curvature. The results can be used to describe spin waves in magnets and ferromagnets. According to the results of this work, research in this direction is entering a new step, which will allow a more detailed study of the construction of different integrable surfaces associated with soliton equations.

Acknowledgments

This work was supported by GF of MES RK Grant Number 0118RK00935.

References

[1] Myrzakulov R, Vijayalakshmi S, Nugmanova G and Lakshmanan M 1999 *Phys. Lett.* A **233** 391–96
[2] Myrzakulov R, Mamyrbekova G, Nugmanova G and Yesmakhanova K 2015 *Phys. Lett.* A **378** 2118–23
[3] Myrzakulov R, Mamyrbekova G, Nugmanova G and Lakshmanan M 2015 *Symmetry* **7** 1352
[4] Yersultanova Z, Zhassybayeva M, Nugmanova G and Myrzakulov R 2015 *Int. J. of Geometric Methods in Modern Phys.* **12** 1550134
[5] Nugmanova G, Sagidullayeva Zh and Myrzakulov R 2017 *J. Phys.: Conf. Series* **804** 012035
[6] Nugmanova G and Sagidullayeva Zh 2017 *Bulletin of the Karaganda university-Mathematics* **86** 91–6
[7] Sagidullayeva Zh, Nugmanova G and Myrzakulov 2019 *AIP Conf. Proc.* **2159** 030029
[8] Sagidullayeva Zh 2018 *Bulletin of the L.N. Gumilyov Eurasian National University. Phys. Astronomy Series* **3(124)** 41–6
[9] Bekova G, Yesmakhanova K and Shaikhova G 2018 *AIP Conf. Proc.* **1997** 020039
[10] Yesmakhanova K, Bekova G, Ybyraiymova S and Shaikhova G 2017 *J. Phys.: Conf. Series* **936** 012098
[11] Bekova G, Yesmakhanova K, Myrzakulov R and Shaikhova G 2017 *AIP Conf. Proc.* **1880** 060022
[12] Lax P 1968 *Commun. on Pure and App. Math.* **21** 467–90
[13] Sym A 1985 *Lecture Notes in Phys.* **239**

Exact solutions for the (3+1)-dimensional Kudryashov-Sinelshchikov equation

G.N. Shaikhova[1], B.B. Kutum[1], A.B. Altaybaeva[1], B.K. Rakhimzhanov[2]

[1]Department of General and Theoretical Physics, L. N. Gumilyov Eurasian National University, Nur-Sultan, 010008, Kazakhstan

[2] JSC "National company "Kazakhstan Gharysh Sapary", Nur-Sultan, 010008, Kazakhstan

E-mail: g.shaikhova@gmail.com

Abstract. In this work, (3+1)-dimensional Kudryashov-Sinelshchikov equation is investigated by using the sine-cosine method and modification of the truncated expansion method. A variety of exact solutions are obtained.

1. Introduction

In this paper, we study the (3+1)-dimensional Kudryashov-Sinelshchikov equation [1]

$$(u_t + \alpha u u_x + \gamma u_{xxx})_x + d u_{yy} + e u_{zz} = 0, \tag{1}$$

where α represents the nonlinearity, γ is dispersion term, while d and e stand for transverse variation of wave in y and z directions. The equation (1) describes the physical characteristics of nonlinear waves in a bubbly liquid. In the case $e = 0$, equation (1) reduces to two-dimensional Korteweg-de Vries equation and for case $d = e = 0$ we obtain the one-dimensional Korteweg-de Vries equation. The equation (1) was studied by the modified tanh-coth method [2], Backlund transformation [3], bifurcation analysis was presented in [4], and density fluctuation symbolic computation in [5]. In one-dimensional and two-dimensional cases the Kudryashov-Sinelshchikov equations were studied by the G'/G expansion method in [6], the first integral method was applied in [7], modification of truncated expansion method [8], the modified exp-function method [9].

The purpose of this work is to find exact solutions for the (3+1)-dimensional Kudryashov-Sinelshchikov equation. The methods for finding exact solution of nonlinear partial differential equations are known. Some of them are the Darboux transformation [10–13], Hirota bilinear method [14–17], Kudryashov method [18], extended tanh method [19, 20], sine-cosine method [20]. To obtain the exact solution for the (3+1)-dimensional Kudryashov-Sinelshchikov equation we use the two methods such as the sine-cosine method [20] and the modification of the truncated method [8].

The organization of the paper is as follows: In Section 2, the description of the sine-cosine method and exact solution are given. In section 3, we study the Kudryashov-Sinelshchikov equation by the modification of the truncated method. Finally, the conclusion is given in Section 4.

Journal of Physics: Conference Series **1416** (2019) 012030 doi:10.1088/1742-6596/1416/1/012030

2. The Sine-cosine method

2.1. Review of the sine-cosine method

In this section we describe sine-cosine method that is presented in [20]. According to method the partial differential equation

$$E_1(u_t, u_x, u_{xx}, u_{xxx}, ...) = 0, \tag{2}$$

can be converted to ODE

$$E_2(u, u', u'', u''', ...) = 0, \tag{3}$$

by using a wave variable

$$u(x,t) = u(\xi), \quad \xi = x - ct. \tag{4}$$

Then equation (3) is integrated as long as all terms contain derivatives where integration constants are considered zeros. The solutions of ODE can be expressed in the form

$$u(x,t) = \lambda \cos^\beta(\mu\xi), \quad |\xi| \le \frac{\pi}{2\mu}, \tag{5}$$

or

$$u(x,t) = \lambda \sin^\beta(\mu\xi), \quad |\xi| \le \frac{\pi}{\mu}, \tag{6}$$

where the parameters λ, μ and β will be determined, and μ is wave number and c is wave speed respectively. The derivatives of (5) become

$$(u^n)_\xi = -n\beta\mu\lambda^n \cos^{n\beta-1}(\mu\xi)\sin(\mu\xi), \tag{7}$$

$$(u^n)_{\xi\xi} = -n^2\mu^2\beta^2\lambda^n \cos^{n\beta}(\mu\xi) + n\mu^2\lambda^n\beta(n\beta-1)\cos^{n\beta-2}(\mu\xi), \tag{8}$$

and the derivatives of (6) have next forms

$$(u^n)_\xi = -n\beta\mu\lambda^n \sin^{n\beta-1}(\mu\xi)\cos(\mu\xi), \tag{9}$$

$$(u^n)_{\xi\xi} = -n^2\mu^2\beta^2\lambda^n \sin^{n\beta}(\mu\xi) + n\mu^2\lambda^n\beta(n\beta-1)\sin^{n\beta-2}(\mu\xi), \tag{10}$$

and so on for the other derivatives. Using (5)-(10) into the reduced ODE gives a trigonometric equation of $\cos^R(\mu\xi)$ or $\sin^R(\mu\xi)$ terms. Then, we determine the parameters by first balancing the exponents of each pair of cosine or sine to determine R. Next, we collect all coefficients of the same power in $\cos^k(\mu\xi)$ or $\sin^k(\mu\xi)$, where these coefficients have to vanish. The system of algebraic equations among the unknown β, λ, and μ will be given and from that, we can determine coefficients.

2.2. Application the sine-cosine method

In this section, we apply sine-cosine method to the (3+1)-dimensional equation (1). By wave variable

$$u(x,y,z,t) = u(\xi), \quad \xi = (x + y + z - ct), \tag{11}$$

the equation (1) can be converted to

$$(-c + d + e)u + \frac{\alpha}{2}u^2 + \gamma u'' = 0. \tag{12}$$

Seeking the solution in (5) and (8)

$$\cos^{\beta}(\mu\xi)\left[(-c+d+e)\lambda-\gamma\mu^2\beta^2\lambda\right]+ \\ +\frac{\alpha}{2}\lambda^2\cos^{2\beta}(\mu\xi)+\mu^2\lambda\beta(\beta-1)\cos^{\beta-2}(\mu\xi)=0. \tag{13}$$

Equating the exponents and the coefficients of each pair of the $\cos(\mu\xi)$ functions we find the following algebraic system:

$$2\beta=\beta-2,\rightarrow\beta=-2. \tag{14}$$

Substituting equation (14) into equation (13) to get

$$\cos^{-2}(\mu\xi)\left[(-c+d+e)\lambda-\gamma\mu^2\beta^2\lambda\right]+ \\ +\frac{\alpha}{2}\lambda^2\cos^{-4}(\mu\xi)+\mu^2\lambda\beta(\beta-1)\cos^{-4}(\mu\xi)=0. \tag{15}$$

Equating the exponents and the coefficients of each pair of the $\cos(\mu\xi)$ functions, we obtain a system of algebraic equations

$$\cos^{-2}(\mu\xi)\ :\ (-c+d+e)\lambda-\gamma\mu^2\beta^2\lambda=0, \tag{16}$$

$$\cos^{-2}(\mu\xi)\ :\ \frac{\alpha}{2}\lambda^2+\mu^2\lambda\beta(\beta-1)=0. \tag{17}$$

Solving the algebraic system (16)-(17), we get:

$$\lambda=\frac{-12\mu^2}{\alpha},\quad \mu=-\frac{1}{2}\sqrt{\frac{d+e-c}{\gamma}}. \tag{18}$$

The result (18) can be easily obtained if we also use the sine method (6) and then we obtain the following exact solutions

$$u_1(x,y,z,t)=\frac{-12\mu^2}{\alpha}\cos^{-2}(\frac{1}{2}\sqrt{\frac{d+e-c}{\gamma}}(x+y+z-ct)), \tag{19}$$

$$u_2(x,y,z,t)=\frac{-12\mu^2}{\alpha}\sin^{-2}(\frac{1}{2}\sqrt{\frac{d+e-c}{\gamma}}(x+y+z-ct)), \tag{20}$$

where $c\neq d+e$.

3. Modification of the truncated expansion method

3.1. Review modification of the truncated expansion method

The partial differential equation

$$E_1(u_t,u_x,u_{xx},u_{xxx},...)=0, \tag{21}$$

can be converted to ODE

$$E_2(u,u',u'',u''',...)=0, \tag{22}$$

by using a wave variable

$$u(x,t)=u(\xi),\quad \xi=x-ct. \tag{23}$$

To find dominant terms we substitute

$$u = \xi^{-p}, \tag{24}$$

into all terms of equation (22). Then we ought to compare degrees of all terms of equations and choose two or more with the highest degree. The maximum value of p is called the pole of the equation (22) and we denote it as N. The method can be applied when N is integer. The exact solution of equation (22) is looked in the form

$$u = a_0 + a_1 Q(\xi) + a_2 Q(\xi)^2 + ... + a_N Q(\xi)^N, \tag{25}$$

where $Q(\xi)$ is the following function

$$Q(\xi) = \frac{1}{1+e^{\xi}}. \tag{26}$$

We can calculate number of derivatives by

$$u_\xi = \sum_{n=0}^{N} a_n n Q^n (Q-1), \tag{27}$$

$$u_{\xi\xi} = \sum_{n=0}^{N} a_n n Q^n (Q-1)[(n+1)Q - n], \tag{28}$$

$$u_{\xi\xi\xi} = \sum_{n=0}^{N} a_n n Q^n (Q-1)[(n^2+3n+2)Q^2 - (2n^2+3n+1)Q + n^2]. \tag{29}$$

3.2. Application to Kudryashov-Sinelshchikov Equation

In this section, we apply modification of truncated expansion method to the (3+1)-dimensional Kudryashov-Sinelshchikov equation (1). By wave variable

$$u(x,y,z,t) = u(\xi), \quad \xi = (x+y+z-ct), \tag{30}$$

the equation (1) can be converted to ODE

$$(-c+d+e)u + \frac{\alpha}{2}u^2 + \gamma u'' = 0. \tag{31}$$

From equation (31) we find $N = 2$ then we look for the solution of equation (31) in the form

$$u = a_0 + a_1 Q(\xi) + a_2 Q(\xi)^2. \tag{32}$$

The second derivative of equation (32) is

$$u_{\xi\xi} = a_1 Q + (4a_2 - 3a_1)Q^2 + (2a_1 - 10a_2)Q^3 + 6a_2 Q^4. \tag{33}$$

Substituting (32)-(33) into (31) we obtain the system of algebraic equations in the form

$$Q^4 : \frac{1}{2}a_2^2\alpha + 6a_2\gamma, \tag{34}$$

$$Q^3 : a_1 a_2 \alpha + 2a_1\gamma - 10a_2\gamma, \tag{35}$$

$$Q^2 : -a_2 c + a_2 d + a_2 e + a_0 a_2 \alpha + \frac{1}{2}a_1^2\alpha - 3a_1\gamma + 4a_2\gamma, \tag{36}$$

$$Q^1 : a_0 a_1 \alpha - a_1 c + a_1 d + a_1 e + a_1\gamma, \tag{37}$$

$$Q^0 : -a_0 c + a_0 d + a_0 e + \frac{1}{2}a_0^2\alpha. \tag{38}$$

From the system (34)-(38) we can find coefficients with two cases as

$$1)\ a_0 = 0,\ a_1 = \frac{12\gamma}{\alpha},\ a_2 = -\frac{12\gamma}{\alpha},\ c = d + e + \gamma, \tag{39}$$

$$2)\ a_0 = -\frac{2\gamma}{\alpha},\ a_1 = \frac{12\gamma}{\alpha},\ a_2 = -\frac{12\gamma}{\alpha},\ c = -\gamma + d + e. \tag{40}$$

Substituting (39)-(40) in (32) we obtain exact solutions of equations (1) in the form

$$u_3(x, y, z, t) = \frac{12\gamma}{\alpha}\frac{1}{1+e^{\xi}} - \frac{12\gamma}{\alpha}\left(\frac{1}{1+e^{\xi}}\right)^2, \tag{41}$$

$$u_4(x, y, z, t) = -\frac{2\gamma}{\alpha} + \frac{12\gamma}{\alpha}\frac{1}{1+e^{\xi}} - \frac{12\gamma}{\alpha}\left(\frac{1}{1+e^{\xi}}\right)^2, \tag{42}$$

where $\xi = (x + y + z - ct)$. The graphical representation of u_3 and u_4 is shown in Fig. 1

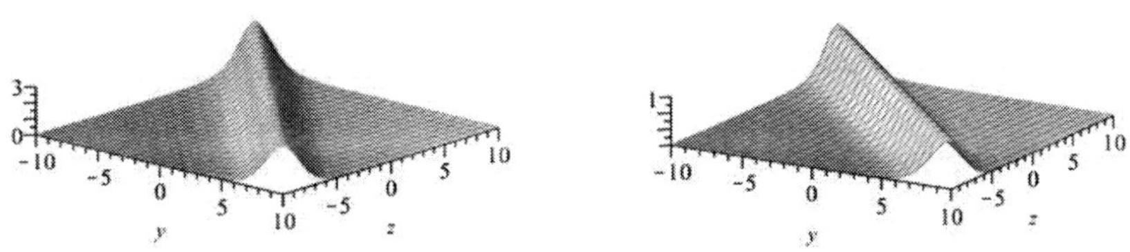

a) b)

Figure 1: Solutions corresponding to u_3 and u_4 for $\alpha = 1, \gamma = 1, d = 0.5, e = 0.5, x = 0, t = 0$.

4. Conclusion

In this paper, the (3+1)-dimensional Kudryashov-Sinelshchikov equation was studied using the two methods such as the sine-cosine method and the modification of the truncated method. The schemes of the two methods were presented. New exact solutions for the Kudryashov-Sinelshchikov equation were obtained. These methods can be applied to other kinds of nonlinear problems.

Acknowledgments

This work is prepared by the frame of the project the Ministry of Education and Science of the Republic of Kazakhstan (IRN: AP08052081).

References

[1] Kudryashov N A, Sinelshchikov D I 2012 *Physica Scripta* **85** 025402
[2] Chukkol Y, Mohamad M N, and Muminov M 2018 *Journal of Applied Mathematics* 1–9
[3] Tu J M, Tian S F, Xu M J, Song X Q and Zhang T T 2016 *Nonlinear Dynamics* **83** 1199–1215

[4] Feng Y, Shan W, Sun W, Zhong H, and Tian B 2014 *Communications in Nonlinear Science and Numerical Simulation* **19** 880–86
[5] Gao X Y 2016 *Modern Physics Letters B* **30** 1650217
[6] He Y, Li Sh, and Yao L 2013 *Mathematical Problems in Engineering Volume* **2013**
[7] Mirzazadeh M, Eslami M 2012 *Modelling and Control* **17** 481–88
[8] Ryabov P N 2010 *Applied Mathematics and Computation* **217** 3585–90
[9] He Y, Yao L 2013 *International Mathematical Forum* **8** 895–902
[10] Bekova G, Shaikhova G, Yesmakhanova K and Ozat N 2018 *J. Phys.: Conf. Series* **965** 012035 (1–10)
[11] Bekova G, Shaikhova G, Yesmakhanova K and Myrzakulov R 2017 *AIP Conf. Proc.* **1880** 060022 (1–7)
[12] Bekova G, Yesmakhanova K, Bekova G, Ybyraiymova S and Shaikhova G 2017 *J. Phys.: Conf. Series* **936** 012098 (1–6)
[13] Bekova G, Yesmakhanova K, Bekova G, Myrzakulov R and Shaikhova G 2017 *J. Phys.: Conf. Series* **804** 012004 (1–8)
[14] Nugmanova G, Sagidullayeva Zh 2017 *Bulletin of the Karaganda University of the-Mathematics* **86**(2) 91-96
[15] Mukhanmedina K, Syzdykova AM and Shaikhova GN 2015 *Bulletin of the Karaganda University-Mathematics* **80** 103–07
[16] Kutum BB, Shaikhova GN 2019 *Bulletin of the Karaganda University-Physics* **3** 22–26
[17] Burdik C, Shaikhova G, Rakhimzhanov B 2019 *arXiv:1909.00826*
[18] Shaikhova G 2018 *Bulletin of the Karaganda University-Mathematics* **92** 94–98
[19] Bekova G, Shaikhova G and Yesmakhanova K 2018 *AIP Conf. Proc.* **1997** 020039 (1–6)
[20] Wazwaz A M 2004 *Mathematical and Computer Modelling* **40** (5) 499–508

Para-Hermitian Geometry and Doubled Aspects of Vaisman Algebroid

Kenta Shiozawa
Department of Physics, Kitasato University, Sagamihara 252-0373, Japan
E-mail: k.shiozawa@sci.kitasato-u.ac.jp

Abstract. The geometry of double field theory (DFT) is described by a para-Hermitian manifold $\mathcal{M}$. A tangent bundle of the para-Hermitian manifold $T\mathcal{M}$ is decomposed into two eigenbundles L and $\tilde{L}$ associated with the eigenvalues of the para-complex structure K. We define a Lie algebroid structure on the eigenbundles L, $\tilde{L}$. The gauge symmetry algebra of DFT is governed by the C-bracket. The algebraic structure based on the C-bracket is not a Courant algebroid, but a metric algebroid proposed by Vaisman (the Vaisman algebroid). We show that the Vaisman algebroid in DFT is naturally defined on $T\mathcal{M}$ by an analogue of the Drinfel'd double of L, $\tilde{L}$. We also find that an algebraic origin of the strong constraint is the condition for $(L, \tilde{L})$ to become a Lie bialgebroid.

1. Introduction

String theory is studied as a candidate of quantum gravity. This theory has fruitful mathematical structures. In particular, duality, which relates various different theories, is an important concept to uncover the whole picture of the theory. For example, we consider a string propagating in a compact space. The strings can wrap around a compactified direction and have energy proportional to the number of windings (the winding mode $w \in \mathbb{Z}$). The Hamiltonian of the string is characterized by the quantized momentum along the compactified direction (the Kaluza-Klein (KK) mode $n \in \mathbb{Z}$) and the winding mode. At this time, the Hamiltonian H is invariant under the interchange of the KK-mode and the winding mode (at the same time, the radius of the compactified direction becomes an inverse number): $H(n, w) \leftrightarrow H(w, n)$. This invariance is called the target space duality, or T-duality [1]. When the compact space is a D-dimensional torus T^D, T-duality is expressed as a symmetry by $O(D, D)$ group. In supergravity, which is a low energy effective theory of strings, T-duality is hidden symmetry.

One of the approaches to deal geometrically with T-duality is the Hitchin's generalized geometry (GG) [2]. This is defined on an extended bundle that combines a tangent bundle and a cotangent bundle of the target space: $\mathbb{T}M = TM \oplus T^*M$. The gauge symmetry with respect to the target space of non-linear sigma models of strings is described by a Courant algebroid appearing in GG. The Courant algebroid accommodates the diffeomorphism of the spacetime and the NSNS B-field gauge symmetry algebra. In contrast to doubling the tangent bundle in GG, there is a formalism that doubles the base space, which is called *double field theory* (DFT) [3]. The action of DFT can be written in the Einstein-Hilbert-like form, *i.e.*, DFT is Einstein's gravity theory with T-duality as a manifest symmetry. DFT is defined in the doubled spacetime $\mathcal{M}^{2D}$, which is locally characterized by the doubled coordinates $X^M = (x^\mu, \tilde{x}_\mu)$. Since

Journal of Physics: Conference Series **1416** (2019) 012031 doi:10.1088/1742-6596/1416/1/012031

DFT includes extra degrees of freedom, it is necessary to impose the *physical section condition* in order to obtain a true physical theory. There are two section conditions: the *weak constraint* derived from the level matching condition of closed strings and the *strong constraint* derived from a condition of the gauge symmetry of DFT. The gauge symmetry algebra of DFT is governed by the *C-bracket* [4]. The C-bracket reduces to the Courant bracket in GG when the strong constraint is imposed.

It is interesting to study the mathematical aspects of DFT because the construction of DFT was done physically rather than mathematically. The *metric algebroid* (we call this the *Vaisman algebroid*) is proposed in [5] as the structure of symmetric algebras based on the C-bracket. The doubled spacetime of DFT is given by a para-Hermitian manifold, and the Vaisman algebroid is an algebraic structure defined by the C-bracket on its tangent bundle [6]. Recently, the Vaisman algebroid was rediscovered as the pre-DFT algebroid in the study of membrane sigma models that corresponds to DFT [7].

It is known that a Courant algebroid is obtained by an analogue of the Drinfel'd double of a Lie bialgebroid [8]. Therefore the Courant algebroid has the doubled structure. Some aspects of the doubled structure of the Vaisman algebroid were studied [9]. In particular, the Vaisman algebroid is obtained as a double of a pair of Lie algebroids [10]. At this time, the key is the derivation condition of Lie bialgebroids.

In this contribution, we show that the Vaisman algebroid in DFT is given by an analogue of the Drinfel'd double of a Lie algebroid pair naturally defined on a para-Hermitian manifold. In section 2, we give a lightning review of DFT for grasping the physical aspects of DFT. We introduce important notions in DFT for the following discussion. In section 3, we review a para-Hermitian manifold. DFT is described as a flat para-Hermitian manifold. In section 4, we calculate the "double" for the Vaisman algebroid in DFT setup. In section 5, we give the conclusion and the future directions. We also discuss the gauge symmetry in DFT. The main results in this contribution are based on the original work [10].

2. A Lightning Review of Double Field Theory

We give a quick introduction to DFT in this section. The T-duality transformation in DFT is expressed as an $O(D,D)$ group action in the doubled spacetime. The doubled spacetime is given by a $2D$-dimensional differential manifold $\mathcal{M}^{2D}$. The coordinates of $\mathcal{M}^{2D}$ is given by

$$X^M = \begin{pmatrix} \tilde{x}_\mu \\ x^\mu \end{pmatrix} \qquad (M = 1, \dots, 2D;\ \mu = 1, \dots, D). \tag{1}$$

Here x^μ is the coordinate of a conventional physical spacetime. $\tilde{x}_\mu$ is the Fourier dual of the winding modes of strings and called the winding coordinate. The doubled index is raised and lowered by the $O(D,D)$ invariant structure (the η-metric):

$$\eta_{MN} = \begin{pmatrix} 0 & \delta^\mu{}_\nu \\ \delta_\mu{}^\nu & 0 \end{pmatrix}, \qquad \eta^{MN} = \begin{pmatrix} 0 & \delta_\mu{}^\nu \\ \delta^\mu{}_\nu & 0 \end{pmatrix}. \tag{2}$$

The dynamical fields in DFT are the generalized metric $\mathcal{H}(g,B)$ and the DFT dilaton $d(g,\phi)$. Here, g is a conventional D-dimensional spacetime metric, B is the NSNS 2-form and ϕ is the dilaton in type II supergravities. The action of DFT is given by the Einstein-Hilbert-like form

$$S_{\mathrm{DFT}} = \int \mathrm{d}^{2D}X\, e^{-2d}\mathcal{R}(\mathcal{H}, d), \tag{3}$$

where $\mathcal{R}(\mathcal{H}, d)$ is called the generalized Ricci scalar. Detailed expressions for $\mathcal{H}$, d and $\mathcal{R}$ is found in [11]. This action is invariant by the $O(D,D)$ group action. Moreover, this action is invariant under the DFT gauge transformation.

The infinitesimal DFT gauge transformation of $O(D,D)$ tensors is given by the generalized Lie derivative $\widehat{\mathcal{L}}_\Xi$. The generalized Lie derivative on a doubled vector V^M is defined by

$$\widehat{\mathcal{L}}_\Xi V^M = \Xi^K \partial_K V^M + (\partial^M \Xi_K - \partial_K \Xi^M) V^K. \tag{4}$$

Here Ξ is a gauge parameter and is a doubled vector. The generalized Lie derivative is also called the D-bracket. The commutator of the generalized Lie derivatives is calculated as

$$[\widehat{\mathcal{L}}_{\Xi_1}, \widehat{\mathcal{L}}_{\Xi_2}] \approx \widehat{\mathcal{L}}_{[\Xi_1,\Xi_2]_\mathsf{C}}. \tag{5}$$

This means that the DFT gauge algebra is closed by the C-bracket. The symbol "$\approx$" in (5) means equality when imposing the strong constraint. The C-bracket is defined by

$$[\Xi_1, \Xi_2]_\mathsf{C}^M = \Xi_1^K \partial_K \Xi_2^M - \Xi_2^K \partial_K \Xi_1^M - \frac{1}{2}\eta_{KL}(\Xi_1^K \partial^M \Xi_2^L - \Xi_2^K \partial^M \Xi_1^L). \tag{6}$$

The strong constraint is a condition for reducing extra degrees of freedom in DFT and obtaining a true physical theory. The strong constraint is given by

$$\eta^{MN} \partial_M * \partial_N * = 0. \tag{7}$$

Here $*$ denotes arbitrary DFT fields and gauge parameters. One of a trivial solution to this constraint is that the derivative of the winding coordinates vanishes $\tilde{\partial}^\mu * = 0$. This solution is called the supergravity frame because it shows that fields in the theory depend only on the coordinate x^μ that are the Fourier dual to the KK-modes. In the supergravity frame $\tilde{\partial}^\mu * = 0$, the DFT action (3) is reduced to the NSNS sector of type II supergravity:

$$S_\text{DFT} \xrightarrow{\tilde{\partial}^\mu *=0} S_\text{NSNS} = \int \mathrm{d}^D x \sqrt{-g} e^{-2\phi}\left[R + 4(\partial\phi)^2 - \frac{1}{12}(H_3)^2\right], \tag{8}$$

where R is the Ricci scalar with the metric g, and $H_3 = \mathrm{d}B$ is the field strength of the B-field. Similarly, the C-bracket is reduced to the Courant bracket in GG:

$$[\Xi_1, \Xi_2]_\mathsf{C} \xrightarrow{\tilde{\partial}^\mu *=0} [A+\alpha, B+\beta]_\mathrm{c} = [A,B]_L + \mathcal{L}_A \beta - \mathcal{L}_B \alpha - \frac{1}{2}\mathrm{d}(\iota_A\beta - \iota_B\alpha). \tag{9}$$

Then, $\Xi_1 = A + \alpha$, $\Xi_2 = B + \beta$ are denoted as a formal sum of vector fields and 1-forms.

We focus on the doubled structure of the C-bracket. The component of the gauge parameters Ξ_i^M and the basis of the doubled vector fields are given by

$$\Xi_1^M = \begin{pmatrix}\alpha_\mu \\ A^\mu\end{pmatrix}, \qquad \Xi_2^M = \begin{pmatrix}\beta_\mu \\ B^\mu\end{pmatrix}, \qquad \partial_M = \begin{pmatrix}\tilde{\partial}^\mu \\ \partial_\mu\end{pmatrix}. \tag{10}$$

Here the doubled vector field is given by $\Xi^M \partial_M$. We call $\alpha_\mu \tilde{\partial}^\mu$ appearing as a component of the doubled vector field as the winding vector field. The Lie bracket and the dual "winding" Lie bracket are defined by

$$[A,B]_L = (A^\nu \partial_\nu B^\mu - B^\nu \partial_\nu A^\mu)\partial_\mu, \qquad [\alpha,\beta]_{\tilde{L}} = (\alpha_\nu \tilde{\partial}^\nu \beta_\mu - \beta_\nu \tilde{\partial}^\nu \alpha_\mu)\tilde{\partial}^\mu. \tag{11}$$

We define naively the exterior derivative d acting on the winding vector fields, the winding exterior derivative $\tilde{\mathsf{d}}$ acting on the vector fields and the interior products $\iota_A, \tilde{\iota}_\alpha$:

$$\mathsf{d}\alpha = \partial_\mu \alpha_\nu \tilde{\partial}^\mu \wedge \tilde{\partial}^\nu, \qquad \tilde{\mathsf{d}} A = \tilde{\partial}^\mu A^\nu \partial_\mu \wedge \partial_\nu, \qquad \iota_A \alpha = \tilde{\iota}_\alpha A = A^\mu \alpha_\mu. \tag{12}$$

Journal of Physics: Conference Series **1416** (2019) 012031 doi:10.1088/1742-6596/1416/1/012031

The Lie derivative $\mathcal{L}_A$ acting on the winding vector fields and the winding Lie derivative $\tilde{\mathcal{L}}_\alpha$ acting on the vector fields are given by analogous formulae of the Cartan's homotopy formula:

$$\mathcal{L}_A\alpha = (\mathsf{d}\iota_A + \iota_A\mathsf{d})\alpha, \qquad \tilde{\mathcal{L}}_\alpha A = (\tilde{\mathsf{d}}\tilde{\iota}_\alpha + \tilde{\iota}_\alpha\tilde{\mathsf{d}})A. \tag{13}$$

Then we can rewrite the C-bracket (6) into a non-$O(D,D)$ covariant form:

$$\begin{aligned}[\Xi_1,\Xi_2]_{\mathsf{C}} = [A,B]_L + \mathcal{L}_A\beta - \mathcal{L}_B\alpha - \frac{1}{2}\mathsf{d}(\iota_A\beta - \iota_B\alpha)\\ + [\alpha,\beta]_{\tilde{L}} + \tilde{\mathcal{L}}_\alpha B - \tilde{\mathcal{L}}_\beta A + \frac{1}{2}\tilde{\mathsf{d}}(\iota_A\beta - \iota_B\alpha).\end{aligned} \tag{14}$$

3. Para-Hermitian Geometry

In this section, we introduce a para-Hermitian manifold as the doubled spacetime of DFT [5, 6, 12, 13]. Vaisman [5] first mentioned that the doubled geometry of DFT is described as a para-Hermitian geometry.

3.1. Almost Para-Hermitian Manifold

An almost para-complex manifold is defined by $(\mathcal{M},K)$ where $\mathcal{M}$ is a differential manifold and K is an endomorphism $K : T\mathcal{M} \to T\mathcal{M}$ satisfying $K^2 = 1$. K is called the almost para-complex structure. The prefix "para" means a real analogue of the (almost) complex structure $J^2 = -1$. By using a para-complex structure K, the tangent bundle is decomposed into $T\mathcal{M} = L \oplus \tilde{L}$. L and $\tilde{L}$ are the eigenbundles associated with the eigenvalues $K = +1, -1$. This decomposition is performed via the projection operators:

$$P = \frac{1}{2}(1+K), \qquad \tilde{P} = \frac{1}{2}(1-K). \tag{15}$$

These can also be viewed as maps $P : T\mathcal{M} \to L$ and $\tilde{P} : T\mathcal{M} \to \tilde{L}$. The eigenbundles L, $\tilde{L}$ are distributions in $T\mathcal{M}$.

An almost para-Hermitian manifold is an almost para-complex manifold $\mathcal{M}$ equipped with a neutral metric $\eta : T\mathcal{M} \times T\mathcal{M} \to \mathbb{R}$. η satisfies the compatibility condition $\eta(K\cdot, K\cdot) = -\eta(\cdot,\cdot)$.

3.2. Integrability

The integrability condition of the para-complex structure is that a real analogue of the Nijenhuis tensor N_K vanishes. Here N_K is given by

$$N_K(X,Y) = \frac{1}{4}\Big\{[K(X),K(Y)] + [X,Y] - K\big([K(X),Y] + [X,K(Y)]\big)\Big\}, \tag{16}$$

where $X, Y \in T\mathcal{M}$. When $\mathcal{M}$ is an almost para-complex manifold, if the integrability condition is imposed on the para-complex structure, $\mathcal{M}$ becomes a para-complex manifold. Similarly, an almost para-Hermitian manifold becomes a para-Hermitian manifold.

On the other hand, the integrability of a distribution is represented by the Frobenius theorem. For any vector field A, B belonging to a distribution L, if the bracket $[A,B]$ also belongs to L, then the distribution L is called involutive. That is, L and $\tilde{L}$ being involutive means that the following tensors are zero:

$$N_P(X,Y) := \tilde{P}[P(X),P(Y)], \qquad N_{\tilde{P}}(X,Y) := P[\tilde{P}(X),\tilde{P}(Y)]. \tag{17}$$

The Frobenius theorem states that a distribution L is integrable if and only if L is involutive. Adding two tensors in (17) gives the Nijenhuis tensor of K: $N_K(X,Y) = N_P(X,Y) + N_{\tilde{P}}(X,Y)$. In contrast to the usual complex case, the integrability of the two distributions $L, \tilde{L}$ is independent each other in the para-complex case.

3.3. Relation between DFT and Generalized Geometry

An alternative representation of the Frobenius theorem states that a subbundle $E \subset T\mathcal{M}$ is integrable if and only if it is defined by a foliation of $\mathcal{M}$. Therefore L and $\tilde{L}$ have foliation structures when they are integrable:

$$L = T\mathcal{F}, \qquad \tilde{L} = T\tilde{\mathcal{F}}. \tag{18}$$

Here the foliation $\mathcal{F}$ is given by the union of leaves $\coprod_{[p]} M_{[p]}$. A leaf M_p is a subspace of $\mathcal{F}$ that pass through a point $p \in \mathcal{M}$. The index space in the union is the leaf space $\mathcal{M}/\mathcal{F}$. For $\mathcal{F}$, the local coordinates x^μ is given along a leaf M_p. The one for the transverse directions to leaves is $\tilde{x}_\mu$. Therefore $\tilde{x}_\mu$ is a constant on a leaf M_p in $\mathcal{F}$. The same discussion for $\tilde{L}$ is true. This discussion means that a leaf M_p is a physical spacetime. In the supergravity frame $\tilde{\partial}* = 0$, the field dependence is fixed on the leaf.

The metric η over $\mathcal{M}$ can be seen as a map $\eta : T\mathcal{M} = L \oplus \tilde{L} \to T^*\mathcal{M} = L^* \oplus \tilde{L}^*$. By using η, the following two isomorphisms $\phi^+ : \tilde{L} \to L^*$ and $\phi^- : L \to \tilde{L}^*$ are defined. They map vectors in $\tilde{L}$ (L) to forms in L^* ($\tilde{L}^*$). Given these maps, the following new isomorphisms are defined:

$$\Phi^+ : T\mathcal{M} \to L \oplus L^* \qquad \text{and} \qquad \Phi^- : T\mathcal{M} \to \tilde{L} \oplus \tilde{L}^*. \tag{19}$$

These maps are called the natural isomorphisms. In particular, Φ^+ is utilized to relate DFT and generalized geometry.

3.4. DFT as A Flat Para-Hermitian Manifold

In the following, we consider DFT in a torus background (we call this the toroidal DFT) without fluxes. The toroidal DFT is given by a flat para-Hermitian manifold $\mathcal{M}$. K and η are given by

$$K = \begin{pmatrix} -1 & 0 \\ 0 & +1 \end{pmatrix}, \qquad \eta = \begin{pmatrix} 0 & 1 \\ 1 & 0 \end{pmatrix}. \tag{20}$$

The integrability condition $N_K = 0$ is satisfied because N_P, $N_{\tilde{P}}$ are naturally zero. The tangent space $T\mathcal{M}$ is spanned by ∂_M. Vector fields on $T\mathcal{M}$ are decomposed by the projection operators $P, \tilde{P}$. Namely, we have

$$\Xi^M \partial_M = A^\mu(x, \tilde{x})\partial_\mu + \alpha_\mu(x, \tilde{x})\tilde{\partial}^\mu, \tag{21}$$

where $\Xi \in T\mathcal{M}$, $A \in L$, $\alpha \in \tilde{L}$. Here $X^M = (x^\mu, \tilde{x}_\mu)$ is the decomposition induced by $P, \tilde{P}$ of the local coordinates on the base space $\mathcal{M}$. L is spanned by ∂_μ while $\tilde{L}$ is spanned by $\tilde{\partial}^\mu$.

4. Doubled Structure of Vaisman Algebroid in DFT Setup

4.1. A Lie Algebroid in DFT

A Lie algebroid is a generalization of a Lie algebra defined over a manifold. Lie algebras are defined by using left invariant vector fields and determined by structure constants. On the other hand, Lie algebroids are defined by using general vector fields and determined by structure functions. We use the following theorem to define the Lie algebroid structures on the para-Hermitian manifold. The theorem by Vaintrob [14] states that a Lie algebroid over a vector bundle $V \to M$ and a Gerstenhaber algebra over multi-vectors $\wedge^\bullet V$ are equivalent. A Gerstenhaber algebra is defined by the multi-vector fields on a manifold through the Schouten-Nijenhuis bracket [15]. Therefore a Lie algebroid structure in DFT is defined by the exterior algebra of multi-vectors with the Schouten-Nijenhuis bracket on the para-Hermitian manifold.

We define a natural exterior algebra of multi-vectors on $T\mathcal{M}$, L and $\tilde{L}$. We introduce a set of the doubled multi-vectors $\hat{\mathcal{A}}^k(\mathcal{M}) = \Gamma(\wedge^k T\mathcal{M})$. Here Γ means the section of a vector

Journal of Physics: Conference Series **1416** (2019) 012031 doi:10.1088/1742-6596/1416/1/012031

bundle. If we define $\mathcal{A}^{r,s}(\mathcal{M})$ as the section of $(\wedge^r L) \wedge (\wedge^s \tilde{L})$, then we obtain the decomposition $\hat{\mathcal{A}}^k(\mathcal{M}) = \bigoplus_{k=r+s} \mathcal{A}^{r,s}(\mathcal{M})$. This decomposition is given by $\pi^{r,s} : \hat{\mathcal{A}}^{r+s}(\mathcal{M}) \to \mathcal{A}^{r,s}(\mathcal{M})$ induced by $P, \tilde{P}$. The exterior derivatives acting on L and $\tilde{L}$ are defined by

$$\tilde{\mathrm{d}} : \mathcal{A}^{r,s}(\mathcal{M}) \to \mathcal{A}^{r+1,s}(\mathcal{M}) \qquad (i.e. \wedge^r L \to \wedge^{r+1} L), \tag{22}$$

$$\mathrm{d} : \mathcal{A}^{r,s}(\mathcal{M}) \to \mathcal{A}^{r,s+1}(\mathcal{M}) \qquad (i.e. \wedge^s \tilde{L} \to \wedge^{s+1} \tilde{L}). \tag{23}$$

$\tilde{\mathrm{d}}$ and d have the following properties: $\tilde{\mathrm{d}}^2 = 0$, $\mathrm{d}^2 = 0$, $\mathrm{d}\tilde{\mathrm{d}} + \tilde{\mathrm{d}}\mathrm{d} = 0$. They are called the para-Dolbeault operators. We also define the interior products and the Lie derivatives (see [10] for details).

We give an explicit form of the Schouten-Nijenhuis bracket and the exterior derivatives. We introduce the "odd coordinates" $\zeta_\mu := \partial_\mu$, then an r-vector is

$$A = \frac{1}{r!} A^{\mu_1 \cdots \mu_r} \partial_{\mu_1} \wedge \cdots \wedge \partial_{\mu_r} = \frac{1}{r!} A^{\mu_1 \cdots \mu_r} \zeta_{\mu_1} \cdots \zeta_{\mu_r}. \tag{24}$$

Here ζ_μ can be treated as a Grassmann number, so the derivative $\partial/\partial\zeta_\mu$ is defined by the right derivative. By using $\partial/\partial\zeta_\mu$, the Schouten-Nijenhuis bracket is explicitly given by

$$[A, B]_{\mathrm{s}} = \left(\frac{\partial}{\partial \zeta_\mu} A \right) \partial_\mu B - (-1)^{(p-1)(q-1)} \left(\frac{\partial}{\partial \zeta_\mu} B \right) \partial_\mu A, \tag{25}$$

where $A \in \wedge^p L$, $B \in \wedge^q L$. The same discussion holds for $[\cdot, \cdot]_{\mathrm{s}}^*$ on $\tilde{L}$ where $\zeta_\mu = \partial_\mu$ is replaced by $\zeta^{*\mu} = \tilde{\partial}^\mu$. By using the local coordinate, we find that the action of $\tilde{\mathrm{d}}$ on an r-vector $A \in \wedge^r L$ is explicitly given by

$$\tilde{\mathrm{d}} A = \frac{1}{r!} \tilde{\partial}^\nu A^{\mu_1 \cdots \mu_r} \partial_\nu \wedge \partial_{\mu_1} \wedge \cdots \wedge \partial_{\mu_r}. \tag{26}$$

This definition is compatible with the dual Schouten-Nijenhuis bracket $[\cdot, \cdot]_{\mathrm{s}}^*$ (see [10] for details). d on $\tilde{L}$ has the same definition with ∂_μ and $\tilde{\partial}^\mu$ interchanged. $\tilde{\mathrm{d}}$ and d defined here are exactly equivalent to the exterior derivative $\tilde{\mathrm{d}}$ and d defined naively (12).

The exterior algebras of multi-vectors are consistently defined. Therefore, we have defined the Lie algebroid $(\wedge^\bullet L, [\cdot, \cdot]_{\mathrm{s}}, \mathrm{d})$ and its dual Lie algebroid $(\wedge^\bullet \tilde{L}, [\cdot, \cdot]_{\mathrm{s}}^*, \tilde{\mathrm{d}})$ in DFT. In the DFT setup, L and $\tilde{L}$ are integrable, so they are tangent Lie algebroids. The anchor maps ρ_L and $\rho_{\tilde{L}}$ are defined by the identity map.

4.2. The Double of Lie Algebroid Pair in DFT

A Lie bialgebroid is defined by the Lie algebroid $(L, [\cdot, \cdot]_L, \rho_L, \mathrm{d})$ and its dual Lie algebroid $(L^*, [\cdot, \cdot]_{L^*}, \rho_{L^*}, \mathrm{d}_*)$ with a compatibility condition between them called the derivation condition:

$$\mathrm{d}_*[A, B]_{\mathrm{s}} = [\mathrm{d}_* A, B]_{\mathrm{s}} + [A, \mathrm{d}_* B]_{\mathrm{s}} \qquad (A, B \in L), \tag{27}$$

where $\mathrm{d} : \wedge^k L^* \to \wedge^{k+1} L^*$ and $\mathrm{d}_* : \wedge^k L \to \wedge^{k+1} L$ are the exterior derivatives. It is known that the Drinfel'd double of a Lie bialgebroid is the Courant algebroid [8].

A double of a pair of Lie algebroids $(L, \tilde{L})$ is given by a direct sum $L \oplus \tilde{L}$ of a vector space with a bilinear form and a skew-symmetric bracket. Now, the term "double" means an analogous notion of the Drinfel'd double. A double of a pair of Lie algebroids does not satisfy some axioms of the Courant algebroid because the derivation condition of Lie bialgebroids is not imposed [10].

In the following, we discuss a double of a pair of Lie algebroids $(L, \tilde{L})$ in DFT setup. We examine the derivation condition in DFT by explicit calculation. We obtain

$$\tilde{\mathrm{d}}[A, B]_{\mathrm{s}} = [\tilde{\mathrm{d}} A, B]_{\mathrm{s}} + [A, \tilde{\mathrm{d}} B]_{\mathrm{s}} + (\partial_\rho A^\mu \tilde{\partial}^\rho B^\nu + \tilde{\partial}^\rho A^\mu \partial_\rho B^\nu) \partial_\mu \wedge \partial_\nu. \tag{28}$$

The last contribution is representing the violation of the derivation condition (27). Therefore, it is shown in DFT that L and $\tilde{L} \simeq L^*$ pairs generally do not form a Lie bialgebroid. We define the two structures required for a double of $(L, \tilde{L})$. The bilinear form $(\cdot, \cdot)$ is given by

$$(A+\alpha, B+\beta) = \frac{1}{2}\Big\{\langle\!\langle \alpha, B\rangle\!\rangle + \langle\!\langle \beta, A\rangle\!\rangle\Big\}, \tag{29}$$

where $\langle\!\langle \cdot, \cdot\rangle\!\rangle$ is the symmetric pairing and is defined by $\langle\!\langle \alpha, A\rangle\!\rangle = A^\mu \alpha_\mu$. The symmetric pairing is regarded as the inner product when $\tilde{L}$ and L^* are identified by Φ^+. The skew-symmetric bracket called the Vaisman bracket is given in the same form as the C-bracket (14). The anchor map is defined as $\rho_{\mathrm{V}} = \rho_L + \rho_{\tilde{L}}$ and the differential operator is defined as $\mathcal{D} = \mathrm{d} + \tilde{\mathrm{d}}$. Then, the quadruple $(L \oplus \tilde{L}, [\cdot, \cdot]_{\mathrm{C}}, \rho_{\mathrm{V}}, (\cdot, \cdot))$ defines a Vaisman algebroid.

We note that the last contribution in (28) is rewritten as

$$\partial_\rho A^\mu \tilde{\partial}^\rho B^\nu + \tilde{\partial}^\rho A^\mu \partial_\rho B^\nu = \eta^{MN} \partial_M A^\mu \partial_N B^\nu. \tag{30}$$

This becomes zero by imposing the strong constraint. Therefore, when the strong constraint is imposed, the derivation condition is satisfied and $(L, \tilde{L})$ becomes a Lie bialgebroid. Then, the double $L \oplus \tilde{L}$ defines a Courant algebroid. This completely agrees with the analysis in [7]. We ascertained that an algebraic origin of the strong constraint in DFT is the derivation condition of Lie bialgebroids.

5. Conclusion and Discussion

In this contribution, we discussed a doubled structure of the Vaisman algebroid in the para-Hermitian geometry. We defined Lie algebroid structures on the two eigenbundles of the para-Hermitian manifolds and examined their doubles. The derivation condition for a pair of Lie algebroids to become a Lie bialgebroid is violated in the form of the strong constraint. Hence, we found an algebraic origin of the strong constraint in DFT.

Based on the doubled structure of the Vaisman bracket in the para-Hermitian manifold, we discuss the physical aspects of the gauge symmetry generated by the C-bracket. Winding vectors in $\tilde{L}$ are identified with 1-forms in L^* by Φ^+, and the Lie bracket $[\cdot, \cdot]_{\tilde{L}}$ in $\tilde{L}$ represents the B-field gauge symmetry algebra. However, $[\cdot, \cdot]_{\tilde{L}}$ in the Vaisman bracket is generally non-zero. Therefore, in DFT, the B-field gauge symmetry should be effectively non-Abelian.

The strong constraint is necessary for the DFT gauge algebra to close. However, in more general setups (*e.g.* gauged DFT [16]), there is no need to impose the strong constraint. The Vaisman algebroid would play even more important roles in applications of DFT. It is interesting in the case of DFT on group manifolds used in the study of the Poisson-Lie T-duality [17]. The Drinfel'd double is also an important notion to understand the Poisson-Lie T-duality [18–20].

In this discussion, we used a para-Hermitian geometry, but in order to introduce the generalized metric in DFT, it is necessary to consider in the framework of Born geometry [21]. A conventional generalized metric can be obtained by performing the B-transformation e^B to a Riemannian metric in Born geometry. This transformation e^B may lead to a twisted Vaisman bracket.

The Vaisman algebroid is an algebra of the infinitesimal gauge transformation in DFT. The finite gauge transformation in DFT is governed by an "integration" of the Vaisman algebroid. We are working on the problem of finding the groupoid-like structure corresponding to the Vaisman algebroid [22]. We believe that the geometric structure of the DFT gauge symmetry leads to an understanding of the string winding effects to spacetimes [23–26].

Acknowledgments

We are grateful to Shin Sasaki and Haruka Mori for the collaboration of the original work and for helpful discussion. We would like to thank the organizers of the XXVIth International

Conference on Integrable Systems and Quantum Symmetries (ISQS-26) for arranged our talk in Prague and for their kind hospitality. The work of K.S. is supported by the Sasakawa Scientific Research Grant (No. 2019-2027) from the Japan Science Society.

References

[1] Giveon A, Porrati M and Rabinovici E 1994 *Phys. Rept.* **244** 77 (*Preprint* hep-th/9401139)
[2] Hitchin N 2003 *Quart. J. Math.* **54** 281 (*Preprint* math/0209099)
[3] Hull C and Zwiebach B 2009 *J. High Energy Phys.* JHEP09(2009)099 (*Preprint* 0904.4664)
[4] Hull C and Zwiebach B 2009 *J. High Energy Phys.* JHEP09(2009)090 (*Preprint* 0908.1792)
[5] Vaisman I 2012 *J. Math. Phys.* **53** 033509 (*Preprint* 1203.0836)
[6] Svoboda D 2018 *J. Math. Phys.* **59** 122302 (*Preprint* 1802.08180)
[7] Chatzistavrakidis A, Jonke L, Khoo F S and Szabo R J 2018 *J. High Energy Phys.* JHEP07(2018)015 (*Preprint* 1802.07003)
[8] Liu Z J, Weinstein A and Xu P 1997 *J. Differential Geom.* **45** 547 (*Preprint* dg-ga/9508013)
[9] Vaisman I 2013 *J. Math. Phys.* **54** 123507 (*Preprint* 1209.0152)
[10] Mori H, Sasaki S and Shiozawa K 2019 *Preprint* 1901.04777
[11] Hohm O, Hull C and Zwiebach B 2010 *J. High Energy Phys.* JHEP08(2010)008 (*Preprint* 1006.4823)
[12] Freidel L, Rudolph F J and Svoboda D 2017 *J. High Energy Phys.* JHEP11(2017)175 (*Preprint* 1706.07089)
[13] Marotta V E and Szabo R J 2019 *Fortsch. Phys.* **67** 1800093 (*Preprint* 1810.03953)
[14] Vaintrob A 1997 *Russian Math. Surveys* **52** 428
[15] Tulczyjew W M 1974 *Bull. Acad. Pol. Sci., Sér. Sci. Math. Astr. Phys.* **22** 937
[16] Grana M and Marques D 2012 *J. High Energy Phys.* JHEP04(2012)020 (*Preprint* 1201.2924)
[17] Hassler F 2017 *Preprint* 1707.08624
[18] Lüst D and Osten D 2018 *J. High Energy Phys.* JHEP05(2018)165 (*Preprint* 1803.03971)
[19] Marotta V E, Pezzella F and Vitale P 2018 *J. High Energy Phys.* JHEP08(2018)185 (*Preprint* 1804.00744)
[20] Marotta V E, Pezzella F and Vitale P 2019 *Preprint* 1903.01243
[21] Freidel L, Rudolph F J and Svoboda D 2018 *Preprint* 1806.05992
[22] Mori H, Sasaki S and Shiozawa K, work in progress
[23] Kimura T and Sasaki S 2013 *Nucl. Phys. B* **876** 493 (*Preprint* 1304.4061), *J. High Energy Phys.* JHEP08(2013)126 (*Preprint* 1305.4439)
[24] Berman D S and Rudolph F J 2015 *J. High Energy Phys.* JHEP05(2015)015 (*Preprint* 1409.6314)
[25] Bakhmatov I, Kleinschmidt A and Musaev E T 2016 *J. High Energy Phys.* JHEP10(2016)076 (*Preprint* 1607.05450)
[26] Kimura T, Sasaki S and Shiozawa K 2018 *J. High Energy Phys.* JHEP07(2018)001 (*Preprint* 1803.11087), *J. High Energy Phys.* JHEP12(2018)095 (*Preprint* 1810.02169)

Journal of Physics: Conference Series **1416** (2019) 012032 doi:10.1088/1742-6596/1416/1/012032

Hidden Supersymmetries of Deformed Supersymmetric Mechanics

Stepan Sidorov

Bogoliubov Laboratory of Theoretical Physics, JINR, 141980 Dubna, Moscow Region, Russia

E-mail: sidorovstepan88@gmail.com

Abstract. We consider quantum models corresponding to superymmetrizations of the two-dimensional harmonic oscillator based on worldline $d=1$ realizations of the supergroup $\mathrm{SU}(\mathcal{N}/2\,|1)$, where the number of supersymmetries $\mathcal{N}$ is arbitrary even number. Constructed models possess the hidden supersymmetry $\mathrm{SU}(\mathcal{N}/2\,|2)$. Degeneracies of energy levels are spanned by representations of the hidden supersymmetry group.

1. Introduction

In [1, 2, 3] there were studied a new class of $\mathcal{N}=4$, $d=1$ supersymmetric quantum mechanics (SQM) models known as "Weak Supersymmetry" models, where the appropriate superalgebra involves, besides Hamiltonian, extra bosonic generators not commuting with supercharges. In [4, 5] we reproduced these models, associated with the multiplets[1] $(\mathbf{1},\mathbf{4},\mathbf{3})$ and $(\mathbf{2},\mathbf{4},\mathbf{2})$, from a superfield approach based on the $d=1$ worldline supersymmetry $\mathrm{SU}(2|1)$. We identified the weak supersymmetry superalgebra with the superalgebra $su(2|1)$, where we considered it as a deformation of the standard $\mathcal{N}=4$, $d=1$ Poincaré superalgebra by a mass parameter m.

The simplest model of weak supersymmetry considered in [1] corresponds to $\mathcal{N}=4$ supersymmetric extension of the one-dimensional harmonic oscillator. One of the distinct features of this model is the non-trivial degeneracy of energy levels (see Figure 1), that was explained in [4] in the framework of the $\mathrm{SU}(2|1)$ representation theory [6]. The first excited level is given by unequal numbers of fermionic and bosonic states forming the fundamental representation of $\mathrm{SU}(2|1)$. All other higher excited levels show up the standard 4-fold degeneracy containing equal numbers of fermionic and bosonic states and corresponding to the simplest typical $\mathrm{SU}(2|1)$ representation.

In [4], we also considered the chiral multiplet $(\mathbf{2},\mathbf{4},\mathbf{2})$ and constructed its general superfield action. One of the simplest models corresponds to $\mathrm{SU}(2|1)$ supersymmetric extension of the two-dimensional harmonic oscillator and possesses the hidden supersymmetry $\mathrm{SU}(2|2)$. Degeneracies at each energy level, except the ground state, are spanned by $\mathrm{SU}(2|2)$ representations characterized by equal numbers of fermionic and bosonic states [7].

In this paper, we consider $\mathrm{SU}(\mathcal{N}/2\,|1)$ supersymmetric extensions of the two-dimensional harmonic oscillator, where the number of supersymmetries $\mathcal{N}$ is even and not restricted from

[1] The numbers $(\mathbf{k},\mathcal{N},\mathcal{N}-\mathbf{k})$ display the field content of $\mathcal{N}$, $d=1$ multiplets. For example, the multiplet $(\mathbf{1},\mathbf{4},\mathbf{3})$ contains **1** physical bosonic field, **4** physical fermionic fields and **3** auxiliary bosonic fields.

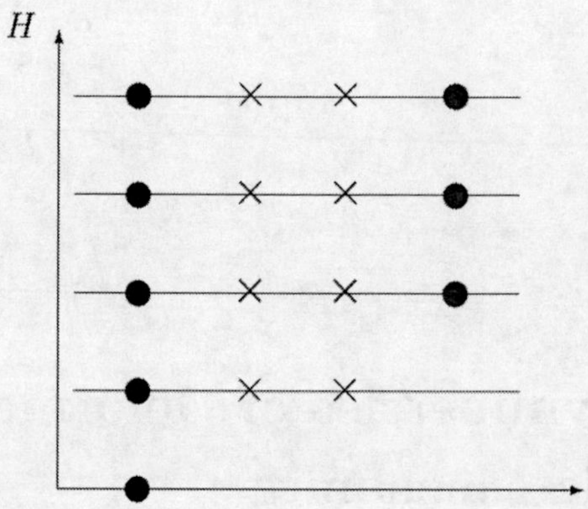

Figure 1. The degeneracy of energy levels of the one-dimensional SU(2|1) supersymmetric harmonic oscillator drawn in [1]. Circles and crosses indicate bosonic and fermionic states, respectively.

above[2]. We show that the presence of hidden supersymmetry is a common feature of such models.

2. Two-dimensional harmonic oscillator

The quantum Hamiltonian of the two-dimensional oscillator reads

$$H_{\text{bos.}} = \frac{1}{2}\sum_{a=1,2}\left(-\,\partial_a^2 + m^2 x_a^2\right) \quad \Rightarrow \quad H_{\text{bos.}} = -\,\partial_z\partial_{\bar z} + m^2 z\bar z\,, \tag{1}$$

where

$$z = \frac{1}{\sqrt{2}}\left(x_1 + i x_2\right), \qquad \bar z = \frac{1}{\sqrt{2}}\left(x_1 - i x_2\right) \tag{2}$$

The Hamiltonian can be rewritten as

$$H_{\text{bos.}} = a^+ a^- + b^+ b^- + m, \tag{3}$$

where creation and annihilation operators are given by

$$\begin{aligned} a^+ &= \frac{i}{\sqrt{2}}\left(\partial_{\bar z} - m z\right), \quad a^- = \frac{i}{\sqrt{2}}\left(\partial_z + m\bar z\right), \qquad \left[a^-, a^+\right] = m\,,\\ b^+ &= \frac{i}{\sqrt{2}}\left(\partial_z - m\bar z\right), \quad b^- = \frac{i}{\sqrt{2}}\left(\partial_{\bar z} + m z\right), \qquad \left[b^-, b^+\right] = m\,. \end{aligned} \tag{4}$$

Wave functions are written via the creation operator a^+ and b^+ acting on the ground state $|0\rangle = \exp\left(-m z\bar z\right)$ as

$$\begin{aligned} &|n_1, n_2\rangle = \left(a^+\right)^{n_1}\left(b^+\right)^{n_2}|0\rangle\,,\\ &|n_1+1, n_2\rangle = a^+\,|n_1, n_2\rangle = \left(a^+\right)^{n_1+1}\left(b^+\right)^{n_2}|0\rangle\,,\\ &|n_1, n_2+1\rangle = b^+\,|n_1, n_2\rangle = \left(a^+\right)^{n_1}\left(b^+\right)^{n_2+1}|0\rangle\,. \end{aligned} \tag{5}$$

[2] SU(4|1) superfield approach for the chiral multiplet $(\mathbf{2}, \mathbf{8}, \mathbf{6})$ was studied in [8].

The operators a^- and b^- annihilate the ground state and reduce the number of creation operators:

$$\begin{aligned} a^- \left|0\right\rangle &= 0\,, \qquad & a^- \left|n_1, n_2\right\rangle &= n_1\, m\, \left|n_1 - 1, n_2\right\rangle , \\ b^- \left|0\right\rangle &= 0\,, \qquad & b^- \left|n_1, n_2\right\rangle &= n_2\, m\, \left|n_1, n_2 - 1\right\rangle . \end{aligned} \tag{6}$$

The spectrum of the Hamiltonian $H_{\text{bos.}}$ is then

$$H_{\text{bos.}} \left|n_1, n_2\right\rangle = (n_1 + n_2 + 1)\, m\, \left|n_1, n_2\right\rangle . \tag{7}$$

2.1. Hidden symmetry

One can introduce SU(2) symmetry generators commuting with the Hamiltonian:

$$J_+ = \frac{1}{m}\, b^+ a^- , \qquad J_- = \frac{1}{m}\, a^+ b^- , \qquad J_3 = \frac{1}{2m}\left(b^+ b^- - a^+ a^-\right) . \tag{8}$$

They commute as

$$[J_+, J_-] = 2 J_3\,, \qquad [J_3, J_\pm] = \pm\, J_\pm\,. \tag{9}$$

It is none other than the well known hidden SU(2) symmetry of the two-dimensional harmonic oscillator. In general, N-dimensional harmonic oscillator possesses SU(N) symmetry (see e.g. [9]).

Since the generators (8) commute with the Hamiltonian, we have a degeneracy of energy levels corresponding to SU(2) representations. As shown in Figure 2, the energy level n is given by $n + 1$ states with the wave function defined as a sum of (5):

$$\left|n\right\rangle = \sum_{n_1=0}^{n} C_{n_1} \left|n_1, n - n_1\right\rangle , \qquad H \left|n\right\rangle = (n+1)\, m\, \left|n\right\rangle , \qquad C_{n_1} = \text{const.} \tag{10}$$

Indeed, the action of $2J_3$ on these states takes the integer eigenvalues from $-n$ to n.

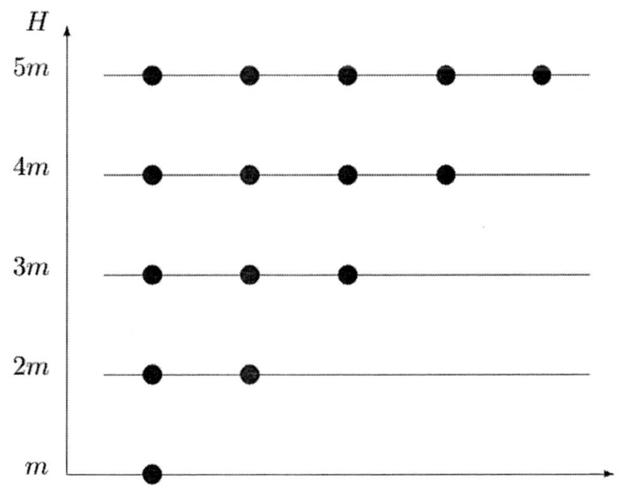

Figure 2. The degeneracy of energy levels of the two-dimensional oscillator. Each energy level is given by $n + 1$ states.

Journal of Physics: Conference Series **1416** (2019) 012032 doi:10.1088/1742-6596/1416/1/012032

3. Supersymmetric two-dimensional harmonic oscillator

Supersymmetric Weyl-ordered Hamiltonian reads

$$H = a^+a^- + b^+b^- + m\,\psi^K\bar{\psi}_K + m\left(1 - \frac{\mathcal{N}}{4}\right), \tag{11}$$

where the fermionic fields ψ^I and $\bar{\psi}_J$ anticommute as

$$\{\psi^I, \bar{\psi}_J\} = \delta^I_J\,. \tag{12}$$

The capital indices I, J, K, L refer to the $\mathrm{SU}(\mathcal{N}/2)$ fundamental and anti-fundamental representations. The supersymmetric Hamiltonian (11) commute with supercharges defined as

$$Q^I = \sqrt{2}\,\psi^I\,a^-, \qquad \bar{Q}_J = \sqrt{2}\,\bar{\psi}_J\,a^+. \tag{13}$$

They close on the centrally-extended superalgebra $\widehat{su}(\mathcal{N}/2\,|1)$ given by the following non-vanishing (anti)commutators:

$$\begin{aligned}
&\{Q^I, \bar{Q}_J\} = 2m\,L^I_J - 2m\,\delta^I_J R + \delta^I_J H, &&[L^I_J, L^K_L] = \delta^K_J L^I_L - \delta^I_L L^K_J\,,\\
&[L^I_J, Q^K] = \delta^K_J Q^I - \frac{2}{\mathcal{N}}\,\delta^I_J Q^K, &&[L^I_J, \bar{Q}_L] = \frac{2}{\mathcal{N}}\,\delta^I_J \bar{Q}_L - \delta^I_L \bar{Q}_J\,,\\
&[R, Q^I] = \left(1 - \frac{2}{\mathcal{N}}\right) Q^I, &&[R, \bar{Q}_J] = -\left(1 - \frac{2}{\mathcal{N}}\right) \bar{Q}_J\,.
\end{aligned} \tag{14}$$

The $\mathrm{SU}(\mathcal{N}/2) \times \mathrm{U}(1)$ subgroup generators L^I_J and R are written as

$$\begin{aligned}
L^I_J &= \psi^I\bar{\psi}_J - \frac{\delta^I_J}{4}\,\psi^K\bar{\psi}_K\,,\\
R &= \frac{1}{2m}\left(b^+b^- - a^+a^-\right) + \left(\frac{1}{2} - \frac{2}{\mathcal{N}}\right)\psi^K\bar{\psi}_K - \frac{\mathcal{N}}{4}\left(\frac{1}{2} - \frac{2}{\mathcal{N}}\right).
\end{aligned} \tag{15}$$

The mass-dimensional generator H, associated with Hamiltonian, is a central charge generator. In the limit $m = 0$, the $\mathrm{SU}(\mathcal{N}/2) \times \mathrm{U}(1)$ generators become automorphism generators of the standard $\mathcal{N}$, $d{=}1$ Poincaré superalgebra.

3.1. Hidden supersymmetry

One can check the hidden SU(2) symmetry generators (8) commutes with the supersymmetric Hamiltonian. However their commutators with (13) give new supercharges

$$[J_-, Q^I] = -\,S^I, \qquad [J_+, \bar{Q}_J] = \bar{S}_J\,. \tag{16}$$

The new supercharges are written as

$$S^I = \sqrt{2}\,\psi^I\,b^-, \qquad \bar{S}_J = \sqrt{2}\,\bar{\psi}_J\,b^+. \tag{17}$$

Thus, the superalgebra (20) is extended by the new supercharges and SU(2) generators.

Splitting the U(1) generator R as

$$R = J_3 + \left(1 - \frac{4}{\mathcal{N}}\right)F \quad\Rightarrow\quad F = \frac{1}{2}\,\psi^K\bar{\psi}_K - \frac{\mathcal{N}}{8}, \qquad J_3 = \frac{1}{2m}\left(b^+b^- - a^+a^-\right), \tag{18}$$

we obtain the centrally-extended superalgebra $\widehat{su}(\mathcal{N}/2\,|2)$ given by

$$
\begin{aligned}
&\{Q^I, \bar{Q}_J\} = 2m\, L^I_J - 2m\,\delta^I_J J_3 + 2m\left(\frac{4}{\mathcal{N}} - 1\right)\delta^I_J F + \delta^I_J H, \qquad \{Q^I, \bar{S}_J\} = 2m\,\delta^I_J J_+\,,\\
&\{S^I, \bar{S}_J\} = 2m\, L^I_J + 2m\,\delta^I_J J_3 + 2m\left(\frac{4}{\mathcal{N}} - 1\right)\delta^I_J F + \delta^I_J H, \qquad \{S^I, \bar{Q}_J\} = 2m\,\delta^I_J J_-\,,\\
&\left[L^i_j, \bar{Q}_K\right] = \frac{2}{\mathcal{N}}\,\delta^I_J \bar{Q}_K - \delta^I_K \bar{Q}_J\,, \qquad \left[L^I_J, Q^K\right] = \delta^K_J Q^I - \frac{2}{\mathcal{N}}\,\delta^I_J Q^K,\\
&\left[L^I_J \bar{S}_K\right] = \frac{2}{\mathcal{N}}\,\delta^I_J \bar{S}_K - \delta^I_K \bar{S}_J\,, \qquad \left[L^I_J, S^k\right] = \delta^K_J S^I - \frac{2}{\mathcal{N}}\,\delta^I_J S^K,\\
&\left[F, Q^I\right] = \frac{1}{2}\,Q^I, \quad \left[F, \bar{Q}_J\right] = -\,\frac{1}{2}\,\bar{Q}_J\,, \quad \left[F, S^I\right] = \frac{1}{2}\,S^I, \quad \left[F, \bar{S}_J\right] = -\,\frac{1}{2}\,\bar{S}_J\,,\\
&\left[J_3, Q^I\right] = \frac{1}{2}\,Q^I, \quad \left[J_3, \bar{Q}_J\right] = -\,\frac{1}{2}\,\bar{Q}_J\,, \quad \left[J_+, \bar{Q}_J\right] = \bar{S}_J\,, \quad \left[J_+, S^I\right] = -\,Q^I,\\
&\left[J_3, S^I\right] = -\,\frac{1}{2}\,S^I, \quad \left[J_3, \bar{S}_J\right] = \frac{1}{2}\,\bar{S}_J\,, \quad \left[J_-, \bar{S}_J\right] = \bar{Q}_J\,, \quad \left[J_-, Q^I\right] = -\,S^I,\\
&\left[L^I_J, L^K_L\right] = \delta^K_J L^I_L - \delta^I_L L^K_J\,, \qquad \left[J_+, J_-\right] = 2J_3\,, \qquad \left[J_3, J_\pm\right] = \pm\,J_\pm\,.
\end{aligned} \tag{19}
$$

3.2. N=4 supersymmetric extension

As instructive example, let us consider in details $\mathcal{N}=4$ supersymmetric extension. Our studies of SU(2|1) supersymmetric mechanics [4] were based on the deformation of the standard $\mathcal{N}=4$, $d=1$ Poincaré superalgebra to the centrally-extended superalgebra $\widehat{su}(2|1)$ given by[3]

$$
\begin{aligned}
&\{Q^i, \bar{Q}_j\} = 2m\, I^i_j - 2m\,\delta^i_j R + \delta^i_j H, \qquad \left[I^i_j, I^k_l\right] = \delta^k_j I^i_l - \delta^i_l I^k_j\,,\\
&\left[I^i_j, \bar{Q}_l\right] = \frac{1}{2}\,\delta^i_j \bar{Q}_l - \delta^i_l \bar{Q}_j\,, \qquad \left[I^i_j, Q^k\right] = \delta^k_j Q^i - \frac{1}{2}\,\delta^i_j Q^k,\\
&\left[R, Q^k\right] = \frac{1}{2}\,Q^k, \qquad \left[R, \bar{Q}_l\right] = -\,\frac{1}{2}\,\bar{Q}_l\,.
\end{aligned} \tag{20}
$$

The indices i, j $(i = 1, 2)$ are SU(2) indices. The generators I^i_j and R correspond to the $SU(2) \times U(1)$ subgroup.

The quantum SU(2|1) generators are written as

$$
\begin{aligned}
&Q^i = \sqrt{2}\,\psi^i a^-, \qquad \bar{Q}_j = \sqrt{2}\,\bar{\psi}_j a^+,\\
&H = a^+a^- + b^+b^- + m\,\psi^k\bar{\psi}_k\,,\\
&R = \frac{1}{2m}\left(b^+b^- - a^+a^-\right) \quad \Rightarrow \quad R \equiv J_3\,,\\
&I^i_j = \psi^i\bar{\psi}_j - \frac{\delta^i_j}{2}\,\psi^k\bar{\psi}_k\,.
\end{aligned} \tag{21}
$$

It is straightforward to check that they satisfy the superalgebra relation (20).

As was shown in [4] the model possesses hidden supersymmetry SU(2|2). The corresponding

[3] The superalgebra relations here differ from those in [4] such that $H \to 2H$.

centrally-extended superalgebra $\widehat{su}(2|2)$ is given by

$$\begin{aligned}
&\{Q^i,\bar{Q}_j\} = \delta^i_j H + 2m\left(I^i_j - \delta^i_j J_3\right), \qquad \{Q^i,\bar{S}_j\} = 2m\,\delta^i_j J_+\,,\\
&\{S^i,\bar{S}_j\} = \delta^i_j H + 2m\left(I^i_j + \delta^i_j J_3\right), \qquad \{S^i,\bar{Q}_j\} = 2m\,\delta^i_j J_-\,,\\
&\left[I^i_j,\bar{Q}_l\right] = \frac{1}{2}\,\delta^i_j\bar{Q}_l - \delta^i_l\bar{Q}_j\,, \qquad \left[I^i_j,Q^k\right] = \delta^k_j Q^i - \frac{1}{2}\,\delta^i_j Q^k,\\
&\left[I^i_j,\bar{S}_l\right] = \frac{1}{2}\,\delta^i_j\bar{S}_l - \delta^i_l\bar{S}_j\,, \qquad \left[I^i_j,S^k\right] = \delta^k_j S^i - \frac{1}{2}\,\delta^i_j S^k,\\
&\left[J_3,Q^i\right] = \frac{1}{2}\,Q^i, \quad \left[J_3,\bar{Q}_j\right] = -\,\frac{1}{2}\,\bar{Q}_j\,, \quad \left[J_+,\bar{Q}_j\right] = \bar{S}_j\,, \quad \left[J_+,S^i\right] = -\,Q^i,\\
&\left[J_3,S^i\right] = -\,\frac{1}{2}\,S^i, \quad \left[J_3,\bar{S}_j\right] = \frac{1}{2}\,\bar{S}_j\,, \quad \left[J_-,\bar{S}_j\right] = \bar{Q}_j\,, \quad \left[J_-,Q^i\right] = -\,S^i,\\
&\left[I^i_j,I^k_l\right] = \delta^k_j I^i_l - \delta^i_l I^k_j\,, \qquad \left[J_+,J_-\right] = 2J_3\,, \qquad \left[J_3,J_\pm\right] = \pm\,J_\pm\,.
\end{aligned} \tag{22}$$

3.2.1. Dual superfield description. Corresponding Lagrangian of the two-dimensional SU(2|1) supersymmetric harmonic oscillator reads

$$\mathcal{L}_{\,\text{on-shell}} = \dot{z}\dot{\bar{z}} + \frac{i}{2}\left(\psi^i\dot{\bar{\psi}}_i - \dot{\psi}^i\bar{\psi}_i\right) - m^2 z\bar{z} - m\,\psi^k\bar{\psi}_k\,. \tag{23}$$

This on-shell Lagrangian can be constructed from superfield approach in two ways, which is exceptional for $\mathcal{N}=4$ supersymmetric extension.

First way is given in the framework of SU(2|1) superfield approach elaborated in [4]. Simple superfield Lagrangian for the chiral multiplet $(\mathbf{2},\mathbf{4},\mathbf{2})$ reads

$$\mathcal{L}_{\,\text{deformed}} = \int d^2\vartheta\, d^2\bar{\vartheta}\left(1 + 2m\,\bar{\vartheta}^k\vartheta_k\right)\Psi\,\bar{\Psi}. \tag{24}$$

The second superfield construction is based on the chiral multiplet of non-deformed (standard) SQM, where the parameter m introduced via superpotential term:

$$\mathcal{L}_{\,\text{non-deformed}} = \int d^2\theta\, d^2\bar{\theta}\,\Phi\,\bar{\Phi} + \frac{m}{2}\left[\int d^2\theta\,\Phi^2 + \int d^2\bar{\theta}\,\bar{\Phi}^2\right]. \tag{25}$$

After elimination of auxiliary fields in both actions, we set equivalence of these Lagrangians. The superalgebra (20) and the standard $\mathcal{N}=4$, $d=1$ superalgebra are realized on the same on-shell set $(z,\bar{z},\psi^i,\bar{\psi}_i)$. The superalgebra (22) is recovered as the closure of these two superalgebras.

Define the new supercharges

$$\Pi^i = \frac{1}{\sqrt{2}}\left(Q^i + \bar{S}^i\right), \qquad \bar{\Pi}_j = \frac{1}{\sqrt{2}}\left(\bar{Q}_j - S_j\right). \tag{26}$$

One can check that they form the standard $\mathcal{N}=4$ superalgebra,

$$\{\Pi^i,\bar{\Pi}_j\} = \delta^i_j H. \tag{27}$$

Thus, it is a subalgebra of (22), while the generators (8) become external automorphism generators:

$$\begin{aligned}
&\left[J_3,\bar{\Pi}_j\right] = -\,\frac{1}{2}\,\bar{\Pi}_j\,, \qquad \left[J_3,\Pi^i\right] = \frac{1}{2}\,\Pi^i,\\
&\left[J_+,\bar{\Pi}_j\right] = \Pi_j\,, \qquad \left[J_-,\Pi^i\right] = \bar{\Pi}^i.
\end{aligned} \tag{28}$$

Journal of Physics: Conference Series **1416** (2019) 012032 doi:10.1088/1742-6596/1416/1/012032

3.2.2. Wave functions. We construct supersymmetric wave functions in terms of (10) satisfying the same spectrum

$$H\,|n\rangle = n\,m\,|n\rangle\,. \tag{29}$$

We impose the standard physical condition on the bosonic states $|n\rangle$ as

$$\bar{\psi}_j\,|n\rangle = 0\,. \tag{30}$$

Then, the fermionic states $\psi_k\,|n\rangle$ have shifted spectrum $(n+1)\,m$:

$$H\,\psi_k\,|n\rangle = (n+1)\,m\,\psi_k\,|n\rangle\,. \tag{31}$$

Additional bosonic states defined as $\psi_k\psi^k\,|n\rangle$ also have shifted spectrum $(n+2)\,m$:

$$H\,\psi_k\psi^k\,|n\rangle = (n+2)\,m\,\psi_k\psi^k\,|n\rangle\,. \tag{32}$$

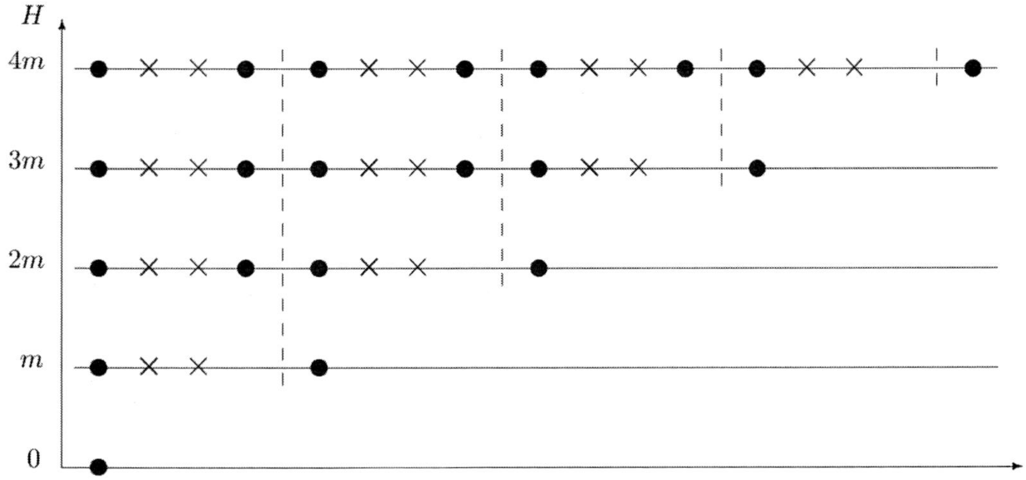

Figure 3. The degeneracy of energy levels of the two-dimensional SU(2|1) supersymmetric oscillator. There are $4n$ states at the excited energy level n. This picture can be interpreted as a multiplication of the pictures drawn in Figures 1 and 2. One can easily see that each tower divided by dashed line is a tower of the same type drawn in Figure 1.

As shown in Figure 3, each excited energy level is given by equal numbers of fermionic and bosonic states. States on the excited level n form a short SU(2|2) representation of the dimension $4n$ [7]. Note that the first excited level corresponds to the fundamental representation of SU(2|2).

4. Conclusions

We considered superextensions of the two-dimensional harmonic oscillator with the relevant supersymmetry group $\mathrm{SU}(\mathcal{N}/2\,|1)$. We performed quantization and constructed the Hilbert space of wave functions in terms of wave functions of bosonic harmonic oscillator. We showed that the hidden on-shell symmetry SU(2) of the two-dimensional harmonic oscillator reveals the hidden supersymmetry $\mathrm{SU}(\mathcal{N}/2\,|2)$ defining degeneracies of energy levels. The case of $\mathcal{N}=4$ superextension is exceptional, since it has dual superfield description.

Another feature of the two-dimensional harmonic oscillator is the presence of conformal SO(2, 1) invariance of the trigonometric type [10], [11], where superconformal Hamiltonian must be an even function of the deformation parameter m [12]. Thus, the Hamiltonian is redefined as

$$H = \mathcal{H}_{\text{conf.}} + 2mF, \qquad \mathcal{H}_{\text{conf.}} = -\,\partial_z\partial_{\bar z} + m^2 z\bar z = a^+a^- + b^+b^- + m. \tag{33}$$

Then the closure of the relevant superconformal group SU($\mathcal{N}/2\,|1,1$) and the hidden symmetry SU(2) yields some extended conformal supersymmetry corresponding to the superalgebra $\sim osp(\mathcal{N}|4)$, where the extra supercharges are written as

$$\begin{aligned} \mathcal{Q}^I &= \sqrt{2}\,\psi^I\,a^+, \qquad & \bar{\mathcal{Q}}_J &= \sqrt{2}\,\bar\psi_J\,a^-,\\ \mathcal{S}^I &= \sqrt{2}\,\psi^I\,b^+, \qquad & \bar{\mathcal{S}}_J &= \sqrt{2}\,\bar\psi_J\,b^-. \end{aligned} \tag{34}$$

It can be interpreted as a spectrum-generating supersymmetry.

It would be interesting to study superextensions of three- or four-dimensional harmonic oscillators for hidden supersymmetries. For example, the multiplet $(\mathbf{4},\mathbf{4},\mathbf{0})$ can describe $\mathcal{N}=4$ superextension of the four-dimensional harmonic oscillator [13].

Acknowledgments

The author thanks the organizers of the conference ISQS26 for the kind hospitality in Prague. He also thanks Evgeny Ivanov and Gor Sarkissian for discussions. This paper was supported by the RFBR Grant No. 18-02-01046, and by the Ministry of Science and Higher Education of Russian Federation, project No. 3.1386.2017.

References

[1] Smilga A V 2004 Weak supersymmetry *Phys. Lett.* B **585** 173 (*Preprint* hep-th/0311023)
[2] Bellucci S and Nersessian A 2004 (Super)Oscillator on CP(N) and Constant Magnetic Field *Phys. Rev.* D **67** 065013 (*Preprint* hep-th/0211070)
[3] Bellucci S and Nersessian A 2004 Supersymmetric Kähler oscillator in a constant magnetic field (*Preprint* hep-th/0401232)
[4] Ivanov E and Sidorov S 2014 Deformed supersymmetric mechanics *Class. Quant. Grav.* **31** 075013 (*Preprint* 1307.7690 [hep-th])
[5] Ivanov E and Sidorov S 2014 Super Kähler oscillator from $SU(2|1)$ superspace *J. Phys.* A **47** 292002 (*Preprint* 1312.6821 [hep-th])
[6] Scheunert M, Nahm W and Rittenberg V 1977 Irreducible Representations of the OSP(2,1) and SPL(2,1) Graded Lie Algebras *J. Math. Phys.* **18** 155
[7] Beisert N 2007 The Analytic Bethe Ansatz for a Chain with Centrally Extended su(2|2) Symmetry *J. Stat. Mech.* **0701** P01017 (*Preprint* nlin/0610017)
[8] Ivanov E, Lechtenfeld O and Sidorov S 2019 Deformed $N = 8$ Supersymmetric Mechanics *Symmetry* **11** 135
[9] Perelomov A 1990 *Integrable systems of classical mechanics and Lie algebras* vol 1 Berlin Birkhäuser Verlag 307 p
[10] Papadopoulos G 2013 New potentials for conformal mechanics *Class. Quant. Grav.* **30** 075018 (*Preprint* 1210.1719 [hep-th])
[11] Holanda N L Toppan F 2014 Four types of (super)conformal mechanics: D-module reps and invariant actions *J. Math. Phys.* **55** 061703 (*Preprint* arXiv:1402.7298 [hep-th])
[12] Ivanov E, Sidorov S and Toppan F 2015 Superconformal mechanics in $SU(2|1)$ superspace *Phys. Rev.* D **91** 085032 (*Preprint* 1501.05622 [hep-th])
[13] Ivanov E and Sidorov S 2016 $SU(2|1)$ mechanics and harmonic superspace *Class. Quant. Grav.* **33** 055001 (*Preprint* 1507.00987 [hep-th])

Journal of Physics: Conference Series **1416** (2019) 012033 doi:10.1088/1742-6596/1416/1/012033

Slightly generalized Maxwell system and longitudinal components of solution

Volodimir Simulik, Igor Gordievich and Taras Zajac

Principal researcher, Institute of Electron Physics, National Academy of Sciences, Uzhgorod 88000, Ukraine;
Researcher, MIRTEK Ltd, Stavropol, Russia;
Docent, Department of Electronic Systems, Uzhgorod National University, Uzhgorod 88000, Ukraine.

E-mail: `vsimulik@gmail.com`

Abstract. We consider slightly generalized Maxwell equations with electric and magnetic currents and charges densities of the gradient type. Among other versions of the Maxwell system these equations differ by the extended symmetry properties. Such system of equations is invariant with respect to a 256-dimensional algebra, and this algebra is not yet the maximum of possible symmetry. The longitudinal components of both vectors of electric and magnetic field strengths, together with two corresponded scalar waves, are found as the exact solution of such generalized Maxwell equations. The longitudinal wave component of the electric field strength vector itself is found as an exact solution of the standard Maxwell equations with fixed current and charge of the gradient type. This wave is corresponded to the scalar wave component, which is propagated in the same direction. The analysis of found solutions demonstrates that longitudinal components are located near the corresponded current and charge densities, which are the sources of such fields. The relationship with modern experiments is considered briefly.

1. Introduction

The theoretical and experimental description of the longitudinal electromagnetic waves is a subject of the different level discussions. There is a lot of indications of the existence of such waves in different media. Note, e.g., the longitudinal waves in plasmons and in plasma. Nevertheless, these oscillations only sometimes are considered as the electromagnetic ones. On the other hand, it is well known fact that the Maxwell equations for the free electromagnetic field contain only the transverse solutions.

Recently in [1] a simple experimental method for generating an intense longitudinal electric field from transverse electromagnetic waves (laser pulses) with radially symmetric polarization has been presented. The laser-generated longitudinal electric field was observed in two dimensions and distinguished from the transverse component using the optical Kerr shutter method. Authors of [2] experimentally demonstrated that a radially polarized field can be focused to a spot size significantly smaller than for linear polarization. For strong focusing, a radially polarized field leads to a longitudinal electric field component at the focus which is sharp and centered at the optical axis. Experimental and theoretical investigations in [3] demonstrates that the longitudinal field component has the maximal amplitude where the "‘usual”’ field component is zero. The high-power industrial CO_2 laser is applied. Authors of [4] studied that, when focused, a radially polarized beam has a net longitudinal field useful for particle acceleration and, perhaps, other unique applications.

The goal of this paper is to contribute in the problem by means of standard classical electrodynamics, e. g., an exact solutions of well defined physical and mathematical systems of equations. Presented below result that the longitudinal wave component of the electric field strength vector is the exact solution of the standard Maxwell equations with specific gradient-type case of electric current and charge densities can be verified by anyone after individual substitution into standard Maxwell system with corresponding sources.

Note that one can find some confusion in using the notions "longitudinal electric wave" and "longitudinal wave component of the electric field strength vector". It leads to the confusion in understanding.

The question about longitudinal electromagnetic waves is the old problem of classical electrodynamics. The first theoretical model has been considered in [5]. Longitudinal electromagnetic waves between parallel plates were investigated.

The hypothesis about the longitudinal electromagnetic waves, which presence in the mathematical formalism follows from the massless Dirac equation (on the basis of a link between the massless Dirac and slightly generalized Maxwell equations), was suggested in [6]. The start in [6] was related to the results [7]. Following [6, 7], in [8, 9] we presented our preliminary point of view on this problem. The general solution of such kind of the Maxwell equations has been found. The conclusion that the longitudinal electric waves can exist within the framework of standard Maxwell electrodynamics in the case, when the electric currents and charges have specific gradient-like form, has been suggested. Such longitudinal electric wave appears [8, 9] in the pair with scalar wave.

However, the problem on the longitudinal electromagnetic waves can be considered independently (without any relation to the massless Dirac equation and the Maxwell equations in the Dirac-like form).

Below (see also [10]) the longitudinal wave component of the electric field strength vector $\overrightarrow{E}$ is found as the exact solution of the standard Maxwell equations with specific partial case of electric current and charge densities.

In the period after our publications [8, 9] an interest to the problem of longitudinal electromagnetic waves in electrodynamics has been arose. Other approaches to the problem should be briefly mentioned as well, see, e. g. [11, 12]. Nevertheless, the generalization of classical electrodynamics to admit a scalar field and longitudinal waves mentioned in [12] was known already from [8, 9].

Note that in [13] scalar potential wave together with a longitudinal electric field $\overrightarrow{E}$ in the direction of propagation has been detected by a ball antenna. Nevertheless, the authors of [14–16] have been doubted in this result both from the theoretical [14, 15] and experimental [16] points of view. The authors of [16] tried to repeat the experiment [13] without any success. In [15] it is proved that the authors of [13] detected classical TEM waves emitted by currents flowing in the Earth and launched by the ball antenna used in the experiment.

Here the contemporary status of our original approach started from [6–9] is presented.

2. Exact solution of the Maxwell system with gradient-type electric sources

As ordinarily in field theory, the system of units $\hbar = c = 1$ is used. Further, in the Minkowski space-time $\mathrm{M}(1,3) = \{x \equiv (x^\mu) = (x^0 = t, \overrightarrow{x} \equiv (x^j))\}; \mu = \overline{0,3}, j = 1,2,3$, the variable x^μ denote the Cartesian (covariant) coordinates of the points of the physical space-time in the arbitrary-fixed inertial reference frame.

Consider the Maxwell equations in the form

$$\partial_0 \overrightarrow{E} - \mathrm{curl}\overrightarrow{H} = -\mathrm{grad}E^0, \quad \partial_0 \overrightarrow{H} + \mathrm{curl}\overrightarrow{E} = 0, \qquad (1)$$

$$\operatorname{div}\vec{E} = -\partial_0 E^0, \quad \operatorname{div}\vec{H} = 0.$$

Compare with the standard Maxwell equations for the free field

$$\partial_0 \vec{E} - \operatorname{curl}\vec{H} = 0, \quad \partial_0 \vec{H} + \operatorname{curl}\vec{E} = 0, \qquad (2)$$
$$\operatorname{div}\vec{E} = 0, \quad \operatorname{div}\vec{H} = 0,$$

and with the standard Maxwell system with electric sources.

From the physical point of view, equations (1) are the partial case of the standard Maxwell equations (the partial case of a standard classical electrodynamics). In this case, the current $\vec{j}(x)$ and charge $\rho(x)$ densities have the form

$$\vec{j}(x) = -\operatorname{grad}E^0(x), \quad \rho(x) = -\partial_0 E^0(x). \qquad (3)$$

The mathematical point of view shows that objects (1) and (2) are the different partial differential equations. Indeed, the general solution of the system (2) is well known and is given by the transverse electromagnetic waves only

$$\vec{E}(x) = \frac{1}{(2\pi)^{\frac{3}{2}}} \int d^3k \sqrt{\frac{\omega}{2}} \left\{ \left[c^1_{\vec{k}} \vec{e}_1 + c^2_{\vec{k}} \vec{e}_2 \right] e^{-ikx} + \left[c^{*1}_{\vec{k}} \vec{e}^{\,*}_1 + c^{*2}_{\vec{k}} \vec{e}^{\,*}_2 \right] e^{ikx} \right\}, \qquad (4)$$
$$\vec{H}(x) = \frac{i}{(2\pi)^{\frac{3}{2}}} \int d^3k \sqrt{\frac{\omega}{2}} \left\{ \left[c^1_{\vec{k}} \vec{e}_1 - c^2_{\vec{k}} \vec{e}_2 \right] e^{-ikx} - \left[c^{*1}_{\vec{k}} \vec{e}^{\,*}_1 - c^{*2}_{\vec{k}} \vec{e}^{\,*}_2 \right] e^{ikx} \right\}.$$

Here $(\vec{E}(x), \vec{H}(x))$ are the real electric and magnetic field strengths, $c^1_{\vec{k}}$, $c^2_{\vec{k}}$ are the complex quantum-mechanical momentum-helicity amplitudes of a photon (the amplitudes of the transverse electromagnetic waves),

$$kx \equiv \omega t - \vec{k}\vec{x}, \quad \omega \equiv \sqrt{\vec{k}^2}, \qquad (5)$$

and the 3-component basis vectors $(\vec{e}_1, \vec{e}_2, \vec{e}_3)$, which, without any loss of generality, can be taken as

$$\vec{e}_1 = \frac{1}{\omega\sqrt{2(k^1k^1 + k^2k^2)}} \begin{vmatrix} \omega k^2 - ik^1k^3 \\ -\omega k^1 - ik^2k^3 \\ i(k^1k^1 + k^2k^2) \end{vmatrix}, \quad \vec{e}_2 = \vec{e}^{\,*}_1, \quad \vec{e}_3 = \frac{\vec{k}}{\omega}, \qquad (6)$$

are the eigen vectors of the quantum-mechanical helicity operator for the spin s = 1.

The general solution (4) is ordinarily found by the Fourier method, and the normalization factor $C \equiv \sqrt{\frac{\omega}{2(2\pi)^3}}$ in (4) is taken from the condition

$$P_0 = \frac{1}{2} \int d^3x \left(\vec{E}^2 + \vec{H}^2 \right) = \int d^3k\, \omega \left(\left| c^1_{\vec{k}} \right|^2 + \left| c^2_{\vec{k}} \right|^2 \right). \qquad (7)$$

It is interesting to find the general solution of the system (1), which is expected to have another (maybe not only transverse) form. The Fourier method in the corresponding rigged Hilbert space leads to the general solution

$$\vec{E}(x) = \frac{1}{(2\pi)^{\frac{3}{2}}} \int d^3k \sqrt{\frac{\omega}{2}} \left\{ \left[c^1_{\vec{k}} \vec{e}_1 + c^2_{\vec{k}} \vec{e}_2 + \alpha_{\vec{k}} \vec{e}_3 \right] e^{-ikx} + \left[c^{*1}_{\vec{k}} \vec{e}^{\,*}_1 + c^{*2}_{\vec{k}} \vec{e}^{\,*}_2 + \alpha^*_{\vec{k}} \vec{e}^{\,*}_3 \right] e^{ikx} \right\},$$
$$\vec{H}(x) = \frac{i}{(2\pi)^{\frac{3}{2}}} \int d^3k \sqrt{\frac{\omega}{2}} \left\{ \left[c^1_{\vec{k}} \vec{e}_1 - c^2_{\vec{k}} \vec{e}_2 \right] e^{-ikx} - \left[c^{*1}_{\vec{k}} \vec{e}^{\,*}_1 - c^{*2}_{\vec{k}} \vec{e}^{\,*}_2 \right] e^{ikx} \right\}, \qquad (8)$$

$$E^0(x) = \frac{1}{(2\pi)^{\frac{3}{2}}} \int d^3k \sqrt{\frac{\omega}{2}} \left(\alpha_{\vec{k}} e^{-ikx} + \alpha^*_{\vec{k}} e^{ikx} \right).$$

It is easy to see that here the electric field strength $\overrightarrow{E}(x)$ contains (together with the ordinary transverse waves) the longitudinal wave as well. This longitudinal electric wave is determined by the amplitude $\alpha_{\vec{k}}$. The scalar function $E^0(x)$ specifies the electric current and charge densities in the Maxwell system (1). The scalar wave $E^0(x)$ is longitudinal as well and is determined by the amplitude $\alpha_{\vec{k}}$.

The validity of the solution (8) can be verified by the direct substitution of (8) into equations (1).

3. Exact solution of the Maxwell system with gradient-type electric and magnetic sources

Consider the Maxwell-like system of equations

$$\begin{gathered} \partial_0 \overrightarrow{E} - \mathrm{curl}\overrightarrow{H} = -\mathrm{grad}E^0, \quad \partial_0 \overrightarrow{H} + \mathrm{curl}\,\overrightarrow{E} = -\mathrm{grad}H^0, \\ \mathrm{div}\,\overrightarrow{E} = -\partial_0 E^0, \quad \mathrm{div}\overrightarrow{H} = -\partial_0 H^0. \end{gathered} \tag{9}$$

Equations (9) contain both electric and magnetic gradient-type current and charge densities. System (9) is directly related to the massless Dirac equation (see e.g. the consideration in [17–19] and the references therein).

The general solution of the system (9) is given by

$$\begin{aligned} \overrightarrow{E}(x) &= \int d^3k\, C \left\{ \left[c^1_{\vec{k}} \vec{e}_1 + c^2_{\vec{k}} \vec{e}_2 + (c^3_{\vec{k}} + c^4_{\vec{k}}) \vec{e}_3 \right] e^{-ikx} + A \right\}, \\ A &\equiv \left[c^{*1}_{\vec{k}} \vec{e}^{\,*}_1 + c^{*2}_{\vec{k}} \vec{e}^{\,*}_2 + (c^{*3}_{\vec{k}} + c^{*4}_{\vec{k}}) \vec{e}^{\,*}_3 \right] e^{ikx}, \\ \overrightarrow{H}(x) &= i \int d^3k\, C \left\{ \left[c^1_{\vec{k}} \vec{e}_1 - c^2_{\vec{k}} \vec{e}_2 + (c^3_{\vec{k}} - c^4_{\vec{k}}) \vec{e}_3 \right] e^{-ikx} - B \right\}, \\ B &\equiv \left[c^{*1}_{\vec{k}} \vec{e}^{\,*}_1 - c^{*2}_{\vec{k}} \vec{e}^{\,*}_2 + (c^{*3}_{\vec{k}} - c^{*4}_{\vec{k}}) \vec{e}^{\,*}_3 \right] e^{ikx}, \\ E^0(x) &= \int d^3k\, C \left[(c^3_{\vec{k}} + c^4_{\vec{k}}) e^{-ikx} + (c^{*3}_{\vec{k}} + c^{*4}_{\vec{k}}) e^{ikx} \right], \\ H^0(x) &= i \int d^3k\, C \left[(c^3_{\vec{k}} - c^4_{\vec{k}}) e^{-ikx} - (c^{*3}_{\vec{k}} - c^{*4}_{\vec{k}}) e^{ikx} \right], \end{aligned} \tag{10}$$

$C \equiv \sqrt{\frac{\omega}{2(2\pi)^3}}$ is the normalization factor. Here both the electric field strength $\overrightarrow{E}(x)$ and the magnetic field strength $\overrightarrow{H}(x)$ contain (together with ordinary transverse waves) the corresponding longitudinal waves as well. These longitudinal electric and longitudinal magnetic wave are given by the amplitudes $c^3_{\vec{k}} + c^4_{\vec{k}}$ and $c^3_{\vec{k}} - c^4_{\vec{k}}$, respectively. The scalar functions $(E^0(x),\, H^0(x))$ specify the electromagnetic currents and charges densities in the Maxwell-like system of equations (9). The scalar waves $(E^0(x),\, H^0(x))$ are longitudinal as well.

In our articles [17–19], the system of equations (9) is called as the slightly generalized Maxwell equations with gradient-type sources. Indeed, at the first step, the system (9) is the generalization of the Maxwell equations because it contains the condition $\mathrm{div}\overrightarrow{H} \neq 0$ and the nonzero magnetic current density in the equation $\partial_0 \overrightarrow{H} + \mathrm{curl}\,\overrightarrow{E} = -\mathrm{grad}H^0$. Nevertheless, at the second step the system (9) is the simplification (specification) of the Maxwell equations. Its electromagnetic currents and charge densities are the partial gradient-type forms of the general form of the electromagnetic sources.

Thus, the system of equations (9) due to the conditions $\partial_0\overrightarrow{H}+\mathrm{curl}\overrightarrow{E}\neq 0$, $\mathrm{div}\overrightarrow{H}\neq 0$ is not the standard Maxwell electrodynamics. Therefore, it is better to start the experimental detection of the longitudinal electromagnetic waves in the experimental modeling of the situation given by the system (1), which is inside the standard Maxwell electrodynamics. This system of equations predicts an existence of the longitudinal component of the vector of electric field strength $\overrightarrow{E}$.

The arguments in the prospect of system (9) are not so evident. Nevertheless, let us mention (i) that system (9) has the maximally possible symmetry properties among the Maxwell and the Maxwell-like systems of equations. In [17] the 256 dimensional algebra of invariance of equations (9) was mentioned. The equations (9) have been considered in the terms of complex functions

$$\mathcal{E}\equiv E-iH=\left|\begin{array}{c}\overrightarrow{E}-i\overrightarrow{H}\\ E^0-iH^0\end{array}\right|. \qquad (11)$$

Taking into account the symmetries found recently in [20–26], the number 256 can be increased. Let us recall that the role of the symmetry principle in electrodynamics is known from the times of Maxwell and Heaviside.

(ii) The system of equation (9) has the property of the Fermi–Bose duality. It can describe both the massless spin 1/2 particle-antiparticle doublet of the fermions and the massless spin (1,0) doublet of bosons (photon and massless spinless boson). In the terms of complex 4-vector (11) equations (9) has the manifestly covariant form

$$\partial_\mu\mathcal{E}_\nu-\partial_\nu\mathcal{E}_\mu+i\varepsilon_{\mu\nu\rho\sigma}\partial^\rho\mathcal{E}^\sigma=0,\quad \partial_\mu\mathcal{E}^\mu=0, \qquad (12)$$

and the form of massless Dirac equation

$$\widetilde{\gamma}^\mu\partial_\mu\mathcal{E}=0, \qquad (13)$$

with specific Clifford–Dirac algebra γ matrix representation

$$\widetilde{\gamma}^0=\left|\begin{array}{cccc}1&0&0&0\\0&1&0&0\\0&0&1&0\\0&0&0&-1\end{array}\right|C,\quad \widetilde{\gamma}^1=\left|\begin{array}{cccc}0&0&0&1\\0&0&-i&0\\0&i&0&0\\-1&0&0&0\end{array}\right|C, \qquad (14)$$

$$\widetilde{\gamma}^2=\left|\begin{array}{cccc}0&0&i&0\\0&0&0&1\\-i&0&0&0\\0&-1&0&0\end{array}\right|C,\quad \widetilde{\gamma}^3=\left|\begin{array}{cccc}0&-i&0&0\\i&0&0&0\\0&0&0&1\\0&0&-1&0\end{array}\right|C,$$

where C is the operator of complex conjugation, $C\mathcal{E}\equiv\mathcal{E}^*$, $\widetilde{\gamma}^\mu\widetilde{\gamma}^\nu+\widetilde{\gamma}^\nu\widetilde{\gamma}^\mu=2g^{\mu\nu}$.

(iii) The system of equations (9) considered in the specific medium [27, 28] can be applied to inneratomic phenomena and describe the hydrogen spectrum [17–19] (another approach to the hydrogen spectrum description on the basis of stationary Maxwell equations in this medium is given in [29, 30]). In [29, 30] and [17–19] the role of longitudinal components is evident.

Therefore, there is some sense in experimental modeling the situation given by the system (9) in the problem of the longitudinal electromagnetic waves investigation. Of course, the advantages of system (1), which is inside the standard electrodynamics, are evident.

Note that solution (8) can be found both by the direct application of the Fourier method and as a partial case of solution (10).

The validity of solution (10) can be verified by the direct substitution of (10) into equations (9).

4. The subsystems of the Maxwell-like equations with gradient-type electric and magnetic sources

Due to the above mentioned properties, the system of equations (1) is the most interesting subsystem of the Maxwell-like equations (9) (in the problem of longitudinal electromagnetic waves consideration). Equations (1) follow from equations (9) after substitution $H^0 = 0 \Rightarrow c^3_{\vec{k}} = c^4_{\vec{k}}$. In this case, in solution (8) the amplitude of the longitudinal wave $\alpha_{\vec{k}} = 2c^3_{\vec{k}}$. This partial case of (9) is presented in Sec. 2 in details.

Another interesting subsystem of (9) follows from (9) after the substitution $E^0(x) = 0$. In this case, the vector $\overrightarrow{H}(x)$ of the magnetic field strength contains the longitudinal wave component. Other properties of the general solution are similar to (8) and are to (8) in evident symmetry. Nevertheless, the interest to this case is lower then for $H^0 = 0$ case. The conditions $\mathrm{div}\overrightarrow{H} \neq 0$ and $\partial_0\overrightarrow{H} + \mathrm{curl}\overrightarrow{E} \neq 0$ are not in the framework of the standard Maxwell electrodynamics.

5. Conclusion

The electric field strength vector $\overrightarrow{E}(x)$ that contains the longitudinal wave component is found as the exact solution of the standard Maxwell equations with the partial gradient case of electric current and charge densities (3). The explicit form of this solution is given in (8). Note that the set of general solutions (8) of such Maxwell equations contains the vector $\overrightarrow{E}(x)$ together with the scalar field $E^0(x)$, which is longitudinal as well. The validity of the solution (8) can be verified by the direct substitution of (8) into equations (1). Thus, the link between the longitudinal electric waves and the standard Maxwell electrodynamics is demonstrated. The author has a hope that corresponding experimental situation can be created by modeling the sources (3) and may be useful. Moreover, the experiments in [1–4] look like to be the physical reality of such longitudinal electric waves.

The analysis of (3) and (8) demonstrates that longitudinal components are located near the current and charge densities. Indeed, the current and charge densities (3) and the longitudinal components in the solution (8) are determined by the same amplitudes $\alpha_{\vec{k}}, \alpha^*_{\vec{k}}$. The best examples of corresponding physical reality are such big charges as the whole water area of closed sea, the Earth in general, their oscillations and corresponding longitudinal electric and scalar waves. The subclass of the functions $\overrightarrow{j}(x)$, $\rho(x)$ from (3) is generated by the oscillating charges in plasma, plasmons and laser beams [1–4]. Longitudinal electric wave in [1–4] appears after the interaction of circularly polarized transverse electromagnetic waves. The authors are considered now the possibility of experiment in solid samples.

Acknowledgements. Volodimir Simulik is much grateful to Prof. N.P. Khvorostenko, who was able many years ago in the far away year 1991 to inspire forever his interest to the problem of longitudinal electromagnetic waves, and especially is grateful to Prof. D.B. Kuryliak for the recent suggestion of possible experiments in solid samples.

References

[1] Miyaji G, Miyanaga N, Tsubakimoto K, Sueda K and Ohbayashi K 2004 *Appl. Phys. Lett.* **84** 3855
[2] Dorn R, Quabis S and Leuchs G 2003 *Phys. Rev. Lett.* **91** 233901
[3] Niziev V G, Yakunin V P and Turkin N G 2009 *Quantum Electronics* **39** 505
[4] Tidwell S C, Ford D H and Kimura W D 1990 *Appl. Opt.* **29** 2234
[5] Ferraro V C A and Flint H T 1941 *Proc. Phys. Soc.* **53** 170
[6] Khvorostenko N P 1992 *Rus. Phys. J.* **35** 223
[7] Simulik V M 1991 *Theor. Math. Phys.* **87** 386
[8] Krivsky I Yu and Simulik V M 1997 General solution of the Maxwell equations with gradient-like sources and longitudinal electromagnetic waves *Proceedings of the First International conference "Non-Euclidean geometry in modern physics. BGL-1."* Uzhgorod, Ukraine 193–199
[9] Krivsky I Yu and Simulik V M 1998 *Uzhgorod University Sci. Herald, Ser. Phys.* **2** 121

[10] Simulik V M 2016 Longitudinal electromagnetic waves in the framework of standard classical electrodynamics *arXiv:* 1606.01738 [physics. class-ph]
[11] Barashenkov V S, Pestov A V and Yuriev M Z 2001 Transmission of information by longitudinal electromagnetic waves *Preprint of the Joint Institute for Nuclear Research. Dubna.* P2-2001-253 (in russian)
[12] K.J. Van Vlaenderen K J and Waser A 2001 *Hadr. J.* **24** 609
[13] Monstein C and Wesley J P 2002 *Europhys. Lett.* **59** 514
[14] Bray J R and Britton M C 2004 *Europhys. Lett.* **66** 153
[15] Rebilas K 2008 *Europhys. Lett.* **83** 60007
[16] Butterworth E J, Allison C B, Cavazos D and Mullen F M 2013 *J. Sci. Explor.* **27** 13
[17] Simulik V M and Krivsky I Yu 2002 *Ann. Fond. L. de Broglie* **27** 303
[18] Simulik V M and Krivsky I Yu 2002 *Rep. Math. Phys.* **50** 315
[19] Simulik V M and Krivsky I Yu 2003 *Electromagnetic Phenomena* **3** 103
[20] Simulik V M and Krivsky I Yu 2009 Extended real Clifford–Dirac algebra and bosonic symmetries of the Dirac equation with nonzero mass *arXiv:* 0908.3106 [math-ph]
[21] Simulik V M and Krivsky I Yu 2010 *Reports of the National Academy of Sciences of Ukraine* **5** 82 (in Ukrainian)
[22] Krivsky I Yu and Simulik V M 2010 *Cond. Matt. Phys.* **13** 43101(1–15)
[23] Simulik V M and Krivsky I Yu 2011 *Phys. Lett.* A **375** 2479
[24] Simulik V M, Krivsky I Yu and Lamer I L 2012 *Cond. Matt. Phys.* **15** 43101(1–10)
[25] Simulik V M, Krivsky I Yu and Lamer I L 2013 *TWMS J. App. Eng. Math.* **3** 46
[26] Simulik V M, Krivsky I Yu and Lamer I L 2013 *Ukr. J. Phys.* **58** 523
[27] Sallhofer H 1978 *Z. Naturforsch.* A **33** 1378
[28] Sallhofer H 1986 *Z. Naturforsch.* A **41** 468
[29] Sallhofer H 1990 *Z. Naturforsch.* A **45** 1361
[30] Sallhofer H 1991 *Z. Naturforsch.* A **46** 869

Journal of Physics: Conference Series **1416** (2019) 012034 doi:10.1088/1742-6596/1416/1/012034

Hidden symmetries of relativistic hydrogen atom

Volodimir Simulik and Igor Gordievich

Principal researcher, Institute of Electron Physics, National Academy of Sciences, Uzhgorod 88000, Ukraine;

Researcher, MIRTEK Ltd, Stavropol, Russia.

E-mail: vsimulik@gmail.com

Abstract. The spin s=(1,0) Bose symmetry of the Dirac equation for the free spinor field, proved recently in our papers, is extended for the Dirac equation interacting with external Coulomb field. Relativistic hydrogen atom is modeling here by such Dirac equation. We are able to present both the fermionic and bosonic symmetries known from our papers about the case of non-interacting spinor field. New symmetry operators were found on the basis of new gamma matrix representations of the Clifford and SO(8) algebras, which were found recently in our papers. Hidden symmetries were found both in the canonical Foldy-Wouthuysen and in the covariant Dirac representations. The symmetry operators, which are simple and graceful in the Foldy-Wouthuysen representation, become non-local in the Dirac model.

1. Introduction

Recently in our papers [1, 2] the bosonic symmetry of the Dirac equation with nonzero mass has been found. The investigation of the symmetries of the Dirac equation in external Coulomb field is the next logical step.

Symmetry of the hydrogen atom was considered at first for the non-relativistic theoretical model. The SO(4) symmetry of the non-relativistic Schrödinger equation for a hydrogen atom was found by V. Fock [3], see also [4].

Relativistic hydrogen atom is modeling here by the Dirac equation in the external Coulomb field

$$\left(i\partial_0 - \widehat{H}\right)\psi(x) = 0; \quad \widehat{H} \equiv \gamma^0\vec{\gamma}\cdot\vec{p} + \gamma^0 m - \frac{Ze^2}{|\vec{x}|}, \tag{1}$$

where $x \in \mathrm{M}(1,3)$, $\partial_\mu \equiv \partial/\partial x^\mu$, $Z = 1$, $\mu = \overline{0,3}$, $j = 1,2,3$, and $\mathrm{M}(1,3) = \{x \equiv (x^\mu) = (x^0 = t,\ \overrightarrow{x} \equiv (x^j))\}$ is the Minkowski space-time, the 4-component function $\psi(x)$ belongs to rigged Hilbert space

$$\mathrm{S}^{3,4} \subset \mathrm{H}^{3,4} \subset \mathrm{S}^{3,4*}. \tag{2}$$

Due to a special role of the time variable $x^0 = t \in (x^\mu)$ (in obvious analogy with non-relativistic theory), in general consideration one can use the quantum-mechanical rigged Hilbert space (2). Here the Schwartz test function space $\mathrm{S}^{3,4}$ is dense in the Schwartz generalized function space $\mathrm{S}^{3,4*}$ and $\mathrm{H}^{3,4}$ is the quantum-mechanical Hilbert space of 4-component functions over $\mathrm{R}^3 \subset \mathrm{M}(1,3)$.

The system of units $\hbar = c = 1$ is chosen, the metric tensor in Minkowski space-time $\mathrm{M}(1,3)$ is given by

$$g^{\mu\nu} = g_{\mu\nu} = g^\mu_\nu, \quad (g^\mu_\nu) = \mathrm{diag}\,(1,-1,-1,-1)\,; \tag{3}$$

$x_\mu = g_{\mu\nu}x^\mu$, and summation over the twice repeated indices is implied. The Dirac γ matrices are taken in the standard Dirac-Pauli representation

$$\gamma^0 = \begin{vmatrix} \mathrm{I} & 0 \\ 0 & -\mathrm{I} \end{vmatrix}, \quad \gamma^\ell = \begin{vmatrix} 0 & \sigma^\ell \\ -\sigma^\ell & 0 \end{vmatrix}, \quad \ell = 1, 2, 3, \tag{4}$$

where the Pauli matrices are given by

$$\sigma^1 = \begin{vmatrix} 0 & 1 \\ 1 & 0 \end{vmatrix}, \sigma^2 = \begin{vmatrix} 0 & -i \\ i & 0 \end{vmatrix}, \sigma^3 = \begin{vmatrix} 1 & 0 \\ 0 & -1 \end{vmatrix}. \tag{5}$$

Below we prove the Dirac equation in the external Coulomb field (1) to possess the symmetry determined by the 31 operators, which form the 31-dimensional algebra SO(6)$\oplus i\gamma^0\cdot$SO(6)$\oplus i\gamma^0$. Two different fermionic D(1/2,0)$\oplus$(0,1/2) representations of the SO(1,3) algebra of the Lorentz group are found. Two different bosonic tensor-scalar D(1,0)$\oplus$(0,0) and vector D(1/2,1/2) representations of this algebra are found as well. The corresponding generators of the above mentioned algebras commute with the operator of the Dirac equation in the external Coulomb field (1), and, therefore, determine the hidden symmetries (algebras of invariance) of such Dirac equation.

At first we consider the known symmetries of the Dirac equation (1), after that we present the mathematical tools, which are necessary for our investigations and, finally, we present the list of different hidden symmetries of the relativistic hydrogen atom.

2. SO(4) symmetry of the Dirac equation in external Coulomb field

The first four constants of motion (symmetry operators, which commutes with the operator of the Dirac equation) for the equation (1) were found by P. Dirac in his paper [5], where the equation (1) has been derived and introduced. They are three components of the vector $\vec{J} = (J^1,\, J^2,\, J^3)$ of the total angular momentum

$$\vec{J} = \vec{L} + \vec{s}, \quad \vec{L} \equiv \vec{x} \times \vec{p}, \quad \vec{s} \equiv \frac{1}{2}\begin{vmatrix} \vec{\sigma} & 0 \\ 0 & \vec{\sigma} \end{vmatrix}, \tag{6}$$

where $\vec{L}$ is the orbital angular momentum, $\vec{s}$ is the spin-1/2 angular momentum, and the found by Dirac additional constant of motion K:

$$K = \gamma^0 \left(2\vec{s}\cdot\vec{L} + 1\right), \quad K^2 = \vec{J}^2 + \frac{1}{4}. \tag{7}$$

Symmetry operator number five is the Johnson–Lippman constant of motion [6]

$$D = 2\vec{s}\cdot\frac{\vec{x}}{|\vec{x}|} + \frac{1}{mZe^2}K\gamma^4\left(\widehat{H} - \gamma^0 m\right), \tag{8}$$

$D^2 = 1 + \left(\frac{\widehat{H}^2}{m^2} - 1\right)\frac{K^2}{Z^2e^4}$, which commutes with the operator $\left(i\partial_0 - \widehat{H}\right)$ of the Dirac equation and anti-commutes with the Dirac symmetry operator K of (7). Here the anti-Hermitian $\gamma^4 = \gamma^0\gamma^1\gamma^2\gamma^3$ instead of Hermitian γ^5 of other authors is used. Note that article [6] about excellent result (8) was published as a brief 1/10 of a journal page remark, containing only single formula for D from (8).

After that the way to SO(4) symmetry of the relativistic hydrogen atom was direct. This symmetry was found in [7] and [8], for the consideration in [9] see the comments of [10]. Thus, the SO(4) symmetry of the Dirac equation (1) for the hydrogen atom is given by the six operators

$$\vec{I} = \vec{J} + \vec{T}, \quad \vec{R} = \vec{J} - \vec{T}, \tag{9}$$

Journal of Physics: Conference Series **1416** (2019) 012034 doi:10.1088/1742-6596/1416/1/012034

where $\vec{J}$ is known from (6) and components of $\vec{T} = (T^1, T^2, T^3)$ have the form

$$T^1 = \frac{D}{2\sqrt{D^2}}, \quad T^2 = \frac{iDK}{2\sqrt{D^2K^2}}, \quad T^3 = \frac{K}{2\sqrt{K^2}}. \tag{10}$$

In the paper [7] the object $\vec{T} = (T^1, T^2, T^3)$ was called as Lentz spin-1/2 vector operator. The notations used in (10) are explained in (7) and (8) above.

Next symmetry is as follows. The Pauli-Gürsey operators [11] and [12]

$$s^{01} = \frac{i}{2}\gamma^2\hat{C}, \quad s^{02} = \frac{1}{2}\gamma^2\hat{C}, \quad s^{12} = -\frac{i}{2}, \tag{11}$$

where $\hat{C}$ is the operator of complex conjugation, $\hat{C}\psi = \psi^*$ (the operator of involution in the space $\mathrm{H}^{3,4}$), determine according to [13] the SO(1,2)⊂SO(1,3) algebra of invariance of the Dirac equation in the form $(i\gamma^\mu\partial_\mu - m + \frac{e^2}{|\vec{x}|})\psi(x) = 0$.

In [14] the symmetry of relativistic hydrogen atom in the form of gl(8,R) algebra has been found. The stationary case of the Dirac equation (1) was considered and the discrete transformation were used.

The author of [15] considered another problem. The quasi-potential two-particle model was presented for the description of spinless relativistic hydrogen atom. The O(4) symmetry and its breaking were investigated.

3. Application of the gamma matrix representations of the real Clifford and SO(8) algebras

In our long time investigations of the mapping of the Maxwell theory on the Dirac theory we also studied the different exotic representations of the Clifford–Dirac algebra, which were useful in the description of the Maxwell theory in the notations taking from the formalism of the Dirac spinor field, see, e.g., [16] and [2]. Step by step we have come to the idea of generalization of the Clifford-Dirac algebra in order to describe the bosonic features of the Dirac theory and fermionic features of the Maxwell theory as well.

Recently, see, e.g., [17], we put into consideration the gamma matrix representations of the Clifford algebra $C\ell^{\mathbb{R}}$(0,6) and the Lie algebra SO(8), which are defined over the field of real numbers. Comparison with well known gamma matrix representations of algebras $C\ell^{\mathbb{C}}$(1,3) (in physical literature the Clifford–Dirac algebra) and SO(1,5), defined over the field of complex numbers, demonstrates that new representations contain much more useful elements and perspectives for application. Appealing to the gamma matrix representations of the Clifford algebra $C\ell^{\mathbb{R}}$(0,6) and the Lie algebra SO(8) made possible, see, e.g., [18] and [19], to find the hidden symmetries of the free (non-interacting) Dirac equation. The Bose symmetries were found as well [1] and [20, 21]. Below we use these new algebraic objects [17] for the problem of finding the hidden symmetries of the relativistic hydrogen atom. The necessary part of consideration from [17] is presented below in a compact version.

Consider the fact that seven γ matrices $\gamma^1, \gamma^2, \gamma^3$,

$$\gamma^4 = \gamma^0\gamma^1\gamma^2\gamma^3, \quad \gamma^5 = \gamma^1\gamma^3\hat{C}, \quad \gamma^6 = i\gamma^1\gamma^3\hat{C}, \quad \gamma^7 = i\gamma^0, \tag{12}$$

where γ^μ matrices are given in (4) and operator $\hat{C}$ is defined after the formulas (11), satisfy the anti-commutation relations

$$\gamma^{\mathrm{A}}\gamma^{\mathrm{B}} + \gamma^{\mathrm{B}}\gamma^{\mathrm{A}} = -2\delta^{\mathrm{AB}}, \quad \mathrm{A,B} = \overline{1,7}, \tag{13}$$

of the Clifford algebra generators over the field of real numbers. Due to the evident fact that only six operators of (12) are linearly independent, $\gamma^4 = -i\gamma^7\gamma^1\gamma^2\gamma^3$, it is the representation of the Clifford algebra $C\ell^{\mathbb{R}}$(0,6) of the dimension $2^6 = 64$.

The first 16 operators are given in the Table 1 of [17], the next 16 are found from them with the help of the multiplication by imaginary unit $i = \sqrt{-1}$. Last 32 are found from first 32 with the help of multiplication by operator $\hat{C}$ of complex conjugation. Thus, if to introduce the notation "stCD" ("st" and CD are taken from standard Clifford–Dirac) for the set of 16 matrices from the Table 1 in [17], then the set of 64 elements of $C\ell^{\mathbb{R}}(0,6)$ algebra representation will be given by

$$\Big\{(\mathrm{stCD}) \cup i\cdot(\mathrm{stCD}) \cup \hat{C}\cdot(\mathrm{stCD}) \cup i\hat{C}\cdot(\mathrm{stCD})\Big\}. \tag{14}$$

As the consequences of the equalities $\gamma^4 \equiv \prod_{\mu=0}^{3}\gamma^{\mu} \rightarrow \prod_{\bar{\mu}=0}^{4}\gamma^{\bar{\mu}} = -\mathrm{I}$, known from the standard Clifford–Dirac algebra $C\ell^{\mathbb{C}}(1,3)$, and the consequences of the anti-commutation relations (13), in $C\ell^{\mathbb{R}}(0,6)$ algebra for the matrices γ^{A} (12) the following extended equalities are valid: $\gamma^7 \equiv -\prod_{\mathrm{A}=1}^{6}\gamma^{\mathrm{A}} \rightarrow \prod_{\mathrm{A}=1}^{7}\gamma^{\mathrm{A}} = \mathrm{I}, \quad \gamma^5\gamma^6 = i$.

Operators (12) generate also the 28 matrices

$$s^{\widetilde{\mathrm{A}}\widetilde{\mathrm{B}}} = \{s^{\mathrm{AB}} = \frac{1}{4}[\gamma^{\mathrm{A}},\gamma^{\mathrm{B}}],\ s^{\mathrm{A}8} = -s^{8\mathrm{A}} = \frac{1}{2}\gamma^{\mathrm{A}}\}, \tag{15}$$

$\widetilde{\mathrm{A}}, \widetilde{\mathrm{B}} = \overline{1,8}$, which satisfy the commutation relations of the Lie algebra SO(8)

$$[s^{\widetilde{\mathrm{A}}\widetilde{\mathrm{B}}}, s^{\widetilde{\mathrm{C}}\widetilde{\mathrm{D}}}] = \delta^{\widetilde{\mathrm{A}}\widetilde{\mathrm{C}}}s^{\widetilde{\mathrm{B}}\widetilde{\mathrm{D}}} + \delta^{\widetilde{\mathrm{C}}\widetilde{\mathrm{B}}}s^{\widetilde{\mathrm{D}}\widetilde{\mathrm{A}}} + \delta^{\widetilde{\mathrm{B}}\widetilde{\mathrm{D}}}s^{\widetilde{\mathrm{A}}\widetilde{\mathrm{C}}} + \delta^{\widetilde{\mathrm{D}}\widetilde{\mathrm{A}}}s^{\widetilde{\mathrm{C}}\widetilde{\mathrm{B}}}. \tag{16}$$

It is evident that here we have the algebra over the field of real numbers as well. Furthermore, it is evident that 28 elements (15) of SO(8) do not form any Clifford algebra and do not form any subalgebra of the Clifford algebra. It is independent from the Clifford algebra mathematical object. Note that here (as in (12) for the gamma matrices) the anti-Hermitian realization of the SO(8) operators is chosen, for the reasons see, e.g., the article [1] and monographs [22] or [23].

The explicit form of the 28 elements of the γ matrix representation of the SO(8) algebra is given in the Table 3 of [17]. The wonderful feature of this gamma matrix representation of the algebra SO(8) is as follows. Here (i) two subsets (s^{23}, s^{31}, s^{12}) and (s^{45}, s^{64}, s^{56}) of operators $s^{\widetilde{\mathrm{A}}\widetilde{\mathrm{B}}}$ from (15) determine two different sets of SU(2) spin 1/2 generators, (ii) commute between each other and (iii) commute with the operator of the Dirac equation in the Foldy–Wouthuysen representation [24].

4. New fermionic and bosonic symmetries of relativistic hydrogen atom

Below we present both the symmetries of the Foldy–Wouthuysen equation [24] in the external Coulomd field

$$(i\partial_0 - \gamma^0\omega + \frac{e^2}{|\vec{x}|})\phi(x) = 0; \quad \omega \equiv \sqrt{-\Delta + m^2}, \tag{17}$$

$x \in \mathrm{M}(1,3)$, $\phi \in \{\mathrm{S}^{3,4} \subset \mathrm{H}^{3,4} \subset \mathrm{S}^{3,4*}\}$, and, of course, the symmetries of the Dirac equation (1) in such external field as well.

The first application of the matrix representations of the algebras $C\ell^{\mathbb{R}}(0,6)$ and SO(8) is the symmetry analysis (the search of groups and algebras with respect to which the equation is invariant). It is easy to understand that the Foldy–Wouthuysen representation [24] is preferable for such analysis. Indeed, in this representation one must calculate the commutation relations of possible pure matrix symmetry operators from (14) (or from (15)) only with two elements of the Foldy–Wouthuysen equation (17) operator: γ^0 and i. After the determining of the symmetries of the Foldy–Wouthuysen equation one can find the symmetries of the Dirac equation on the basis of the inverse Foldy–Wouthuysen transformation [24]. Note that after such transformation only the small part of symmetry operators will be pure matrix, the main part of operators will contain the nonlocal operator $\omega \equiv \sqrt{-\Delta + m^2}$ and the functions of it.

Journal of Physics: Conference Series **1416** (2019) 012034 doi:10.1088/1742-6596/1416/1/012034

Now we can start the consideration of the new symmetries of relativistic hydrogen atom. The fundamental assertions are as follows.

(i) The gamma matrix representation of the subalgebra SO(6) of the algebra SO(8), which is formed by the operators

$$\{s^{\breve{A}\breve{B}}\} = \{s^{\breve{A}\breve{B}} \equiv \frac{1}{4}[\gamma^{\breve{A}}, \gamma^{\breve{B}}]\}, \quad \breve{A}, \breve{B} = \overline{1,6}, \tag{18}$$

determines the algebra of invariance of the Dirac equation in the Foldy–Wouthuysen representation $(\partial_0 + i\gamma^0\omega - \frac{e^2}{|\vec{x}|})\phi(x) = 0$ (in (18) the six matrices $\{\gamma^{\breve{A}}\} = \{\gamma^1, \gamma^2, \gamma^3, \gamma^4, \gamma^5, \gamma^6,\}$ are known from (12)).

(ii) On the basis of SO(6) (18) the 31-dimensional gamma matrix representation of the Lie algebra SO(6)$\oplus i\gamma^0$SO(6)$\oplus i\gamma^0$ is constructed, which is formed by the elements from $C\ell^{\mathbb{R}}(0,6)$ and is the maximal pure matrix algebra of invariance of the Dirac equation in the Foldy–Wouthuysen representation $(\partial_0 + i\gamma^0\omega - \frac{e^2}{|\vec{x}|})\phi(x) = 0$.

(iii) The Dirac equation in external Coulomb field (1) is invariant with respect to the 31-dimensional gamma matrix representation of the algebra $\widetilde{\mathrm{SO}}(6) \oplus i\widetilde{\gamma}^0\widetilde{\mathrm{SO}}(6) \oplus i\widetilde{\gamma}^0$, where the representation of the algebra $\widetilde{\mathrm{SO}}(6)$ is given in the form of (18) with gamma operators found from matrices (12) by the inverse Foldy–Wouthuysen transformation [24]. Resulting operators are given by

$$\vec{\widetilde{\gamma}} = \vec{\gamma}\frac{-\vec{\gamma}\cdot\nabla + m}{\omega} + \vec{p}\frac{-\vec{\gamma}\cdot\nabla + \omega + m}{\omega(\omega+m)},$$

$$\widetilde{\gamma}^4 = \gamma^4\frac{-\vec{\gamma}\cdot\nabla + m}{\omega}, \quad \widetilde{\gamma}^5 = \widetilde{\gamma}^1\widetilde{\gamma}^3\widetilde{C}, \quad \widetilde{\gamma}^6 = i\widetilde{\gamma}^1\widetilde{\gamma}^3\widetilde{C}, \tag{19}$$

$$\widetilde{\gamma}^7 = i\widetilde{\gamma}^0, \quad \widetilde{\gamma}^0 = \gamma^0\frac{-\vec{\gamma}\cdot\nabla + m}{\omega},$$

where $\widetilde{C} = (\mathrm{I} + 2\frac{i\gamma^1\partial_1 + i\gamma^2\partial_2}{\sqrt{2\omega(\omega+m)}})\hat{C}$, and $\omega \equiv \sqrt{-\Delta + m^2}$. These formulas give the images of gamma matrices (12) in the Dirac representation after fulfilling the inverse Foldy–Wouthuysen transformation.

Thus, the $\widetilde{\mathrm{SO}}(6) \oplus i\widetilde{\gamma}^0\widetilde{\mathrm{SO}}(6) \oplus i\widetilde{\gamma}^0$ algebra is found from SO(6)$\oplus i\gamma^0$SO(6)$\oplus i\gamma^0$ on the basis of the inverse Foldy–Wouthuysen transformation [24]. For the Dirac equation only the part of this algebra is pure matrix, other elements contain the operator $\omega \equiv \sqrt{-\Delta + m^2}$.

Consider the symmetries of the relativistic hydrogen atom with respect to the Lorentz group. On the basis of $C\ell^{\mathbb{R}}(0,6)$ and SO(8) we can determine two different realizations of the $\mathrm{D}(0,\frac{1}{2}) \oplus (\frac{1}{2},0)$ representation of the Lie algebra of *universal covering* $\mathcal{L}$ = SL(2,C) of the proper ortochronous Lorentz group $\mathrm{L}^{\uparrow}_{+}$ = SO(1,3)=$\{\Lambda = (\Lambda^{\mu}_{\nu})\}$, with respect to which the equation $(\partial_0 + i\gamma^0\omega - \frac{e^2}{|\vec{x}|})\phi(x) = 0$ is invariant:

$$s^{\mu\nu}_{\mathrm{I}} = \{s^{0k}_{\mathrm{I}} = \frac{i}{2}\gamma^k\gamma^4, \quad s^{km}_{\mathrm{I}} = \frac{1}{4}[\gamma^k, \gamma^m]\}; \quad \gamma^4 \equiv \gamma^0\gamma^1\gamma^2\gamma^3, \quad (k, m = \overline{1,3}), \tag{20}$$

$$s^{01}_{\mathrm{II}} = -\frac{i}{2}\gamma^2\hat{C}, \, s^{02}_{\mathrm{II}} = -\frac{1}{2}\gamma^2\hat{C}, \, s^{03}_{\mathrm{II}} = \frac{1}{2}\gamma^0, \, s^{23}_{\mathrm{II}} = -\frac{1}{2}\gamma^0\gamma^2\hat{C}, \, s^{31}_{\mathrm{II}} = \frac{i}{2}\gamma^0\gamma^2\hat{C}, \, s^{12}_{\mathrm{II}} = -\frac{i}{2}. \tag{21}$$

Taking the combinations of operators (20), (21) we construct the generators of bosonic representations:

$$s^{\mu\nu}_{\mathrm{TS}} = \{s^{0k}_{\mathrm{TS}} = s^{0k}_{\mathrm{I}} + s^{0k}_{\mathrm{II}}, \quad s^{mn}_{\mathrm{TS}} = s^{mn}_{\mathrm{I}} + s^{mn}_{\mathrm{II}}\}; \quad s^{\mu\nu}_{\mathrm{V}} = \{s^{0k}_{\mathrm{V}} = -s^{0k}_{\mathrm{I}} + s^{0k}_{\mathrm{II}}, \quad s^{mn}_{\mathrm{V}} = s^{mn}_{\mathrm{TS}}\}, \tag{22}$$

where $s^{\mu\nu}_{\rm TS}$ and $s^{\mu\nu}_{\rm V}$ are the generators of the tensor-scalar $D(1,0)\oplus(0,0)$ and irreducible vector $D(\frac{1}{2},\frac{1}{2})$ representations of the Lie algebra SO(1,3) of the Lorentz group $\mathcal{L}$ respectively, with respect to which the Foldy–Wouthuysen equation $(\partial_0 + i\gamma^0\omega - \frac{e^2}{|\vec{x}|})\phi(x) = 0$ is invariant.

Anti-Hermitian operators of every set (20), (21), or (22), satisfy the commutation relations of the Lie algebra SO(1,3) of the Lorentz group $\mathcal{L}$:

$$[s^{\mu\nu}, s^{\rho\sigma}] = -g^{\mu\rho}s^{\nu\sigma} - g^{\rho\nu}s^{\sigma\mu} - g^{\nu\sigma}s^{\mu\rho} - g^{\sigma\mu}s^{\rho\nu}. \tag{23}$$

For the Dirac equation in the space of Dirac spinors $\{\psi\}$ (i.e. in the Pauli–Dirac representation) the form of the generators of the tensor-scalar D(1,0)⊕(0,0) and irreducible vector $D(\frac{1}{2},\frac{1}{2})$ representations of the Lie algebra SO(1,3) of the Lorentz group $\mathcal{L}$ is similar to (22) (with (20), (21)) but the gamma operators are in this case too much complicated and are given by (19). The images of the operators (20), (21), or (22), in the Dirac representation satisfy the commutation relations (23) as well.

In [2] and [19] for the free Dirac and Foldy–Wouthuysen equations we used also the *evident bosonic representation* of (22), in which the Casimir operators are diagonal and the proof of Bose properties is most convenient.

5. Conclusion

The generalization of our results on the Bose symmetries of the free non-interacting Dirac equation (see, e.g., [2] and the references therein) for the case of the presence of external Coulomb field is suggested. The hidden Fermi and Bose symmetries of the relativistic hydrogen atom are found.

Note that physical picture of the hydrogen atom as the electron in the external Coulomb field is related to the spin 1/2 Fermi symmetries. On the other hand the physical picture of the hydrogen atom as a compound system of proton and electron is related to the total spin 1 (or zero) Bose symmetries.

The main result of this paper is putting into consideration of two different bosonic tensor-scalar D(1,0)⊕(0,0) and vector D(1/2,1/2) representations of the Lie algebra SO(1,3) of the Lorentz group $\mathcal{L}$, with respect to which the Dirac equation in the external Coulomb field is invariant. Two fermionic symmetries of this equation, given by different D(1/2,0)⊕(0,1/2) representations of the SO(1,3) algebra of the Lorentz group, are found as well. The maximal pure matrix symmetry of the Foldy–Wouthuysen equation for hydrogen atom is found and the explicit forms of operators of corresponding symmetry of the Dirac model are calculated.

References

[1] Simulik V M and Krivsky I Yu 2011 *Phys. Lett.* A **375** 2479
[2] Simulik V M, Krivsky I Yu and Lamer I L 2013 *Ukr. J. Phys.* **58** 523
[3] Fock V 1935 *Z. Phys.* **98** 145
[4] Bargmann V 1936 *Z. Phys.* **99** 576
[5] Dirac P A M 1928 *Proc. Roy. Soc. Lond.* A **117** 610
[6] Johnson M H and Lippmann B A 1950 *Phys. Rev.* **78** 329
[7] De Groot E 1982 *Am. J. Phys.* **50** 1141
[8] Stahlhofen A A 1997 *Helv. Phys. Acta* **70** 372
[9] Chen J-L, Deng D-L and Hu M-G 2008 *Phys. Rev.* A **77** 034102
[10] Stahlhofen A A 2008 *Phys. Rev.* A **78** 036101
[11] Pauli W 1957 *Nuov. Cim.* **6** 204
[12] Gürsey F 1958 *Nuov. Cim.* **7** 411
[13] Krivsky I Yu and Simulik V M 1992 *Theor. Math. Phys.* **90** 265
[14] Nikitin A G 1999 In *Photon and Poincare group* Nova Science, New York 74
[15] Ruijgrok Th W 1976 *Acta Phys. Pol.* **87** 43
[16] Simulik V M and Krivsky I Yu 1997 *Adv. Appl. Cliff. Algebras* **7** 25

Journal of Physics: Conference Series **1416** (2019) 012034 doi:10.1088/1742-6596/1416/1/012034

[17] Simulik V M 2018 *Adv. Appl. Cliff. Algebras* **28** 93
[18] Simulik V M and Krivsky I Yu 2010 *Reports of the National Academy of Sciences of Ukraine* **5** 82 (in ukrainian)
[19] Krivsky I Yu and Simulik V M 2010 *Cond. Matt. Phys.* **13** 43101
[20] Simulik V M, Krivsky I Yu and Lamer I L 2013 *TWMS J. App. Eng. Math.* **3** 46
[21] Simulik V M 2019 *J. Phys: Conf. Ser.* **1194** 012099
[22] J. Elliott, P. Dawber, *Symmetry in physics.* vol.1, London, 1979.
[23] B. Wybourne, *Classical groups for physicists.* New York, 1974.
[24] Foldy L L and Wouthuysen S A 1950 *Phys. Rev.* **78** 29

Spin transport in a rippled graphene periodic chain

Jan Smotlacha[1], **Michal Pudlak**[2], **and Rashid G. Nazmitdinov**[1,3]
[1]BLTP, Joint Institute for Nuclear Research, 141980 Dubna, Moscow region, Russia
[2]Institute of Experimental Physics, 04001 Kosice, Slovak Republic
[3]Dubna State University, 141982 Dubna, Moscow region, Russia

E-mail: smota@centrum.cz

Abstract. We discuss the solution of the scattering problem of ballistic electrons on the finite periodic chain. Each element of this chain is composed of a combination of a planar and rippled graphene elements. The analysis is carried out in the effective mass approximation when only the interaction between nearest neighbour atoms is taken into account. We find that, with the increase of number of elements, the chain becomes more transparent for electrons with spin up travelling in one direction and for electrons with spin down travelling in the opposite direction.

1. Introduction

In quantum mechanics the study of electron motion in an infinite or half-infinite one-dimensional periodic chain provides a valuable tool in understanding and prediction various physical properties of real materials. The most famous example is the Kronig-Penney model [1]. Nowadays, the problem of scattering with a finite number of scattering centres attracts a considerable interest in view of an ability of modern technology to create periodic rippled graphene structures. For example, the chemical vapor deposition provides quite promising way to create periodic nanoripples [2]. On the other hand, it is found that ripples or wrinkles act as potential barriers for charged carriers leading to their localization [3].

In this communication we discuss the electron scattering in the graphene based superlattice (figure 1). Each element of the periodic chain consists of the flat graphene piece continuously connected to the ripple that is connected again to the flat graphene piece. In our consideration a ripple consists of a curved graphene surface in the form of an arc of a circle. We analyze the transmission and reflection of an electron flow through a rippled system in the effective mass approximation. In this case only the interaction between nearest neighbour atoms is taken into account (see details in [4]).

Our study is based on the results discussed in Refs.[5, 6]. We suppose that the incident electrons move from the left planar part (L) to the right planar part (R) along the x axis (figure 1) and its energy is an integral of motion. Hereafter, we consider a wide enough graphene sheet $W \gg M$, where W and M being, respectively, as the width along the y axis and the length along x axis of the graphene sheet. It means that we keep the translational invariance along the y axis and neglect the edge effects.

In our system the scattering mechanism is due to the curvature-induced spin-orbit couplings characterized by the coefficients λ_x and λ_y in carbon nanotubes (see details in Refs.[4, 5, 6]). Considering the curved surface as part of the armchair nanotube, it was shown that the curvature-induced spin-orbit coupling [4] yields backscattering with spin inversion. This spin

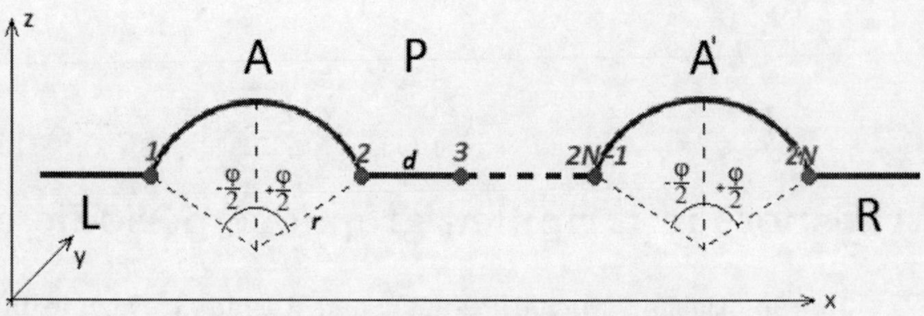

Figure 1. The system structure: planar parts (P) and arc pieces (A). It is bounded both from the left and the right side by planar graphene (L and R, respectively). Here, r is the radius of arc, d is the length of the inner planar part. The unit element is composed of the one arc (A) and the one planar piece (P). All the structure is composed of $N-1$ units. An additional arc (A') is added to connect the structure with the right planar part. The particular parts are connected in the connection points (1 to $2N$).

inversion is caused by the spin-orbit term that breaks spin symmetry in the effective Hamiltonian of the armchair carbon nanotube [5, 6]. The same mechanism has been used recently to study Klein collimation by rippled graphene superlattice [7].

In Section 2, we describe the base functions of the subsystems, introduce the relations between the energy and momenta and clarify the spin properties of the base functions. In Section 3, we provide a detailed description of the calculation of the transmission of the states with different spin properties. Section 4 is aimed to derive the analytical expression through our one-dimensional rippled graphene superlattice. Summary is presented in Section 5.

2. Wave functions

The particular subsystems of the investigated structure can be characterized by the base wave function sets [5]. Each of these sets is composed of the four wave functions that depend on the momentum $\vec{k} = (k_x, k_y)$ or angular momentum m, respectively. In the calculations, we assume that $k_y = 0$. In other words, we consider the one-dimensional problem.

For the planar parts, the eigen energy depends on the momentum as

$$E = \pm\gamma k_x \tag{1}$$

and the base wave functions have the form

$$\psi_j^P = \frac{1}{2}\exp(\pm \mathrm{i}k_x x)w_j, \qquad j \in \{1,2,3,4\} \tag{2}$$

(signs "+" or "-", correspond to the indices "1,2" or "3,4", respectively), where $\exp(\pm \mathrm{i}k_x x)$ are the eigenfunctions of the momentum and w_1, w_2, w_3, w_4 are the energy eigenvectors from the set

$$S_P = \left\{ \begin{pmatrix} 1 \\ \mathrm{i} \\ 1 \\ \mathrm{i} \end{pmatrix}, \begin{pmatrix} 1 \\ -\mathrm{i} \\ 1 \\ -\mathrm{i} \end{pmatrix}, \begin{pmatrix} -1 \\ -\mathrm{i} \\ 1 \\ \mathrm{i} \end{pmatrix}, \begin{pmatrix} -1 \\ \mathrm{i} \\ 1 \\ -\mathrm{i} \end{pmatrix} \right\}. \tag{3}$$

For the planar piece, the wave function can be written in the form

$$\psi = \frac{1}{2}\sum_{j=1}^{4} C_{P,j}\exp(\pm \mathrm{i}k_x x)w_j, \tag{4}$$

Journal of Physics: Conference Series **1416** (2019) 012035 doi:10.1088/1742-6596/1416/1/012035

where $C_{P,j}$ are the coefficients of the linear combination. The matrix form of this wave function has the following form

$$M_0 \exp(\mathrm{i}\hat{K}x)C_P, \tag{5}$$

where the columns of M_0 are created by the vectors from S_P and the components of C_P are created by the coefficients $C_{P,j}$. The expression $\exp(\mathrm{i}\hat{K}x)$ is a diagonal matrix

$$\exp(\mathrm{i}\hat{K}x) = \mathrm{diag}\{\exp(\mathrm{i}k_x x), \exp(\mathrm{i}k_x x), \exp(-\mathrm{i}k_x x), \exp(-\mathrm{i}k_x x)\}. \tag{6}$$

For the arc pieces, there are four different values of the angular momentum m_1, m_2, m_3, m_4 since the angular momentum is not conserved in this case. They are connected with the eigenenergy as (see figure 2 and [5]):

$$E = \mathcal{E}_1 = \lambda_x + \sqrt{\lambda_y^2 + t_{m_1}^2}, \qquad E = \mathcal{E}_2 = \lambda_x - \sqrt{\lambda_y^2 + t_{m_2}^2}, \tag{7}$$

$$E = \mathcal{E}_3 = -(\lambda_x - \sqrt{\lambda_y^2 + t_{m3}^2}), \qquad E = \mathcal{E}_4 = -(\lambda_x + \sqrt{\lambda_y^2 + t_{m4}^2}), \tag{8}$$

where $t_{m_j} = \frac{\gamma}{R} m_j$. For the purpose of the simplification, we can write

$$t_{m1,2} = \pm t_{m+}, \qquad t_{m3,4} = \pm t_{m-}. \tag{9}$$

Then,

$$t_{m+} = \sqrt{(E-\lambda_x)^2 - \lambda_y^2}, \qquad t_{m-} = \sqrt{(E+\lambda_x)^2 - \lambda_y^2}. \tag{10}$$

The base eigenfunctions have the form

$$\psi_j^A = \exp(\pm \mathrm{i} m_\pm \theta) \exp(\mathrm{i}k_y y) v, \tag{11}$$

where v is an energy eigenvector from the list

$$S_A = \left\{ \begin{pmatrix} -\mathrm{i}t_m \\ \lambda_y - \sqrt{\lambda_y^2 + t_m^2} \\ \mathrm{i}(\lambda_y - \sqrt{\lambda_y^2 + t_m^2}) \\ -t_m \end{pmatrix}, \begin{pmatrix} -\mathrm{i}t_m \\ \lambda_y + \sqrt{\lambda_y^2 + t_m^2} \\ \mathrm{i}(\lambda_y + \sqrt{\lambda_y^2 + t_m^2}) \\ -t_m \end{pmatrix}, \begin{pmatrix} -\mathrm{i}t_m \\ \lambda_y + \sqrt{\lambda_y^2 + t_m^2} \\ -\mathrm{i}(\lambda_y + \sqrt{\lambda_y^2 + t_m^2}) \\ t_m \end{pmatrix}, \begin{pmatrix} -\mathrm{i}t_m \\ \lambda_y - \sqrt{\lambda_y^2 + t_m^2} \\ \mathrm{i}(-\lambda_y + \sqrt{\lambda_y^2 + t_m^2}) \\ t_m \end{pmatrix} \right\}, \tag{12}$$

but here the choice of the appropriate vector in S_A is more complicated.

The electron eigenfunction, associated with the arc piece, can be (analogously to (5)) written as

$$M_A(\theta) \exp(\mathrm{i}\hat{m}\theta) C_A, \tag{13}$$

where $M_A(\theta) = U(-\theta)M_A(0)$. The structure of the matrix $M_A(0)$ depends on the chosen interval of energy (figure 2). If we denote the particular vectors from S_A by v_1, v_2, v_3, v_4, it follows that:

- $E > \lambda_x \Rightarrow E = \mathcal{E}_1 \vee E = \mathcal{E}_3$; the base eigenfunctions are $\exp(\pm \mathrm{i}m_+\theta)v_1(\pm t_{m+})$ and $\exp(\pm \mathrm{i}m_-\theta)v_3(\pm t_{m-})$. The columns of $M_A(0)$ are created by $v_1(t_{m+})$, $v_3(t_{m-})$, $v_1(-t_{m+})$, $v_3(-t_{m-})$, adjusted by the normalization constants; the case $E < \lambda_x + \lambda_y$ corresponds to the evanescent modes (m is imaginary);
- $-\lambda_x < E < \lambda_x \Rightarrow E = \mathcal{E}_2 \vee E = \mathcal{E}_3$; the base eigenfunctions are $\exp(\pm \mathrm{i}m_+\theta)v_2(\pm t_{m+})$ and $\exp(\pm \mathrm{i}m_-\theta)v_3(\pm t_{m-})$. The columns of $M_A(0)$ are created by $v_2(t_{m+})$, $v_3(t_{m-})$, $v_2(-t_{m+})$, $v_3(-t_{m-})$ adjusted by the normalization constants; the case when $|E \pm \lambda_x| \le \lambda_y$ corresponds to the evanescent modes;
- $E < -\lambda_x \Rightarrow E = \mathcal{E}_2 \vee E = \mathcal{E}_4$; the base eigenfunctions are $\exp(\pm \mathrm{i}m_+\theta)v_2(\pm t_{m+})$ and $\exp(\pm \mathrm{i}m_-\theta)v_4(\pm t_{m-})$. The columns of $M_A(0)$ are created by $v_2(t_{m+})$, $v_4(t_{m-})$, $v_2(-t_{m+})$, $v_4(-t_{m-})$ adjusted by the normalization constants; the case when $E > -\lambda_x - \lambda_y$ corresponds to the evanescent modes.

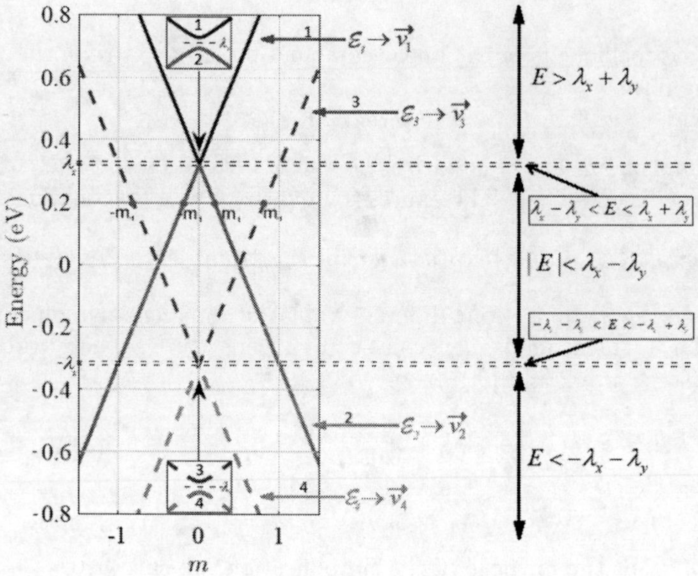

Figure 2. Dependence of the energy on the quantum number m for rippled graphene (purple line). There are two anticrossings (top and the bottom of the left panel) due to the presence of nonzero term $\lambda_y = 0.0043$ eV, which yields the energy gaps $2\lambda_y$. These gaps contain the evanescent modes. The shifts of these gaps with the respect to the zero energy are given by the term $\lambda_x = 0.32$ eV, so they are ± 0.32 eV. The values of λ_x, λ_y follow from the values of the parameters: $R = 10$ Å, $\delta = 0.01$, $p = 0.1$, $\gamma = (4.5 \cdot 1.42)\,\mathrm{eV}\cdot$ Å, $\gamma' = \frac{8}{3}\gamma$.

Next,

$$\exp(\mathrm{i}\hat{m}\theta) = \mathrm{diag}\left(\exp(\mathrm{i}m_+\theta), \exp(\mathrm{i}m_-\theta), \exp(-\mathrm{i}m_+\theta), \exp(-\mathrm{i}m_-\theta)\right) \tag{14}$$

where the angular momenta

$$m_\pm = \frac{R}{\gamma} t_{m\pm} = \frac{R}{\gamma}\sqrt{(E \mp \lambda_x)^2 - \lambda_y^2} \tag{15}$$

and C_A is a vector of the coefficients of the linear combination.

3. Spin scattering

In the investigated system, we define the unit structure (figure 1) that is $(N-1)-$times repeated (plus the connection arc A'); and we denote $2N$ connection points. In these connection points, the eigenfunctions of the neighboring parts are equal. This fact leads to the system of $2N$ equations

$$M_0 \begin{pmatrix} \Phi_{in} \\ r \end{pmatrix} = M_A(-\frac{\varphi}{2}) \cdot \exp(-\mathrm{i}\hat{m}\frac{\varphi}{2}) C_{1A}, \tag{16}$$

$$M_A(+\frac{\varphi}{2}) \cdot \exp(+\mathrm{i}\hat{m}\frac{\varphi}{2}) C_{1A} = M_0 \cdot \exp(-\mathrm{i}\hat{K}d/2) C_{1P}, \tag{17}$$

$$M_0 \cdot \exp(+\mathrm{i}\hat{K}d/2) C_{1P} = M_A(-\frac{\varphi}{2}) \cdot \exp(-\mathrm{i}\hat{m}\frac{\varphi}{2}) C_{2A}, \tag{18}$$

$$\vdots$$

$$M_0 \cdot \exp(+i\hat{K}d/2)C_{N-1,P} = M_A(-\frac{\varphi}{2}) \cdot \exp(-i\hat{m}\frac{\varphi}{2})C_{N,A}, \tag{19}$$

$$M_A(+\frac{\varphi}{2}) \cdot \exp(+i\hat{m}\frac{\varphi}{2})C_{N,A} = M_0 \begin{pmatrix} t \\ 0 \end{pmatrix}. \tag{20}$$

$-\varphi, \varphi$ are the edge angles of the arcs, d is the length of the planar part. The form of the eigenfunctions for the particular parts follows from (13) and (5), respectively. The symbols C_{1A}, C_{1P},...,$C_{N-1,A}$, $C_{N-1,P}$, $C_{N,A}$ denote the vectors of the linear combination at planar part (C_{iP}) and rippled part (C_{iA}) of the graphene. Their forms for the left and the right planar part are $\begin{pmatrix} \Phi_{in} \\ r \end{pmatrix}$ and $\begin{pmatrix} t \\ 0 \end{pmatrix}$, respectively; where $\Phi_{in}, r, t, 0$ represent two component vectors with $\Phi_{in} = \frac{1}{\sqrt{2}}\begin{pmatrix} 1 \\ 1 \end{pmatrix}$ and "0" represents two zeros. These forms follow from the idea that the electron flux moves from the left to the right, and the both components of the incoming state, corresponding to the positive momentum in Φ_{in}, are equal. The factor $\frac{1}{\sqrt{2}}$ stands for the normalization.

The system (16)-(20) can be rewritten in the matrix form

$$\begin{pmatrix} \Phi_{in} \\ r \end{pmatrix} = {M_0}^{-1} A M_0 \begin{pmatrix} t \\ 0 \end{pmatrix} = M_C \begin{pmatrix} t \\ 0 \end{pmatrix}, \tag{21}$$

where

$$A = \left(\left[M_A(-\frac{\varphi}{2})\exp(-i\,\hat{m}\varphi){M_A}^{-1}(\frac{\varphi}{2})\right] \cdot \left[M_0 \exp(-i\,\hat{K}d){M_0}^{-1}\right]\right)^{N-1} \cdot$$
$$\cdot \left[M_A(-\frac{\varphi}{2})\exp(-i\,\hat{m}\varphi){M_A}^{-1}(\frac{\varphi}{2})\right]. \tag{22}$$

Introducing the notation

$$B = M_0^{-1}\left[M_A(-\frac{\varphi}{2})\exp(-i\,\hat{m}\varphi){M_A}^{-1}(\frac{\varphi}{2})\right] M_0, \tag{23}$$

we have

$$A = M_0(B\exp(-i\,\hat{K}d))^{N-1}BM_0^{-1}. \tag{24}$$

As a result, it follows that

$$M_C = (B\exp(-i\,\hat{K}d))^{N-1}B. \tag{25}$$

The schematic form of B is

$$\begin{pmatrix} z_1 & 0 & 0 & z_1' \\ 0 & z_2 & z_2' & 0 \\ 0 & z_2'^* & z_2^* & 0 \\ z_1'^* & 0 & 0 & z_1^* \end{pmatrix}, \tag{26}$$

where

$$z_1 = e^{i\frac{\varphi}{2}}\left[\cos(m_-\varphi) - i\,\mathrm{sgn}(E+\lambda_x)\frac{\sqrt{\lambda_y^2 + t_{m_-}^2}}{t_{m_-}}\sin(m_-\varphi)\right], \tag{27}$$

$$z_2 = e^{-i\frac{\varphi}{2}}\left[\cos(m_+\varphi) - i\,\mathrm{sgn}(E-\lambda_x)\frac{\sqrt{\lambda_y^2 + t_{m_+}^2}}{t_{m_+}}\sin(m_+\varphi)\right], \tag{28}$$

$$z_1' = -\frac{\lambda_y \sin(m_-\varphi)}{t_{m_-}}, \qquad z_2' = -\frac{\lambda_y \sin(m_+\varphi)}{t_{m_+}}. \tag{29}$$

Here, the application of the function "sgn" (signum) is a consequence of different forms of $M_A(0)$ for different intervals of energies.

It can be easily proven that in the general case, if z_1, z_2, z_1', z_2' are complex numbers, an arbitrary product of the matrices of such a kind, including diagonal matrices, has the same structure. Because M_C is created by a product of the mentioned kind of matrices, (21) can be transformed to the following form

$$\begin{pmatrix} \Phi_{in} \\ r \end{pmatrix} = \begin{pmatrix} \alpha_1 & 0 & 0 & \beta_1 \\ 0 & \alpha_2 & \beta_2 & 0 \\ 0 & \beta_2^* & \alpha_2^* & 0 \\ \beta_1^* & 0 & 0 & \alpha_1^* \end{pmatrix} \begin{pmatrix} t \\ 0 \end{pmatrix}, \tag{30}$$

where $\alpha_1, \alpha_2, \beta_1, \beta_2$ are complex numbers. Because

$$\begin{pmatrix} \Phi_{in} \\ r \end{pmatrix} = \frac{1}{\sqrt{2}} \begin{pmatrix} 1 \\ 1 \\ r_1 \\ r_2 \end{pmatrix}, \qquad \begin{pmatrix} t \\ 0 \end{pmatrix} = \frac{1}{\sqrt{2}} \begin{pmatrix} t_1 \\ t_2 \\ 0 \\ 0 \end{pmatrix}, \tag{31}$$

we can immediately write

$$t_1 = 1/\alpha_1, \qquad t_2 = 1/\alpha_2. \tag{32}$$

Note, that within our model, the transmission coefficient t_1 corresponds to spin up states, while the coefficient t_2 corresponds to spin down states; the both coefficients correspond to the positive pseudospin quantum number.

4. Results

By using the mathematical induction, we can show that in the case of N samples in the ripple,

$$\alpha_{1,2}^{(N)} = \frac{z_{1,2}\left((\lambda_{1,2}^{(+)})^N - (\lambda_{1,2}^{(-)})^N\right) - \exp(ik_x d)\left((\lambda_{1,2}^{(+)})^{N-1} - (\lambda_{1,2}^{(-)})^{N-1}\right)}{\lambda_{1,2}^{(+)} - \lambda_{1,2}^{(-)}} \tag{33}$$

and

$$\beta_{1,2}^{(N)} = z_{1,2}' \frac{(\lambda_{1,2}^{(+)})^N - (\lambda_{1,2}^{(-)})^N}{\lambda_{1,2}^{(+)} - \lambda_{1,2}^{(-)}}, \tag{34}$$

where

$$\lambda_{1,2}^{(\pm)} = \mathrm{Re}\,(z_{1,2}\exp(-ik_x d)) \pm \sqrt{[\mathrm{Re}\,(z_{1,2}\exp(-ik_x d))]^2 - 1}. \tag{35}$$

Furthermore,

$$|z_{1,2}|^2 = 1 + |z_{1,2}'|^2 \tag{36}$$

and at the same time,

$$|\alpha_{1,2}^{(N)}|^2 = 1 + |\beta_{1,2}^{(N)}|^2. \tag{37}$$

Then, the final formula for the transmission probabilities is

$$|t_{1,2}|^2 = \frac{1}{1 + |\beta_{1,2}^{(N)}|^2} = \frac{1}{1 + |z_{1,2}'|^2 \left(\frac{(\lambda_{1,2}^{(+)})^N - (\lambda_{1,2}^{(-)})^N}{\lambda_{1,2}^{(+)} - \lambda_{1,2}^{(-)}}\right)^2}. \tag{38}$$

Taking into account (15), (1), (27) and (28), we obtain the following result

$$|t_1(E)|^2 = |t_2(-E)|^2. \tag{39}$$

It follows from (38) and (29) that the perfect transmission ($|t_{1,2}|^2 = 1$) is ensured in the case when

$$m_{\mp}\varphi = \pi j,\ j \in \mathcal{Z}. \tag{40}$$

We can show that the corresponding value of the angle equals

$$\varphi = \frac{j\gamma}{R}\frac{\pi}{\sqrt{(E \pm \lambda_x)^2 - \lambda_y^2}}. \tag{41}$$

In [5], it is demonstrated that in the case of one sample, the transmission has a minimum close to point $E = \mp\lambda_x$. So, it is reasonable to suppose that for a suitable value of the angle, the state corresponding to the opposite spin has a maximum in this range of values. Due to (39), we can conclude that the state with the minimum of the transmission in the point $E = \mp\lambda_x$ has the appropriate maximum in the point $E = \pm\lambda_x$. Then, in the case $j = 1$, the value of the corresponding angle is

$$\varphi_c = \frac{j\gamma}{R}\frac{\pi}{\sqrt{4\lambda_x^2 - \lambda_y^2}} \sim 0.996\pi. \tag{42}$$

So, let us do a choice $\varphi_c = \pi$.

We show the plots of the transmission probabilities $|t_1|^2$, $|t_2|^2$ for the cases of N=2 and 200 samples in the investigated structure for $d = 16$ Å and 100 Å. The results are displayed in figure 3. The values of the parameters are specified in the description of figure 2 and $\varphi = \pi$.

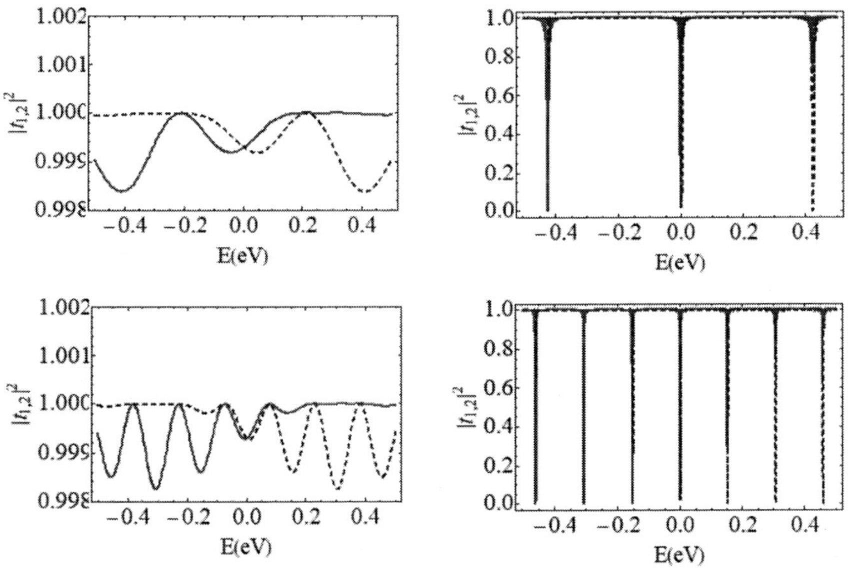

Figure 3. Transmission probabilities $|t_1|^2$ (red) and $|t_2|^2$ (blue) for various lengths of the planar parts: $d = 16$ Å (up), $d = 100$ Å (bottom). The number N of elements in the superlattice: $N = 2$ (left), $N = 200$ (right).

In the case of N=2 elements in the system under consideration, the transmission change is almost negligible. With the increasing number of the elements in the superlattice, the effect strengthens but the interval of significance is getting narrower. If we increase the distance between the arcs in the structure, the number of the peaks in the corresponding plots increases but the interval of significance remains the same.

5. Conclusion

We have considered the electron scattering in the superlattice created by periodically repeated the semi-rippled and at graphene structures. One element of the superlattice acts as a semipermeable membrane in some energy range which is transparent for electrons with spin up travelling in one direction and for electrons with spin down travelling in the opposite direction. The results show that this effect is negligible for a few semi-ripples. However, it can acquire considerable values in the case of a few hundreds of the elements. Finally, we have derived the analytical formula for the transmission of electrons with different spin orientation through our one-dimensional superlattice. It would be useful to analyse further the curvature induced spin-orbit effects in various graphene based nanostructures for the purposes of spintronics.

References

[1] Kronig R and Penney W G 1930 *Proc. R. Soc.* A **130** 499.
[2] Ni G-X et al 2012 *ACS Nano* **6** 1158.
[3] Vasić B, Zurutuza A and Gajić R 2016 *Carbon* **102** 304.
[4] Ando T 2000 *J. Phys. Soc. Jpn.* **69** 1757.
[5] Pichugin K N, Pudlak M and Nazmitdinov R G 2014 *Eur. Phys. J.* B **87: 124** 1.
[6] Pudlak M, Pichugin K N and Nazmitdinov R G 2015 *Phys. Rev.* B **92** 205432.
[7] Pudlak M and Nazmitdinov R G 2019 *J. Phys.: Condens. Matter* **31** 495301.

Journal of Physics: Conference Series 1416 (2019) 012036 doi:10.1088/1742-6596/1416/1/012036

A new concept of QM-models and truth in quantum mechanics (QM), the new hybrid-epistemic model of QM and its observer independence, the proof of the invalidity of no-go theorems

Jiří Souček

Charles University, Faculty od Arts; Prague, Czech Republic, U Kříže 8, 15800

E-mail: jiri.soucek@ff.cuni.cz

Abstract. We define a concept of a QM-model. The true theorem in QM is a statement which is true in all QM-models. In some QM-models the Bell's theorem can be proved (e.g. in the standard model of QM) while in other QM-models the Bell's theorem cannot be proved (e.g. in the hybrid-epistemic model of QM). The same situation is true when other no-go theorems are considered. Thus no-go theorems are not true in QM and the proof of the non-locality of QM is invalid.

Then the axiomatic definition of the hybrid-epistemic (HE) model of QM is presented in all details. At the end the recent proof of the inconsistency of the standard QM-model is discussed.

Our main program is to start the axiomatic study of QM (in the sense of the Hilbert's sixth problem), to prove the invalidity of no-go theorems and to identify the right (acceptable) QM-model.

1. Introduction

The situation in the foundations of quantum mechanics (QM) is now quite confused (see the recent inconsistency proofs ([1], [2]). This situation requires the exact study which can be done only by using the axiomatic method. We propose to start the systematic study of different models and different interpretations of QM and comparisons among them using systematically the axiomatic method. Each proposed model/interpretation should be transformed into the exactly defined axiomatic theory. This direction of study makes a part of the famous sixth Hilbert's problem – this problem proposes the axiomatization of physical theories in general (the axiomatization of QM is a special, but very important, part of this problem). Here we present the first steps in this program and the first results:

- The concept of an axiomatic QM-model
- The new concept of truth in QM
- The discovery of the hybrid-epistemic (HE) model and its axiomatic formulation
- The impossibility to prove the no-go theorems in the HE model
- The impossibility to prove no-go theorems in QM and the non-locality of QM
- The proof of the locality of EPR correlations in the HE model of QM

We think that the exact axiomatic study of these questions is completely necessary for the true understanding of the contemporary situation in the foundations of QM.

2. QM-models, truth in QM

We shall assume the standard QM model formalized by von Neumann ([3]) – its formulation has not the exact axiomatic form but it is an "almost" axiomatic theory. Moreover we shall assume that

- QM is a true description of the physical world
- We shall consider only finite dimensional quantum systems, i.e. quantum systems which Hilbert space has finite dimension (for simplicity)

Definition 2.1. The *QM-model* is the axiomatic system such that the set of its empirical predictions is identical to the set of empirical predictions of the standard model of QM.

This means that all QM-models are empirically equivalent among them. But this **does not imply** that there cannot exist theorems which are provable in one model and non-provable in another model.

Let us assume that there exists an axiomatic theory T such that (we shall show below that such a theory T exists)

- T is a QM-model
- no-go theorems (and thus the non-locality of QM) cannot be proved in T.

This is a quite new situation which requires the new definition of truth in QM.

Definition 2.2. The true statement in QM is the statement which is true in all QM-models.

For example, **all empirical predictions** of the standard QM are the true QM statement. The negative reformulation of this definition is the following theorem.

Theorem 2.1. Let us assume that a statement S is not proved in a certain QM-model R. Then the statement S cannot be proved as a true statement of QM.

Proof 2.1. We are not able to find which QM-model is the "true" model representing the Nature. Thus R is a possible candidate for a "true" model describing the Nature. Then S cannot be a true statement of QM.

The exact statement is the following: the empirical content of QM does not imply the statement S.

Theorem 2.2. The following statements are not the true statements in QM:

(i) No-go theorems cited above

(ii) The non-locality of QM

The basic fact is that such a theory T (mentioned above) exists and, in fact, it is, for example, the hybrid-epistemic model (*HE model*) defined below.

The intuition under this theorem is the fact that all QM-models are (from our empirical point of view) equally possible candidates for the "true" model of the Nature. We usually prefer the standard model of QM but there is no real reason to do this. From the empirical point of view all QM-models are equally good candidates.

The true meaning of the Bell's theorem is not a non-locality of QM but the inconsistency of the standard (von Neumann's) model of QM (see below).

Remark 2.1. The exact content of Theorem 2.2 can be illustrated by following statements.

Journal of Physics: Conference Series **1416** (2019) 012036 doi:10.1088/1742-6596/1416/1/012036

- "The locality of QM is incompatible with the empirical content of QM". This statement is false (since the HE model is local and respects the empirical content of QM).
- "The non-locality of QM is a consequence of the empirical content of QM together with the assumption of the standard model of QM". This statement is true.
- Some people assert that the "local realism" implies the Bell's inequality. It seems that this statement is true. But the "local realism" is not the same as "locality". The violation of the "local realism" does not imply the violation of "locality".

3. The hybrid-epistemic (HE) model of QM

This model was for the first time published in [4]. The presentation given here is slightly different but equivalent to the formulation given in [4]. The main point is the meaning of the wave function, i.e. of the quantum state (we shall not consider in this paper mixed states since this is not necessary). There are two possibilities:

- The wave function describes the state of the individual system – this is the main idea of the standard model of QM (called the ontic model in [4]).
- The wave function describes only states of an ensembles of systems – this is the basic postulate of the epistemic model of QM (this assumption is also a base of the so-called statistical interpretation of QM).

The HE model of QM is a certain mixture of these two positions: each wave function can be a state of an ensemble but only some wave functions in some special situations can be considered as states of individual systems (they are then called individual states).[1] On the other hand the properties are associated with individual systems. The concept of properties is the fundamental new element in the HE model of QM.

Definition 3.1. The structured *Hilbert space* is a couple $(\mathbf{H}, \mathbf{L})$ where $\mathbf{H}$ is a (finite-dimensional) Hilbert space and $\mathbf{L}$ is an orthogonal decomposition of $\mathbf{H}$, i.e. $\mathbf{H} = \oplus\{\mathbf{L}^{(v)} | v \in I\}$.

Axiom 3.1. We shall assume that to each system S there is associated a complex n – dimensional structured Hilbert space $(\mathbf{H}_S, \mathbf{L}_S)$, $n \geq 2$, (together with its decomposition $\mathbf{H}_S = \oplus\{\mathbf{L}_S^{(v)} | v \in I_S\}$).

Definition 3.2. An ensemble is a set of systems $\mathbf{E} = \{S_1, \ldots, S_N\}$ which are created by some preparation procedure and satisfy conditions $\mathbf{H}(S_1) \cong \ldots \cong \mathbf{H}(S_N) \cong \mathbf{H}(\mathbf{E}), \mathbf{L}(S_1) \cong \ldots \cong \mathbf{L}(S_N) \cong \mathbf{L}(E)$, where $\cong$ denotes the isomorphisms. It is assumed that the isomorphism between $\mathbf{H}(S_i)$ and $\mathbf{H}(S_j)$ generates the isomorphism between $\mathbf{L}(S_i)$ and $\mathbf{L}(S_j)$. Moreover $I(S_1) \cong \ldots \cong I(S_N) \cong I(E)$. It is also assumed that systems $S_1, \ldots, S_N$ are independent.

Axiom 3.2.

(i) The state of an ensemble (in a given moment of time) is described by some element of $\mathbf{H}_E$ – more precisely by a ray $[\Psi] = \{\alpha\Psi | |\alpha| = 1, \alpha \in C\}$, where $||\Psi|| = 1$. The state of $\mathbf{E}$ is denoted by $St(\mathbf{E})$.

(ii) The time evolution of the state $[\Psi(t)]$ of an ensemble $\mathbf{E}$ is given by the unitary group $\{U_t\}$ in $\mathbf{H}_E$ and we have $[\Psi(t)] = [U_t(\Psi(0))]$.

[1] Each individual state is also a state of an ensemble. This ensemble consists of systems where each of these systems is in a given individual state (this ensemble is called the homogeneous ensemble). In this sense each individual state is also a collective state.

Axiom 3.3. Let ensembles $\mathbf{E}$ and $\mathbf{F}$ be such that for some system S we have $S \in \mathbf{E}, S \in \mathbf{F}$. We assume that $St(\mathbf{E}) = [\Psi], \Psi \in \mathbf{L}_S^{(v)}$ for some $v \in I_S = I_E$ and that $St(\mathbf{F}) = [\Phi], \Phi \in \mathbf{L}_S^{(w)}$ for some $w \in I_S = I_F$. Then $v = w$. (I.e. the property of the individual system S depends only on S.)

Axiom 3.4. For each system S there exists the ensemble $\mathbf{E}$ and $v \in I_S$ such that $S \in \mathbf{E}$, $St(\mathbf{E}) = [\Psi]$ and $\Psi \in \mathbf{L}_S^{(v)}$.

Theorem 3.1. For each system S there exists a unique $v \in I_S$ such that for each ensemble $\mathbf{E}$ satisfying $S \in \mathbf{E}$ it is true that $St(\mathbf{E}) = [\Psi]$ implies $\Psi \in \mathbf{L}_S^{(v)}$.

Definition 3.3. For each system S we denote $v_S \in I_S$ such v which satisfies the conditions from Theorem 3.1 Then we also define a function $v : I_S \to \{1, 0\}$ by $v(S) = 1$ if $v = v_S$ and $v(S) = 0$ if $v \neq v_S$.

Definition 3.4. Let us consider an ensemble $\mathbf{E} = \{S_1, \ldots, S_N\}$ and $v \in I_E$. Then we define

$$p(v|\mathbf{E}) = \lim 1/N \sum_{i=1}^{N} v(S_i) = \lim 1/N |\{S \in \mathbf{E} | v(S) = 1\}|^2$$

Axiom 3.5. Let $\mathbf{E}$ and $\mathbf{F}$ be two ensembles such that $St(\mathbf{E}) = St(\mathbf{F}) = [\Psi], \Psi \in \mathbf{H}_E$. Then for each $v \in I_E$ we have $p(v|\mathbf{E}) = p(v|\mathbf{F})$. This means that the probability of observing the property v depends only on the state of an ensemble.

On the basis of Axiom 3.5 the probability $p(v|\mathbf{E})$ depends only on the state of $\mathbf{E}$. Thus we can define

Definition 3.5. We set $p(v|[\Psi]) = p(v|\mathbf{E})$ where $St(\mathbf{E}) = [\Psi]$.

Thus we have defined the probability distribution $p(.|[\Psi])$ on I_E depending only on $[\Psi]$.

Axiom 3.6. (The composed systems.) Let us consider two ensembles $\mathbf{M} = \{M_1, \ldots, M_N\}, \mathbf{S} = \{S1, \ldots, S_N\}$, where $St(\mathbf{M}) = [\Psi], \Psi \in \mathbf{H_M}, St(\mathbf{S}) = [\Phi], \Phi \in \mathbf{H_S}$. Let us consider the composed ensemble $\mathbf{E} = \mathbf{M} \oplus \mathbf{S} = \{M_1 \oplus S_1, \ldots, M_N \oplus S_N\}$. Then

(i) $\mathbf{H_E} = \mathbf{H_M} \otimes \mathbf{H_S}$

(ii) $I_E = I_M \times I_S, \mathbf{L_{M \oplus S(v,w)}}^{(v,w)} = \mathbf{L_M}^{(v)} \otimes \mathbf{L_S}(w)$

(iii) If $\mathbf{M}$ and $\mathbf{S}$ are independent ensembles and if $St(\mathbf{M}) = [\Psi], \Psi \in \mathbf{H_M}, St(\mathbf{S}) = [\Psi], \Psi \in \mathbf{H_S}$, then $St(\mathbf{M} \oplus \mathbf{S}) = [\Psi \otimes \Phi]$.

Definition 3.6. Let $P : \mathbf{H_M} \to \mathbf{H_M}$ be a projection and $Id(\mathbf{H_S})$ be an identity map on $\mathbf{H_S}$. Then the projection $P \otimes Id(\mathbf{H_S}) : \mathbf{H_E} \to \mathbf{H_E}$ is defined by

$$(P \otimes Id(\mathbf{H_S}))(\Psi \otimes \Phi) = P(\Psi) \otimes \Phi \text{ for each } \Psi \in \mathbf{H_M}, \Phi \in \mathbf{H_S}.$$

Axiom 3.7. (The Born's rule.) Let $\mathbf{E} = \mathbf{M} \oplus \mathbf{S} = \{M_1 \oplus S_1, \ldots, M_N \oplus S_N\}$ and let $St(\mathbf{E}) = [\Psi], \Psi \in \mathbf{H_{M \oplus S}} = \mathbf{H_M} \otimes \mathbf{H_S}$. Let $P_M^{(v)} : \mathbf{H_M} \to \mathbf{L_M^{(v)}}, v \in I_M$ be a standard projection operator. Let us assume that the individual system $M_k \oplus S_k$ be a composition of two individual systems. Then the probability that the individual system M_k will have the property v (i.e. $v(Mk) = 1$) will be

$$p(v|[\Psi]) = ||(P(v)_M \otimes Id(\mathbf{H_S}))(\Psi)||^2.$$

[2] We assume the stability (i.e. the existence of the limit) of this relative frequency.

Axiom 3.8. (The up-date rule.) Let $\mathbf{E} = \mathbf{M} \oplus \mathbf{S} = M_1 \oplus S_1, \ldots, M_N \oplus S_N$ and let $St(\mathbf{E}) = [\Psi], \Psi \in \mathbf{H_{M \oplus S}} = \mathbf{H_M} \otimes \mathbf{H_S}$. Let $P_M^{(v)} : \mathbf{H_M} \to \mathbf{L}_M^{(v)}, v \in I_M$ be a standard projection operator. Let us assume that the individual system $M_k \oplus S_k$ be a composition of two individual systems. Let us assume that the observation M_k gives that $v(M_k) = 1$, i.e. that the subsystem M_k has that property v. Then the individual system $M_k \oplus S_k$ will be an element of an up-dated ensemble $\mathbf{E}^{(v)} = \{M_I \oplus S_i | v(M_i) = 1\}$ which is in the state

$$||(P_M^{(v)} \otimes Id(\mathbf{H}_S))(\Psi)||^{-1}.(P_M^{(v)} \otimes Id(\mathbf{H}_S))(\Psi).$$

Axiom 3.9. (The existence of measuring systems.) For each $n \geq 2$ there exists at least one ensemble $\mathbf{E} = \{M_1, \ldots, M_N\}$ such that $|I(\mathbf{E})| = n$.

In the reference [4] there are proved the basic properties of the HE model:

(i) The HE model is empirically equivalent to the standard model of QM (see [4], sect. 7, p.36, where the standard model of QM is called the ontic model of QM)

(ii) The measurement process is the standard process in QM and uses the concept of an observation of the fact that the measuring system has certain property v.

(iii) The EPR correlations are explicitly local in the HE model of QM.

Thus the HE model of QM satisfies the requirements for the model T considered in the preceding section and thus theorems stated in the preceding section are proved.

The existence of different QM-models is the basic new fact presented in [4]. The invalidity of no-go theorems in QM and invalidity of the non-locality of QM are consequences of this discovery.

4. The impossibility to prove no-go theorems (and the non-locality) in the HE model of QM. The inconsistences in the standard model of QM.

The simplest sub-case of the HE model is the hybrid model (introduced in [5] under the name of the modified QM). The hybrid system is such system S for which $dim\mathbf{L}_S^(v) = 1$ for each $v \in I_S$. The hybrid model of QM is the model such that it contains only hybrid systems.

In the hybrid model each state $[\Psi], \Psi \in \mathbf{L}_S^{(v)}$ can be considered as an individual state since it can be associated with the individual system S (through the property $v(S) = 1$).

Each proof of any no-go theorem is based on the existence and the use of individual states. In general at least at one part (Alice's or Bob's) the set of considered individual states must contain at least two orthogonal bases. I.e. at least two settings of measuring apparatus is needed. But in the hybrid model we have in a disposition only one base composed of individual states for a given system. Thus the standard way of proving the no-go theorem cannot be applied. Thus the no-go theorems cannot be proved in the hybrid model. In the HE model the situation with individual states is the same. As a consequence also the proof of the non-locality of QM is impossible in the HE model of QM (details can be found in [4]).

The essential feature of the HE model of QM is its observer independence. This means that the concept of the observer does not take a part in the formulation of the model. In the HE model the measurement process is the one of standard processes in QM. Instead of the concept of a measurement there is a concept of an observation that the individual measuring system has (or does not have) a given property. The detailed description of the internal measurement process in the HE model can be found in [4].

The recently discovered **inconsistency of the standard model** ([1], [2]) was proved by constructing a rather sophisticated example based on the paradox of "Wigner's friend". Such an example cannot be constructed in the HE model since the HE model is observer independent and the paradox of "Wigner's friend" does not exists in this model. The observer independence creates the basic difference between the standard model and the HE model of QM (i.e. the HE model is a completely new type of a model of QM).

The inconsistency mentioned above implies that the HE model is the **unique realistic local possibility** for QM (the Bohm's model of QM is explicitly non-local).

5. Conclusion

(i) There exist different QM-models, e.g., the standard model, the HE model and others. The true theorem of QM must be true in all QM models.

(ii) The no-go theorems (and consequently the non-locality of QM) are not the consequences of the empirical content of QM. The standard proof of no-go theorems requires the existence of "individual states" which do not exist in the HE model of QM (or, at least, do not exist in the sufficiently big amount).

(iii) It can be proved that the EPR correlations are explicitly local in the HE-model. All details of formulations and proofs can be found in [4]

(iv) The HE model of QM is defined in the exact form of the axiomatic theory

(v) The HE model does not contain the axiomatic definition of the measurement process (in contrast to the standard model). Thus this model is observer independent. The concept of an observer does not enter into this model.

(vi) As a consequence, the Wigner's friend paradox does not exists in the HE model of QM (since there is no role of an observer in the HE model of QM).

(vii) In the light of the recent proofs of the inconsistency of the standard model ([**?**], [2]), the HE model of QM is the unique surviving local possibility for QM.

References

[1] Frauchiger D and Renner R 2016 Quantum theory cannot consistently describe the use of itself *Preprint* arXiv:1604.07422
[2] Frauchiger D and Renner R 2018 *Nature Communications* **9** 3711
[3] von Neumann J 1932 *Mathematische Grundlagen der Quantenmechanik* (Berlin: Springer-Verlag)
[4] Soucek J 2018 *A new model for quantum mechanics and the invalidity of no-go theorems* (Berlin: LAP LAMBERT Academic Publishing)
[5] Soucek J 2014 The principle of anti-superposition in QM and the local solution of the Bell's inequality problem *Preprint* http://www.nusl.cz/ntk/nusl-177617

Journal of Physics: Conference Series **1416** (2019) 012037 doi:10.1088/1742-6596/1416/1/012037

The NSVZ relation and the NSVZ scheme for $\mathcal{N}=1$ non-Abelian supersymmetric theories, regularized by higher covariant derivatives

Konstantin Stepanyantz

Moscow State University, Faculty of Physics, Department of Theoretical Physics, 119991, Moscow, Russia

E-mail: stepan@m9com.ru

Abstract. We discuss, how the exact NSVZ β-function appears in $\mathcal{N}=1$ supersymmetric non-Abelian gauge theories, regularized by higher covariant derivatives. In particular, we demonstrate that the renormalization group functions defined in terms of the bare couplings satisfy the NSVZ relation in the case of using this regularization. This occurs, because the loop integrals giving the β-function are integrals of double total derivatives with respect to loop momenta. It is also shown that for the renormalization group functions standardly defined in terms of the renormalized couplings the NSVZ scheme can be obtained if the theory is regularized by higher covariant derivatives and only powers of $\ln\Lambda/\mu$ are included into the renormalization constants. These statements are confirmed by the explicit calculations in the three-loop approximation, where the scheme dependence is essential.

1. Introduction

The exact Novikov, Shifman, Vainshtein, and Zakharov (NSVZ) β-function [1, 2, 3, 4] is an all-order expression relating the β-function of $\mathcal{N}=1$ supersymmetric gauge theories to the anomalous dimension of the matter superfields,

$$\beta(\alpha,\lambda) = -\frac{\alpha^2\Big(3C_2 - T(R) + C(R)_i{}^j\gamma_j{}^i(\alpha,\lambda)/r\Big)}{2\pi(1-C_2\alpha/2\pi)}. \qquad (1)$$

In this equation α and λ denote the gauge and Yukawa couplings, respectively, r is the dimension of a simple gauge group G, $\mathrm{tr}\,(T^A T^B) \equiv T(R)\,\delta^{AB}$, $(T^A)_i{}^k (T^A)_k{}^j \equiv C(R)_i{}^j$, and $C_2 = T(Adj)$. If the matter superfields are absent, then the NSVZ relation gives the all-order β-function of the $\mathcal{N}=1$ supersymmetric Yang–Mills (SYM) theory, which appears to be the geometric series.

Three- and four-loop calculations made for $\mathcal{N}=1$ supersymmetric theories with the help of dimensional reduction supplemented by the modified minimal subtractions demonstrate [5, 6, 7, 9] that the NSVZ relation in the $\overline{\mbox{DR}}$-scheme is valid only for the one- and two-loop β-function. In these approximations the renormalization group functions (RGFs) entering the NSVZ equation are scheme independent. (The scheme-independent two-loop β-function is related to the scheme-independent one-loop anomalous dimension of the matter superfields.) However, it turned out that in the three- and four-loop approximations the NSVZ relation can be restored by a specially tuned finite renormalization of the gauge coupling constant. It should be noted that the possibility of this tuning is highly nontrivial. This implies that the disagreement can be explained by the scheme-dependence of the NSVZ equation. Thus, the NSVZ relation takes place only in certain (NSVZ) subtraction schemes, which do not include the $\overline{\text{DR}}$-scheme.

Here we will discuss how to derive the NSVZ relation and construct the NSVZ scheme in all loops with the help of the higher covariant derivative (HD) regularization proposed by A.A.Slavnov in [10, 11] in subsequently generalized to the supersymmetric case in [12, 13].

2. The NSVZ relation with the higher derivative regularization in $\mathcal{N}=1$ SQED

First, let us briefly discuss the simplest case of the $\mathcal{N}=1$ supersymmetric electrodynamics (SQED) with N_f flavours, for which the exact NSVZ equation [14, 15] relates the β-function in a certain loop to the anomalous dimension in the previous loop.

With the HD regularization the NSVZ relation is valid in all orders [16, 17] for the (scheme-independent for a fixed regularization) RGFs defined in terms of the bare coupling constant

$$\beta(\alpha_0) \equiv \frac{d\alpha_0(\alpha,\Lambda/\mu)}{d\ln\Lambda}\Big|_{\alpha=\text{const}}; \qquad \gamma(\alpha_0) \equiv -\frac{d\ln Z(\alpha,\Lambda/\mu)}{d\ln\Lambda}\Big|_{\alpha=\text{const}}, \tag{2}$$

where α_0 and α are the bare and renormalized coupling constants, respectively. Λ is the dimensionful regularization parameter playing the role of the ultraviolet cut-off, and μ denotes the renormalization point. For RGFs (2) the equation

$$\beta(\alpha_0) = \frac{\alpha_0^2 N_f}{\pi}\Big(1-\gamma(\alpha_0)\Big) \tag{3}$$

is valid for an arbitrary renormalization prescription in the case of using the HD regularization. Note that in the case of using the regularization by dimensional reduction it is not so, see [18].

The NSVZ equation for RGFs defined in terms of the bare couplings appears in the perturbation theory because the loop integrals giving $\beta(\alpha_0)$ are integrals of total [19] and even double total derivatives [20] in the momentum space in the case of using the HD regularization. For example, in the two-loop approximation (see, e.g., [21])

$$\frac{\beta(\alpha_0)}{\alpha_0^2} = N_f \frac{d}{d\ln\Lambda}\Big\{2\pi\sum_I c_I \int \frac{d^4q}{(2\pi)^4}\frac{\partial}{\partial q^\mu}\frac{\partial}{\partial q_\mu}\frac{\ln(q^2+M_I^2)}{q^2} + 4\pi\int\frac{d^4q}{(2\pi)^4}\frac{d^4k}{(2\pi)^4}$$
$$\times\frac{e_0^2}{k^2R_k}\frac{\partial}{\partial q^\mu}\frac{\partial}{\partial q_\mu}\Big(\frac{1}{q^2(k+q)^2} - \sum_I c_I\frac{1}{(q^2+M_I^2)((k+q)^2+M_I^2)}\Big) + O(e_0^4)\Big\}. \tag{4}$$

The double total derivatives cut matter lines producing contributions to the anomalous dimension, while by attaching two gauge lines in all possible ways we obtain contributions to the β-function [20], see Fig. 1 as an illustration.

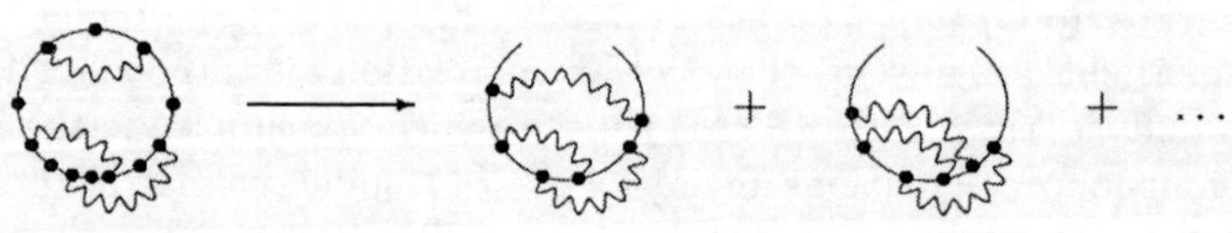

Figure 1. Cutting internal lines in a vacuum supergraph we obtain superdiagrams contributing to the anomalous dimension. From the other side, attaching two external gauge lines in all possible way to the points denoted by disks, we obtain superdiagrams contributing to $\beta(\alpha_0)$.

Note that RGFs (2) should be distinguished from RGFs standardly defined in terms of the renormalized coupling constant,

$$\widetilde{\beta}(\alpha) \equiv \frac{d\alpha_0(\alpha,\Lambda/\mu)}{d\ln\mu}\Big|_{\alpha_0=\text{const}}; \qquad \widetilde{\gamma}(\alpha) \equiv \frac{d\ln Z(\alpha_0,\Lambda/\mu)}{d\ln\mu}\Big|_{\alpha_0=\text{const}}. \tag{5}$$

RGFs defined by these equations appear to be scheme-dependent and satisfy the NSVZ equation only for special (NSVZ) renormalization prescriptions. One of these prescriptions (so-called HD+MSL) is obtained with the HD regularization, when Minimal Subtractions of Logarithms are used for renormalization. In this case only powers of $\ln\Lambda/\mu$ are included into the renormalization constants and all finite constants are set to 0. Then according to [22, 23, 24] both definitions of RGFs give the same result up to the change of the argument,

$$\beta(\alpha_0\to\alpha) = \widetilde{\beta}(\alpha); \qquad \gamma(\alpha_0\to\alpha) = \widetilde{\gamma}(\alpha). \tag{6}$$

Taking into account that RGFs (2) satisfy Eq. (3) in all orders, we see that the HD+MSL prescription gives the NSVZ scheme in all loops. However, this NSVZ scheme is not unique. Various NSVZ schemes are related by the finite renormalizations of the form $\alpha' = \alpha'(\alpha)$, $Z'(\alpha',\Lambda/\mu) = z(\alpha)Z(\alpha,\Lambda/\mu)$ which satisfy the equation [25]

$$\frac{1}{\alpha'(\alpha)} - \frac{1}{\alpha} - \frac{N_f}{\pi}\ln z(\alpha) = B, \tag{7}$$

where a constant B appears due to an arbitrariness in choosing the renormalization point μ. According to [26], another example of an all-loop NSVZ prescription is the on-shell scheme. This implies that both the HD+MSL and on-shell schemes belong to the class of the NSVZ schemes.

3. The NSVZ relation in the non-Abelian case

For non-Abelian $\mathcal{N}=1$ supersymmetric theories the NSVZ equation (1) relates the β-function to the matter superfield anomalous dimension in all previous loops due to the coupling constant dependent denominator. Moreover, it has not a qualitative interpretation similar to the Abelian case, because the cuts of internal lines can also give superdiagrams contributing to the anomalous dimensions of the Faddeev–Popov ghosts and of the quantum gauge superfields. However, it is possible to rewrite the NSVZ equation in a different (but equivalent) form. This can be done with the help of the non-renormalization theorem for the triple gauge-ghost vertices [27]. This theorem states that the three-point vertices with two ghost legs and a single leg of the quantum gauge superfield are finite in all loops. There are 4 such vertices corresponding to $\bar{c}Vc$, $\bar{c}^+Vc$, $\bar{c}Vc^+$, and $\bar{c}^+Vc^+$ external lines. All of them have the same renormalization constant $Z_\alpha^{-1/2}Z_cZ_V$. Therefore, their finiteness leads to the identity

$$\frac{d}{d\ln\Lambda}(Z_\alpha^{-1/2}Z_cZ_V) = 0, \tag{8}$$

where the renormalization constants are defined by the equations $Z_\alpha = \alpha/\alpha_0$, $V = Z_VZ_\alpha^{-1/2}V_R$, $\bar{c}c = Z_cZ_\alpha^{-1}\bar{c}_Rc_R$, and $\phi_i = (\sqrt{Z_\phi})_i{}^j(\phi_R)_j$. Therefore, it is possible to choose such a renormalization prescription that $Z_\alpha^{-1/2}Z_cZ_V = 1$.

For RGFs defined in terms of the bare couplings (see Eq. (18) below) the NSVZ equation can be equivalently presented in the form

$$\frac{\beta(\alpha_0,\lambda_0)}{\alpha_0^2} = -\frac{3C_2 - T(R) + C(R)_i{}^j(\gamma_\phi)_j{}^i(\alpha_0,\lambda_0)/r}{2\pi} + \frac{C_2}{2\pi}\cdot\frac{\beta(\alpha_0,\lambda_0)}{\alpha_0}. \tag{9}$$

The β-function in the right hand side can be expressed in terms of the anomalous dimensions of the quantum gauge superfield (γ_V) and of the Faddeev–Popov ghosts (γ_c) with the help of the non-renormalization theorem for the triple gauge-ghost vertices,

$$\beta(\alpha_0,\lambda_0) = -\alpha_0\frac{d\ln Z_\alpha}{d\ln\Lambda}\Big|_{\alpha,\lambda=\text{const}} = -2\alpha_0\frac{d\ln(Z_c Z_V)}{d\ln\Lambda}\Big|_{\alpha,\lambda=\text{const}} = 2\alpha_0\Big(\gamma_c(\alpha_0,\lambda_0)+\gamma_V(\alpha_0,\lambda_0)\Big). \quad (10)$$

Substituting this expression into the right hand side of Eq. (9) one obtains the equivalent form of the NSVZ relation

$$\frac{\beta(\alpha_0,\lambda_0)}{\alpha_0^2} = -\frac{1}{2\pi}\Big(3C_2 - T(R) - 2C_2\gamma_c(\alpha_0,\lambda_0) - 2C_2\gamma_V(\alpha_0,\lambda_0) + C(R)_i{}^j(\gamma_\phi)_j{}^i(\alpha_0,\lambda_0)/r\Big). \quad (11)$$

This equation relates the β-function in a certain loop to the anomalous dimensions of the quantum superfields in the previous loop and does not contain the coupling dependent denominator in the right hand side. Its graphical interpretation is similar to the Abelian case, because cuts of internal lines in vacuum supergraphs produce superdiagrams contributing to γ_V, γ_c, and $(\gamma_\phi)_j{}^i$. This allows suggesting that it is this equation that follows from the perturbative calculations. This guess was confirmed by an explicit calculation in the two-loop approximation, see [28].

4. Outline of the all-loop derivation

The new form of the NSVZ equation for RGFs defined in terms of the bare couplings (11) can be derived in all orders in the case of using the higher covariant derivative regularization. The proof is based on the method proposed in Refs. [16, 17]. Here we briefly outline its main steps.

1. Let us consider the two-point Green function of the background gauge superfield $\boldsymbol{V}$

$$\Gamma^{(2)}_{\boldsymbol{V}} = -\frac{1}{8\pi}\text{tr}\int\frac{d^4p}{(2\pi)^4}\,d^4\theta\,\boldsymbol{V}(-p,\theta)\partial^2\Pi_{1/2}\boldsymbol{V}(p,\theta)\,d^{-1}(\alpha_0,\lambda_0,\Lambda/p) \quad (12)$$

and make the formal replacement $\boldsymbol{V}^A \to \theta^4 v^A$, where v^A is a function slowly decreasing at a scale $R\to\infty$. This allows extracting the β-function defined in terms of the bare couplings,

$$\frac{d\Delta\Gamma^{(2)}_{\boldsymbol{V}}}{d\ln\Lambda} = \frac{\mathcal{V}_4}{2\pi}\frac{d}{d\ln\Lambda}\Big(d^{-1}(\alpha_0,\lambda_0,\Lambda/p)-\alpha_0^{-1}\Big)\Big|_{\alpha,\lambda=\text{const};\ p\to 0} = \frac{\mathcal{V}_4}{2\pi}\cdot\frac{\beta(\alpha_0,\lambda_0)}{\alpha_0^2}, \quad (13)$$

where

$$\Delta\Gamma^{(2)}_{\boldsymbol{V}} \equiv \Gamma^{(2)}_{\boldsymbol{V}} - S^{(2)}_{\boldsymbol{V}} \qquad \text{and} \qquad \mathcal{V}_4 \equiv \int d^4x\,(v^A)^2. \quad (14)$$

2. Using the rules for calculating supergraphs and the identity [17]

$$\theta^2 AB\theta^2 + 2(-1)^{P_A+P_B}\theta^a A\theta^2 B\theta_a - \theta^2 A\theta^2 B - A\theta^2 B\theta^2 = O(\theta), \quad (15)$$

where $(-1)^{P_A}$ and $(-1)^{P_B}$ are the Grassmannian parities of A and B (which are constructed from supersymmetric and usual derivatives and superspace δ-functions), the expression for the β-function can be reduced to a formally vanishing integral of double total derivatives [29]. This integral is similar to

$$\int\frac{d^4q}{(2\pi)^4}\frac{\partial}{\partial q^\mu}\frac{\partial}{\partial q_\mu}\Big(\frac{f(q^2)}{q^2}\Big) = \frac{1}{8\pi^4}\int\limits_{S^3_\varepsilon} dS\,\frac{f(q^2)}{q^3} = \frac{1}{4\pi^2}f(0), \quad (16)$$

where $f(q^2)$ is a non-singular function rapidly decreasing at infinity. This implies that $\beta(\alpha_0,\lambda_0)$ beyond the one-loop approximation is given by the sum of singular contributions.

Journal of Physics: Conference Series **1416** (2019) 012037 doi:10.1088/1742-6596/1416/1/012037

3. The sum of various singular contributions was found to be

$$\frac{\beta(\alpha_0,\lambda_0)}{\alpha_0^2}-\frac{\beta_{\text{1-loop}}(\alpha_0)}{\alpha_0^2}=\frac{1}{\pi}C_2\gamma_V(\alpha_0,\lambda_0)+\frac{1}{\pi}C_2\gamma_c(\alpha_0,\lambda_0)-\frac{1}{2\pi r}C(R)_i{}^j(\gamma_\phi)_j{}^i(\alpha_0,\lambda_0). \qquad (17)$$

(The paper describing the details of this calculation is in preparation.) After substituting the one-loop contribution to the β-function $\beta_{\text{1-loop}}(\alpha_0)=-\alpha_0^2(3C_2-T(R))/2\pi$ we obtain the new form of the NSVZ relation.

5. The NSVZ equation and various definitions of RGFs

The reasoning sketched above allows deriving the NSVZ equation for RGFs defined in terms of the bare couplings

$$\beta(\alpha_0,\lambda_0)\equiv\frac{d\alpha_0}{d\ln\Lambda};\qquad (\gamma_\phi)_i{}^j(\alpha_0,\lambda_0)\equiv-\frac{d\ln(Z_\phi)_i{}^j}{d\ln\Lambda};$$
$$\gamma_V(\alpha_0,\lambda_0)\equiv-\frac{d\ln Z_V}{d\ln\Lambda};\qquad \gamma_c(\alpha_0,\lambda_0)\equiv-\frac{d\ln Z_c}{d\ln\Lambda}. \qquad (18)$$

For a fixed regularization they do not depend on the renormalization prescription. Therefore, the NSVZ equation for these RGFs is valid in an arbitrary subtraction scheme in the case of using the higher covariant derivative regularization. However, RGFs defined in terms of the renormalized couplings

$$\widetilde\beta(\alpha,\lambda)\equiv\frac{d\alpha}{d\ln\mu};\qquad (\widetilde\gamma_\phi)_i{}^j(\alpha,\lambda)\equiv\frac{d\ln(Z_\phi)_i{}^j}{d\ln\mu};$$
$$\widetilde\gamma_V(\alpha,\lambda)\equiv\frac{d\ln Z_V}{d\ln\mu};\qquad \widetilde\gamma_c(\alpha,\lambda)\equiv\frac{d\ln Z_c}{d\ln\mu} \qquad (19)$$

are scheme dependent and satisfy the NSVZ equation only in certain (NSVZ) schemes.

Similarly to the $\mathcal{N}=1$ SQED case, one of the NSVZ schemes for non-Abelian $\mathcal{N}=1$ supersymmetric theories is given by the HD+MSL prescription [27],

$$\text{HD+MSL}=\text{NSVZ}, \qquad (20)$$

because in this case both definitions of RGFs give the same result,

$$\beta(\alpha_0\to\alpha,\lambda_0\to\lambda)=\widetilde\beta(\alpha,\lambda);\qquad (\gamma_\phi)_i{}^j(\alpha_0\to\alpha,\lambda_0\to\lambda)=(\widetilde\gamma_\phi)_i{}^j(\alpha,\lambda);$$
$$\gamma_V(\alpha_0\to\alpha,\lambda_0\to\lambda)=\widetilde\gamma_V(\alpha,\lambda);\qquad \gamma_c(\alpha_0\to\alpha,\lambda_0\to\lambda)=\widetilde\gamma_c(\alpha,\lambda). \qquad (21)$$

6. Explicit perturbative three-loop calculation for terms containing Yukawa couplings

The above results have been verified by the explicit calculation of terms containing the Yukawa couplings in the three-loop β-function [30, 31]. They are generated by the supergraphs presented in Fig. 2 to which one should attach two external lines of the background gauge superfield in all possible ways. From the other side, cuts of the internal lines in these graphs produce various two-loop contributions to the anomalous dimensions of the quantum gauge superfield and of the matter superfields. (For the supergraphs under consideration superdiagrams contributing to the anomalous dimension of the Faddeev–Popov ghosts do not appear.)

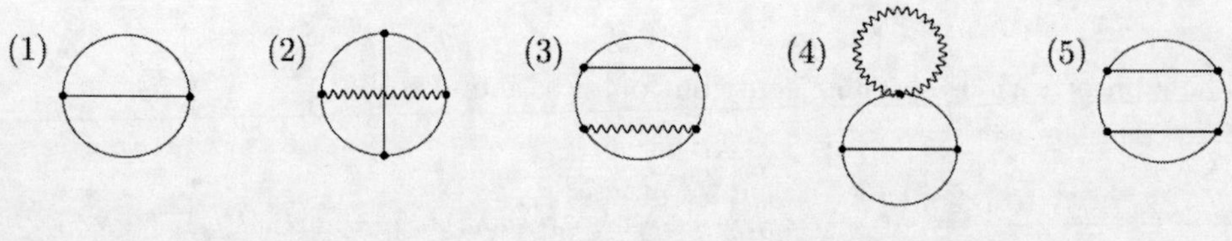

Figure 2. Supergraphs generating the three-loop contributions to the β-function containing the Yukawa couplings.

Expressions for all graphs obtained after the direct calculation of the two-point superdiagrams and summing the results turned out to be given by integrals of double total derivatives. This makes possible the analytic evaluation of one of the loop integrals, which has lead to the relations

$$\Delta_A\Big(\frac{\beta(\alpha_0,\lambda_0)}{\alpha_0^2}\Big) = \frac{1}{\pi}C_2\Delta_A\gamma_V(\alpha_0,\lambda_0) - \frac{1}{2\pi r}C(R)_i{}^j(\Delta_A\gamma_\phi)_j{}^i(\alpha_0,\lambda_0) \qquad (22)$$

for all considered graphs. For instance, the terms quartic in the Yukawa couplings (first found in [30]) are written as

$$\frac{\Delta\beta(\alpha_0,\lambda_0)}{\alpha_0^2} = -\frac{2\pi}{r}C(R)_i{}^j\frac{d}{d\ln\Lambda}\int\frac{d^4k}{(2\pi)^4}\frac{d^4q}{(2\pi)^4}\,\lambda_0^{imn}\lambda^*_{0jmn}\frac{\partial}{\partial q_\mu}\frac{\partial}{\partial q^\mu}\Big(\frac{1}{k^2F_k\,q^2F_q\,(q+k)^2F_{q+k}}\Big)$$
$$+\frac{4\pi}{r}C(R)_i{}^j\frac{d}{d\ln\Lambda}\int\frac{d^4k}{(2\pi)^4}\frac{d^4l}{(2\pi)^4}\frac{d^4q}{(2\pi)^4}\Big(\lambda_0^{iab}\lambda^*_{0kab}\lambda_0^{kcd}\lambda^*_{0jcd}\Big(\frac{\partial}{\partial k_\mu}\frac{\partial}{\partial k^\mu} - \frac{\partial}{\partial q_\mu}\frac{\partial}{\partial q^\mu}\Big) + 2\lambda_0^{iab}\lambda^*_{0jac}$$
$$\times\lambda_0^{cde}\lambda^*_{0bde}\frac{\partial}{\partial q_\mu}\frac{\partial}{\partial q^\mu}\Big)\frac{1}{k^2F_k^2\,q^2F_q\,(q+k)^2F_{q+k}\,l^2F_l\,(l+k)^2F_{l+k}} = -\frac{1}{2\pi r}C(R)_i{}^j\Delta\gamma_\phi(\lambda_0)_j{}^i. \qquad (23)$$

For all supergraphs such integrals have been calculated for the higher derivative regulators $F(x) = 1 + x^n$; $R(x) = 1 + x^m$. The result for RGFs defined in terms of the bare couplings appear to be

$$\frac{\beta(\alpha_0,\lambda_0)}{\alpha_0^2} = -\frac{1}{2\pi}\Big(3C_2 - T(R)\Big) - \frac{1}{2\pi r}C(R)_j{}^i\Big(\frac{1}{4\pi^2}\lambda^*_{0imn}\lambda_0^{jmn} + \frac{\alpha_0}{8\pi^3}\lambda^*_{0imn}\lambda_0^{jmn}C_2$$
$$-\frac{\alpha_0}{8\pi^3}\lambda^*_{0lmn}\lambda_0^{jmn}C(R)_i{}^l\Big(1-\frac{1}{n}\Big) + \frac{\alpha_0}{4\pi^3}\lambda^*_{0imn}\lambda_0^{jml}C(R)_l{}^n\Big(1+\frac{1}{n}\Big) - \frac{1}{16\pi^4}\lambda^*_{0iac}\lambda_0^{jab}\lambda^*_{0bde}\lambda_0^{cde}\Big)$$
$$+O(\alpha_0^2\lambda_0^2,\alpha_0\lambda_0^4,\lambda_0^6) + \text{terms without the Yukawa couplings;} \qquad (24)$$

$$(\gamma_\phi)_i{}^j(\alpha_0,\lambda_0) = -\frac{\alpha_0}{\pi}C(R)_i{}^j + \frac{1}{4\pi^2}\lambda^*_{0imn}\lambda_0^{jmn} - \frac{\alpha_0}{8\pi^3}\lambda^*_{0lmn}\lambda_0^{jmn}C(R)_i{}^l\Big(1-\frac{1}{n}\Big)$$
$$+\frac{\alpha_0}{4\pi^3}\lambda^*_{0imn}\lambda_0^{jml}C(R)_l{}^n\Big(1+\frac{1}{n}\Big) - \frac{1}{16\pi^4}\lambda^*_{0iac}\lambda_0^{jab}\lambda^*_{0bde}\lambda_0^{cde} + O(\alpha_0^2,\alpha_0\lambda_0^4,\lambda_0^6); \qquad (25)$$

$$\gamma_V(\alpha_0,\lambda_0) = -\frac{\alpha_0}{4\pi}\Big(3C_2 - T(R)\Big) - \frac{\alpha_0}{16\pi^3 r}\lambda^*_{0jmn}\lambda_0^{imn}C(R)_i{}^j + O(\alpha_0^2,\alpha_0\lambda_0^4). \qquad (26)$$

RGFs defined in terms of the renormalized couplings are

$$\frac{\widetilde{\beta}(\alpha,\lambda)}{\alpha^2} = -\frac{1}{2\pi}\Big(3C_2 - T(R)\Big) - \frac{1}{2\pi r}C(R)_j{}^i\Big(\frac{1}{4\pi^2}\lambda^*_{imn}\lambda^{jmn} + \frac{\alpha}{8\pi^3}\lambda^*_{imn}\lambda^{jmn}C_2$$
$$+\frac{\alpha}{4\pi^3}\lambda^*_{lmn}\lambda^{jmn}C(R)_i{}^l\Big[b_2 - g_{11} - \frac{1}{2}\Big(1-\frac{1}{n}\Big)\Big] + \frac{\alpha}{2\pi^3}\lambda^*_{imn}\lambda^{jml}C(R)_l{}^n\Big[b_2 - g_{11} + \frac{1}{2}\Big(1+\frac{1}{n}\Big)\Big]$$

$$-\frac{1}{8\pi^4}\lambda^*_{iac}\lambda^{jab}\lambda^*_{bde}\lambda^{cde}\Big[b_2-g_{12}+\frac{1}{2}\Big]+\frac{1}{16\pi^4}\lambda^*_{iab}\lambda^{kab}\lambda^*_{kcd}\lambda^{jcd}\Big[g_{12}-b_2\Big]\Big)+O(\alpha^2\lambda^2,\alpha\lambda^4,\lambda^6)$$

$$+\text{terms without the Yukawa couplings;} \qquad (27)$$

$$(\widetilde{\gamma}_\phi)_i{}^j(\alpha,\lambda)=-\frac{\alpha}{\pi}C(R)_i{}^j+\frac{1}{4\pi^2}\lambda^*_{imn}\lambda^{jmn}+\frac{\alpha}{4\pi^3}\lambda^*_{lmn}\lambda^{jmn}C(R)_i{}^l\Big[g_{12}-g_{11}-\frac{1}{2}\Big(1-\frac{1}{n}\Big)\Big]$$

$$+\frac{\alpha}{2\pi^3}\lambda^*_{imn}\lambda^{jml}C(R)_l{}^n\Big[g_{12}-g_{11}+\frac{1}{2}\Big(1+\frac{1}{n}\Big)\Big]-\frac{1}{16\pi^4}\lambda^*_{iac}\lambda^{jab}\lambda^*_{bde}\lambda^{cde}+O(\alpha^2,\alpha\lambda^4,\lambda^6); \qquad (28)$$

$$\widetilde{\gamma}_V(\alpha,\lambda)=-\frac{\alpha}{4\pi}\Big(3C_2-T(R)\Big)-\frac{\alpha}{16\pi^3 r}\lambda^*_{jmn}\lambda^{imn}C(R)_i{}^j+O(\alpha^2,\alpha\lambda^4). \qquad (29)$$

The finite constants b_1, b_2, g_{11}, and g_{12} entering these expressions are defined by equations

$$\frac{1}{\alpha}-\frac{1}{\alpha_0}=-\frac{1}{2\pi}\Big(3C_2-T(R)\Big)\Big[\ln\frac{\Lambda}{\mu}+b_1\Big]-\frac{1}{8\pi^3 r}C(R)_j{}^i\lambda^*_{imn}\lambda^{jmn}\Big[\ln\frac{\Lambda}{\mu}+b_2\Big]+O(\alpha,\alpha\lambda^2,\lambda^4);$$

$$(Z_\phi)_i{}^j=\delta_i^j+\frac{\alpha}{\pi}\Big(\ln\frac{\Lambda}{\mu}+g_{11}\Big)C(R)_i{}^j-\frac{1}{4\pi^2}\lambda^*_{imn}\lambda^{jmn}\Big(\ln\frac{\Lambda}{\mu}+g_{12}\Big)+O(\alpha^2,\alpha\lambda^2,\lambda^4) \qquad (30)$$

and (partially) fix the subtraction scheme in the lowest approximation. The HD+MSL prescription evidently corresponds to $g_{11}=0$, $g_{12}=0$, $b_1=b_2=0$.

Comparing Eqs. (27), (28), and (29) we see that for the considered terms in the considered approximation the new form of the NSVZ relation

$$\frac{\widetilde{\beta}(\alpha,\lambda)}{\alpha^2}=-\frac{1}{2\pi}\Big(3C_2-T(R)-2C_2\widetilde{\gamma}_c(\alpha,\lambda)-2C_2\widetilde{\gamma}_V(\alpha,\lambda)+C(R)_i{}^j(\widetilde{\gamma}_\phi)_j{}^i(\alpha,\lambda)/r\Big) \qquad (31)$$

is not valid for a general renormalization prescription. However, in the HD+MSL renormalization scheme (for which $g_{12}-g_{11}=0$, $b_2-g_{11}=0$, $b_2-g_{12}=0$) Eq. (31) is really valid. Moreover, it is easy to see that in this case both definition of RGFs give the same functions, see Eq. (21). It is possible to verify that the original form of the NSVZ relation is also valid in this case. Thus, for the considered class of superdiagrams HD + MSL = NSVZ.

7. Conclusion

The NSVZ equation naturally appears in the perturbation theory if the higher covariant derivative method is used for regularization. However, it is obtained in the new form, which relates the β-function to the anomalous dimensions of the quantum gauge superfield, of the Faddeev–Popov ghosts, and of the matter superfields. With the HD regularization the NSVZ relation is satisfied by RGFs defined in terms of the bare couplings independently of the prescription used for renormalization. This relation originates from the factorization of integrals giving the β-function into integrals of double total derivatives. If RGFs are defined in terms of the renormalized couplings, then the NSVZ scheme in all orders is given by the HD+MSL prescription, when the theory is regularized by higher covariant derivatives and the renormalization constants include only powers of $\ln\Lambda/\mu$.

All these general statements concerning the NSVZ relation and the NSVZ scheme have been confirmed by some explicit calculations in the three-loop approximation, where the scheme dependence is essential.

Acknowledgments

This research is supported by Foundation for Advancement of Theoretical Physics and Mathematics 'BASIS', grant No. 19-1-1-45-1.

References

[1] Novikov V A, Shifman M A, Vainshtein A I and Zakharov V I 1983 Exact Gell-Mann-Low Function of Supersymmetric Yang-Mills Theories from Instanton Calculus *Nucl. Phys.* B **229** 381.
[2] Jones D R T 1983 More on the Axial Anomaly in Supersymmetric Yang-Mills Theory *Phys. Lett.* **123B** 45.
[3] Novikov V A, Shifman M A, Vainshtein A I and Zakharov V I 1986 Beta Function in Supersymmetric Gauge Theories: Instantons Versus Traditional Approach *Phys. Lett.* **166B** 329.
[4] Shifman M A and Vainshtein A I 1986 Solution of the Anomaly Puzzle in SUSY Gauge Theories and the Wilson Operator Expansion *Nucl. Phys.* B **277** 456.
[5] Avdeev L V and Tarasov O V 1982 The Three Loop Beta Function in the $N = 1$, $N = 2$, $N = 4$ Supersymmetric Yang-Mills Theories *Phys. Lett.* **112B** 356.
[6] Jack I, Jones D R T and North C G 1996 N=1 supersymmetry and the three loop gauge Beta function *Phys. Lett.* B **386** 138.
[7] Jack I, Jones D R T and North C G 1997 Scheme dependence and the NSVZ Beta function *Nucl. Phys.* B **486** 479.
[8] Jack I, Jones D R T and Pickering A 1998 The Connection between DRED and NSVZ *Phys. Lett.* B **435** 61.
[9] Harlander R V, Jones D R T, Kant P, Mihaila L and Steinhauser M 2006 Four-loop beta function and mass anomalous dimension in dimensional reduction *JHEP* **0612** 024.
[10] Slavnov A A 1971 Invariant regularization of nonlinear chiral theories *Nucl. Phys.* B **31** 301.
[11] Slavnov A A 1972 Invariant regularization of gauge theories *Theor.Math.Phys.* **13** 1064.
[12] Krivoshchekov V K 1978 Invariant Regularizations for Supersymmetric Gauge Theories *Theor. Math. Phys.* **36** 745.
[13] West P C 1986 Higher Derivative Regulation of Supersymmetric Theories *Nucl. Phys.* B **268** 113.
[14] Vainshtein A I, Zakharov V I and Shifman M A 1985 Gell-mann-low Function In Supersymmetric Electrodynamics *JETP Lett.* **42** 224.
[15] Shifman M A, Vainshtein A I and Zakharov V I 1986 Exact Gell-mann-low Function In Supersymmetric Electrodynamics *Phys. Lett.* **166B** 334.
[16] Stepanyantz K V 2011 Derivation of the exact NSVZ β-function in N=1 SQED, regularized by higher derivatives, by direct summation of Feynman diagrams *Nucl. Phys.* B **852** 71.
[17] Stepanyantz K V 2014 The NSVZ β-function and the Schwinger-Dyson equations for $\mathcal{N} = 1$ SQED with N_f flavors, regularized by higher derivatives *JHEP* **1408** 096.
[18] Aleshin S S, Goriachuk I O, Kataev A L and Stepanyantz K V 2017 The NSVZ scheme for $\mathcal{N} = 1$ SQED with N_f flavors, regularized by the dimensional reduction, in the three-loop approximation *Phys. Lett.* B **764** 222.
[19] Soloshenko A A and Stepanyantz K V 2004 Three loop beta function for N=1 supersymmetric electrodynamics, regularized by higher derivatives *Theor. Math. Phys.* **140** 1264.
[20] Smilga A V and Vainshtein A 2005 Background field calculations and nonrenormalization theorems in 4-D supersymmetric gauge theories and their low-dimensional descendants *Nucl. Phys.* B **704** 445.
[21] Stepanyantz K V 2012 Derivation of the exact NSVZ beta-function in N=1 SQED regularized by higher derivatives by summation of Feynman diagrams *J. Phys. Conf. Ser.* **343** 012115.
[22] Kataev A L and Stepanyantz K V 2013 NSVZ scheme with the higher derivative regularization for $\mathcal{N} = 1$ SQED *Nucl. Phys.* B **875** 459.
[23] Kataev A L and Stepanyantz K V 2014 Scheme independent consequence of the NSVZ relation for N=1 SQED with N_f flavors *Phys. Lett.* B **730** 184.
[24] Kataev A L and Stepanyantz K V 2014 The NSVZ beta-function in supersymmetric theories with different regularizations and renormalization prescriptions *Theor. Math. Phys.* **181** 1531.
[25] Goriachuk I O, Kataev A L and Stepanyantz K V 2018 A class of the NSVZ renormalization schemes for $\mathcal{N} = 1$ SQED *Phys. Lett.* B **785** 561.
[26] Kataev A L, Kazantsev A E and Stepanyantz K V 2019 On-shell renormalization scheme for $\mathcal{N} = 1$ SQED and the NSVZ relation *Eur. Phys. J.* C **79** 477.
[27] Stepanyantz K V 2016 Non-renormalization of the $V\bar{c}c$-vertices in $\mathcal{N} = 1$ supersymmetric theories *Nucl. Phys.* B **909** 316.
[28] Shakhmanov V Y and Stepanyantz K V 2018 New form of the NSVZ relation at the two-loop level *Phys. Lett.* B **776** 417.
[29] Stepanyantz K V 2019 The β-function of $\mathcal{N} = 1$ supersymmetric gauge theories regularized by higher covariant derivatives as an integral of double total derivatives *Preprint* arXiv:1908.04108 [hep-th].
[30] Shakhmanov V Y and Stepanyantz K V 2017 Three-loop NSVZ relation for terms quartic in the Yukawa couplings with the higher covariant derivative regularization *Nucl. Phys.* B **920** 345.
[31] Kazantsev A E, Shakhmanov V Y and Stepanyantz K V 2018 New form of the exact NSVZ β-function: the three-loop verification for terms containing Yukawa couplings *JHEP* **1804** 130.

Drinfeld Yangian of the queer Lie superalgebra sq_1

Vladimir Stukopin

South Mathematical Institute and Moscow Institute of Physics and Technology (MIPT), Institutsky, 9, Dolgoprudny, Moscow region, Russian Federation

E-mail: stukopin@mail.ru

Abstract. Drinfeld Yangian of a queer Lie superalgebra sq_1 is defined as the quantization of a Lie bisuperelgebra of twisted polynomial currents. An analogue of the new system of generators of Drinfeld is being constructed. It is proved for the case Lie superalgebra sq_1 that this so defined Yangian and the Yangian, introduced earlier by M. Nazarov using the Faddeev-Reshetikhin-Takhtadzhjan approach, are isomorphic.

Mathematical Subject Classification (2000). Primary 17B37; Secondary 81R50, 13F60.

Keywords Super Yangian, Queer Lie Superalgebra, Drinfel'd Yangian, Hopf superalgebra.

1. Introduction

The Yangians of simple Lie superalgebras were defined by Drinfeld in the mid-1980s (see, for example, [3], [4], [5], [6], [1]), as quantizations (or non-commutative deformations in the Hopf algebra class) of Lie polynomial Lie bisuperalgebras with a bracket, determined by a rational r-matrix (the Yang matrix). But the Yangian of general linear algebra appeared earlier in the works of the mathematical physicists of the Leningrad school L.D. Faddeev. In the second case, it was determined using the defining relations given by the Yang quantum R-matrix. V.Drinfeld established the equivalence of these two different definitions of the Yangian for a special linear algebra, but he did not publish proof of this fact. The first published evidence appeared much later (first for sl_2-case and later, almost 20 years after the first works of Drinfeld and for the case sl_n (see [10])).

Later, in the early 90s of the last century, L.D. Faddeev, N.Yu. Reshetikhin and L. Takhtadzhjan published a detailed description of the approach to the definition of quantum groups based on the use of the quantum R- matrix for defining a system of defining relations ([10]). Such an approach was naturally related to the quantum method of the inverse scattering problem developed by them, which is now more often called the algebraic Bethe ansatz. Because, it is natural to call their name the traditional definition for the mathematical physicists of the Yangian, in contrast to V. Drinfeld's definition. In the 90s of the last century, the Yangians of Lie superalgebras began to be investigated (see the works [12], [18]). Moreover, in the first paper the definition of the Yangian of the general linear Lie superalgebra was given by the Faddeev-Reshetikhin-Takhdazhjan approach (the FRT approach), and in the second work, the Yangian of the special linear superalgebra was defined in accordance with Drinfeld's approach. The Yangian thus defined we call the Drinfeld Yangian. We note that the Yangians of Lie superalgebras have now found numerous applications in quantum field theory, in particular, in

Journal of Physics: Conference Series **1416** (2019) 012038 doi:10.1088/1742-6596/1416/1/012038

the quantum theory of superstrings ([2], [17]).

For definition Yangian in accordance with the Drinfeld approach it is necessary to have a nondegenerate invariant bilinear form on the original Lie algebra (or Lie superalgebra), which is used in the definition of the Young matrix or for definition cobracket in Lie bisuperalgebra. But such a form is absent in the queer Lie superalgebra, which prevents the extension of the Drinfeld definition of Yangian on this case. Nevertheless, as will be seen below, instead of deforming the superalgebra of polynomial currents, one can consider the deformation of a twisted superalgebra of currents. The Yangians associated with the queer Lie superalgebra (see [9], [7]) were first constructed by M. Nazarov ([13]), using approach Faddeev-Reshetikhin-Takhdazhjan. Later in the paper by M. Nazarov and A. Sergeev ([14]) was given definition of Yangian of the queer Lie superalgebra using a general approach based on the centralizer construction of G. Olshansky. This definition was later used in [15] to establish the connection between the Yangian of queer Lie superalgebra and W-algebras, introduced in [16].

In the paper [23], the Yangian of the queer Lie superalgebra was introduced, following the Drinfeld approach as a quantization of the Lie bisuperalgebra of twisted currents. There was also obtained his description in terms of the current system of generators and defining relations (a new system of generators in the terminology of V. Drinfel'd, [5]). It should be noted that the Yangian of a queer Lie superalgebra is defined as the deformation of a Lie bisuperalgebra of twisted polynomial currents in which the structure of a Lie superalgebra is given by a rational r-matrix (see [20]).

We note that this approach can not be fully extended to quantizations of other twisted superalgebras of currents, since, generally speaking, they can not be endowed with the structure of a Lie bisuperalgebra (see [24]). Nevertheless, with a small modification, it can be used for definition of twisted Yangians, which, generally speaking, are not Hopf superalgebras.

In this paper, which is the first part of the work devoted to the study of the Yangian of the queer Lie superalgebra (see also [29], [28]), we define the Yangian of the queer Lie superalgebra based on the Drinfeld approach, and establish a connection with the Yangian Nazarov of the queer Lie superalgebra in the particular case of the Lie superalgebra q_1. A few words about the organization of this work. In the second section we recall the definitions of the queer Lie superalgebra and the twisted current Lie bisuperelgebra. We describe the deformation of the twisted current Lie bisupergalgebra and introduce the main actor, the Yangian of the queer Lie superalgebra. In the section 3 we recall the definition of the Yangian of the queer Lie superalgebras given by M. Nazarov and introduce the current system of generators for Drinfeld Yangian in section 4. We prove in the section 5 the theorem on the isomorphism of the Drinfeld Yangian of the queer Lie superalgebra and the Yangian introduced by M. Nazarov for the partial case Q_1. Our proof is based on well-known ideas and essentially uses the triangular decomposition of the transfer matrix used in definition of the defining relations in the Yangian introduced by M. Nazarov and the construction of the quasi-determinants that was introduced by I.M. Gelfand and V.S. Retakh (see [8]).

In this paper we give the sketch of idea of the proof of general case using quasi-determinant theory (see [8], [11]) and uses some features of the root system of the queer Lie superalgebra (see also [29]).

We note that simplifying the proof of the theorem on isomorphism, we somewhat changed the current system of generators in this paper in comparison with the papers [22], [23], [25], [28] which led to a somewhat simpler and more natural form of defining relations. Let's note that exist isomorphism between super Yangian and quantum loop superalgebra for special linear superalgebra, which is twin of queer Lie superalgebra ([26], [27]). To construct such isomorphism for Yangian of queer Lie superalgebra is an open problem.

2. Queer Lie superalgebra q_1, graded by involution Lie superalgebra $gl(1,1)$ and Yangian

In this case we are dealing not with the basic Lie superalgebra. Nevertheless, the general construction of quantization considered in the previous subsection can also be used in this particular case. We recall the definition of the Lie superalgebra q_1. This superalgebra is generated by the generators h_0, h_1, and $p(h_0) = 0$, $p(h_1) = 1$, which satisfy the following relations

$$[h_0, h_0] = 0, \quad [h_0, h_1] = 0, \quad [h_1, h_1] = 2h_0. \tag{1}$$

The relations for a twisted superalgebra of currents with values in $gl(1,1)$ have the following form.

$$[h_0, h_0^1] = [h_0^1, h_0^1] = 0, \quad [h_0^1, h_1^1] = -[h_1^1, h_0^1] = \tfrac{1}{2}h_1, \tag{2}$$
$$[h_1^1, h_1^1] = -\tfrac{1}{2}h_0, \quad [h_0^1, h_1] = 2h_1^1, \quad [h_1^1, h_1] = 2h_1^1. \tag{3}$$

The Casimir element corresponding to the invariant scalar product in $gl(1,1)$ is defined by the formula:

$$t_0 = h_1^1 \otimes h_1 + h_0^1 \otimes h_0 - h_1 \otimes h_1^1 + h_0 \otimes h_0^1. \tag{4}$$

We note that in this case the structure of the bisuperalgebra on the twisted current algebra $gl(1,1)^{tw}[t]$ is defined by a bracket

$$\delta : gl(1,1)^{tw}[t] \to gl(1,1)^{tw}[t] \hat{\otimes} gl(1,1)^{tw}[t],$$

which is given by formula:

$$\delta : a(u) \to [a(u) \otimes 1 + 1 \otimes a(v), r_\sigma(u,v)], \quad r_\sigma(u,v) = \frac{1}{2}\frac{t_0 + t_1}{u - v} + \frac{1}{2}\frac{t_0 - t_1}{u + v}, \tag{5}$$

where

$$t_0 = h_0 \otimes h_0^1 - h_1 \otimes h_1^1, \quad t_1 = h_0^1 \otimes h_0 - h_1^1 \otimes h_1.$$

Next we will be interested in the Yangian of the Lie superalgebra sq_1, that is, $Y(sq_1)$, and also Yangian $Y(Q_1)$.

Definition 2.1. *Yangian $Y(sq_1)$ is an associative superalgebra generated by generators $h_{0,0}, h_{1,0}, h_{1,0}, h_{1,1}$ which satisfy the following defining relations:*

$$[h_{0,i}, h_{0,j}] = 0, \quad i, j \in \{0,1\}, \qquad [h_{1,0}, h_{1,0}] = 2h_{0,0}, \tag{6}$$
$$[h_{0,0}, h_{i,k}] = 0, \quad i, j \in \{0,1\}, \qquad [h_{0,1}, h_{1,0}] = 2h_{1,1}, \tag{7}$$
$$[h_{1,0}, h_{1,1}] = 0. \tag{8}$$

Yangian $Y(sq_1)$ is a Hopf superalgebra with comultiplication on generators which is given by the following formulas:

$$\Delta(h_{i,0}) = h_{i,0} \otimes 1 + 1 \otimes h_{i,0}, \quad i = 0,1, \tag{9}$$
$$\Delta(h_{0,1}) = h_{0,1} \otimes 1 + 1 \otimes h_{0,1} + h_{1,0} \otimes h_{1,0}, \tag{10}$$
$$\Delta(h_{1,1}) = h_{1,1} \otimes 1 + 1 \otimes h_{1,1} - h_{1,0} \otimes h_{0,0} + h_{0,0} \otimes h_{1,0}. \tag{11}$$

We odtain system generators and defining relations for $Y(q_1)$ as in papers [25], [28]. Further we need in some modification of system generators, namely in so-called new (or current) system generators (see section 4).

3. Nazarov Yangian $Y_N(q_1)$

Recall that Yangian $Y_N(q_1)$ introduced by M. Nazarov ([13], [14]) is the associative unital superalgebra over C with countable set of generators

$$t_{i,j}^m, \quad i,j = \pm 1, \quad m = 1, 2, \ldots.$$

The Z_2 – grading of the $Y_N(q_1)$ is defined as follows $p(t_{1,1}^m) = p(t_{-1,-1}^m) = 0$, $p(t_{1,-1}^m) = p(t_{-1,1}^m) = 1$.

To write down the defining relations for these generators we employ the formal series in $Y_N(q_1)[[u^{-1}]]$:

$$t_{i,j}(u) = \delta_{i,j} \cdot 1 + t_{i,j}^1 u^{-1} + t_{i,j}^2 u^{-2} + \ldots.$$

Then for all possible indices i, j, k, l we have the relations:

$$(u^2 - v^2)([t_{i,j}(u), t_{k,l}(v)]) \cdot (-1)^{p(i)p(k)+p(i)p(l)+p(k)p(l)} =$$
$$(u+v)(t_{k,j}(u)t_{i,l}(v) - t_{k,j}(v)t_{i,l}(u)) - (u-v)(t_{-k,j}(u)t_{-i,l}(v) - t_{k,-j}(v)t_{i,-l}(u))(-1)^{p(k)+p(l)}, \quad (12)$$

where v is a formal parameter independent for u, and we have an equality in the algebra of formal Laurent series in u^{-1}, v^{-1} with coefficients in $Y(q_1)$. Also we have the relations

$$t_{i,j}(-u) = t_{-i,-j}(u). \quad (13)$$

Note, that the relations (12) and (13) are equivalent to the following defining relations:

$$([t_{i,j}^{m+1}, t_{k,l}^{r-1}] - [t_{i,j}^{m-1}, t_{k,l}^{r+1}]) \cdot (-1)^{p(i)p(k)+p(i)p(l)+p(k)p(l)} =$$
$$t_{k,j}^m t_{i,l}^{r-1} + t_{k,j}^{m-1} t_{i,l}^r - t_{k,j}^{r-1} t_{i,l}^m - t_{k,j}^r t_{i,l}^{m-1} +$$
$$(-1)^{p(k)+p(l)}(-t_{-k,j}^m t_{-i,l}^{r-1} + t_{-k,j}^{m-1} t_{-i,l}^r + t_{k,-j}^{r-1} t_{i,-l}^m - t_{k,-j}^r t_{i,-l}^{m-1}), \quad , \quad (14)$$

$$t_{-i,-j}^m = (-1)^m t_{i,j}^m, \quad (15)$$

where $m, r = 1, 2, \ldots$ and $t_{i,j}^0 = \delta_{i,j}$.

Recall that $Y_N(q_1)$ is a Hopf algebra with comultiplication given by the formula

$$\Delta(t_{i,j}^m) = \sum_{s=0}^m \sum_k (-1)^{(p(i)+p(k))(p(j)+p(k))} t_{i,k}^s \otimes t_{k,j}^{m-s}. \quad (16)$$

Let's note that this definition can be rewrite using generating functions as follows. Let, $I = \{1, 2 \ldots, n$, $I_1 = I \bigcup (-I) = \{\pm 1, \pm 2, \ldots, \pm n\}$ and as above

$$J = \sum_{i \in I_1} E_{i,-i}(-1)^{p(i)},$$

$$P = \sum_{i,j \in I_1} E_{i,j} \otimes E_{j,i}(-1)^{p(j)}, \quad P_1 = P \otimes 1, \quad P_2 = 1 \otimes P,$$

$$T(u) = \sum_{i,j \in I_1} E_{i,j} \otimes t_{i,j}(u), \quad T_1(u) = T(u) \otimes 1, \quad T_2(u) = 1 \otimes T(u).$$

We define quantum R-matrix by formula:

$$R(u,v) = 1 - \frac{P}{u-v} + \frac{PJ_1J_2}{u+v}. \quad (17)$$

Journal of Physics: Conference Series **1416** (2019) 012038 doi:10.1088/1742-6596/1416/1/012038

Then defining relations can be presented in the following form

$$(R(u,v)\otimes 1)T_(u)T_2(v) = T_2(v)T_1(u)(R(u,v)\otimes 1). \quad (18)$$

We write explicitly the relations for the Yangian $Y_N(q_1)$:

$$[t^m_{-1,1}, t^k_{-1,1}] = \sum_{r=1}^{m-1}\left(t^{k+r-1}_{-1,1}t^{m-r}_{-1,1} - t^{m-r}_{-1,1}t^{k+r-1}_{-1,1}\right) +$$

$$\sum_{r=1}^{m-1}(-1)^r\left((-1)^{k+m}t^{k+r-1}_{1,1}t^{m-r}_{1,1} - t^{m-r}_{1,1}t^{k+r-1}_{1,1}\right). \quad (19)$$

Similarly, we obtain, that

$$[t^m_{1,1}, t^k_{1,1}] = -\sum_{r=1}^{m-1}\left[t^{k+r-1}_{1,1}, t^{m-r}_{1,1}\right] +$$

$$\sum_{r=1}^{m-1}(-1)^r\left((-1)^{k+m}t^{k+r-1}_{-1,1}t^{m-r}_{-1,1} - t^{m-r}_{-1,1}t^{k+r-1}_{-1,1}\right). \quad (20)$$

Last relation has the following form:

$$[t^m_{1,1}, t^k_{-1,1}] = \sum_{r=1}^{m-1}\left(t^{k+r-1}_{1,1}t^{m-r}_{-1,1} - t^{m-r}_{1,1}t^{k+r-1}_{1,1}\right) +$$

$$\sum_{r=1}^{m-1}(-1)^r\left((-1)^{k+m}t^{k+r-1}_{-1,1}t^{m-r}_{1,1} - t^{m-r}_{-1,1}t^{k+r-1}_{1,1}\right). \quad (21)$$

4. Yangian $Y(sq_1)$. Current generators.

Let

$$\{a,b\} := a\cdot b + (-1)^{p(a)p(b)}b\cdot a$$

be an anticommutator.

Definition 4.1. *Yangian $\bar{Y}(sq_1)$ is the associative unital superalgebra over C with countable set of generators $\tilde{h}_{0,k}$, $\tilde{h}_{1,k}$, $k \in Z_{k\geq 0}$, which satisfy the following system of defining relations.*

$$[\tilde{h}_{0,2k}, \tilde{h}_{0,2r}] = [\tilde{h}_{2k+1}, \tilde{h}_{2r+1}] = 0, \quad [\tilde{h}_{0,0}, \tilde{h}_{0,k}] = 0,$$
$$[\tilde{h}_{0,1}, \tilde{h}_{1,k}] = 2\tilde{h}_{1,k} + \tfrac{1}{2}(\tilde{h}_{0,0}\tilde{h}_{1,k} + \tilde{h}_{1,k}\tilde{h}_{0,0}),$$
$$[\tilde{h}_{1,0}, \tilde{h}_{1,k}] = \tilde{h}_{0,k},$$
$$[\tilde{h}_{0,k+1}, \tilde{h}_{1,r}] = [\tilde{h}_{0,k}, \tilde{h}_{1,r+1}] + \{\tilde{h}_{0,k}, \tilde{h}_{1,r}\} + \{\tilde{h}_{1,k}, \tilde{h}_{0,r}\},$$
$$[\tilde{h}_{1,k+1}, \tilde{h}_{1,r}] - [\tilde{h}_{1,k}, \tilde{h}_{1,r+1}] = \{\tilde{h}_{1,k}, \tilde{h}_{1,r}\} + \{\tilde{h}_{0,k}, \tilde{h}_{0,r}\},$$

Theorem 4.1. *([25], [28], [29]) Yangian $Y(sq_1)$ is isomorphic as an associative algebra to $\bar{Y}(sq_1)$.*

Let's define the new generators in $Y(sq_1)$ by the following formulas:

$$h_{1,k+1} := [h_{0,1}, h_{1,k}], \quad h_{0,2k} := [h_{1,0}, h_{1,2k}]. \tag{22}$$

Easy to check that the following relations are satisfied:

$$[h_{1,k+2}, h_{0,2r}] = [h_{1,k}, h_{0,2(r+1)}] + \dots \tag{23}$$

We define the system of generators and defining relations of Drinfeld Yangian of the Lie superalgebra sq_1.

Theorem 4.2. *Yangian $Y_D(sq_1)$ is an associative Hopf superalgebra over C generated by generators $h_{0,2k}$, $h_1 1, k$, $k \in Z_{k\geq 0}$, which satisfy the following defining relations:*

$$[h_{2k}, h_{0,2l}] = 0,$$
$$[h_{0,2k+2}, h_{1,l}] = [h_{1,2k+1}, h_{0,l+1}] + [h_{0,2k}, h_{1,l+2}] + \{h_{0,2k,h_{1,l+1}}\} + (-1)^l\{h_{1,2k}, h_{0,l}\},$$
$$[h_{1,k+2}, h_{1,l}] = [h_{1,k}, h_{1,l+2}] = \{h_{1,k+1}, h_{1,l}\} + \{h_{0,k}, h_{0,l+1}\}.$$

5. Isomorphism between Nazarov Yangian $Y_N(q_1)$ and Drinfeld Yangian $Y(q_1)$

I recall the triangular decomposition construction for Yangian $Y_N(q_1)$. Let $T(u) = (t_{i,j}(u))_{i,j=\pm 1}$. Then we have the following decomposition

$$\begin{pmatrix} t_{1,1}(u) & t_{1,-1}(u) \\ t_{-1,1}(u) & t_{-1,-1}(u) \end{pmatrix} = \begin{pmatrix} 1 & 0 \\ h_{-1}(u) & 0 \end{pmatrix} \cdot \begin{pmatrix} d_1(u) & 0 \\ 0 & d_2(u) \end{pmatrix} \cdot \begin{pmatrix} 1 & h_1(u) \\ 0 & 1 \end{pmatrix}. \tag{24}$$

So, we have the following equality

$$\begin{pmatrix} t_{1,1}(u) & t_{1,-1}(u) \\ t_{-1,1}(u) & t_{-1,-1}(u) \end{pmatrix} = \begin{pmatrix} d_1(u) & d_1(u)\cdot h_1(u) \\ h_{-1}(u)\cdot d_1(u) & d_2(u) + h_{-1}(u) d_1(u) h_1(u). \end{pmatrix} \tag{25}$$

We can formulate the main result of this subsection.

Theorem 5.1. *The correspondence*

$$t_{1,1}(u) \to h_0(u), \quad t_{1,1}(u)^{-1} t_{-1,1}(u) \to h_1(u) \tag{26}$$

define the epimorphism

$$F : Y_N(q_1) \to Y_D(sq_1), \tag{27}$$

where $KerF = \langle h_{0,3}, h_{0,5}, \dots \rangle$ the ideal generated by elements $h_{0,2k+1}$ for $k \geq 1$.

Proof. Let's check that map F respects the defining relations of $Y_N(q_1)$ and $Y_D(sq_1)$. We have that

$$t_{1,-1} = d_1(u)\cdot h_1(u).$$

Therefore

$$t_{1,-1} = \sum_{k=1}^{\infty} t_{1,-1}^k u^{-k} = \left(1 + \sum_{k=0}^{\infty} h_{0,k} u^{-k-1}\right)\left(\sum_{r=0}^{\infty} h_{1,r} u^{-r-1}\right) = \sum_{n=0}^{\infty}\left(\sum_{k=-1}^{n-1} h_{0,k}\cdot h_{n-k}\right) u^{-n-1}.$$

Here $h_{0,-1} = 1$. Therefore

$$t_{1,-1}^{n+1} = \sum_{k=-1}^{n-1} h_{0,k}\cdot h_{n-k} = -h_{1,n} + h_{0,0} h_{1,n-1} + \dots + h_{0,n-1} h_{0,0}.$$

Also, we obtain

$$t_{1,1} = 1 + \sum_{n=1}^{\infty} t_{1,1}^{n} u^{-n} = d_1(u) = h_0(u) = 1 + \sum_{n=0}^{\infty} h_{0,n} u^{-n-1}.$$

Therefore

$$t_{1,1}^{n} = h_{0,n-1}.$$

□

As a corollary of proved theorem we obtain a definition of Yangian

$$Y_N(sq_1) := Y_N(q_1)/KerF.$$

In Nazarov Yangian $Y_N(sq_1)$ we'll use only even generators t_{11}^{2k}, because only this generators corresponds to generators $h_{0,2k}$ in Drinfel'd Yangian $Y_D(sq_1)$. We have the following lemma is hold.

Lemma 5.1. *We have the following relation*

$$[t_{11}^{2k}, t_{11}^{2r}] = 0. \tag{28}$$

for all $r, k > 0$.

Proof. We give a sketch of proof. For $r, k = 0, 1$ the formula (28) is evidently satisfied. Easy to check also that $[t_{11}^{0}, t_{11}^{2r}] = 0$ and $[t_{11}^{0}, t_{-11}^{2r}] = 0$ for all $r \geq 0$.

We suggest that $[t_{11}^{2k}, t_{11}^{2r}] = 0$ and let's prove that $[t_{11}^{2(k+1)}, t_{11}^{2(r+1)}] = 0$ is also true. □

The work is supported by grant "Domestic Research in Moscow"' of Interdisciplinary Scientific Center J.-V. Poncelet (CNRS UMI 2615) and Scoltech Center for Advanced Study.

References

[1] Chari V, Pressley A 1995 A guide to quantum groups *Cambridge* Camb. Univ. Press 622
[2] Dolan L, Nappi Ch, Witten E 2004 Yangian Symmetry in D=4 Superconformal Yang-Mills theory *Preprint arXiv: hep-th/0401243*
[3] Drinfeld V 1988 Quantum groups. *Proc. Int. Cong. Math.* 789 – 820
[4] Drinfel'd V 1985 Hopf algebras and quantum Yang-Baxter equation *Doklady Ac. of Sc. USSR* **283** 1060– 64
[5] Drinfel'd V 1988 New realization of Yangians and quantum affine algebras *Doklady Ac of Sc. USSR* **36** 212–16
[6] Drinfel'd V 1986 Degenerate affine Hecke algebras and Yangians *Funct. Anal. Appl.* **20** 69 – 70
[7] Frappat L, Sciarrino A, Sorba P 2000 Dictionary on Lie Superalgebras *London* Academic Press
[8] Gelfand I, Gelfand S, Retakh V 2005 Quasideterminants. *Adv. Math.* **193** 1 56 – 141 (*Preprint* math.QA/0208146)
[9] Kac V 1977 A Sketch of Lie Superalgebra Theory *Commun.Math.Phys.* **53** 31 – 64
[10] Molev A 2002 Yangians and their applications *Preprint arxiv: math.* QA/0211288
[11] Molev A and Retakh V 2004 Quasideterminants and Casimir elements for the general Lie superalgebras. *Intern. Math. Res. Notes*, no. 13, 611 – 19.
[12] Nazarov M 1991 Quantum Berezinian and the classical Capelly identity. *Lett. Math. Phys.* **21** 121 – 31 pp.
[13] Nazarov M 1999 Yangian of the queer Lie superalgebra *Comm. Math. Phys.* **208** 195 – 223
[14] Nazarov M, Sergeev A 2006 Centralizer construction of the Yangian of the queer Lie superalgebra *Studies in Lie Theory, in Progr. Math.* **243** 417 – 41
[15] Poletaeva E, Serganova V 2016 On Konstant's theorem for Lie superalgebra Q(n) *Advances in Mathematics* **300** 320 – 359
[16] Premet 2002 A Special transverse slices and their enveloping algebras *Advances in Mathematics* **170** 1 – 55
[17] Spill F, Torrielli A 2008 On Drinfeld's second realization of the *AdS/CFT* $su(2|2)$ Yangian. *Preprint arxiv:* hep-th/0803.3194

[18] Stukopin V 1994 Yangians of Lie superalgebras $A(m, n)$ type. *Funct. Anal. Appl.* **28** 3 217– 19
[19] Stukopin V 2006 The Yangian Double of the Lie Superalgebra $A(m, n)$ *Funct. Anal. Appl.* **40** 2 155– 58
[20] Stukopin V 2007 The quantum double of the Yangian of the Lie superalgebra $A(m, n)$ and computation of the universal R-matrix *Journal of Math. Sci.* **142** 2 1989 – 2006
[21] Stukopin V 2013 Representations of the Yangian of a Lie superalgebra of type $A(m, n)$ *Izvestija Mathematics* **77** 5 1021– 43
[22] Stukopin V 2013 The Yangian of the strange Lie superalgebra and its quantum double *Theoret. and Math. Phys.* **174** 1 122 – 33
[23] Stukopin V 2007 Yangian of the strange Lie Superalgebra Q_{n-1}, Drinfel'd approach *SIGMA* **3** 069
[24] Stukopin V 2009 Twisted Yangians, Drinfel'd approach *Journal of Math. Sci.* **161**(2009) 1 143– 62
[25] Stukopin V 2013 *Journal of Physics. C. S.* **422** 1 012027 1– 14
[26] Stukopin V 2019 Isomorphism of the Yangian of $Y_\hbar(A(m, n))$ of special linear superalgebra and quantum loop superalgebra $U_\hbar(LA(m, n))$ *Theoret. and Math. Phys.* **198** 1 129 – 44
[27] Stukopin V 2019 *Journal of Physics. C. S.* **1194** 012103 1– 14
[28] Stukopin V 2018 *Journal of Physics. C. S.* **965** 012049, 1 – 14
[29] Stukopin 2018 Drinfeld Yangian of the queer Lie superalgebra. I *Preprint arxiv:* math.QA/1807.00919

Journal of Physics: Conference Series **1416** (2019) 012039 doi:10.1088/1742-6596/1416/1/012039

Features of the twistor formulation of the massless superparticle on $AdS_5 \times S^5$ superbackground

D V Uvarov
NSC Kharkov Institute of Physics and Technology, 61108 Kharkov, Ukraine
E-mail: d_uvarov@hotmail.com

Abstract. We study supertwistor formulations of the $D = 10$ massless superparticle model on $AdS_5 \times S^5$ superbackground of IIB supergravity. Product structure of the background suggests using $Spin(1,4)$ variables to express momentum components tangent to AdS_5 and $Spin(5)$ variables to express momentum components tangent to S^5 that yields eight-supertwistor formulation of the superparticle's Lagrangian. We find incidence relations connecting supertwistor components with the $AdS_5 \times S^5$ superspace coordinates and the set of the constraints that supertwistors satisfy. Solving the constraints for the $Spin(1,4)$ and $Spin(5)$ variables it is possible to reduce eight-supertwistor formulation to the four-supertwistor one. Respective supertwistors agree with those introduced previously in other models. The advantage of the four-supertwistor formulation is the presence only of the first-class constraints that facilitates analysis of the superparticle model.

1. Introduction

AdS_5/CFT_4 correspondence is commonly formulated and studied in (super)space-time but $D = 4$ $N = 4$ superYang-Mills theory can also be formulated in *supertwistor space* with *manifest* and *linearly realized* superconformal symmetry at the level of supermultiplet, field equations and Lagrangian, free and including interactions, as well as the scattering amplitudes [1], [2], [3], [4], [5], [6], [7], [8]. Basic ingredients are Penrose-Ferber supertwistors [9], ambitwistors and in the case of the scattering amplitudes – momentum twistors [7]. Thus it is natural to wonder whether IIB superstring/supergravity theories admit a twistor formulation? In Refs. [10], [11] $SU(2)$-doublet of Penrose twistors was extended to the Poincare patch of AdS_5 and found the twistor formulation of the massive particle's action. Recently it was extended to the family of $Sp(4,\mathbb{K}) \simeq Spin(2, 2 + \dim\mathbb{K})$ ($\mathbb{K} = \mathbb{R}, \mathbb{C}, \mathbb{H}$) twistor formulations of particle models in AdS_D with $D = 3 + \dim\mathbb{K}$ [12], [13]. Another definition of AdS_5 twistors was given in [14], there were also performed calculations reproducing some simple gauge theory correlators in the framework of AdS_5/CFT_4 duality using twistor methods in the bulk. Generalization of these results to $AdS_5 \times S^5$ superspace is not obvious since it is not superconformally flat [15] unlike Poincare patch of AdS_D. Independently I. Bars in the framework of his 2T approach proposed 'parent' superparticle and tensionless string models [16], [17] extended gauge symmetries of which can be fixed in a variety of ways. Massless superparticle on $AdS_5 \times S^5$ in the superspace and supertwistor formulations arises from the 'parent' superparticle upon choosing different gauge conditions. In the supertwistor gauge dynamical variables of the model are two $SU(2)$-doublets of supertwistors $\mathcal{Z}_a^A$ and Ψ_q^A. In the 'parent' superparticle model by I. Bars superspace coordinates and supertwistors appear to be unrelated to one another so the incidence relations cannot be derived. This motivated us to follow in [18], [19] the traditional root of deriving supertwistor formulation for massless superparticle on $AdS_5 \times S^5$ superbackground described in this note that starts with the first-order form of the superspace Lagrangian and relies on the superparticle's momentum representation in terms of the product of the spinor variables.

2. Massless superparticle on $AdS_5 \times S^5$ superbackground: eight-supertwistor formulation

Classical action of the massless superparticle on $AdS_5 \times S^5$ superbackground can be presented in the first-order form

$$S_{AdS_5\times S^5} = \int d\tau \mathscr{L}_{AdS_5\times S^5}, \quad \mathscr{L}_{AdS_5\times S^5} = p_{m'}E^{m'}_\tau + p_{I'}E^{I'}_\tau - \frac{g}{2}(p_{m'}p^{m'} + p_{I'}p_{I'}), \tag{1}$$

where $E^{m'}_\tau$ and $E^{I'}_\tau$ are world-line pullbacks of the $D = 10$ supervielbein bosonic components tangent to AdS_5 and S^5. To pass to the supertwistor formulation we need to express null 10-momentum $(p_{m'}, p_{I'})$ in terms of the product of spinor variables. Momentum components tangent to AdS_5 can be presented as

$$p_{m'} = -\frac{1}{2}v^{\mathrm{T}\boldsymbol{\lambda}}{}_{\boldsymbol{\alpha}}\gamma_{m'}{}^{\boldsymbol{\alpha}}{}_{\boldsymbol{\beta}}v^{\boldsymbol{\beta}}_{\boldsymbol{\nu}}\gamma^{(0)\boldsymbol{\nu}}{}_{\boldsymbol{\lambda}} = -\frac{1}{2}(v^b_{\boldsymbol{\alpha}}\gamma_{m'}{}^{\boldsymbol{\alpha}}{}_{\boldsymbol{\beta}}v^{\boldsymbol{\beta}}_b - v^{\dot b}_{\boldsymbol{\alpha}}\gamma_{m'}{}^{\boldsymbol{\alpha}}{}_{\boldsymbol{\beta}}v^{\boldsymbol{\beta}}_{\dot b}), \tag{2}$$

where

$$v^{\boldsymbol{\alpha}}_{\boldsymbol{\mu}} = (-v^{\boldsymbol{\alpha}}_b, v^{\boldsymbol{\alpha}\dot b}) \tag{3}$$

is the 4×4 matrix of the $Spin(1,4)$ variables. Its index from the beginning of the Greek alphabet is acted by the left $Spin(1,4)_L$ transformations resulting in the Lorentz rotations of the momentum components and index from the middle part of the Greek alphabet, that in (3) was decomposed into the $SU(2)_R\times SU(2)_R$ indices labelled by small letters from the beginning of the Latin alphabet, is acted by the right local $Spin(1,4)_R$ transformations, whose $SU(2)_R\times SU(2)_R$ subgroup leaves the momentum intact and enters the set of the gauge symmetries of the superparticle's action. In (2) the two sets of γ-matrices in $D = 1+4$ dimensions are invariant under left and right $Spin(1,4)$ transformations respectively and satisfy the defining relations

$$\begin{gathered}\gamma_{m'}{}^{\boldsymbol{\alpha}}{}_{\boldsymbol{\delta}}\gamma_{n'}{}^{\boldsymbol{\delta}}{}_{\boldsymbol{\beta}} + \gamma_{n'}{}^{\boldsymbol{\alpha}}{}_{\boldsymbol{\delta}}\gamma_{m'}{}^{\boldsymbol{\delta}}{}_{\boldsymbol{\beta}} = -2\eta_{m'n'}\delta^{\boldsymbol{\alpha}}_{\boldsymbol{\beta}}, \quad \eta_{m'n'} = \mathrm{diag}(-,+,+,+,+)\\ \gamma^{(k')\boldsymbol{\mu}}{}_{\boldsymbol{\lambda}}\gamma^{(l')\boldsymbol{\lambda}}{}_{\boldsymbol{\nu}} + \gamma^{(l')\boldsymbol{\mu}}{}_{\boldsymbol{\lambda}}\gamma^{(k')\boldsymbol{\lambda}}{}_{\boldsymbol{\nu}} = -2\eta^{(k')(l')}\delta^{\boldsymbol{\mu}}_{\boldsymbol{\nu}}, \quad (k') = (0,\hat I),\ (l') = (0,\hat J).\end{gathered} \tag{4}$$

In the last equality in (2) there has been used the diagonal realization of $\gamma^{(0)\boldsymbol{\nu}}{}_{\boldsymbol{\lambda}}$

$$\gamma^{(0)\boldsymbol{\nu}}{}_{\boldsymbol{\lambda}} = \begin{pmatrix} -\delta^b_a & 0 \\ 0 & \delta^{\dot a}_{\dot b}\end{pmatrix}. \tag{5}$$

Matrix (3) obeys the reality conditions

$$\gamma^{(0)\boldsymbol{\lambda}}{}_{\boldsymbol{\nu}}(v^{\boldsymbol{\alpha}}_{\boldsymbol{\nu}})^\dagger\gamma^{0\boldsymbol{\alpha}}{}_{\boldsymbol{\beta}} = v^{\mathrm{T}\boldsymbol{\lambda}}{}_{\boldsymbol{\beta}} \tag{6}$$

ensuring that $p_{m'}$ is real. In terms of the 4×2 rectangular blocks $v^{\boldsymbol{\alpha}}_b$ and $v^{\boldsymbol{\alpha}\dot b}$ they acquire the form of the $SU(2)$-Majorana conditions

$$(v^{\boldsymbol{\alpha}}_b)^\dagger\gamma^{0\boldsymbol{\alpha}}{}_{\boldsymbol{\beta}} = v^{\mathrm{T}b}{}_{\boldsymbol{\beta}}, \quad (v^{\boldsymbol{\alpha}\dot b})^\dagger\gamma^{0\boldsymbol{\alpha}}{}_{\boldsymbol{\beta}} = v^{\mathrm{T}}{}_{\boldsymbol{\beta}\dot b}. \tag{7}$$

In addition matrix (3) satisfies the constraints

$$v^{\boldsymbol{\alpha}}_a v^{\dot a}_{\boldsymbol{\alpha}} - 0 \tag{8}$$

so that

$$p_{m'}p^{m'} = -\frac{1}{4}(v_a^{\boldsymbol{\alpha}} v_{\boldsymbol{\alpha}}^a + v_{\dot{a}}^{\boldsymbol{\alpha}} v_{\boldsymbol{\alpha}}^{\dot{a}})^2. \tag{9}$$

The transformation and reality properties of the spinor variables (3) are the same as those of $D = 1+4$ spinor Lorentz harmonics parametrizing the $Spin(1,4)/(SU(2) \times SU(2))$ coset [19][1]. Also Eq. (2) coincides with the general relation that connects the time-like component (first-column) of the matrix of the vector Lorentz harmonics and the spinor Lorentz harmonics [22], [23], [24] up to the overall normalization chosen here to produce canonical form of the superparticle's kinetic term in the supertwistor formulation. The difference is that the constraints (8) are weaker than those imposed on the $D = 1+4$ spinor Lorentz harmonics [19]. This is necessary to correctly balance the degrees of freedom of the null 10-momentum and spinor variables.

Further introduce 4×4 matrix of the $Spin(5)$ variables

$$\ell_N^A = (\ell_p^A,\ \ell^{A\dot{p}}). \tag{10}$$

It transforms under left $Spin(5)_L \simeq USp(4)_L$ transformations acting on the capital index from the beginning of the Latin alphabet and local $Spin(5)_R \simeq USp(4)_R$ transformations that act on the capital index from the middle part of the Latin alphabet. $SU(2)_R \times SU(2)_R \subset Spin(5)_R$ subgroup will enter the set of the superparticle's gauge symmetries, accordingly the index N have been split into $SU(2)_R \times SU(2)_R$ indices labelled by small letters from the middle part of the Latin alphabet. $Spin(5)$ variables satisfy the reality conditions

$$(\ell_N^A)^\dagger = \ell^{\mathrm{T}\,N}{}_A: \quad (\ell_p^A)^\dagger = \ell^{\mathrm{T}\,p}{}_A, \quad (\ell^{A\dot{p}})^\dagger = \ell^{\mathrm{T}}_{A\dot{p}} \tag{11}$$

and are constrained by the relations

$$\ell_p^A \ell^{\dot{p}}_A = 0. \tag{12}$$

Let us note that the properties of the $Spin(5)$ variables are very close to those of the $USp(4)/(SU(2) \times SU(2))$ harmonics [25]. They have the same transformation properties and satisfy the same reality conditions but the harmonics obey extra constraints that adjust the number of their independent degrees of freedom to the dimension of the $USp(4)/(SU(2) \times SU(2))$ coset. In analogy with (2) momentum components tangent to S^5 can be expressed in terms of the $Spin(5)$ variables as

$$p_{I'} = \frac{1}{2}\ell^{\mathrm{T}L}{}_A \gamma_{I'}{}^A{}_B \ell_N^B \gamma^{(5)N}{}_L = -\frac{1}{2}(\ell^q_A \gamma_{I'}{}^A{}_B \ell_q^B + \ell^{\dot{q}}_A \gamma_{I'}{}^A{}_B \ell_{\dot{q}}^B), \tag{13}$$

where the two sets of $D = 5$ γ-matrices are invariant under $Spin(5)_L$ and $Spin(5)_R$ transformations respectively and obey

$$\begin{gathered} \gamma_{I'}{}^A{}_C \gamma_{J'}{}^C{}_B + \gamma_{J'}{}^A{}_C \gamma_{I'}{}^C{}_B = 2\delta_{I'J'}\delta_B^A, \\ \gamma_{(I')}{}^N{}_L \gamma_{(J')}{}^L{}_M + \gamma_{(J')}{}^N{}_L \gamma_{(I')}{}^L{}_M = 2\delta_{(I')(J')}\delta_M^N, \quad (I') = (\hat{I}', 5),\ (J') = (\hat{J}', 5). \end{gathered} \tag{14}$$

In the last equality in (13) there has been chosen the diagonal realization of $\gamma^{(5)N}{}_L$

$$\gamma^{(5)N}{}_L = \begin{pmatrix} -\delta_q^p & 0 \\ 0 & \delta_{\dot{p}}^{\dot{q}} \end{pmatrix}. \tag{15}$$

[1] $D = 1+4$ Lorentz harmonics parametrizing other cosets were considered in [20], [21].

Then the square of the momentum components tangent to S^5 equals

$$p_{I'}p_{I'} = \frac{1}{4}(\ell^A_q \ell^q_A - \ell^A_{\dot q}\ell^{\dot q}_A)^2 \tag{16}$$

and the null-momentum condition in 10 dimensions acquires the form

$$(v^{\boldsymbol{\alpha}}_a v^a_{\boldsymbol{\alpha}} + v^{\boldsymbol{\alpha}}_{\dot a} v^{\dot a}_{\boldsymbol{\alpha}})^2 = (\ell^A_q \ell^q_A - \ell^A_{\dot q}\ell^{\dot q}_A)^2. \tag{17}$$

At this point there is a discrepancy between the number of the independent degrees of freedom since null 10-momentum $(p_{m'}, p_{I'})$ has 9 independent components, whereas spinors $(v^{\boldsymbol{\alpha}}_a, v^{\boldsymbol{\alpha}}_{\dot a})$ and $(\ell^A_q, \ell^A_{\dot q})$ have $6+6-1=11$ independent components. Two additional constraints for the spinor variables will be obtained below.

To pass to the eight-supertwistor formulation of the $D=10$ massless superparticle on the $AdS_5 \times S^5$ superbackground it is convenient to introduce diagonal supermatrix

$$\mathrm{V}^{\mathcal{A}}{}_{\mathcal{N}} = \begin{pmatrix} v^{\boldsymbol{\alpha}}_{\nu} & 0 \\ 0 & \ell^A_N \end{pmatrix}, \tag{18}$$

whose upper-diagonal block constitutes the matrix of the $Spin(1,4)$ variables (3), while in the lower-diagonal block there is the matrix of the $Spin(5)$ variables (10). Reality conditions (6) and (11) in the supermatrix form read

$$\mathcal{H}^{\mathcal{L}}{}_{\mathcal{N}}(\mathrm{V}^{\mathcal{A}}{}_{\mathcal{N}})^{\dagger}\mathcal{H}^{\mathcal{A}}{}_{\mathcal{B}} = \mathrm{V}^{\mathrm{T}\mathcal{L}}{}_{\mathcal{B}} = \begin{pmatrix} v^{\mathrm{T}\lambda}_{\boldsymbol{\beta}} & 0 \\ 0 & \ell^{\mathrm{T}L}_{B} \end{pmatrix}, \tag{19}$$

where

$$\mathcal{H}^{\mathcal{A}}{}_{\mathcal{B}} = \begin{pmatrix} \gamma^{0\boldsymbol{\alpha}}{}_{\boldsymbol{\beta}} & 0 \\ 0 & \delta^A_B \end{pmatrix}, \quad \mathcal{H}^{\mathcal{L}}{}_{\mathcal{N}} = \begin{pmatrix} \gamma^{(0)\boldsymbol{\lambda}}{}_{\nu} & 0 \\ 0 & \delta^L_N \end{pmatrix}. \tag{20}$$

Next introduce the $PSU(2,2|4)/(SO(1,4)\times SO(5))$ representative [26], [27]

$$\mathcal{G}^{\mathcal{A}}{}_{\mathcal{B}} = \begin{pmatrix} G^{\boldsymbol{\alpha}}{}_{\boldsymbol{\beta}} & G^{\boldsymbol{\alpha}}{}_{B} \\ G^{A}{}_{\boldsymbol{\beta}} & G^{A}{}_{B} \end{pmatrix} \in PSU(2,2|4)/(SO(1,4)\times SO(5)) \tag{21}$$

that satisfies defining relations

$$(\mathcal{G}^{\mathcal{A}}{}_{\mathcal{B}})^{\dagger} = \mathcal{H}^{\mathcal{B}}{}_{\mathcal{C}}\mathcal{G}^{-1\mathcal{C}}{}_{\mathcal{D}}\mathcal{H}^{\mathcal{D}}{}_{\mathcal{A}}. \tag{22}$$

Now we can define the twistor supermatrix

$$\mathcal{Z}^{\mathcal{A}}{}_{\mathcal{N}} = \mathcal{G}^{\mathcal{A}}{}_{\mathcal{B}}\mathrm{V}^{\mathcal{B}}{}_{\mathcal{N}} = \left(\mathcal{Z}^{\mathcal{A}}_{\nu}\ \Psi^{\mathcal{A}}_{N}\right) = \left(-\mathcal{Z}^{\mathcal{A}}_{b}\ \mathcal{Z}^{\mathcal{A}\dot b}\ \Psi^{\mathcal{A}}_{p}\ \Psi^{\mathcal{A}\dot p}\right) \tag{23}$$

and its dual

$$\bar{\mathcal{Z}}^{\mathcal{L}}{}_{\mathcal{B}} = \mathcal{H}^{\mathcal{L}}{}_{\mathcal{N}}(\mathcal{Z}^{\mathcal{A}}{}_{\mathcal{N}})^{\dagger}\mathcal{H}^{\mathcal{A}}{}_{\mathcal{B}} = \mathrm{V}^{\mathrm{T}\mathcal{L}}{}_{\mathcal{C}}\mathcal{G}^{-1\mathcal{C}}{}_{\mathcal{B}} = \begin{pmatrix} \bar{\mathcal{Z}}^{\lambda}_{\mathcal{B}} \\ \bar{\Psi}^{L}_{\mathcal{B}} \end{pmatrix} = \begin{pmatrix} \bar{\mathcal{Z}}^{b}_{\mathcal{B}} \\ \bar{\mathcal{Z}}_{\mathcal{B}\dot b} \\ \bar{\Psi}^{q}_{\mathcal{B}} \\ \bar{\Psi}_{\mathcal{B}\dot q} \end{pmatrix}. \tag{24}$$

Let us note that our choice of the γ-matrices in $D=2+4$ and $D=1+4$ dimensions is such that γ^0 equals 'metric tensor' connecting **4** and $\bar{\mathbf{4}}$ representations of $SU(2,2)$. As explained in detail in [18] and [19] $\gamma^{0\boldsymbol{\alpha}}{}_{\boldsymbol{\beta}}$ and $\gamma^{(0)\boldsymbol{\lambda}}{}_{\boldsymbol{\nu}}$ correspond to its off-diagonal (twistor) and diagonal (oscillator) realizations.

Supertwistors $\mathcal{Z}_b^A$, $\mathcal{Z}^{A\dot b}$ and their duals are conventional ones and were named c-type supertwistors in [18] because their $SU(2,2)_L$ components $Z_b^{\boldsymbol{\alpha}}$, $Z^{\boldsymbol{\alpha}\dot b}$ are even and $SU(4)_L$ components η_b^A, $\eta^{A\dot b}$ are odd. On the contrary supertwistors Ψ_p^A, $\Psi^{A\dot p}$ and their duals were named a-type supertwistors since their $SU(2,2)_L$ components $\xi_p^{\boldsymbol{\alpha}}$, $\xi^{\boldsymbol{\alpha}\dot p}$ are odd but $SU(4)_L$ components L_p^A, $L^{A\dot p}$ are even. Appearance of such supertwistors with unconventional statistics of their components in the superparticle model on $AdS_5 \times S^5$ superbackground was discussed in [16] and [17].

The 1-form that defines kinetic term of the superparticle's Lagrangian (1) can be expressed in terms of the introduced supertwistors

$$\begin{aligned} p_{m'}E^{m'}(d) + p_{I'}E^{I'}(d) =& \ \tfrac{\mathrm{i}}{2}\left(\bar{\mathcal{Z}}^{\mathcal{L}}{}_{\mathcal{A}} d\mathcal{Z}^{\mathcal{A}}{}_{\mathcal{N}} - d\bar{\mathcal{Z}}^{\mathcal{L}}{}_{\mathcal{A}}\mathcal{Z}^{\mathcal{A}}{}_{\mathcal{N}}\right)\Gamma^{\mathcal{N}}{}_{\mathcal{L}} \\ =& \ \tfrac{\mathrm{i}}{2}(\bar{\mathcal{Z}}^c_{\mathcal{A}} d\mathcal{Z}^{\mathcal{A}}_c - d\bar{\mathcal{Z}}^c_{\mathcal{A}}\mathcal{Z}^{\mathcal{A}}_c) + \tfrac{\mathrm{i}}{2}(\bar{\mathcal{Z}}_{\mathcal{A}\dot c} d\mathcal{Z}^{\mathcal{A}\dot c} - d\bar{\mathcal{Z}}_{\mathcal{A}\dot c}\mathcal{Z}^{\mathcal{A}\dot c}) \\ +& \ \tfrac{\mathrm{i}}{2}(\bar{\Psi}^q_{\mathcal{A}} d\Psi^{\mathcal{A}}_q - d\bar{\Psi}^q_{\mathcal{A}}\Psi^{\mathcal{A}}_q) - \tfrac{\mathrm{i}}{2}(\bar{\Psi}_{\mathcal{A}\dot q} d\Psi^{\mathcal{A}\dot q} - d\bar{\Psi}_{\mathcal{A}\dot q}\Psi^{\mathcal{A}\dot q}) \end{aligned} \tag{25}$$

with the diagonal supermatrix

$$\Gamma^{\mathcal{N}}{}_{\mathcal{L}} = \begin{pmatrix} \gamma^{(0)\boldsymbol{\nu}}{}_{\boldsymbol{\lambda}} & 0 \\ 0 & -\gamma^{(5)N}{}_L \end{pmatrix} = \begin{pmatrix} \begin{matrix} -\delta^b_a & 0 \\ 0 & \delta^{\dot a}_{\dot b} \end{matrix} & \mathbf{0} \\ \mathbf{0} & \begin{matrix} \delta^p_q & 0 \\ 0 & -\delta^{\dot q}_{\dot p} \end{matrix} \end{pmatrix}. \tag{26}$$

In its upper-diagonal block there is $\gamma^{(0)\boldsymbol{\nu}}{}_{\boldsymbol{\lambda}}$ matrix that, as discussed in [19], not just breaks $SU(2,2)_R$ symmetry but actually 'switches' it to the $SU(4)_R$ symmetry. Similarly $\gamma^{(5)N}{}_L$ 'switches' $SU(4)_R$ symmetry to the $SU(2,2)_R$ one. So the supertwistor 1-form (25) is invariant under 'twisted' $SU(2,2|4)_R$ symmetry with $SU(4)_R$ parameters in the upper-diagonal block and $SU(2,2)_R$ parameters in the lower-diagonal block. This hints at another definition of the dual twistor supermatrix

$$\tilde{\mathcal{Z}}^{\mathcal{N}}{}_{\mathcal{A}} = \Gamma^{\mathcal{N}}{}_{\mathcal{L}}\bar{\mathcal{Z}}^{\mathcal{L}}{}_{\mathcal{A}} = \tilde{\mathcal{H}}^{\mathcal{N}}{}_{\mathcal{L}}(\mathcal{Z}^{\mathcal{B}}{}_{\mathcal{L}})^{\dagger}\mathcal{H}^{\mathcal{B}}{}_{\mathcal{A}}, \tag{27}$$

where

$$\tilde{\mathcal{H}}^{\mathcal{N}}{}_{\mathcal{L}} = \begin{pmatrix} \delta^{\boldsymbol{\nu}}_{\boldsymbol{\lambda}} & 0 \\ 0 & -\gamma^{(5)N}{}_L \end{pmatrix}, \tag{28}$$

resulting in the following expression for the 1-form (25)

$$\frac{\mathrm{i}}{2}\left(\tilde{\mathcal{Z}}^{\mathcal{N}}{}_{\mathcal{A}} d\mathcal{Z}^{\mathcal{A}}{}_{\mathcal{N}} - d\tilde{\mathcal{Z}}^{\mathcal{N}}{}_{\mathcal{A}}\mathcal{Z}^{\mathcal{A}}{}_{\mathcal{N}}\right). \tag{29}$$

This 'twisted' $SU(2,2|4)_R$ symmetry will, however, be broken by the supertwistor constraints that we discuss below. So let us write kinetic term of the superparticle's Lagrangian in the eight-supertwistor formulation in the form with manifest $(SU(2)_R)^4$ invariance

$$\begin{aligned} \mathscr{L}_{\text{kin}} =& \ \tfrac{\mathrm{i}}{2}(\bar{\mathcal{Z}}^c_{\mathcal{A}}\dot{\mathcal{Z}}^{\mathcal{A}}_c - \dot{\bar{\mathcal{Z}}}^c_{\mathcal{A}}\mathcal{Z}^{\mathcal{A}}_c) + \tfrac{\mathrm{i}}{2}(\bar{\mathcal{Z}}_{\mathcal{A}\dot c}\dot{\mathcal{Z}}^{\mathcal{A}\dot c} - \dot{\bar{\mathcal{Z}}}_{\mathcal{A}\dot c}\mathcal{Z}^{\mathcal{A}\dot c}) \\ +& \ \tfrac{\mathrm{i}}{2}(\bar{\Psi}^q_{\mathcal{A}}\dot{\Psi}^{\mathcal{A}}_q - \dot{\bar{\Psi}}^q_{\mathcal{A}}\Psi^{\mathcal{A}}_q) - \tfrac{\mathrm{i}}{2}(\bar{\Psi}_{\mathcal{A}\dot q}\dot{\Psi}^{\mathcal{A}\dot q} - \dot{\bar{\Psi}}_{\mathcal{A}\dot q}\Psi^{\mathcal{A}\dot q}). \end{aligned} \tag{30}$$

As is common supertwistors are constrained dynamical variables. The constraints are contained in the $SU(2,2|4)_L$-invariant supermatrix

$$\bar{\mathcal{Z}}^{\mathcal{L}}{}_{\mathcal{A}}\mathcal{Z}^{\mathcal{A}}{}_{\mathcal{N}}. \tag{31}$$

It is possible to decompose it on the constituents irreducible under all four $SU(2)_R$ subgroups of $SU(2,2|4)_R$ and calculate their Dirac bracket (D.B.) relations using the basic D.B. relations that follow from (30)

$$\begin{aligned} \{Z_a^{\mathcal{A}}, \bar{Z}_{\mathcal{B}}^b\}_{D.B.} = \mathrm{i}\delta_a^b\delta_{\mathcal{B}}^{\mathcal{A}}, \quad \{Z^{\mathcal{A}\dot{a}}, \bar{Z}_{\mathcal{B}\dot{b}}\}_{D.B.} = \mathrm{i}\delta_{\dot{b}}^{\dot{a}}\delta_{\mathcal{B}}^{\mathcal{A}}, \\ \{\Psi_q^{\mathcal{A}}, \bar{\Psi}_{\mathcal{B}}^p\}_{D.B.} = \mathrm{i}\delta_q^p\delta_{\mathcal{B}}^{\mathcal{A}}, \quad \{\Psi^{\mathcal{A}\dot{q}}, \bar{\Psi}_{\mathcal{B}\dot{p}}\}_{D.B.} = -\mathrm{i}\delta_{\dot{p}}^{\dot{q}}\delta_{\mathcal{B}}^{\mathcal{A}}. \end{aligned} \tag{32}$$

In such a way one identifies 15 bosonic

$$L^a{}_b = \bar{Z}_{\mathcal{A}}^a Z_b^{\mathcal{A}} - \tfrac{1}{2}\delta_b^a \bar{Z}_{\mathcal{A}}^c Z_c^{\mathcal{A}} \approx 0, \quad M_{\dot{a}}{}^{\dot{b}} = \bar{Z}_{\mathcal{A}\dot{a}} Z^{\mathcal{A}\dot{b}} - \tfrac{1}{2}\delta_{\dot{a}}^{\dot{b}} \bar{Z}_{\mathcal{A}\dot{c}} Z^{\mathcal{A}\dot{c}} \approx 0, \tag{33}$$

$$R^q{}_p = \bar{\Psi}_{\mathcal{A}}^q \Psi_p^{\mathcal{A}} - \tfrac{1}{2}\delta_p^q \bar{\Psi}^r{}_{\mathcal{A}} \Psi_r^{\mathcal{A}} \approx 0, \quad S_{\dot{q}}{}^{\dot{p}} = \bar{\Psi}_{\mathcal{A}\dot{q}} \Psi^{\mathcal{A}\dot{p}} - \tfrac{1}{2}\delta_{\dot{q}}^{\dot{p}} \bar{\Psi}_{\mathcal{A}\dot{r}} \Psi^{\mathcal{A}\dot{r}} \approx 0, \tag{34}$$

$$C + E = \bar{Z}^c Z_c - \bar{Z}_{\dot{c}} Z^{\dot{c}} + \bar{\Psi}^r \Psi_r + \bar{\Psi}_{\dot{r}} \Psi^{\dot{r}} \approx 0, \tag{35}$$

$$T = \bar{Z}^c Z_c + \bar{Z}_{\dot{c}} Z^{\dot{c}} + \bar{\Psi}^r \Psi_r - \bar{\Psi}_{\dot{r}} \Psi^{\dot{r}} \approx 0, \tag{36}$$

$$U = \bar{Z}^c Z_c + \bar{Z}_{\dot{c}} Z^{\dot{c}} - \bar{\Psi}^r \Psi_r + \bar{\Psi}_{\dot{r}} \Psi^{\dot{r}} \approx 0 \tag{37}$$

and 16 fermionic first-class constraints

$$\bar{Z}_{\mathcal{A}}^b \Psi_q^{\mathcal{A}} \approx 0, \quad \bar{\Psi}_{\mathcal{A}}^p Z_a^{\mathcal{A}} \approx 0, \quad \bar{Z}_{\mathcal{A}\dot{b}} \Psi^{\mathcal{A}\dot{q}} \approx 0, \quad \bar{\Psi}_{\mathcal{A}\dot{p}} Z^{\mathcal{A}\dot{a}} \approx 0. \tag{38}$$

The constraint (35) upon substitution of the incidence relations for supertwistors (23) and (24) equals the square root of the constraint (17), while (36) and (37) are two extra sought for constraints needed to reduce the number of independent components of the $Spin(1,4)$ and $Spin(5)$ variables to that of the null momentum in 10 dimensions. Besides that there are 16 bosonic

$$\bar{Z}_{\mathcal{A}}^a Z^{\mathcal{A}\dot{b}} \approx 0, \quad \bar{Z}_{\mathcal{A}\dot{b}} Z_a^{\mathcal{A}} \approx 0, \quad \bar{\Psi}_{\mathcal{A}}^q \Psi^{\mathcal{A}\dot{p}} \approx 0, \quad \bar{\Psi}_{\mathcal{A}\dot{p}} \Psi_q^{\mathcal{A}} \approx 0 \tag{39}$$

and 16 fermionic second-class constraints

$$\bar{Z}_{\mathcal{A}}^b \Psi^{\mathcal{A}\dot{q}} \approx 0, \quad \bar{\Psi}_{\mathcal{A}\dot{p}} Z_a^{\mathcal{A}} \approx 0, \quad \bar{Z}_{\mathcal{A}\dot{b}} \Psi_q^{\mathcal{A}} \approx 0, \quad \bar{\Psi}_{\mathcal{A}}^p Z^{\mathcal{A}\dot{a}} \approx 0. \tag{40}$$

Thus the number of the physical degrees of freedom in the eight-supertwistor formulation of the superparticle precisely matches that in the superspace formulation. The first-class constraints can be added to the Lagrangian with appropriate Lagrange multipliers so that the complete Lagrangian in the eight-supertwistor formulation equals

$$\mathscr{L}_{8-\mathrm{stwistor}} = \mathscr{L}_{\mathrm{kin}} + \mathscr{L}_{\mathrm{constr}}, \tag{41}$$

where expression for the kinetic term is given in (30) and

$$\begin{aligned} \mathscr{L}_{\mathrm{constr}} = &\ \Lambda^b{}_a L^a{}_b + \Lambda_{\dot{b}}{}^{\dot{a}} M_{\dot{a}}{}^{\dot{b}} + \Lambda^p{}_q R^q{}_p + \Lambda_{\dot{p}}{}^{\dot{q}} S_{\dot{q}}{}^{\dot{p}} + \Lambda_{C+E}(C+E) + \Lambda_T T + \Lambda_U U \\ &+ \mathrm{i}\Lambda_p^a \bar{\Psi}_{\mathcal{A}}^p Z_a^{\mathcal{A}} + \mathrm{i}\bar{\Lambda}_a^p \bar{Z}_{\mathcal{A}}^a \Psi_p^{\mathcal{A}} + \mathrm{i}\Lambda_{\dot{a}}^{\dot{p}} \bar{\Psi}_{\mathcal{A}\dot{p}} Z^{\mathcal{A}\dot{a}} + \mathrm{i}\bar{\Lambda}_{\dot{p}}^{\dot{a}} \bar{Z}_{\mathcal{A}\dot{a}} \Psi^{\mathcal{A}\dot{p}}. \end{aligned} \tag{42}$$

3. Massless superparticle on $AdS_5 \times S^5$ superbackground: four-supertwistor formulation

The fact that in the eight-supertwistor formulation of the massless superparticle model there are 32 second-class constraints essentially hampers further Hamiltonian analysis, requiring introduction of the D.B. for these constraints. Since the presence of the second-class constraints signals redundancy of the degrees of freedom of the model, it is possible to remove some of them that will simplify the algebra of the remaining constraints. The form of the incidence relations (23) and (24) shows that reduction of the $Spin(1,4)$ and $Spin(5)$ variables will lead to reduction of the supertwistors. In the case of the $Spin(1,4)$ variables (3) using the constraints (8) and (35)-(37) one can show that in (2) contributions of the first and the second summands equal so that one can exclude, e.g. spinor variables with dotted $SU(2)$ indices

$$p_{m'} = -\frac{1}{2} v^{b}_{\alpha} \gamma_{m'}{}^{\alpha}{}_{\beta} v^{\beta}_{b}, \tag{43}$$

where emerging factor of two has been absorbed in the definition of remaining $Spin(1,4)$ variables. Reduction of the $Spin(5)$ variables proceeds analogously. Resulting expression for the momentum components tangent to S^5 is

$$p_{I'} = -\frac{1}{2} \ell^{q}_{B} \gamma_{I'}{}^{B}{}_{A} \ell^{A}_{q}. \tag{44}$$

Expressions for the momentum components (43) and (44) were the starting point to derive the four-supertwistor formulation of the massless superparticle on $AdS_5 \times S^5$ superbackground in [18]. Resulting four-supertwistor representation of the superparticle's Lagrangian is

$$\begin{aligned} \mathscr{L}_{4-\text{stwistor}} = & \ \tfrac{\mathrm{i}}{2}\left(\bar{\mathcal{Z}}^{a}_{A}\dot{\mathcal{Z}}^{A}_{a} - \dot{\bar{\mathcal{Z}}}^{a}_{A}\mathcal{Z}^{A}_{a}\right) + \tfrac{\mathrm{i}}{2}(\bar{\Psi}^{q}_{A}\dot{\Psi}^{A}_{q} - \dot{\bar{\Psi}}^{q}_{A}\Psi^{A}_{q}) \\ + & \ \Lambda^{b}{}_{a}L^{a}{}_{b} + \Lambda^{p}{}_{q}R^{q}{}_{p} + \Lambda(\bar{\mathcal{Z}}^{a}_{A}\mathcal{Z}^{A}_{a} + \bar{\Psi}^{q}_{A}\Psi^{A}_{q}) \\ + & \ \mathrm{i}\Lambda^{a}_{q}\bar{\Psi}^{q}_{A}\mathcal{Z}^{A}_{a} + \mathrm{i}\bar{\Lambda}^{q}_{a}\bar{\mathcal{Z}}^{a}_{A}\Psi^{A}_{q}. \end{aligned} \tag{45}$$

Lagrangian (45) coincides with that obtained in Ref. [17] by partial gauge fixing $2T$ superparticle model in 2+10 dimensions. In our approach it became possible also to find the incidence relations for the supertwistors

$$\mathcal{Z}^{A}_{a} = G^{A}{}_{B}\begin{pmatrix} v^{\beta}_{a} \\ 0 \end{pmatrix} = \begin{pmatrix} Z^{\alpha}_{a} \\ \eta^{A}_{a} \end{pmatrix}, \quad \bar{\mathcal{Z}}^{a}_{A} = (\mathcal{Z}^{B}_{a})^{\dagger}\mathcal{H}^{B}{}_{A} = (\bar{Z}^{a}_{\alpha}\ \bar{\eta}^{a}_{A}) \tag{46}$$

and

$$\Psi^{A}_{q} = G^{A}{}_{B}\begin{pmatrix} 0 \\ \ell^{B}_{q} \end{pmatrix} = \begin{pmatrix} \xi^{\alpha}_{q} \\ L^{A}_{q} \end{pmatrix}, \quad \bar{\Psi}^{q}_{A} = (\Psi^{B}_{q})^{\dagger}\mathcal{H}^{B}{}_{A} = (\bar{\xi}^{q}_{\alpha}\ \bar{L}^{q}_{A}) \tag{47}$$

that satisfy seven bosonic

$$L^{a}{}_{b} = \bar{\mathcal{Z}}^{a}_{A}\mathcal{Z}^{A}_{b} - \frac{1}{2}\delta^{a}_{b}\bar{\mathcal{Z}}^{c}_{A}\mathcal{Z}^{A}_{c} \approx 0, \quad R^{q}{}_{p} = \bar{\Psi}^{q}_{A}\Psi^{A}_{p} - \frac{1}{2}\delta^{q}_{p}\bar{\Psi}^{r}{}_{A}\Psi^{A}_{r} \approx 0, \quad \bar{\mathcal{Z}}^{a}\mathcal{Z}_{a} + \bar{\Psi}^{q}\Psi_{q} \approx 0 \tag{48}$$

and eight fermionic constraints

$$\bar{\Psi}^{q}_{A}\mathcal{Z}^{A}_{a} \approx 0, \quad \bar{\mathcal{Z}}^{a}_{A}\Psi^{A}_{q} \approx 0 \tag{49}$$

generating the superalgebra of the $SU(2|2)_R$ color supergroup, whose notion was introduced in [16]. These constraints are the generators of the gauge symmetries of the superparticle's action based on the Lagrangian (45) discussed in [18].

The fact that the constraints and the Lagrangian (45) are quadratic in supertwistors clears the way to the Dirac quantization of the model. In our previous work [18] we considered the simplest case when the superparticle propagates only within the AdS_5 subspace of the $AdS_5 \times S^5$ superspace, i.e. its momentum components in directions tangent to S^5 vanish. In supertwistor formulation this means that Ψ^{A}_{q} and $\bar{\Psi}^{q}_{A}$ vanish so that only the c-type supertwistors $\mathcal{Z}^{A}_{a}$ and $\bar{\mathcal{Z}}^{a}_{A}$ remain as dynamical variables. It was demonstrated in [18] that the states of quantized superparticle in this case coincide with the $D = 5$ $N = 8$ gauged supergravity multiplet [28] both in the supertwistor and superoscillator approaches. Next task is to take into account Ψ^{A}_{q} and $\bar{\Psi}^{q}_{A}$ supertwistors that corresponds to the case when superparticle moves both within the AdS_5 and S^5 subspaces of the $AdS_5 \times S^5$ superspace.

References

[1] Shirafuji T 1983 Lagrangian mechanics of massless particles with spin *Prog. Theor. Phys.* **70** 18
[2] Witten E 1978 An interpretation of classical Yang-Mills theory *Phys. Lett.* B **77** 394
[3] Witten E 2004 Perturbative gauge theory as a string theory in twistor space *Comm. Math. Phys.* **252** 189 (*Preprint* hep-th/0312171)
[4] Mason L J 2005 Twistor actions for non-self-dual fields; a new foundation for twistor-string theory *J. High Energy Phys.* JHEP10(2005)009 (*Preprint* hep-th/0507269)
[5] Mason L J and Skinner D 2006 An ambitwistor Yang-Mills Lagrangian *Phys. Lett.* B **636** 60 (*Preprint* hep-th/0510262)
[6] Boels R, Mason L and Skinner D 2007 Supersymmetric gauge theories in twistor space *J. High Energy Phys.* JHEP02(2007)014 (*Preprint* hep-th/0604040)
[7] Hodges A 2013 Eliminating spurious poles from gauge-theoretic apmlitudes *J. High Energy Phys.* JHEP05(2013)135 (*Preprint* 0905.1473 [hep-th])
[8] Mason L and Skinner D 2009 Dual superconformal invariance, momentum twistors and Grassmannians *J. High Energy Phys.* JHEP11(2009)045 (*Preprint* 0909.0250 [hep-th])
[9] Ferber A 1978 Supertwistors and conformal supersymmetry *Nucl. Phys.* B **132** 55
[10] Claus P, Rahmfeld J and Zunger Y 1999 A simple particle action from a twistor parametrization of AdS_5 *Phys. Lett.* B **466** 181 (*Preprint* hep-th/9906118)
[11] Claus P, Kallosh R and Rahmfeld J 1999 BRST quantization of a particle in AdS_5 *Phys. Lett.* B **462** 285 (*Preprint* hep-th/9906195)
[12] Arvanitakis A S, Barns-Graham A E and Townsend P K 2017 Anti-de Sitter particles and manifest (super)isometries *Phys. Rev. Lett.* **118** 141601 (*Preprint* 1608.04380 [hep-th])
[13] Arvanitakis A S, Barns-Graham A E and Townsend P K 2018 Twistor description of spinning paricles in *AdS* *J. High Energy Phys.* JHEP01(2018)059 (*Preprint* 1710.09557 [hep-th])
[14] Adamo T, Skinner D and Williams J 2016 Twistor methods for AdS_5 *J. High Energy Phys.* JHEP08(2016)167 (*Preprint* 1607.03763 [hep-th])
[15] Bandos I, Ivanov E, Lukierski J and Sorokin D 2002 On the superconformal flatness of *AdS* superspaces, *J. High Energy Phys.* JHEP06(2002)040 (*Preprint* hep-th/0205104)
[16] Bars I 2004 Twistor superstring in 2T-physics *Phys. Rev.* D **70** 104022 (*Preprint* hep-th/0407239)
[17] Bars I 2005 Twistors and 2T-physics *AIP Conf. Proc.* **767** 3 (*Preprint* hep-th/0502065)
[18] Uvarov D V 2018 Supertwistor formulation for massless superparticle in $AdS_5 \times S^5$ superspace *Nucl. Phys.* B **936** 690 (*Preprint* 1807.08318 [hep-th])
[19] Uvarov D V 2019 Multitwistor mechanics of massless superparticle on $AdS_5 \times S^5$ superbackground *Preprint* 1907.13613 [hep-th]
[20] Sokatchev E 1989 An off-shell formulation of $N = 4$ supersymmetric Yang-Mills theory in twistor harmonic superspace *Phys. Lett.* B **217** 489
[21] Uvarov D V 2016 Spinor description of $D = 5$ massless low-spin gauge fields *Class. Quantum Grav.* **33** 135010 (*Preprint* 1506.01881 [hep-th])
[22] Delduc F, Galperin A S and Sokatchev E 1992 Lorentz-harmonic (super)fields and (super)particles *Nucl. Phys.* B **368** 143
[23] Galperin A S, Howe P S and Stelle K S 1992 The superparticle and the Lorentz group *Nucl. Phys.* B **368** 248 (*Preprint* hep-th/9201020)
[24] Bandos I A and Zheltukhin A A 1992 Green-Schwarz superstrings in spinor moving frame formalism, *Phys. Lett.* B **288** 77
[25] Ivanov E, Kalitsyn S, Nguyen Ai Viet and Ogievetsky V 1985 Harmonic superspaces of extended supersymmetry. The calculus of harmonic variables *J. Phys. A: Math. Gen.* **18** 3433
[26] Metsaev R R and Tseytlin A A 1998 Type IIB superstring action in $AdS_5 \times S^5$ background *Nucl. Phys.* B **533** 109 (*Preprint* hep-th/9805028)
[27] Kallosh R, Rahmfeld J and Rajaraman A 1998 Near horizon superspace *J. High Energy Phys.* JHEP09(1998)002 (*Preprint* hep-th/9805217)
[28] Gunaydin M and Marcus N 1985 The spectrum of the S^5 compactification of the chiral $N = 2$, $D = 10$ supergravity and the unitary supermultiplets of $U(2, 2|4)$ *Class. Quantum Grav.* **2** L11

Journal of Physics: Conference Series **1416** (2019) 012040 doi:10.1088/1742-6596/1416/1/012040

On modeling neutrinos oscillations by geometry methods in the frames of the theory for a fermion with three mass parameters

Ya.A. Voynova[1], E A Ovsiyuk[2], V V Kisel [3], V A Pletyukhov[4], V V Gilewsky [5], V M Red'kov[1]

[1] B.I. Stepanov Institute of Physics, NAS of Belarus, Independence ave. 68-2, 220072 Minsk
[2] Mozyr State Pedagogical University, Studencheskaya st. 28, Mozyr 247760, Belarus
[3] Belarus State University of Informatics and Radio-electronics, P. Brovki st 6, 220013 Minsk
[4] Brest State University, Kosmonavtov st 21, 224016 Brest, Belarus
[5] JINP-Sosny, NAS of Belarus, Minsk

E-mail: e.ovsiyuk@mail.ru , voinovayanina@mail.ru , vasiliy-bspu@mail.ru , terphys@brsu.brest.by , valentin.gilewsky@gmail.com , v.redkov@ifanbel.bas-net.by

Abstract. Starting from the general Gel'fand-Yaglom approach, we develop the theory of a new wave equation for a spin 1/2 fermion, which is characterized by three mass parameters. On the base of a 20-component wave function, three auxiliary bispinors are introduced, in absence of external fields these bispinors obey three separate Dirac-like equations with different masses. In presence of external electromagnetic fields or gravitational non-Euclidean background with a non-vanishing Ricci scalar curvature, the main equation is not split into separated three equations, instead a quite definite mixing of three Dirac-like equations arises. It is shown that for neutral Majorana particle, a generalized equation with three mass parameters exists as well. Such a generalized Majorana equation is not split into three separated equations in the curved space-time background, if the Ricci scalar of that space-time does not vanish. We have studied in detail the Majorana case, assuming approximation when an external cosmological background is taken into account by a constant Ricci parameter, $R = \text{const}$, and the Cartesian coordinates are used. With the help of a special linear transformation, the system of three linked Majorana equations transforms into three separate ones, with modified mass parameters, the last are solvable in the usual way. The spectrum of arising mass parameters is studied analytically and numerically.

1. Fermion with 3 mass parameter, theory in absence of external fields

Existence of different and more general wave equations than commonly used ones is well known [?, ?, ?, ?]. In the context of existence of the similar neutrinos of different masses, we examine a possibility within the theory of relativistic wave equations to describe a spin 1/2 particle with three mass parameters. Such a generalized equation for fermion with 3 mass parameters is based on the use of 20-component wave function (bispinor Ψ_0 and vector bispinor Ψ_μ). Omitting many details of this theory in the frames of Gel'fand–Yaglom approach [?], we start with a system of equations in spin-tensor form (first, consider the model in absence of external fields)

$$c_1 \hat{\partial}(\gamma_\mu \Psi_\mu) + \frac{c_3}{\sqrt{6}}[\hat{\partial}(\gamma_\mu \Psi_\mu) - 4(\partial_\mu \Psi_\mu)] + M(\gamma_\mu \Psi_\mu) = 0, \quad (1)$$

Journal of Physics: Conference Series **1416** (2019) 012040 doi:10.1088/1742-6596/1416/1/012040

$$c_2\hat{\partial}\Psi_0 - i\frac{4c_4}{\sqrt{6}}[-\frac{1}{4}\hat{\partial}(\gamma_\mu\Psi_\mu) + (\partial_\mu\Psi_\mu)] + M\Psi_0 = 0, \tag{2}$$

$$-\frac{2fc_3^*}{\sqrt{6}}[\partial_\lambda(\gamma_\mu\Psi_\mu) - \frac{1}{4}\gamma_\lambda\hat{\partial}(\gamma_\mu\Psi_\mu)] + i\frac{2gc_4^*}{\sqrt{6}}(\partial_\lambda\Psi_0 - \frac{1}{4}\gamma_\lambda\hat{\partial}\Psi_0) + M[\Psi_\lambda - \frac{1}{4}\gamma_\lambda(\gamma_\mu\Psi_\mu)] = 0. \tag{3}$$

Numerics c_1, c_2 are real, and c_3, c_4 are complex; $f, g \in \{\pm 1\}$; we use notation $\hat{\partial} = \gamma_\mu\partial_\mu$. Physical sense of the numerical parameters will be clarified later. The system can be transformed to a the form of equations with respect to three bispinors $\gamma_\mu\Psi_\mu$, Ψ_0, $\partial_\mu\Psi_\mu$:

$$\frac{1}{c_2(c_1 + c_3/\sqrt{6})}\left\{c_1c_2(c_1 + c_3/\sqrt{6}) + fc_2|c_3|^2 - \frac{g}{\sqrt{6}}c_3|c_4|^2\right\}\hat{\partial}(\gamma_\mu\Psi_\mu)$$
$$-ig\frac{c_3c_4^*}{c_2}\hat{\partial}\Psi_0 - \frac{4c_3}{M\sqrt{6}}\frac{fc_2|c_3|^2 + gc_1|c_4|^2}{c_2(c_1 + c_3/\sqrt{6})}\hat{\partial}(\partial_\mu\Psi_\mu) + M(\gamma_\mu\Psi_\mu) = 0,$$
$$\frac{i}{\sqrt{6}}\frac{c_4}{c_2(c_1 + c_3/\sqrt{6})}(\sqrt{6}fc_2c_3^* - g|c_4|^2)\hat{\partial}(\gamma_\mu\Psi_\mu) + \frac{c_2^2 + g|c_4|^2}{c_2}\hat{\partial}\Psi_0$$
$$-i\frac{4c_4}{M\sqrt{6}}\frac{fc_2|c_3|^2 + gc_1|c_4|^2}{c_2(c_1 + c_3/\sqrt{6})}\hat{\partial}(\partial_\mu\Psi_\mu) + M\Psi_0 = 0,$$
$$\frac{M}{4}\frac{1}{c_2(c_1 + c_3/\sqrt{6})}\left\{\sqrt{6}fc_2c_3^* - g|c_4|^2 - c_2(c_1 + c_3/\sqrt{6})\right\}\hat{\partial}(\gamma_\mu\Psi_\mu)$$
$$-igM\frac{\sqrt{6}}{4}\frac{c_4^*}{c_2}\hat{\partial}\Psi_0 - \frac{fc_2|c_3|^2 + gc_1|c_4|^2}{c_2(c_1 + c_3/\sqrt{6})}\hat{\partial}(\partial_\mu\Psi_\mu) + M(\partial_\mu\Psi_\mu) = 0.$$

In brief, it is presented in a matrix form as follows

$$K\hat{\partial}\begin{vmatrix}\gamma_\mu\Psi_\mu\\ \Psi_0\\ \partial_\mu\Psi_\mu\end{vmatrix} = M\begin{vmatrix}\gamma_\mu\Psi_\mu\\ \Psi_0\\ \partial_\mu\Psi_\mu\end{vmatrix}, \quad K = \begin{vmatrix}A_1 & B_1 & R_1\\ A_2 & B_2 & R_2\\ A_3 & B_3 & R_3\end{vmatrix}. \tag{4}$$

The numerical matrix K can be diagonalized, there arises a cubic equation for diagonal elements λ_i:

$$\Psi' = S\Psi, \quad SKS^{-1} = K' = \begin{vmatrix}\lambda_1 & 0 & 0\\ 0 & \lambda_2 & 0\\ 0 & 0 & \lambda_3\end{vmatrix}, \quad S = \begin{vmatrix}a_1 & a_2 & a_3\\ b_1 & b_2 & b_3\\ r_1 & r_2 & r_3\end{vmatrix},$$
$$\lambda^3 - \lambda^2(c_1 + c_2) + \lambda(c_1c_2 - f|c_3|^2 - g|c_4|^2) + fc_2|c_3|^2 + gc_1|c_4|^2 = 0, \tag{5}$$

The rows of the matrix S satisfy to corresponding subsystems:

$$a_1A_1 + a_2A_2 + a_3A_3 = \lambda_1a_1, \quad a_1B_1 + a_2B_2 + a_3B_3 = \lambda_1a_2, \quad a_1R_1 + a_2R_2 + a_3R_3 = \lambda_1a_3;$$

$$b_1A_1 + b_2A_2 + b_3A_3 = \lambda_2b_1, \quad b_1B_1 + b_2B_2 + b_3B_3 = \lambda_2b_2, \quad b_1R_1 + b_2R_2 + b_3R_3 = \lambda_2b_3;$$
$$r_1A_1 + r_2A_2 + r_3A_3 = \lambda_3r_1, r_1B_1 + r_2B_2 + r_3B_3 = \lambda_3r_2, r_1R_1 + r_2R_2 + r_3R_3 = \lambda_3r_3.$$

Their solutions have similar structure:

$$a_2 = a_1\frac{-igc_3c_4^*(c_1 + \frac{c_3}{\sqrt{6}})\lambda_1}{\lambda_1c_2(\lambda_1 - c_2) + (\lambda_1 - c_2)[fc_2|c_3|^2 + gc_1|c_4|^2] - \lambda_1(c_1 + \frac{c_3}{\sqrt{6}})g|c_4|^2},$$

$$a_3 = -a_1\frac{4a_1}{\sqrt{6}M}\frac{c_3(\lambda_1 - c_2)[fc_2|c_3|^2 + gc_1|c_4|^2]}{\lambda_1c_2(\lambda_1 - c_2) + (\lambda_1 - c_2)[fc_2|c_3|^2 + gc_1|c_4|^2] - \lambda_1(c_1 + \frac{c_3}{\sqrt{6}})g|c_4|^2},$$

and results for b_j, r_j are similar: $b_1, b_2, b_3 \Longrightarrow \lambda_2$, $r_1, r_2, r_3 \Longrightarrow \lambda_3$.

In this way we transform the system into three unlinked Dirac-like equations with different masses: $M_1 = M/\lambda_1, M_2 = M/\lambda_2, M_3 = M/\lambda_3$. We are to examine possible values of the roots $\lambda_1, \lambda_2, \lambda_3$ of characteristic cubic equations (5). It is readily proved that real positive values are possible if $c_1 > 0,\ c_2 > 0,\ f = -1\,,\ g = -1$. Using simplifying notations

$$|c_4|^2 = a^2, \quad |c_3|^2 = b^2, \quad \Gamma = \frac{c_1 a^2 + c_2 b^2}{(c_1 + c_2)^3}, \tag{6}$$

we can introduce simple parametrization for the roots:

$$M_3 = \frac{M}{\lambda_3} = \frac{\mu}{\cos\alpha}, \quad \mu = \frac{M}{(c_1 + c_2)}, \quad \alpha \in (0, \frac{\pi}{2}), \tag{7}$$

$$M_{1,2} = \frac{M}{\lambda_2} = \frac{\mu}{\sin^2(\alpha/2) \pm \sqrt{\sin^4(\alpha/2) - \Gamma/\cos\alpha}}\,. \tag{8}$$

Thus, in absence of external electromagnetic fields, the initial system is reducible to the form of three separated Dirac-like equations for bispinors:

$$\begin{aligned}
\Phi_1 &= a_1(\gamma_\mu \Psi_\mu) + a_2 \Psi_0 + a_3(\partial_\mu \Psi_\mu), \quad (\hat{\partial} + M_1)\Phi_1 = 0\,, \quad M_1 = M/\lambda_1\,; \\
\Phi_2 &= b_1(\gamma_\mu \Psi_\mu) + b_2 \Psi_0 + b_3(\partial_\mu \Psi_\mu), \quad (\hat{\partial} + M_2)\Phi_2 = 0, \quad M_2 = M/\lambda_2\,; \\
\Phi_3 &= r_1(\gamma_\mu \Psi_\mu) + r_2 \Psi_0 + r_3(\partial_\mu \Psi_\mu), \quad (\hat{\partial} + M_3)\Phi_3 = 0\,, \quad M_3 = M/\lambda_3\,.
\end{aligned} \tag{9}$$

2. The presence of electromagnetic field

Now, let us take unto account external electromagnetic fields. To this end, we are to turn back to the system (1)—(3) and modify the derivative operator, $D_\mu = \partial_\mu - ieA_\mu(x)$:

$$c_2 \hat{D}\Psi_0 - \frac{4ic_4}{\sqrt{6}} \left[(D_\mu \Psi_\mu) - \frac{1}{4}\hat{D}(\gamma_\mu \Psi_\mu) \right] + M\Psi_0 = 0, \tag{10}$$

$$(c_1 + \frac{c_3}{\sqrt{6}})\hat{D}(\gamma_\mu \Psi_\mu) - \frac{4c_3}{\sqrt{6}}(D_\mu \Psi_\mu) + M(\gamma_\mu \Psi_\mu) = 0, \tag{11}$$

$$\begin{aligned}
&-\frac{2fc_3^*}{\sqrt{6}} \left[\frac{3}{4}D^2 - \frac{1}{4}(-ieF_{\lambda\rho})\,\sigma_{\lambda\rho} \right] (\gamma_\mu \Psi_\mu) \\
&+\frac{2igc_4^*}{\sqrt{6}} \left[\frac{3}{4}D^2 - \frac{1}{4}(-ieF_{\lambda\rho})\sigma_{\lambda\rho} \right] \Psi_0 - \frac{M}{4}\hat{D}(\gamma_\mu \Psi_\mu) + M\,(D_\lambda \Psi_\lambda) = 0.
\end{aligned} \tag{12}$$

Equation in (12) contains a 2-nd order operator $D^2 = D_\nu D_\nu$. With the help of two first equations, we may exclude such an operator from the third equation. Further, we can reduce the system to a different (but equivalent) form

$$\begin{aligned}
&A_1\hat{D}\,\bar{\Phi}_1 + B_1\hat{D}\,\bar{\Phi}_2 + R_1\hat{D}\,\bar{\Phi}_3 + M\,\bar{\Phi}_1 + \frac{4}{3}\frac{f|c_3|^2}{M}\Sigma\,\bar{\Phi}_1 - \frac{4i}{3}\frac{gc_3c_4^*}{M}\Sigma\,\bar{\Phi}_2 = 0, \\
&A_2\hat{D}\,\bar{\Phi}_1 + B_2\hat{D}\,\bar{\Phi}_2 + R_2\hat{D}\,\bar{\Phi}_3 + M\,\bar{\Phi}_2 + \frac{4i}{3}\frac{fc_3^*c_4}{M}\Sigma\,\bar{\Phi}_1 + \frac{4}{3}\frac{g|c_4|^2}{M}\Sigma\,\bar{\Phi}_2 = 0, \\
&A_3\hat{D}\,\bar{\Phi}_1 + B_3\hat{D}\,\bar{\Phi}_2 + R_3\hat{D}\,\bar{\Phi}_3 + M\,\bar{\Phi}_3 + \frac{2}{\sqrt{6}}fc_3^*\,\Sigma\,\bar{\Phi}_1 - i\frac{2}{\sqrt{6}}\,gc_4^*\Sigma\,\bar{\Phi}_2 = 0,
\end{aligned} \tag{13}$$

where $\bar{\Phi}_1 = \gamma_\mu \Psi_\mu$, $\bar{\Phi}_2 = \Psi_0$, $\bar{\Phi}_3 = D_\mu \Psi_\mu$, $\Sigma = -ieF_{\lambda\rho}\sigma_{\lambda\rho}$. The system (13) differs from similar one (4) in free case: it contains additional mixing terms depending on the tensor $F_{\lambda\rho}$ of the external electromagnetic field.

From equations (13), for three new bispinors

$$\Phi_1 = a_1\bar{\Phi}_1 + a_2\bar{\Phi}_2 + a_3\bar{\Phi}_3,\ \Phi_2 = b_1\bar{\Phi}_1 + b_2\bar{\Phi}_2 + b_3\bar{\Phi}_3,\ \Phi_3 = r_1\,\bar{\Phi}_1 + r_2\,\bar{\Phi}_2 + r_3\,\bar{\Phi}_3, \tag{14}$$

we derive a system with a more symmetric structure

$$\begin{aligned}
&(\lambda_1\hat{D} + M)\Phi_1 + \frac{4c_2c_3}{3M}\,\lambda_1(\lambda_1 - c_2)\,\Sigma(x)\,[fc_3^*\,\bar{\Phi}_1 - igc_4^*\bar{\Phi}_2)] = 0,\\
&(\lambda_2\hat{D} + M)\Phi_2 + \frac{4c_2c_3}{3M}\,\lambda_2(\lambda_2 - c_2)\,\Sigma(x)\,[fc_3^*\,\bar{\Phi}_1 - igc_4^*\bar{\Phi}_2] = 0,\\
&(\lambda_3\hat{D} + M)\Phi_3 + \frac{4c_2c_3}{3M}\,\lambda_3(\lambda_3 - c_2)\,\Sigma(x)\,[fc_3^*\,\bar{\Phi}_1 - igc_4^*\bar{\Phi}_2] = 0.
\end{aligned} \tag{15}$$

Expressing $\bar{\Phi}_j$ through Φ_j, we get

$$\begin{aligned}
&\hat{D}\Phi_1(x) + M_1\Phi_1(x) + Y_1\,\Sigma(x)\Phi(x) = 0\,,\\
&\hat{D}\Phi_2(x) + M_2\Phi_2(x) + Y_2\,\Sigma(x)\Phi(x) = 0\,,\\
&\hat{D}\Phi_3(x) + M_3\Phi_3(x) + Y_3\,\Sigma(x)\Phi(x) = 0\,.
\end{aligned} \tag{16}$$

In the system (16) the notations are used

$$\begin{aligned}
&\Phi(x) = L_1\Phi_1(x) + L_2\Phi_2(x) + L_3\Phi_3(x)\,,\\
&L_1 = \frac{-L|c_4|^2 - L|c_3|^2 + c_2{}^2 - c_2(\lambda_2 + \lambda_3) + \lambda_2\lambda_3}{Lc_2\,c_3(\lambda_1 - \lambda_2)(\lambda_1 - \lambda_3)}\,,\\
&L_2 = \frac{-L|c_4|^2 - L|c_3|^2 + c_2{}^2 - c_2(\lambda_3 + \lambda_1) + \lambda_3\,\lambda_1}{Lc_2\,c_3(\lambda_2 - \lambda_3)(\lambda_2 - \lambda_1)}\,,\\
&L_3 = \frac{-L|c_4|^2 - L|c_3|^2 + c_2{}^2 - c_2(\lambda_1 + \lambda_2) + \lambda_1\,\lambda_2}{Lc_2\,c_3(\lambda_3 - \lambda_1)(\lambda_3 - \lambda_2)}\,.\\
&Y_i = \frac{4c_3}{3M}c_2(\lambda_i - c_2), \quad i = 1,2,3\,, \quad L = c_1 + \frac{c_3}{\sqrt{6}}\,,
\end{aligned} \tag{17}$$

Equations (16) can be presented in a matrix form

$$\begin{vmatrix} \hat{D}+M_1 & 0 & 0 \\ 0 & \hat{D}+M_2 & 0 \\ 0 & 0 & \hat{D}+M_3 \end{vmatrix} \begin{vmatrix} \Phi_1 \\ \Phi_2 \\ \Phi_3 \end{vmatrix} + \Sigma(x) \begin{vmatrix} Y_1L_1 & Y_1L_2 & Y_1L_3 \\ Y_2L_1 & Y_2L_2 & Y_2L_3 \\ Y_3L_1 & Y_3L_2 & Y_3L_3 \end{vmatrix} \begin{vmatrix} \Phi_1 \\ \Phi_2 \\ \Phi_3 \end{vmatrix} = 0\,. \tag{18}$$

Let it be $|c_4| = a, |c_3| = b$; taking in mind that according the relations (8) parameter $(c_1 + c_2)$ may be taken up by arbitrary M, we set $c_1 = 1, c_2 = 1$. Then the the elements of the mixing matrix read

$$\begin{aligned}
&Y_1L_1 = \frac{4}{3M}(\lambda_1 - 1)\frac{-L(a^2 + b^2) + 1 - (\lambda_2 + \lambda_3) + \lambda_2\lambda_3}{L(\lambda_1 - \lambda_2)(\lambda_1 - \lambda_3)}\,,\\
&Y_1L_2 = \frac{4}{3M}(\lambda_1 - 1)\frac{-L(a^2 + b^2) + 1 - (\lambda_3 + \lambda_1) + \lambda_3\,\lambda_1}{L(\lambda_2 - \lambda_3)(\lambda_2 - \lambda_1)}\,,\\
&Y_1L_3 = \frac{4}{3M}(\lambda_1 - 1)\frac{-L(a^2 + b^2) + 1 - (\lambda_1 + \lambda_2) + \lambda_1\,\lambda_2}{L(\lambda_3 - \lambda_1)(\lambda_3 - \lambda_2)}\,,\\
&Y_2L_1 = \frac{4}{3M}(\lambda_2 - 1)\frac{-L(a^2 + b^2) + 1 - (\lambda_2 + \lambda_3) + \lambda_2\lambda_3}{L(\lambda_1 - \lambda_2)(\lambda_1 - \lambda_3)}\,,
\end{aligned}$$

$$
\begin{aligned}
Y_2L_2 &= \frac{4}{3M}(\lambda_2-1)\frac{-L(a^2+b^2)+1-(\lambda_3+\lambda_1)+\lambda_3\lambda_1}{L(\lambda_2-\lambda_3)(\lambda_2-\lambda_1)},\\
Y_2L_3 &= \frac{4}{3M}(\lambda_2-1)\frac{-L(a^2+b^2)+1-(\lambda_1+\lambda_2)+\lambda_1\lambda_2}{L(\lambda_3-\lambda_1)(\lambda_3-\lambda_2)},\\
Y_3L_1 &= \frac{4}{3M}(\lambda_3-1)\frac{-L(a^2+b^2)+1-(\lambda_2+\lambda_3)+\lambda_2\lambda_3}{L(\lambda_1-\lambda_2)(\lambda_1-\lambda_3)},\\
Y_3L_2 &= \frac{4}{3M}(\lambda_3-1)\frac{-L(a^2+b^2)+1-(\lambda_3+\lambda_1)+\lambda_3\lambda_1}{L(\lambda_2-\lambda_3)(\lambda_2-\lambda_1)},\\
Y_3L_3 &= \frac{4}{3M}(\lambda_3-1)\frac{-L(a^2+b^2)+1-(\lambda_1+\lambda_2)+\lambda_1\lambda_2}{L(\lambda_3-\lambda_1)(\lambda_3-\lambda_2)}.
\end{aligned}
$$

3. Riemannian space-time geometry

Let us extend the model to a curved space-time. Instead of *ict*-metric, now we use the metric tensor $g_{\alpha\beta}(x)$ and slightly other Dirac matrices, also we apply more complicated derivative operations

$$
\gamma^0 = \begin{vmatrix} 0 & I \\ I & 0 \end{vmatrix}, \quad \gamma^i = \begin{vmatrix} 0 & -\sigma^i \\ \sigma^i & 0 \end{vmatrix}, D_\alpha(x) = \nabla_\alpha + \Gamma_\alpha(x) + ieA_\alpha(x), \hat{D} = \gamma^\alpha(x) D_\alpha(x), \qquad (19)
$$

where $\Gamma_\alpha(x)$ is the conventional bispinor connection, and tetrad based local Dirac matrices $\gamma^\alpha(x) = \gamma^a e^\alpha_{(a)}(x)$. The final system of equations for a fermion with three mass parameters in Riemannian space-time has the form

$$
\begin{aligned}
&i\gamma^\alpha(x)\left[\partial_\alpha + \Gamma_\alpha(x) + ieA_\alpha(x)\right]\Phi_1(x) - M_1\Phi_1(x) + Y_1\,\Sigma(x)\Phi(x) = 0\,,\\
&i\gamma^\alpha(x)\left[\partial_\alpha + \Gamma_\alpha(x) + ieA_\alpha(x)\right]\Phi_2(x) - M_2\Phi_2(x) + Y_2\,\Sigma(x)\Phi(x) = 0\,,\\
&i\gamma^\alpha(x)\left[\partial_\alpha + \Gamma_\alpha(x) + ieA_\alpha(x)\right]\Phi_3(x) - M_3\Phi_3(x) + Y_3\,\Sigma(x)\Phi(x) = 0\,, \qquad (20)
\end{aligned}
$$

where $\Sigma(x) = -ieF_{\alpha\beta}\sigma^{\alpha\beta}(x) + \frac{1}{4}R(x)$, $R(x)$ is a Ricci scalar. It should be noticed that for geometrical models with non-vanising Ricci scalar, $R(x) \neq 0$, even in absence of the electromagnetic interaction, equations (20) link three bispinors in the unified system. And what is more, because in any Majorana the following properties of Dirac matrices and bispinor connection hold

$$
[i\gamma^\alpha(x)]^* = i\gamma^\alpha(x)\,, \quad \Gamma^*_\alpha(x) = \Gamma_\alpha(x)\,,
$$

the above system (20) describes Majorana-type neutral fermions with three mass parameters as well/ The system remains the same only with change: $\partial_\alpha + \Gamma_\alpha(x) + ieA_\alpha(x) \Longrightarrow \partial_\alpha + \Gamma_\alpha(x)$.

Therefore, this theory permits us to take into account the effects of cosmological geometry background on such a complicated fermion. The most simple physically interpretable example is as follows: we may employ an approximation when locally the use of Cartesian coordinates is justified, and an external cosmological background is described by a non-vanishing constant Ricci scalar.

4. Model example

For simplicity, let us follow a simple 1-dimensional case $(t, x, y = 0, z = 0)$. So we start with the system

$$
\begin{aligned}
&(i\gamma^0\partial_t + i\gamma^1\partial_1 - M_1)\Phi_1 + d_1(L_1\Phi_1 + L_2\Phi_2 + L_3\Phi_3) = 0\,,\\
&(i\gamma^0\partial_t + i\gamma^1\partial_1 - M_2)\Phi_2 + d_2(L_1\Phi_1 + L_2\Phi_2 + L_3\Phi_3) = 0\,,\\
&(i\gamma^0\partial_t + i\gamma^1\partial_1 - M_3)\Phi_3 + d_3(L_1\Phi_1 + L_2\Phi_2 + L_3\Phi_3) = 0\,; \qquad (21)
\end{aligned}
$$

where $d_i = Y_i \frac{R}{4}$, $i = 1, 2, 3$. The system (21) is transformed to the matrix form

$$(i\gamma^0\partial_t + i\gamma^1\partial_1 - M)\begin{vmatrix} \Phi_1 \\ \Phi_2 \\ \Phi_3 \end{vmatrix} = -\begin{vmatrix} M - M_1 + d_1L_1 & d_1L_2 & d_1L_3 \\ d_2L_1 & M - M_2 + d_2L_2 & d_2L_3 \\ d_3L_1 & d_3L_2 & M - M_3 + d_3L_3 \end{vmatrix}\begin{vmatrix} \Phi_1 \\ \Phi_2 \\ \Phi_3 \end{vmatrix},$$

or in brief $\Delta\ \Phi = T\ \Phi$, $\Delta = -(i\gamma^0\partial_t + i\gamma^1\partial_1 - M)$. The 3-column Φ is subject to a linear transformation to diagonalize the mixing matrix T:

$$\bar{\Phi} = S\Phi\,, \quad STS^{-1} = T_0 = \begin{vmatrix} \mu_1 & 0 & 0 \\ 0 & \mu_2 & 0 \\ 0 & 0 & \mu_3 \end{vmatrix}. \tag{22}$$

After that we will have three separate Dirac-like equations with new mass parameters $\bar{M_1}$:

$$\begin{aligned} &(\,i\gamma^0\partial_t + i\gamma^1\partial_1 - \bar{M}_1\,)\,\bar{\Phi}_1 = 0\,, \quad \bar{M}_1 = M + \mu_1\,; \\ &(\,i\gamma^0\partial_t + i\gamma^1\partial_1 - \bar{M}_2\,)\,\bar{\Phi}_2 = 0\,, \quad \bar{M}_2 = M + \mu_2\,; \\ &(\,i\gamma^0\partial_t + i\gamma^1\partial_1 - \bar{M}_3\,)\,\bar{\Phi}_3 = 0\,, \quad \bar{M}_3 = M + \mu_3\,, \end{aligned} \tag{23}$$

by physical reason we assume the real-valuedness of μ_i, and positiveness of $M + \mu_i$. To find the transformation S, we should solve the equation $ST = T_0S$. It leads to three linear subsystems. For diagonal elements in T_0 we get a cubic algebraic equation. To get a more simple form of this equation we are to make several steps. First, we apply the substitution $M_i = \frac{M}{\lambda_i}$, where M is arbitrary. We may simplify the task without loss of generality by setting $c_1 = c_2 = 1$, then the cubic equation for λ_i give simple roots

$$\lambda^3 - 2\lambda^2 + (1+k)\lambda - k = 0\,,\ \lambda_3 = 1\,, \lambda_{1,2} = \frac{1}{2} \mp \frac{1}{2}\sqrt{1-4k}\,,\ k = (a^2 + b^2) \in (0, \frac{1}{4})\,. \tag{24}$$

Correspondingly, the masses M_i equal to

$$M_3 = M\,, \quad M_{1,2} = \frac{2M}{1 \mp \sqrt{1-4k}}\,, \quad k \in (0, \frac{1}{4})\,. \tag{25}$$

It is convenient to introduce dimensionless parameter r: $R = 6r\ M^2 \implies d_i = Mrb(\lambda_i - \frac{1}{2}) = MD_i$, and the notation $L = 1 + \frac{b}{\sqrt{6}}, 0 < 2b < 1$. Besides, the roots may be done dimensionless as well $\mu_i = M\ \Delta_i$.

In this way, we arrive at the following cubic equation for Δ_i:

$$\Delta^3 + \frac{2k-1}{k}\Delta^2 + \left[1 + \left(1 - \frac{\sqrt{6}}{2(\sqrt{6}+b)}\frac{1-4k}{2k}\right)r\right]\Delta + \left(1 + \frac{\sqrt{6}}{2(\sqrt{6}+b)}\frac{1-4k}{2k}\right)r = 0\,, \tag{26}$$

where $0 < k < \frac{1}{4}$, $0 < b < 2$. Taking in mind relationship $R = 6rM^2$, we expect dimensionless parameter r is small, because effect of geometry in the model under consideration should be small. Besides, there exist two physically different possibilities: $r > 0$ at positive curvature, and $r < 0$ at negative curvature. We have followed several case of weak and strong gravitation of different curvature sings:

$$\begin{aligned} &r = +10^{-30}\,, r = +10^{-5}, r = +10^{-3}, r = +10^{-2}, r = +1\,; \\ &r = -10^{-30}\,, r = -10^{-5}, r = -10^{-3}, r = -10^{-2}, r = -1\,. \end{aligned} \tag{27}$$

The cases $r = \pm 10^{-2}, \pm 1$ correspond to very strong curvature of space. Numerical study showed that dependence of the roots Δ_i upon parameter $b \in (0, 2)$ is very small, by this reason below we take the value $b = 0$.

Numerical study shows that at the values of curvature in the region $r = \pm 10^{-2}, \pm 1$ the model becomes non-interpretable, because there appear complex-valued roots.

Acknowledgments
The present work was developed within the framework of the project BRFFR A18-RA-015, set within the framework of the agreement between the National Academy of Sciences of Belarus and the Romanian Academy, and of the project BRFFR F18Y-009.

Special thanks to EU Project MOST, Grant R-f26C-50842, for financial support of attending the 26-th Internatinal Conference on Integrable Systems and Quantum Symmetries, July 8–12, 2019, Prague.

Table 1. The roots at$b = 0,\ r = -10^{-5}$.

k	Δ_1	Δ_2	Δ_3
0.24	0.0000104	0.667	1.500
0.20	0.0000125	0.382	2.618
0.16	0.0000156	0.250	4.000
0.12	0.0000208	0.162	6.171
0.08	0.0000313	0.096,	10.404
0.04	0.0000626	0.043	22.956

Table 2. The roots at$b = 0,\ r = +10^{-5}$.

k	Δ_1	Δ_2	Δ_3
0.24	0.667	1.500	-0.00001
0.20	0.382	2.618	-0.00001
0.16	0.250	3.999	-0.00002
0.12	0.162	6.171	-0.00002
0.08	0.096	10.404,	-0.00003
0.04	0.044	22.956	-0.00006

Table 3. The roots at$b = 0,\ r = -10^{-2}$.

k	Δ_1	Δ_2	Δ_3
0.24	0.011	0.637	1.519
0.20	0.013	0.363	2.624
0.16	0.017	0.231	4.002
0.12	0.025	0.137	6.172
0.08	10.403	0.485+0.256 i	0.485-0.256 i
0.04	22.955	0.226+0.470 i	0.226-0.470 i

Table 4. The roots at$b = 0$, $r = +10^{-2}$.

k	Δ_1	Δ_2	Δ_3
0.24	0.697	1.480	-0.010
0.20	0.399	2.613	-0.012
0.16	0.267	3.998	-0.015
0.12	0.181	6.171	-0.019
0.08	0.120	10.405	-0.025
0.04	0.077	22.958	=-0.035

Table 5. The roots at$b = 0$, $r = +1$.

k	Δ_1	Δ_2	Δ_3
0.24	1.265+1.13 i	-0.362	1.265-1.13 i
0.18	1.173	2.804	-0.422
0.12	0.695	6.128	-0.489
0.06	0.503,	=14.727	-0.563

Table 6. The roots at$b = 0$, $r = -1$.

k	Δ_1	Δ_2	Δ_3
0.24	2.339	-0.86+0.66 i	-0.86-0.66 i
0.18	=3.556	-0.247+0.63 i	-0.24-0.63 i
0.12	6.213	0.60+0.58 i	0.60-0.58 i
0.06	14.468	0.996+0.53 i	0.996-0.53 i

References

[1] Gel'fand I M and Yaglom A M 1948 *Zh. Eksp. Teor. Fiz.* **18** 703–733
[2] Red'kov V M 2009 *Fields in Riemannian space and the Lorentz group* (Minsk: Belarusian Science)
[3] Pletjukhov V A, Red'kov V M and Strazhev V I 2015 *Relativistic wave equations and intrinsic degrees of freedom* (Minsk: Belarusian Science)
[4] Kisel V V, Ovsiyuk E M, Balan V, Veko O V and Red'kov V M 2018 *Elementary particles with internal structure in externalfield. Vol. I. General theory; Vol. II. Physical problems* (New York: Nova Science Publishers Inc.)

Journal of Physics: Conference Series **1416** (2019) 012041 doi:10.1088/1742-6596/1416/1/012041

Topological Bounds from Label Translation Symmetry of Non-Barotropic MHD

Asher Yahalom
Ariel University, Ariel 40700, Israel
E-mail: asya@ariel.ac.il

Abstract. The theorem of Noether dictates that for every continuous symmetry group of an Action the system must possess a conservation law. In this paper we discuss some subgroups of Arnold's labelling symmetry diffeomorphism related to non-barotropic magnetohydrodynamics (MHD) and the conservations laws associated with them. Furthermore, we will study the dynamical bounds resulting from the topological Noether currents associated with label translation symmetry groups.

1. Introduction

The theorem of Noether dictates that for every continuous symmetry group of an Action the system must possess a conservation law. For example time translation symmetry results in the conservation of energy, while spatial translation symmetry results in the conservation of linear momentum and rotation symmetry in the conservation of angular momentum to list some well known examples. But sometimes the conservation law is discovered without reference to the Noether theorem by using the equations of the system. In that case one is tempted to inquire what is the hidden symmetry associated with this conservation law and what is the simplest way to represent it.

The concept of metage as a label for fluid elements along a vortex line in ideal fluids was first introduced by Lynden-Bell & Katz [1]. A translation group of this label was found to be connected to the conservation of Moffat's [2] helicity by Yahalom [3]. The concept of metage was later generalized by Yahalom & Lynden-Bell [4] for barotropic MHD, but now as a label for fluid elements along magnetic field lines which are comoving with the flow in the case of ideal MHD. Yahalom & Lynden-Bell [4] has also shown that the translation group of the magnetic metage is connected to Woltjer [5, 6] conservation of cross helicity for barotropic MHD. Recently the concept of metage was generalized also for non barotropic MHD in which magnetic field lines lie on entropy surfaces [7]. This was later generalized by dropping the entropy condition on magnetic field lines [8].

Cross Helicity was first described by Woltjer [5, 6] and is give by:

$$H_C \equiv \int \vec{B} \cdot \vec{v} d^3x, \tag{1}$$

in which $\vec{B}$ is the magnetic field, $\vec{v}$ is the velocity field and the integral is taken over the entire flow domain. H_C is conserved for barotropic or incompressible MHD and is given a topological interpretation in terms of the knottiness of magnetic and flow field lines. A generalization of barotropic fluid dynamics conserved quantities including helicity to non barotropic flows including topological constants of motion is given by Mobbs [9]. However, Mobbs did not discuss the MHD case.

Both conservation laws for the helicity in the fluid dynamics case and the barotropic MHD case were shown to originate from a relabelling symmetry through the Noether theorem [3, 10, 11, 4]. Webb et al. [13] have generalized the idea of relabelling symmetry to non-barotropic MHD and derived their generalized cross helicity conservation law by using Noether's theorem but without using the simple representation which is connected with the metage variable. The conservation law deduction involves a divergence symmetry of the action. These conservation laws were written as Eulerian conservation laws of the form $D_t + \vec{\nabla} \cdot \vec{F} = 0$ where D is the conserved density and F is the conserved flux. Webb et al. [15] discuss the cross helicity conservation law for non-barotropic MHD in a multi-symplectic formulation of MHD. Webb et al. [12, 13] emphasize that the generalized cross helicity conservation law, in MHD and the generalized helicity conservation law in non-barotropic fluids are non-local in the sense that they depend on the auxiliary nonlocal variable σ, which depends on the Lagrangian time integral of the temperature $T(x, t)$. Notice that a potential vorticity conservation equation for non-barotropic MHD is derived by Webb, G. M. and Mace, R.L. [16] by using Noether's second theorem.

It should be mentioned that Mobbs [9] derived a helicity conservation law for ideal, non-barotropic fluid dynamics, which is of the same form as the cross helicity conservation law for non-barotropic MHD, except that the magnetic field induction is replaced by the generalized fluid helicity $\vec{\Omega} = \vec{\nabla} \times (\vec{v} - \sigma\vec{\nabla}s)$. Webb et al. [12, 13] also derive the Eulerian, differential form of Mobbs [9] conservation law (although they did not reference Mobbs [9]). Webb and Anco [14] show how Mobbs conservation law arises in multi-symplectic, Lagrangian fluid mechanics.

Variational principles for magnetohydrodynamics were introduced by previous authors both in Lagrangian and Eulerian form. Sturrock [17] has discussed in his book a Lagrangian variational formalism for magnetohydrodynamics. Vladimirov and Moffatt [22] in a series of papers have discussed an Eulerian variational principle for incompressible magnetohydrodynamics. However, their variational principle contained three more functions in addition to the seven variables which appear in the standard equations of incompressible magnetohydrodynamics which are the magnetic field $\vec{B}$ the velocity field $\vec{v}$ and the pressure P. Kats [23] has generalized Moffatt's work for compressible non barotropic flows but without reducing the number of functions and the computational load. Sakurai [21] has introduced a two function Eulerian variational principle for force-free magnetohydrodynamics and used it as a basis of a numerical scheme, his method is discussed in a book by Sturrock [17]. Yahalom & Lynden-Bell [4] combined the Lagrangian of Sturrock [17] with the Lagrangian of Sakurai [21] to obtain an **Eulerian** Lagrangian principle for barotropic magnetohydrodynamics which will depend on only six functions. The variational derivative of this Lagrangian produced all the equations needed to describe barotropic magnetohydrodynamics without any additional constraints. The equations obtained resembled the equations of Frenkel, Levich & Stilman [24] (see also [25]). Yahalom [29] have shown that for the barotropic case four functions will suffice. Moreover, it was shown that the cuts of some of those functions [30] are topological local conserved quantities.

Previous work was concerned only with barotropic magnetohydrodynamics. Variational principles of non barotropic magnetohydrodynamics can be found in the work of Bekenstein & Oron [26] in terms of 15 functions and V.A. Kats [23] in terms of 20 functions. Morrison [27] has suggested a Hamiltonian approach but this also depends on 8 canonical variables (see table 2 [27]). The variational principle introduced in [18, 19] show that only five functions will suffice to describe non barotropic MHD in the case that we enforce a Sakurai [21] representation for the magnetic field.

The variational principle was find as an efficient tool for deducing a generalized cross helicity using the translation symmetry groups of labels to both the global non barotropic cross helicity conservation law and the conservation law of circulations of topological velocity along magnetic field lines. The latter were shown to be equivalent to the amount of non barotropic cross helicity per unit of magnetic flux [31, 32]. Further more we have shown that two additional cross helicity conservation laws exist: the χ and η cross helicities [33]. Possible applications for MHD constraints of the generalized cross helicity and magnetic helicity are described in [32].

The plan of this paper is as follows: First we introduce the basic quantities and equations of non-barotropic MHD. Then we discuss some translation subgroups of Arnold's labelling symmetry diffeomorphism related to non-barotropic MHD and the conservations laws associated with them. Finally we introduce dynamical bounds associated with those Noether currents.

2. Basic Equations

Consider the equations of non-barotropic MHD [17, 18]:

$$\frac{\partial \vec{B}}{\partial t} = \vec{\nabla} \times (\vec{v} \times \vec{B}), \tag{2}$$

$$\vec{\nabla} \cdot \vec{B} = 0, \tag{3}$$

$$\frac{\partial \rho}{\partial t} + \vec{\nabla} \cdot (\rho \vec{v}) = 0, \tag{4}$$

$$\rho \frac{d\vec{v}}{dt} = \rho(\frac{\partial \vec{v}}{\partial t} + (\vec{v} \cdot \vec{\nabla})\vec{v}) = -\vec{\nabla} p(\rho, s) + \frac{(\vec{\nabla} \times \vec{B}) \times \vec{B}}{4\pi}, \tag{5}$$

$$\frac{ds}{dt} = \frac{\partial s}{\partial t} + (\vec{v} \cdot \vec{\nabla})s = 0. \tag{6}$$

In the above the following notations are utilized: $\frac{\partial}{\partial t}$ is the temporal derivative, $\frac{d}{dt}$ is the temporal material derivative and $\vec{\nabla}$ has its standard meaning in vector calculus. ρ is the fluid density and s is the specific entropy. Finally $p(\rho, s)$ is the pressure which depends on the density and entropy (the non-barotropic case). Equation (2) describes the fact that the magnetic field lines are moving with the fluid elements ("frozen" magnetic field lines), equation (3) describes the fact that the magnetic field is solenoidal, equation (4) describes the conservation of mass and equation (5) is the Euler equation for a fluid in which both pressure and Lorentz magnetic forces apply. Equation (6) describes the fact that heat is not created (zero viscosity, zero resistivity) in ideal non-barotropic MHD and is not conducted, thus only convection occurs. The number of independent variables for which one needs to solve is eight ($\vec{v}, \vec{B}, \rho, s$) and the number of equations (2,4,5,6) is also eight. Notice that equation (3) is a condition on the initial $\vec{B}$ field and is satisfied automatically for any other time due to equation (2).

3. The labelling symmetry group and its subgroups

It is obvious that the choice of fluid labels is quite arbitrary. However, when enforcing the χ, η, μ coordinate system (see [33] for details) such that:

$$\rho = \frac{\partial(\chi, \eta, \mu)}{\partial(x, y, z)}. \tag{7}$$

The choice is restricted to $\tilde{\chi}, \tilde{\eta}, \tilde{\mu}$ such that:

$$\frac{\partial(\tilde{\chi}, \tilde{\eta}, \tilde{\mu})}{\partial(\chi, \eta, \mu)} = 1. \tag{8}$$

Moreover the Euler potential magnetic field representation:

$$\vec{B} = \vec{\nabla}\chi \times \vec{\nabla}\eta, \tag{9}$$

reduces the choice further to:

$$\frac{\partial(\tilde{\chi},\tilde{\eta})}{\partial(\chi,\eta)} = 1. \tag{10}$$

3.1. Metage translations

In what follows we consider the transformation:

$$\tilde{\chi} = \chi, \tilde{\eta} = \eta, \tilde{\mu} = \mu + a(\chi,\eta) \tag{11}$$

Hence a is a label displacement which may be different for each magnetic field line, as the field line is closed one need not worry about edge difficulties. This transformation satisfies trivially the conditions (8,10). If $a = \delta\mu$ is small we can calculate the associated fluid element displacement with this relabelling (see [33] for details).

$$\vec{\xi} = -\frac{\partial \vec{r}}{\partial \mu}\delta\mu = -\delta\mu\frac{\vec{B}}{\rho}. \tag{12}$$

Inserting this expression into the boundary term in equation (32) of [33] will result in:

$$\delta A_B = \int dt \oint d\vec{S} \cdot [\vec{B}(\frac{1}{2}\vec{v}^2 - w(\rho,s)) - \vec{v}((\vec{v} - \sigma\vec{\nabla}s)\cdot\vec{B})]\delta\mu = 0, \tag{13}$$

which is the condition for magnetic cross helicity conservation. Inserting equation (12) into equation (34) of [33] we obtain the conservation law:

$$\delta J = \int d^3x\rho(\vec{v} - \sigma\vec{\nabla}s)\cdot\vec{\xi} = -\int d^3x\delta\mu(\vec{v} - \sigma\vec{\nabla}s)\cdot\vec{B} \tag{14}$$

In the simplest case we may take $\delta\mu$ to be a small constant, hence:

$$\delta J = -\delta\mu \int d^3x(\vec{v} - \sigma\vec{\nabla}s)\cdot\vec{B} = -\delta\mu H_{CNB} \tag{15}$$

Where H_{CNB} is the non barotropic global cross helicity [12, 31, 32] defined as:

$$H_{CNB} \equiv \int d^3x(\vec{v} - \sigma\vec{\nabla}s)\cdot\vec{B} = \int d^3x\vec{v}_t\cdot\vec{B} \tag{16}$$

in which $\vec{v}_t \equiv \vec{v} - \sigma\vec{\nabla}s$ is the topological velocity field. We thus obtain the conservation of non-barotropic cross helicity using the Noether theorem and the symmetry group of metage translations. Of course one can perform a different translation on each magnetic field line, in this case one obtains:

$$\delta J = -\int d^3x\delta\mu\vec{v}_t\cdot\vec{B} = -\int d\chi d\eta\delta\mu \oint_{\chi,\eta} d\mu\rho^{-1}\vec{v}_t\cdot\vec{B} \tag{17}$$

Now since $\delta\mu$ is an arbitrary (small) function of χ,η it follows that:

$$I = \oint_{\chi,\eta} d\mu\rho^{-1}\vec{v}_t\cdot\vec{B} \tag{18}$$

is a conserved quantity for each magnetic field line. Along a magnetic field line the following equations hold:

$$d\mu = \vec{\nabla}\mu \cdot d\vec{r} = \vec{\nabla}\mu \cdot \hat{B} dr = \frac{\rho}{B} dr \tag{19}$$

in the above $\hat{B}$ is an unit vector in the magnetic field direction and equation (15) of [33] is used. Inserting equation (19) into equation (18) we obtain:

$$I = \oint_{\chi,\eta} dr \vec{v}_t \cdot \hat{B} = \oint_{\chi,\eta} d\vec{r} \cdot \vec{v}_t. \tag{20}$$

which is just the circulation of the topological velocity along the magnetic field lines. This quantity can be written in terms of the generalized Clebsch representation of the velocity [18]:

$$\vec{v} = \vec{\nabla}\nu + \alpha\vec{\nabla}\chi + \beta\vec{\nabla}\eta + \sigma\vec{\nabla}s, \qquad \vec{v}_t = \vec{\nabla}\nu + \alpha\vec{\nabla}\chi + \beta\vec{\nabla}\eta \tag{21}$$

as:

$$I = \oint_{\chi,\eta} d\vec{r} \cdot \vec{v}_t = \oint_{\chi,\eta} d\vec{r} \cdot \vec{\nabla}\nu = [\nu]. \tag{22}$$

$[\nu]$ is the discontinuity of ν. This was shown to be equal to the amount of non barotropic cross helicity per unit of magnetic flux [31, 32].

$$I = [\nu] = \frac{dH_{CNB}}{d\Phi}. \tag{23}$$

3.2. Transformations of magnetic surfaces

Consider the following transformations:

$$\tilde{\eta} = \eta + \delta\eta(\chi,\eta), \qquad \tilde{\chi} = \chi + \delta\chi(\chi,\eta), \qquad \tilde{\mu} = \mu \tag{24}$$

in which $\delta\eta, \delta\chi$ are considered small in some sense. Inserting the above quantities into equation (10) and keeping only first order terms we arrive at:

$$\partial_\eta \delta\eta + \partial_\chi \delta\chi = 0. \tag{25}$$

This equation can be solved as follows:

$$\delta\eta = \partial_\chi \delta f, \qquad \delta\chi = -\partial_\eta \delta f, \tag{26}$$

in which $\delta f = \delta f(\chi,\eta)$ is an arbitrary small function. In this case we obtain a particle displacements of the form:

$$\vec{\xi} = -\frac{\partial\vec{r}}{\partial\chi}\delta\chi - \frac{\partial\vec{r}}{\partial\eta}\delta\eta = -\frac{1}{\rho}\left(\vec{\nabla}\eta \times \vec{\nabla}\mu \, \delta\chi + \vec{\nabla}\mu \times \vec{\nabla}\chi \, \delta\eta\right) = \frac{\vec{\nabla}\mu}{\rho} \times (\vec{\nabla}\eta\delta\chi - \vec{\nabla}\chi\delta\eta) \tag{27}$$

A special case that satisfies equation (25) is the case of a constant $\delta\chi$ and $\delta\eta$, those two independent displacements lead to two new topological conservation laws:

$$\delta J_\chi = \delta\chi \int d^3x \vec{v}_t \cdot \vec{\nabla}\mu \times \vec{\nabla}\eta = \delta\chi \; H_{CNB\chi}, \quad \delta J_\eta = \delta\eta \int d^3x \vec{v}_t \cdot \vec{\nabla}\chi \times \vec{\nabla}\mu = \delta\eta \; H_{CNB\eta}. \tag{28}$$

Where the new non barotropic global cross helicities are defined as:

$$H_{CNB\chi} \equiv \int d^3x \vec{v}_t \cdot \vec{\nabla}\mu \times \vec{\nabla}\eta, \quad H_{CNB\eta} \equiv \int d^3x \vec{v}_t \cdot \vec{\nabla}\chi \times \vec{\nabla}\mu \tag{29}$$

We will find it useful to introduce the abstract "magnetic fields" as follows:

$$\vec{B}_\chi \equiv \vec{\nabla}\mu \times \vec{\nabla}\eta, \qquad \vec{B}_\eta \equiv \vec{\nabla}\chi \times \vec{\nabla}\mu \tag{30}$$

In terms of which we obtain the new helicities in a more conventional form:

$$H_{CNB\chi} = \int d^3x \vec{v}_t \cdot \vec{B}_\chi, \qquad H_{CNB\eta} = \int d^3x \vec{v}_t \cdot \vec{B}_\eta \tag{31}$$

We remark that those topological constants will only be conserved under special boundary conditions satisfying:

$$\oint d\vec{S} \cdot \left[-\rho\vec{\xi}(\frac{1}{2}\vec{v}^2 - w(\rho, s)) + \rho\vec{v}(\vec{v}_t \cdot \vec{\xi}) + \frac{1}{4\pi}\vec{B} \times (\vec{\xi} \times \vec{B}) \right] = 0 \tag{32}$$

for:

$$\vec{\xi}_\chi = \frac{\vec{B}_\chi}{\rho}\delta\chi, \qquad \vec{\xi}_\eta = \frac{\vec{B}_\eta}{\rho}\delta\eta \tag{33}$$

This is more plausible for magnetic field lines which lie on topological torii. In this case η is non single valued [4] and thus the translation in this direction resembles moving fluid elements along closed loops. Both those helicities suffer a topological interpretation in terms of the knottiness of the abstract magnetic field lines and the flow lines. Finally we remark that for barotropic MHD $\vec{v}_t$ can be replaced with $\vec{v}$.

4. Possible Application

In his important review paper "Physics of magnetically confined plasmas" A. H. Boozer [34] states that: "A spiky current profile causes a rapid dissipation of energy relative to magnetic helicity. If the evolution of a magnetic field is rapid, then it must be at constant helicity." Usually topological conservation laws are used in order to deduce lower bounds on the "energy" of the flow. Those bounds are only approximate in non ideal flows but due to their topological nature simulations show that they are approximately conserved even when the "energy" is not. For example it is easy to show that the "energy" is bounded from below by the non-barotropic cross helicity as follows (see [33]):

$$H_{CNB} = \int \vec{B} \cdot \vec{v_t} d^3x \leq \frac{1}{2}\int \left(\vec{B}^2 + \vec{v_t}^2\right) d^3x, \tag{34}$$

$$H_{CNB} = \int \vec{B} \cdot \vec{v_t} d^3x \leq \sqrt{\int \vec{v_t}^2 d^3x}\sqrt{\int \vec{B}^2 d^3x}, \tag{35}$$

the second equation is a result of the Cauchy-Schwartz inequality. In this sense a configuration with a highly complicated topology is more stable since its energy is bounded from below. It is a simple thing to show that similar bounds occur also for the χ and η helicities:

$$H_{CNB\chi} = \int \vec{B_\chi} \cdot \vec{v_t} d^3x \leq \frac{1}{2}\int \left(\vec{B_\chi}^2 + \vec{v_t}^2\right) d^3x, \tag{36}$$

$$H_{CNB\chi} = \int \vec{B_\chi} \cdot \vec{v_t} d^3x \leq \sqrt{\int \vec{v_t}^2 d^3x}\sqrt{\int \vec{B_\chi}^2 d^3x}, \tag{37}$$

$$H_{CNB\eta} = \int \vec{B_\eta} \cdot \vec{v_t} d^3x \leq \frac{1}{2}\int \left(\vec{B_\eta}^2 + \vec{v_t}^2\right) d^3x, \tag{38}$$

$$H_{CNB\eta} = \int \vec{B_\eta} \cdot \vec{v_t} d^3x \leq \sqrt{\int \vec{v_t}^2 d^3x}\sqrt{\int \vec{B_\eta}^2 d^3x}, \tag{39}$$

Hence the kinetic energy is bounded by three differen bounds and so it the "total" energy. The importance of each of those bounds is dependent on the flow.

Journal of Physics: Conference Series **1416** (2019) 012041 doi:10.1088/1742-6596/1416/1/012041

5. Conclusion

We have shown the connection of the translation symmetry groups of labels to both the global non barotropic cross helicity conservation law and the conservation law of circulations of topological velocity along magnetic field lines. The latter were shown to be equivalent to the amount of non barotropic cross helicity per unit of magnetic flux [31, 32, 33]. Further more we have shown that two additional cross helicity conservation laws exist the χ and η cross helicities. Those lead to new bounds on MHD flows in addition to the bounds of the standarad non-barotropic cross helicity discussed in [32] for ideal non-barotropic MHD. The importance of constants of motion for stability analysis is also discussed in [35]. The significance of those constraints for non-ideal MHD and for plasma physics in general remains to be studied in future works.

References

[1] Lynden-Bell D and Katz J 1981 *Proceedings of the Royal Society of London. Series A, Mathematical and Physical Sciences*, **378**, No. 1773, 179
[2] Moffatt H K 1969 *J. Fluid Mech.* **35** 117
[3] Yahalom A 1995 *J. Math. Phys.* **36** 1324
[4] Yahalom A and Lynden-Bell D 2008 *Journal of Fluid Mechanics* **607** 235
[5] Woltjer L 1958a *Proc. Nat. Acad. Sci. U.S.A.* **44** 489
[6] Woltjer L 1958b *Proc. Nat. Acad. Sci. U.S.A.* **44** 833
[7] Yahalom A 2017 *Proceedings of the Chaotic Modeling and Simulation International Conference CHAOS* ed Christos H. Skiadas , 859
[8] Yahalom A 2018 *Quantum Theory and Symmetries with Lie Theory and Its Applications in Physics Volume 2* (Springer Proceedings in Mathematics & Statistics) **255** 387
[9] Mobbs S D 1981 *Journal of Fluid Mechanics* **108** 475
[10] Padhye N and Morrison P J 1996 *Phys. Lett. A* **219** 287
[11] Padhye N and Morrison P J 1996 *Plasma Phys. Rep.* **22** 869
[12] Webb et al 2014 *J. Phys. A, Math. and Theoret* **47** 095501
[13] Webb et al 2014 *J. Phys A, Math. and Theoret* **47** 095502
[14] Webb G M and Anco S. C. 2016 *J. Phys A, Math. and Theoret.* **49** 075501
[15] Webb G M, McKenzie J F and Zank G P 2015 *J. Plasma Phys.* **81** 905810610
[16] Webb G M and Mace R L 2015 *J. Plasma Phys.***81**, Issue 1, 905810115
[17] Sturrock P A 1994 *Plasma Physics* (Cambridge University Press, Cambridge)
[18] Yahalom A 2016 *J. Plasma Phys.* **82** 905820204
[19] Yahalom A 2016 *International Journal of Geometric Methods in Modern Physics* **13** 1650130
[20] Yahalom A 2016 *International Journal of Mechanics* **10** 336
[21] Sakurai T 1979 *Pub. Ast. Soc. Japan* **31** 209
[22] Vladimirov V A and Moffatt H K 1995 *J. Fluid. Mech* **283** 125
[23] Kats A V 2003 *JETP Lett.* **77** 657
[24] Frenkel A , Levich E and Stilman L 1982 *Phys. Lett. A* **88** 461
[25] Zakharov V E and Kuznetsov E A 1997 *Usp. Fiz. Nauk* **40** 1087
[26] Bekenstein J D and Oron A 2000 *Physical Review E* **62** 5594
[27] Morrison P J 1982 *AIP Conference proceedings* **88**, Table 2, 13
[28] Yahalom A and Lynden-Bell D 2014 *Geophysical & Astrophysical Fluid Dynamics* **108** 6
[29] Yahalom A 2010 *EPL* **89** 34005
[30] Yahalom A 2013 *Physics Letters A* **377** 1898
[31] Yahalom A 2017 *Journal of Geophysical & Astrophysical Fluid Dynamics* **111** 131
[32] Yahalom A 2017 *Fluid Dynamics Research* **50** 011406
[33] Yahalom A 2019 Journal of Physics: Conf. Series **1194** 012113
[34] Boozer A. H. 2004 *Rev. Mod. Phys.* **76** 1071
[35] Katz J, Inagaki S and Yahalom A 1993 *Pub. Astro. Soc. Japan* **45** 421

Soliton surfaces induced by the Fokas-Lenells equation

Kuralay Yesmakhanova, Meruyert Zhassybayeva, Ratbay Myrzakulov
L.N. Gumilyov Eurasian National University, Nur-Sultan, Kazakhstan
E-mail: kryesmakhanova@gmail.com

Abstract. In this paper, we study the application of the theory of solitons in differential geometry. The recently proposed soliton equation, which is Fokas-Lenells equation, has been investigated, and its two-dimensional soliton surface in the three-dimensional Euclidean space ($\mathbb{R}^2 \to \mathbb{R}^3$) has been constructed. Thus the connection between the Fokas-Lenells equation and the surface was established by using the Sym-Tafel formula. We find the first and the second quadratic forms, surface area, and Gaussian curvature. The obtained results have various applications in mathematical physics, the geometry of curves and the theory of surfaces.

1. Introduction

The study of properties as geometry [1]-[4], integrability [5], and exact [6]-[10] solutions of nonlinear equation have an impotant role in the application on of physics. The theory of surfaces in three-dimensional Euclidean space is widely used in various fields of science, in particular in mathematics, theoretical physics, etc. In this paper, we apply it to the theory of integrable system. To this end, the integrable Fokas-Lenells (FL) equation, which describes the propagation of ultrashort nonlinear light pulses in optical fibers and looks as follows, is investigated:

$$iq_{xt} - iq_{xx} + 2q_x - |q|^2 q_x + iq = 0, \tag{1}$$

$$ir_{xt} - ir_{xx} - 2r_x + |q|^2 r_x + ir = 0, \tag{2}$$

where $q(x,t)$ is the complex envelope of a field, the indices x, t denote partial derivatives with respect to the arguments x, t and i are an imaginary unit. In physical applications, two natural reductions are used ($r = \pm\bar{q}$), which are of great interest in physics, one of them is a of defocusing case, with $r = -\bar{q}$ [11]-[13]

$$FL_- : \quad iq_{xt} - iq_{xx} + 2q_x - |q|^2 q_x + iq = 0,$$

and second, the of focusing case, with $r = \bar{q}$

$$FL_+ : \quad iq_{xt} - iq_{xx} + 2q_x + |q|^2 q_x + iq = 0.$$

In this paper, we consider the first case, that is

$$iq_{xt} - iq_{xx} + 2q_x - |q|^2 q_x + iq = 0. \tag{3}$$

Since, the equation under consideration is integrable, it has the Lax pair (LP), which plays an important role in the theory of integrable systems. It allows you to apply the inverse scattering method to construct exact solutions and to study the asymptotics of problems with initial conditions. The LP for equation (3) has next form

$$\Phi_x(x,t,\lambda) = U(x,t,\lambda)\Phi(x,t,\lambda), \tag{4}$$

$$\Phi_t(x,t,\lambda) = V(x,t,\lambda)\Phi(x,t,\lambda), \tag{5}$$

where $\Phi = (\Phi_1, \Phi_2)^T$ is called 2×2 the matrix eigenfunction of the eigenvalue λ (or spectral parameter) and the matrix operators U and V are given in the following form:

$$U = -i\lambda^2\sigma_3 + \lambda Q, \tag{6}$$

$$V = -i\lambda^2\sigma_3 + \lambda Q + V_0 + \frac{1}{\lambda}V_{-1} - \frac{i}{4\lambda^2}\sigma_3. \tag{7}$$

Here

$$Q = \begin{pmatrix} 0 & q_x \\ \bar{q}_x & 0 \end{pmatrix}, \quad V_0 = i\sigma_3 - \frac{i|q|^2}{2}\sigma_3, \quad V_{-1} = \frac{i}{2}\begin{pmatrix} 0 & q \\ -\bar{q} & 0 \end{pmatrix}, \quad \sigma_3 = \begin{pmatrix} 1 & 0 \\ 0 & -1 \end{pmatrix}.$$

2. The first fundamental form of the surface

The first fundamental form (1FF) plays an important role in the theory of surfaces and serves primarily to measure infinitely small arcs on a surface. 1FF of a smooth surface P is called a scalar square the radius of the vector $d\mathbf{r} = \mathbf{r}_x dx + \mathbf{r}_t dt$, that is,

$$I = d\mathbf{r}^2 = (\mathbf{r}_x dx + \mathbf{r}_t dt)^2 = \mathbf{r}_x^2 dx^2 + 2\mathbf{r}_x\mathbf{r}_t dx dt + \mathbf{r}_t^2 dt^2. \tag{8}$$

The equation (8) at each point of the surface P represents the quadratic form of the differentials dx and dt.

For coefficients 1FF, the following notation is often used:

$$E = \mathbf{r}_x^2, \quad F = \mathbf{r}_x\mathbf{r}_t, \quad G = \mathbf{r}_t^2, \tag{9}$$

then the equation (8) is rewritten as

$$I = Edx^2 + 2Fdxdt + Gdt^2, \tag{10}$$

where $EG - F^2 > 0$.

Two-dimensional surfaces and integrable equations are relate by the Lax representation. This is done with the help of the so-called Sym-Tafel formula (STF), which has the form [12]

$$r = \Phi^{-1}\Phi_\lambda, \tag{11}$$

and defines a λ-family of surfaces parametrized by coordinates x and t. From equation (11) we can define 1FF. The surfaces are immersed in the linear space of the matrices from the equation (11) [14].

It is very convenient to apply the STF because of the very simple rule for calculating the derivatives with respect to r

$$r_x = (\Phi^{-1})_x\Phi_\lambda + \Phi^{-1}\Phi_{\lambda x} = -\Phi^{-1}\Phi_x\Phi^{-1}\Phi_\lambda + \Phi^{-1}(U\Phi)_\lambda =$$

$$= -\Phi^{-1}U\Phi_\lambda + \Phi^{-1}U_\lambda\Phi + \Phi^{-1}U\Phi_\lambda = \Phi^{-1}U_\lambda\Phi, \tag{12}$$

and similarly,

$$r_t = \Phi^{-1} V_\lambda \Phi. \tag{13}$$

Scalar product is standard, that is, proportional to the trace of the matrix. For example: $r_x \cdot r_t = c\,\mathrm{tr}(U_\lambda V_\lambda)$, where c is a constant. Then, in terms of matrix operators, the equation (8) is rewritten as follows:

$$I = \frac{1}{2}\left(\mathrm{tr}(U_\lambda^2)dx^2 + 2\,\mathrm{tr}(U_\lambda V_\lambda)dxdt + \mathrm{tr}(V_\lambda^2)dt^2\right). \tag{14}$$

Thus, it became obvious that the STF serves as a bridge for establishing a connection with the soliton equation and the surface.

Now, let's differentiate the matrix operators U and V by the spectral parameter λ, we take them in the square and find their trace, that is,

$$\begin{aligned}
\mathrm{tr}(U_\lambda^2) &= 2(|q_x|^2 - 4\lambda^2), &(15)\\
\mathrm{tr}(U_\lambda V_\lambda) &= 2\left(|q_x|^2 - 4\lambda^2 + \frac{1}{\lambda^2}\right) + \frac{i}{2\lambda^2}(\bar{q}q_x - q\bar{q}_x), &(16)\\
\mathrm{tr}(V_\lambda^2) &= 2\left(|q_x|^2 - 4\lambda^2 + \frac{i}{2\lambda^2}(\bar{q}q_x - q\bar{q}_x) + \frac{2}{\lambda^2} + \frac{1}{4\lambda^4}|q|^2 - \frac{1}{4\lambda^6}\right). &(17)
\end{aligned}$$

Substituting the equations (15)-(17) into the equation (14), we get

$$I = (|q_x|^2 - 4\lambda^2)dx^2 + \left(2(|q_x|^2 - 4\lambda^2 + \frac{1}{\lambda^2}) + \frac{i}{2\lambda^2}(\bar{q}q_x - q\bar{q}_x)\right)dxdt +$$

$$+\left(|q_x|^2 - 4\lambda^2 + \frac{i}{2\lambda^2}(\bar{q}q_x - q\bar{q}_x) + \frac{2}{\lambda^2} + \frac{1}{4\lambda^4}|q|^2 - \frac{1}{4\lambda^6}\right)dt^2, \tag{18}$$

or as equation (10)

$$I = Edx^2 + 2Fdxdt + Gdt^2,$$

where

$$\begin{aligned}
E &= |q_x|^2 - 4\lambda^2, &(19)\\
F &= E + \frac{1}{\lambda^2} + \frac{i}{4\lambda^2}(\bar{q}q_x - q\bar{q}_x), &(20)\\
G &= E + \frac{i}{2\lambda^2}(\bar{q}q_x - q\bar{q}_x) + \frac{2}{\lambda^2} + \frac{1}{4\lambda^4}|q|^2 - \frac{1}{4\lambda^6}. &(21)
\end{aligned}$$

Thus, we found 1FF surfaces of the FL equation.

3. The second fundamental form of the surface

The second fundamental form (2FF) describes the surface in the second approximation. It shows how the surface deviates from the tangent plane and completely determines the curvature of the surface. 2FF regular surface f is the scalar product of vectors $d^2\mathbf{r}$ and $\mathbf{n}$:

$$II = (d^2\mathbf{r}\mathbf{n})^2, \tag{22}$$

where the unit normal vector is defined as

$$\mathbf{n} = \frac{\mathbf{r}_x \wedge \mathbf{r}_t}{|\mathbf{r}_x \wedge \mathbf{r}_t|} = \frac{\mathbf{r}_x \wedge \mathbf{r}_t}{\sqrt{EG - F^2}}$$

and the second differential of the vector function $\mathbf{r}(x,t)$

$$d^2\mathbf{r} = \mathbf{r}_{xx}dx^2 + 2\mathbf{r}_{xt}dxdt + \mathbf{r}_{tt}dt^2. \tag{23}$$

Then, the equation (22) will has written as

$$II = (\mathbf{r}_{xx}\cdot\mathbf{n})\,dx^2 + 2(\mathbf{r}_{xt}\cdot\mathbf{n})\,dxdt + (\mathbf{r}_{tt}\cdot\mathbf{n})\,dt^2. \tag{24}$$

For coefficients 2FF, the following notation is used:

$$L = \mathbf{r}_{xx}\cdot\mathbf{n}, \qquad M = \mathbf{r}_{xt}\cdot\mathbf{n}, \qquad N = \mathbf{r}_{tt}\cdot\mathbf{n}.$$

This allows you to rewrite the equation (24) as

$$II = Ldx^2 + 2Mdxdt + Ndt^2, \tag{25}$$

where the coefficients L, M and N are calculated as

$$L = \frac{1}{2}\operatorname{tr}(r_{xx}n), \qquad M = \frac{1}{2}\operatorname{tr}(r_{xt}n), \qquad N = \frac{1}{2}\operatorname{tr}(r_{tt}n), \tag{26}$$

where

$$n = \pm\frac{\Phi^{-1}[U_\lambda, V_\lambda]\Phi}{\sqrt{\frac{1}{2}\operatorname{tr}([U_\lambda, V_\lambda]^2)}}, \tag{27}$$

and the second derivatives of the equation (12)-(13) can be easily calculated

$$r_{xx} = (\Phi^{-1}U_\lambda\Phi)_x = -\Phi^{-1}\Phi_x\Phi^{-1}U_\lambda\Phi + \Phi^{-1}U_{\lambda x}\Phi + \Phi^{-1}U_\lambda\Phi_x =$$
$$= -\Phi^{-1}UU_\lambda\Phi + \Phi^{-1}U_{\lambda x}\Phi + \Phi^{-1}U_\lambda U\Phi = \Phi^{-1}(-UU_\lambda + U_{\lambda x} + U_\lambda U)\Phi = \Phi^{-1}(U_{\lambda x} + [U_\lambda, U])\Phi, \tag{28}$$

and, by same way we can find,

$$r_{xt} = \Phi^{-1}(U_{\lambda t} + [U_\lambda, V])\Phi, \tag{29}$$
$$r_{tt} = \Phi^{-1}(V_{\lambda t} + [V_\lambda, V])\Phi. \tag{30}$$

Thus, substituting the equations (28)-(30) into the equation (26) we get the final formula for calculating the coefficients 2FF

$$L = \frac{1}{2}\frac{\operatorname{tr}((U_{\lambda x} + [U_\lambda, U])[U_\lambda, V_\lambda])}{\sqrt{\frac{1}{2}\operatorname{tr}([U_\lambda, V_\lambda]^2)}}, \tag{31}$$

$$M = \frac{1}{2}\frac{\operatorname{tr}((U_{\lambda t} + [U_\lambda, V])[U_\lambda, V_\lambda])}{\sqrt{\frac{1}{2}\operatorname{tr}([U_\lambda, V_\lambda]^2)}}, \tag{32}$$

$$N = \frac{1}{2}\frac{\operatorname{tr}((V_{\lambda t} + [V_\lambda, V])[U_\lambda, V_\lambda])}{\sqrt{\frac{1}{2}\operatorname{tr}([U_\lambda, V_\lambda]^2)}}. \tag{33}$$

So, we derive the equations for calculating the coefficients of 2FF and next step is to find it. For this purpose we find the commutators

$$[U_\lambda, U] = 2i\lambda^2\begin{pmatrix} 0 & -q_x \\ \bar{q}_x & 0 \end{pmatrix}, \tag{34}$$

$$[U_\lambda, V] = i\alpha\begin{pmatrix} 0 & -q_x \\ \bar{q}_x & 0 \end{pmatrix} + \frac{i}{2\lambda}\begin{pmatrix} -(|q|^2)_x & 0 \\ 0 & (|q|^2)_x \end{pmatrix} + 2\begin{pmatrix} 0 & q \\ \bar{q} & 0 \end{pmatrix}, \tag{35}$$

$$[V_\lambda, V] = i(\alpha - \frac{1}{\lambda^2})\begin{pmatrix} 0 & -q_x \\ \bar{q}_x & 0 \end{pmatrix} + \frac{i}{\lambda}\begin{pmatrix} -(|q|^2)_x & 0 \\ 0 & (|q|^2)_x \end{pmatrix} + \beta\begin{pmatrix} 0 & q \\ \bar{q} & 0 \end{pmatrix}, \tag{36}$$

$$[U_\lambda, V_\lambda] = \frac{i}{\lambda^3}\begin{pmatrix} 0 & -q_x \\ \bar{q}_x & 0 \end{pmatrix} - \frac{i}{2\lambda^2}\begin{pmatrix} -(|q|^2)_x & 0 \\ 0 & (|q|^2)_x \end{pmatrix} - \frac{2}{\lambda}\begin{pmatrix} 0 & q \\ \bar{q} & 0 \end{pmatrix}, \tag{37}$$

Journal of Physics: Conference Series **1416** (2019) 012042 doi:10.1088/1742-6596/1416/1/012042

where

$$\alpha = 2\lambda^2 + 2 - |q|^2 - \frac{1}{2\lambda^2},$$
$$\beta = 3 - \frac{1}{\lambda^2} + \frac{|q|^2}{2\lambda^2} - \frac{1}{4\lambda^4}.$$

Then, substituting the equations (34)-(37) into the equations (31)-(33) we can obtain

$$L = \frac{\frac{i}{2\lambda^2}(\bar{q}_x q_{xx} - q_x \bar{q}_{xx}) + 2|q_x|^2 - \bar{q}q_{xx} - q\bar{q}_{xx} + 2i\lambda^2(\bar{q}q_x - q\bar{q}_x)}{\sqrt{\frac{1}{\lambda^4}|q_x|^2 + \frac{2i}{\lambda^2}(\bar{q}q_x - q\bar{q}_x) - \frac{1}{4\lambda^2}((|q|^2)_x)^2 + 4|q|^2}}, \tag{38}$$

$$M = \frac{\frac{i}{2\lambda^2}(\bar{q}_x q_{xt} - q_x \bar{q}_{xt}) + \frac{a}{\lambda^2}|q_x|^2 - ia(q\bar{q}_x - \bar{q}q_x) + \frac{1}{4\lambda^2}((|q|^2)_x)^2 - \bar{q}q_{xt} - q\bar{q}_{xt} - 4|q|^2}{\sqrt{\frac{1}{\lambda^4}|q_x|^2 + \frac{2i}{\lambda^2}(\bar{q}q_x - q\bar{q}_x) - \frac{1}{4\lambda^2}((|q|^2)_x)^2 + 4|q|^2}}, \tag{39}$$

$$N = \frac{M + \frac{1}{4\lambda^4}(q_t\bar{q}_x + \bar{q}_t q_x) - \frac{1}{\lambda^4}|q_x|^2 + \frac{i\beta}{2\lambda^2}(q\bar{q}_x - \bar{q}q_x) + \frac{1}{4\lambda^2}((|q|^2)_x)^2 + \frac{i}{2\lambda^2}(\bar{q}q_t - q\bar{q}_t) - 2(\beta - 2)|q|^2}{\sqrt{\frac{1}{\lambda^4}|q_x|^2 + \frac{2i}{\lambda^2}(\bar{q}q_x - q\bar{q}_x) - \frac{1}{4\lambda^2}((|q|^2)_x)^2 + 4|q|^2}}, \tag{40}$$

where $a = \alpha - 1/\lambda^2$.

Thus, we found a 2FF surface, which is defined by the equation (25), where the coefficients are equal to the expressions (38)-(40).

4. Surface area

If a surface in a Euclidean space is given by a parametrically smooth function $r(x,t)$, where the parameters x,t change in the D domain on the x,t plane, then the surface area S can be expressed by a double integral

$$S = \int\int_D |\mathbf{r}_x \wedge \mathbf{r}_t| dx dt, \tag{41}$$

where the module the vector product of the vectors $\mathbf{r}_x$ and $\mathbf{r}_t$ is equal to

$$|\mathbf{r}_x \wedge \mathbf{r}_t| = \frac{1}{2}\,\mathrm{tr}(r_{xx}).$$

It is known that r_{xx} is calculated by the formula (28), then the surface area S is calculated by the following formula:

$$S = \int\int_D \sqrt{\frac{1}{2}\,\mathrm{tr}\,(U_{\lambda x} + [U_\lambda, U])^2} dx dt, \tag{42}$$

then our required surface area S look as

$$S = \int\int_D \sqrt{|q_{xx}|^2 + 2i\lambda^2(\bar{q}_x q_{xx} - q_x \bar{q}_{xx}) + 4\lambda^4|q_x|^2} dx dt. \tag{43}$$

5. Gaussian surface curvature

The Gaussian curvature of a surface is the product of the principal curvatures of a regular surface at a given point and calculate by the ratio of the discriminants of the first and second fundamental forms [15]

$$K = \frac{LN - M^2}{EG - F^2}, \tag{44}$$

that is, by substituting equations (38)-(40) and (19)-(21) into the equation (44), we can obtain the Gaussian curvature.

6. Conclusion

Thus, in this paper, for the Fokas-Lenells equation using the Sym-Tafel formula, 1FF and 2FF surfaces with corresponding coefficients are found. 2FF is an effective tool for studying the geometric properties of a regular surface. Through this form, can enter important geometric characteristics that measure the degree and type of surface deviation from the tangent plane. Using these forms, gave a formula for calculating the Gaussian curvature of a surface, which is convenient to use for the classification of points of a regular surface: the sign of which at a given point indicates the nature of the surface behavior at this point ($K > 0$ is elliptic, $K < 0$ is hyperbolic, $K = 0$ is parabolic). Also found the surface area.

Acknowledgments The work was carried out with the financial support of the Ministry of Education and Science of the Republic of Kazakhstan, Grant No. 0118RK00935.

References

[1] Myrzakulov R, Vijayalakshmi S, Nugmanova G and Lakshmanan M 1997 A (2+1)-dimensional integrable spin model: Geometrical and gauge equivalent counterparts, solitons and localized coherent structures *J. Phys. Lett. A* **233** 391

[2] Bekova G, Yesmakhanova K, Shaikhova G, Nugmanova G and Myrzakulov R 2018 Integrable flows of curves/surfaces, generalized Heisenberg ferromagnet equation and complex coupled dispersionless equation *Preprint gr-qc/1812.02152*

[3] Martina L, Myrzakul K, Myrzakulov R and Soliani G 2001 Deformation surfaces, integrable systems and Chern-Simons theory *J.Math.Phys.* **42** 1397

[4] Myrzakul A and Myrzakulov R 2017 Integrable motion of two interacting curves, spin systems and the Manakov system *Geom. Methods Mod. Phys.* **14** 1750115

[5] Bekova G, Yesmakhanova K and Shaikhova G 2018 Travelling wave solutions for the two-dimensional Hirota system of equations *AIP Conf. Proc.* **1997** 020039

[6] Bekova G, Yesmakhanova K, Myrzakulov R and Shaikhova G 2017 Exact solutions for the (2+1)-dimensional Hirota-Maxwell-Bloch system *AIP Conf. Proc.* **1880** 060022

[7] Bekova G, Yesmakhanova K, Myrzakulov R and Shaikhova G 2017 Lax representation and soliton solutions for the (2+1) -dimensional two-component complex modified Korteweg-de Vries equations *Journal of Physics: Conference Series* **804** 012004

[8] Bekova G, Yesmakhanova K, Myrzakulov R and Shaikhova G 2017 Darboux transformation and soliton solution for the (2+1)-dimensional complex modifed Korteweg-de Vries equations *Journal of Physics: Conference Series* **936** 012045

[9] Nugmanova G, Sagidullayeva Zh and Myrzakulov R 2017 Hirota's method for a spin model with self-consistent potential *J. Phys.: Conf. Series* **804** 012035

[10] Nugmanova G and Sagidullayeva Zh 2017 Generalized spin model with vector potential and its solution *Bulletin of the Karaganda university-Mathematics* **86** 91–6

[11] Jingsong H, Shuwei X and Kuppuswamy P 2012 Rogue Waves of the Fokas-Lenells Equation *J. Phys. Soc. Jpn* **81** 12

[12] Zhao P and Fan E 2013 Reality problems for the algebro-geometric solutions of Fokas-Lenell hierarchy *Preprint gr-qc/1309.236*

[13] Yesmakhanova K and Zhassybayeva M 2018 On the equivalence of the Myrzakulov-LXXII equation and the Fokas-Lenells equation *Preprint rg*

[14] Halifax N S 1999 *Backlund and Darboux transformations. The geometry of solitons* (Canada: AARMS-CRM Workshop)

[15] Pressley A 2001 *Elementary differential geometry* (Springer-Verlag)

Journal of Physics: Conference Series **1416** (2019) 012043 doi:10.1088/1742-6596/1416/1/012043

A Quantum Electromagnetic Theory of the Pions, Muons and Their Emitting Particles (I)

J X Zheng-Johansson

Institute of Fundamental Physics Research, 611 93 Nyköping, Sweden

Abstract. In direct accordance to the overall relevant experimental demonstrations, we represent the charged pion π^- as a heavy electron h^- in precessional-orbital (P-O) motion at essentially the light speed c about $\bar{\nu}_e$-orbit of a normal at quantised angle $\pi - \theta_{1/2} = -\arccos(1/\sqrt{3})$ to the z axis. h^- is the level $N = 1$ oscillation of charge $-e$ and its electromagnetic radiation originally generated in the weak potential field of another particle. The P-O kinetic energy current and two additional opposite ones created upon π^- decay represent confined neutrinos $\bar{\nu}_e$, $\bar{\nu}_\mu$, ν_μ. The muon μ^- is a xy-projected h^- in two superposing P-O motions along $\bar{\nu}_e$-, ν_μ-orbits of normals at angles $\pi - \theta_{1/2}$, $\theta_{1/2}$ to z. The μ^- (rest) mass is a geometric projection of the reduced π^- mass, $M_\mu \doteq (M_\pi - M_{\nu_\mu})\sqrt{\cos\theta_{1/2}} = 105.86$ MeV. The μ^- mass is fundamentally predetermined by the mixed two states $m_l = -1, +1$ of level $n = 2$ of a double heavy positronium in the CM frame produced in a relativistic e^-, e^+ collision, and is *ab initio* predicted to be $M_{\mu^-} = (\frac{3}{4}\frac{2}{\alpha} + 1)M_e = 105.549$ MeV, where $\alpha = e^2/4\pi\epsilon_0\hbar c$. The un-projected $n = 2$ level gives the bound π^- mass containing a friction term $\mathcal{O}_\eta$, $M_\pi + \mathcal{O}_\eta = (\frac{2}{\alpha} + 1)M_e = 140.525$ MeV. Their antiparticles π^+, μ^+ and the tauons $\tau^\mp$ can be similarly represented. The remaining unstable elementary particles can be constructed as composites of two or more single charged ones in certain spatial quantised P-O motions.

1. Introduction

In the Standard Model (SM) description[1, 2], elementary particles are divided into 2×6 leptons, baryons / mesons made of 2×6 quarks, and five or so force mediators. Regarding their weak decays, such as the neutron β decay $n \to e^- p \bar{\nu}_e$, the SM quark model assumes that the decay product particles $e^-, p, \bar{\nu}_e$ do not exist until n decays, and that they are instantaneously produced and emitted upon the n decay. The SM has been successful in reproducing the charges, spins, C, P, T symmetries of the hundreds of observational elementary particles, and predicting their decay rates - absolute or comparative. But the SM has unnatural aspects, concerning the quarks and weak decays in particular. The baryons and mesons are made of quarks rather than the particles they decay into, which is an abrupt departure from the atomic and nuclear descriptions. Free quarks are never observed in experiment. The SM weak interaction Hamiltonian $H = G/(\frac{4\pi}{3}r^3)$ is a phenomenological construction. The weak force is not predicted. The weak, strong, gravitational and electromagnetic forces are not unified. The basic questions common to the regular particles e^-, p (being the two only permanent particles - why?) remain outstanding, including the origin of mass, the nature of matter waves, and the cause of gravity.

The Internally Electrodynamic (IED) particle model (see a review and the original references given in [3]) is a complementary approach to the SM and is beyond the SM. The IED model was developed using overall observational properties of particles as input information, and aimed from the beginning to achieve a unification of the three basic mechanics and of the four fundamental forces. In terms of this model, the electron e^- and the proton p are composed of oscillatory charges $-e$ and $+e$ and their electromagnetic radiation (EMR) fields produced in a polarisable dielectric vacuum filled of vacuuons[4, 5]. Their distinct masses are uniquely determined [6] by the quantum oscillation levels of $-e, +e$ in their vacuum potentials that are asymmetric to $-e, +e$. The masses of their antiparticles $e^+, \bar{p}$ are determined accordingly by pair productions. The remaining single charged particles, including the six manifestly free forms $\pi^\mp$, $\mu^\mp$, $\tau^\mp$ and ones in (confined) higher excited states such as comprising $\rho^0(770)$, are the main subject of investigation of this paper. We shall show that, in accordance to the overall experimental demonstrations, these can be generally described each as a charge $-e$ or $+e$ oscillation and the resulting EMR originally generated in the weak potential field of another particle, that are as a whole in certain precessional orbital (P-O) motions. The remaining unstable elementary matter particles such as n, Λ, π^0, $\rho(770)$ are composites of two or more single charged ones. A formal e^-, p model of the neutron n and first principles solutions for spin, magnetic -hence weak- interaction force, coupling constant, and (unpublished) mixing angle have been achieved in [7]. The Λ hyperon is the simplest baryon emitting a pion π^-, but otherwise a direct analogue of n, and will serve a prototype system for the model constructions in this paper.

The formal representation divides in three sections. In Sec 2, using the relevant observations as input information we propose the existence of mass states "heavy electrons (or positrons)", originally generated by charges $-e$'s (or $+e$'s) in the weak potential fields of other particles. In terms of this we propose the structures of the pions $\pi^\mp$ and muon $\mu^\mp$. In Sec 3, we present a generalised first principles description of the masses of the single charged particles such as $\pi^\mp, \mu^\mp$, e^-, p, and the composite particles such as n, Λ. In Sec 4, we predict the existence of a double heavy positronium preceding the rudimentary productions of $\mu^\mp$, $\pi^\mp$, etc. and, based on its eigen level $n = 2$ solutions, we *ab initio* predict the masses of $\mu^\mp$, $\pi^\mp$. Representation of the remaining unstable composite elementary particles will be described in a separate paper.

2. Structures of the charged pions and the muons

2.1 The charged pions π^-, π^+ The simplest composite baryon emitting a charged pion π^-, and secondarily a muon μ^-, is the Λ hyperon. Observationally, Λ has an identical charge (0), spin ($\frac{1}{2}$), and analogous decay reaction ($\Lambda \to \pi^- p \to \mu^- \bar{\nu}_\mu p$) to those of the neutron n (0, $\frac{1}{2}$, $n \to e^- \bar{\nu}_e p$), but Λ has a heavier mass 1115.6 MeV. These suggest an analogous structure of Λ to n [7] (Fig 1a, Inset), except in place of e^- of n, Λ apparently has a "heavy electron" (h^-) in spatial quantised P-O motion relative to p (Fig 1a). In accordance to the implication of Λ, the observational π^- charge $-e$, spin $s_\pi = 0$, mass $M_\pi = 139.569$ MeV, decay reaction $\pi^- \to \mu^- \bar{\nu}_\mu \to (e^- \bar{\nu}_e \nu_\mu) \bar{\nu}_\mu$ and rudimentary direct production reaction $e^- e^+ \to \rho^0(770) \to \pi^- \pi^+$, we propose a two-step description of the π^- structure: (1) There presents a (confined) particle state called heavy electron, h^-, that has a charge $-e$, spin $\frac{1}{2}$ as the electron e^-, and that is yet a heavier mass state, the charge $-e$ oscillation and its resultant EMR of a mass M_h, (originally) generated in the weak potential field $V_{\frac{1}{2}}$ of another particle. (2) The pion π^- is a heavy electron h^-, of spin $S_{hz} = \frac{1}{2}\hbar$ in z direction, in P-O (precessional-orbital) motion at essentially the light speed, $v_{\frac{1}{2}} \doteq c$, along $\bar{\nu}_e$-orbit of a normal $(-z')$ at quantised angle $\pi - \theta_{\frac{1}{2}}$ to the z axis (Fig 1b). The P-O kinetic energy current along $\bar{\nu}_e$-orbit resembles a confined antineutrino $\bar{\nu}_e$, of an apparent rest mass M_{ν_e}. Two equal but opposite P-O momentum currents along $\bar{\nu}_\mu$-, ν_μ-orbits (of confined neutrinos $\bar{\nu}_\mu$ ν_μ) are generated momentarily upon π^- decay in an explosive collision (Fig 1b, Inset), these do not contribute to the π^- dynamical variables. By virtual of its anti-symmetric properties, π^+ is a heavy positron h^+ in P-O motion along orbit ν_e. The rest mass of π^- (or π^+) is formally $M_\pi = M_h + M_{\nu_e}$.

The total and the z component angular momenta of the P-O motion of h^- in the lab frame are given by the eigen solutions of bound state $n = 2$, $l = 1$, $j = l - l^{TP} = \frac{1}{2}$ in the original $V_{\frac{1}{2}}$ field (e.g. of p as in Λ) similarly as for n ([7]), $J_{\frac{1}{2}} = |\mathbf{r}_{\frac{1}{2}\pi} \times (m_\pi \mathbf{v}_{\frac{1}{2}})| (= |\mathbf{r}_{\frac{1}{2}\Lambda} \times (\mathscr{M}_\Lambda \mathbf{v}_{\frac{1}{2}\Lambda})|) = \sqrt{j(j+1)}|_{j=1/2}\hbar = \frac{\sqrt{3}}{2}\hbar$, $J_{\frac{1}{2}z} = J_{\frac{1}{2}} \cos(\pi - \theta_{\frac{1}{2}}) = -j\hbar = -\frac{1}{2}\hbar$, where $\cos(\pi - \theta_{\frac{1}{2}}) = -1/\sqrt{3}$, $m_\pi = \gamma_\pi M_\pi$, $\gamma_\pi = (1 - (v_{\frac{1}{2}}/c)^2)^{-1/2}$ ($\mathscr{M}_\Lambda$, is the reduced mass and $r_{1\Lambda}$ the orbit radius of π^-, p comprising Λ). Those of the total system $\pi^-(e^-_h[\bar{\nu}_e])$ are $S_{\pi z} = J_{\frac{1}{2}z} + S_{hz} = (-\frac{1}{2} + \frac{1}{2})\hbar = s_\pi \hbar = 0$, $s_\pi = 0$, and $S_\pi = \sqrt{s_\pi(s_\pi + 1)}\hbar = 0$. The kinetic energy is $T_\pi = \frac{\gamma}{\gamma+1} m_\pi v_{\frac{1}{2}}^2 = \frac{\gamma}{\gamma+1} J_{\frac{1}{2}} c/r_1$.

2.2 The muons μ^-, μ^+ In direct accordance to the observational charge $-e$, spin $s_\mu = \frac{1}{2}$, mass $M_\mu = 105.658$ MeV and decay reaction $\mu^- \to e^- \bar{\nu}_e \nu_\mu$ of μ^-, to its specific production from $\pi^- \to \mu^- \bar{\nu}_\mu$ and the π^- structure of Sec 2.1, we propose: By an apparent loss of mass energy projected in the $y'z$ plane(s), h^- transforms to a xy-projected mass state h^-_{xy} of mass $M_{h_{xy}}$, and of an unchanged charge $-e$ and spin $s_{h_{xy}} = \frac{1}{2} = S_{h_{xy}z}/\hbar$ in $+z$ direction here. The muon μ^- is a h^-_{xy} in rotational motion along two coinciding elliptics $\nu_{y'z}$, $\nu_{y''z}$ projected from the $\bar{\nu}_e$-, ν_μ-orbits in a $y'z$ (or $y''z$) plane. The $\nu_{y'z}$- (or $\nu_{y''z}$-) motion is the supposition of two P-O motions at speed $v_{\frac{1}{2}} \doteq c$ each along $\bar{\nu}_e$-, ν_μ-orbits of equal radius $r_{1\mu}$, and of normals z', z'' at quantised angles $\pi - \theta_{\frac{1}{2}}$ and $\theta_{\frac{1}{2}}$ to the z axis (Fig 2a).

The projections of the $\bar{\nu}_e$-,ν_μ- orbits in the $y'z$ plane, $\nu_{y'z}$,$\nu_{y''z}$, or in the xy plane, ν_{xy}'s, necessarily coincide. The component kinetic motions of $-e$ along ν_{xy}'s are thus equal and opposite and cancel out, leaving h^-_{xy} at rest on ν_{xy} at a random angle φ to the x axis; and the motions along $\nu_{y'z}$,$\nu_{y''z}$ add up. Similarly as for n[7] and π^- (Sec 2.1), the angular momenta of the P-O motions along $\bar{\nu}_e$, ν_μ in the lab frame are each given by the eigen solutions in the states $n = 2$, $l = 1$, $j = l - l^{TP} = \frac{1}{2}$ in the original $V_{\frac{1}{2}}$ field: $J_{\frac{1}{2}} = r_{1\mu}\gamma_\mu M_\mu c = \sqrt{j(j+1)}\hbar = \frac{\sqrt{3}}{2}\hbar$. The projections along z (with $m'_j, m''_j = -\frac{1}{2}, \frac{1}{2}$) are $J'_{\frac{1}{2}z} = -J''_{\frac{1}{2}z} = J_{\frac{1}{2}} \cos(\pi - \theta_{\frac{1}{2}}) = -\frac{1}{2}\hbar$, and in the xy plane are $J'_{\frac{1}{2}xy} = J''_{\frac{1}{2}xy} = J_{\frac{1}{2}} \sin\theta_{\frac{1}{2}} = \frac{\sqrt{2}}{2}\hbar$, where $\cos\theta_{\frac{1}{2}} = 1/\sqrt{3}$, $\sin\theta_{\frac{1}{2}} = \sqrt{2/3}$. For the total P-O kinetic motion, $J_{\frac{1}{2}z,\mu} = J'_{\frac{1}{2}z} + J''_{\frac{1}{2}z} = -\frac{1}{2}\hbar + \frac{1}{2}\hbar = 0$ along z, and $J_{\frac{1}{2}xy,\mu} = J'_{\frac{1}{2}xy} + J''_{\frac{1}{2}xy} = \sqrt{2}\hbar$ at angle φ to x in the xy plane, with a time average $\langle J_{\frac{1}{2}xy,\mu}\rangle = 0$. The spin (angular momentum) of the total system $\mu^-(h^-_{xy}[\bar{\nu}_e, \nu_\mu])$ is $S_{\mu z} = S_{h_{xy},z} + J_{\frac{1}{2}z,\mu} = \frac{1}{2}\hbar$ in $+z$ direction.

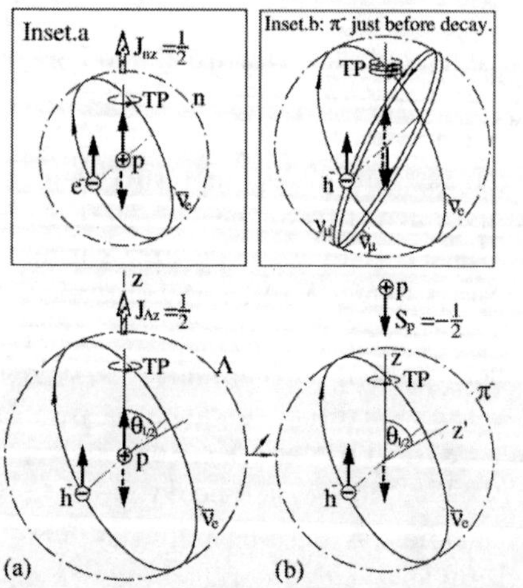

Figure 1. Schematic structures of (a) Λ in analogy to n (Inset.a), except in place of e^-, in Λ is a heavy electron h^- in P-O motion shown in the p rest frame, and (b) stationary π^-; (Inset.b) shows π^- just before decay.

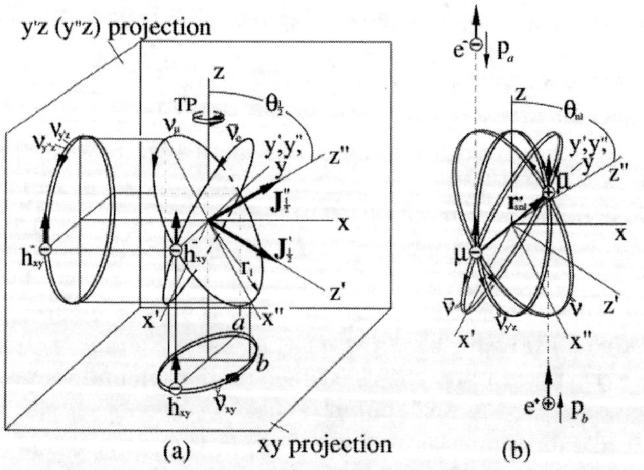

Figure 2. (a) Schematic structure of μ^-, which is a h^-_{xy} of charge $-e$, spin $\frac{1}{2}$, and in two superposing P-O motions along orbits $\nu_\mu, \bar{\nu}_e$ of normals z', z'' at angles $\pi - \theta_{1/2}$, $\theta_{1/2}$ to the z axis. (b) e^-, e^+ near head-on collision at separation $r^*_{s2,1}$ to form a double heavy positronium, preceding to $\mu^-\mu^+$ production.

3. A generalised IED model extending to weak potential field

In the IED picture, the mass of a single charged permanent particle α (e.g. e^-) is generated by the oscillation of its charge $q(=-e)$, of a displacement u_q (at the CM of the minute but extensive q), in the vacuum potential field V_v (mainly the electrostatic field produced by the surrounding vacuuon-dipoles) [3, 6]. In absence of other particles, $V_v = \widetilde{V}_v + V_{v0} \doteq \frac{1}{2}\beta_v u_q^2 + V_{v0}$, where $\beta_v = \mathfrak{M}_v \Omega_v^2$ is the force constant, $\mathfrak{M}_v$ is the (sub-vacuum) mass of the charge q, and Ω_v is the angular frequency; $\mathfrak{M}_v$ represents a friction against the q motion ordinarily in its own V_v field. Assume q oscillates about a fixed position and hence α is at rest here. If another particle α' of charge q' presents in the vicinity and exerting on the vacuum a potential $\widetilde{V}_v'$, then $\mathfrak{M}_v$ ($\propto$ friction), hence β_v, $\widetilde{V}_v$ of q of α would in general be modified, by a factor η_v to $\mathfrak{M}_v^\dagger$, $\beta_v^\dagger$, and

$$\widetilde{V}_v^\dagger(u_q) = \tfrac{1}{2}\beta_v^\dagger u_q^2 = \tfrac{1}{2}\beta_v u_q^2 + \tfrac{1}{2}\mathcal{O}_\eta c^2, \quad \beta_v^\dagger = \beta_v\left(1 + \tfrac{\mathcal{O}_\eta c^2}{2\langle \widetilde{V}_v\rangle}\right), \quad \mathcal{O}_\eta = \tfrac{\eta_v D\langle \widetilde{V}_{v'}\rangle}{D' c^2}, \tag{1}$$

where D, D' are the dimensions of the q, q' oscillations; $\langle\ \rangle$ indicates the time average.

Consider a particle α of charge q is in P-O motion along orbit $\bar{\nu}$ (of a normal z') relative to α' in the central magnetic or weak potential field of α', $V_{\frac{1}{2}}(r_1) = -G/(\frac{4}{3}\pi r_1^3)$, at an equilibrium separation $r_1 (\sim 10^{-18}$ m), G being the coupling constant. Attach the local co-ordinates ξ, ζ to the moving α, which origin is fixed at r_1 and ξ, ζ axes are parallel to r, z' (the radius and normal of orbit $\bar{\nu}$). The oscillation of (the CM of) q, of displacement $\mathbf{u}_q = \mathbf{r} - \mathbf{r}_1 = \boldsymbol{\xi}_q + \boldsymbol{\zeta}_q$, in $\widetilde{V}_v$ is now in the (instantaneous) ξ, ζ plane of dimension $D = 2$, at a random angle ϑ to ζ, and of a rest amplitude $\mathcal{A}_{v\vartheta} = \mathcal{A}_v|\sin\vartheta\hat{\xi} + \cos\vartheta\hat{\zeta}| = |\mathcal{A}_{v\xi}\hat{\xi} + \mathcal{A}_{v\zeta}\hat{\zeta}|$; ϑ varies randomly in 2π over time. Consider further q is at initial time driven by an external force ($\mathbf{F}_{ext}$) into displacement ξ_q away from r_1 along radial r- or ξ- direction. After a relaxation time, q maintains an oscillation of displacement $\xi_q(t) = r - r_1 = \mathcal{A}_h\psi_q(t)$, expressed for a free α at rest here, under the restoring force $\mathbf{F}_h = -\frac{\partial V_h}{\partial \xi_q}\hat{\xi}$ produced by one half of the difference weak potential,

$$V_h(\xi_q) = \frac{1}{2}[V_{\frac{1}{2}}(r_1 + \xi_q) - V_{\frac{1}{2}}(r_1)] \doteq \frac{9G|\xi_q|}{8\pi r_1^4} \approx \frac{9G\xi_q^2}{8\pi r_1^4 \mathcal{A}_h} = \frac{1}{2}\beta_h\xi_q^2, \quad \beta_h = \frac{9G}{4\pi r_1^4\mathcal{A}_h} = \frac{2K_h c^2}{\mathcal{A}_h}, \tag{2}$$

where the second of Eqs (2a) is given for $\xi/r_1 << 1$; the third is a quadratic approximation but gives the exact energy level at $\xi_q = \mathcal{A}_h$. The other half $V_{h'}(\xi_{q'}) = V_h(\xi_q)$ is shared by α'. α' similarly presents to V_h a friction term $\mathcal{O}_\eta = \frac{\eta_v D_h\langle\widetilde{V}_{v'}\rangle}{D'c^2}$, $V_h^\dagger = V_h + \frac{1}{2}\mathcal{O}_\eta c^2$, $\beta_h^\dagger = \beta_h(1 + \frac{\mathcal{O}_\eta c^2}{2\langle V_h\rangle})$.

Setting the friction effect aside, the total restoring forces on q in ξ, ζ directions are thus $F(\xi_q) = F_v(\xi_q) + F_h(\xi_q) = -(\beta_v + \beta_h)\xi_q$, $F(\zeta_q) = F_v(\zeta_q) = -\beta_v\zeta_q$. The Newtonian equations of motion (eom's) of the CM of q along ξ, ζ are thus $-(\beta_v + \beta_h)\xi_q = (\mathfrak{M}_v + \mathfrak{M}_h)\frac{\partial^2\xi_q}{\partial t^2}$, $-\beta_v\zeta_q = (\mathfrak{M}_v)\frac{\partial^2\zeta_q}{\partial t^2}$. The general solutions are $\xi_q^{(c)} = |\mathcal{A}_v e^{i\Omega_v t}\hat{e}_0 + \mathcal{A}_h e^{i\Omega_h t}\hat{e}_h|$, $\zeta_q^{(c)} = \mathcal{A}_v e^{i\Omega_v t}$, where $\Omega_v^2 = \beta_v/\mathfrak{M}_v$, $\Omega_h^2 = \beta_h/\mathfrak{M}_h$; $\hat{e}_0, \hat{e}_h$ are unit vectors for the regular and excited h mass states and assumed orthogonal. $\mathcal{A}_v$ is the time average of $\mathcal{A}_{v\vartheta}$, $\mathcal{A}_v = \frac{\mathcal{A}_{v\vartheta}}{2\pi} = \frac{1}{2}\mathcal{A}_{v\xi}$. The total mechanical energies of the q oscillations in $\widetilde{V}_v^\dagger$, $V_h^\dagger$ including friction are $\mathcal{E}_v^\dagger = \mathcal{E}_v + \mathcal{O}_\eta c^2$, $\mathcal{E}_v = \frac{1}{2}\mathfrak{M}_v\dot{u}_q^2 + \widetilde{V}_v = \frac{1}{2}\beta_v|u_q^{(c)}|^2 = \frac{1}{2}\beta_v\mathcal{A}_v^2$, $\mathcal{E}_h^\dagger = \mathcal{E}_h + \mathcal{O}_\eta c^2$, $\mathcal{E}_h = \frac{1}{2}\mathfrak{M}_h\dot{\xi}_q^2 + V_h = \frac{1}{2}\beta_h|\xi_q^{(c)}|^2 = \frac{1}{2}\beta_h\mathcal{A}_h^2$. For simplicity of discussion we consider the extreme case that no (internal) radiation is being generated by the charge oscillation; $\mathcal{E}_v^\dagger$,$\mathcal{E}_h^\dagger$ thus equal the originally imparted energies.

The charge q is point like to its EMR field and yet is extensive at the scale 10^{-18} m and hence in $V_{\frac{1}{2}}(r)$[6] and V_v[7]. Denote by $\rho_c = |\psi_c(\mathbf{u}, t)|^2$ the dimensionless charge density; ψ_c is a complex function, and $\mathbf{u} = \mathbf{r} - \mathbf{r}_1 = \boldsymbol{\xi} + \boldsymbol{\zeta}$ is the displacement a volume element on q makes at time t. The ρ_c current is thus $j_c = \rho_c\frac{d\mathbf{u}}{dt} = -D_q[\psi_c^*(\nabla\psi_c) - (\nabla\psi_c^*)\psi_c]$; $D = \frac{i\hbar}{2\mathfrak{M}(\mathbf{u}_q)}$ is an imaginary diffusion constant. The stationary state is described by the continuity equation

$\partial_t \rho_c + \nabla j_c + \mathcal{O}\rho_c = 0$, where $\mathcal{O} = \frac{V(\mathbf{u})-V(\mathbf{u})}{i\hbar} \equiv 0$. This decomposes to two equations of a Schrödinger form, given for ψ_c as $i\hbar\partial_t\psi_c = \left(-(\hbar^2/2\mathfrak{M})\nabla^2 + V(\mathbf{u})\right)\psi_c$. Using in it (1a) for $V(\zeta) = \widetilde{V}_{\mathrm{v}}^{\dagger} = \frac{1}{2}\beta_{\mathrm{v}}^{\dagger}\zeta^2$ and (2a) for $V(\xi) = V_h(\xi) = \frac{1}{2}\beta_h\xi^2$ for the two pure orthogonal oscillation states along ζ, ξ, one obtains the standard solutions as for a usual harmonic oscillator: The eigen functions ψ_c are hermit polynomials. The eigen energies are quantised,

$$\mathcal{E}_{\mathrm{v}1}^{\dagger} = \hbar\Omega_{\mathrm{v}}^{\dagger} = \tfrac{1}{2}\beta_{\mathrm{v}}\mathcal{A}_{\mathrm{v}1}^2 + \mathcal{O}_\eta c^2 = M_\alpha^{\dagger}c^2, \quad \mathcal{E}_{hN}^{\dagger} = N\mathcal{E}_h + \mathcal{O}_\eta c^2, \quad \mathcal{E}_h = \hbar\Omega_h = \tfrac{1}{2}\beta_h\mathcal{A}_{h1}^2 = M_h c^2, \quad (3)$$

where based on comparison with experiment [6] only the $n_{\mathrm{v}} = 1$th excited state is permissible in $\widetilde{V}_{\mathrm{v}}$, and $N = 0, 1, 2, \ldots$ in $V_{\frac{1}{2}}$; the terms $\frac{1}{2}\hbar\Omega_{\mathrm{v}}$, $\frac{1}{2}\hbar\Omega_h$ are judged unphysical and dropped. The second of Eqs (3a),(c) identify with the classical energies, except that $\mathcal{A}_{\mathrm{v}1}$, $\mathcal{A}_{hN} = \sqrt{N}\mathcal{A}_{h1}$ are here quantised. The last equate $\mathcal{E}_{\mathrm{v}1}^{\dagger}$,$\mathcal{E}_{hN}$ with the rest-mass energies, and $M_\alpha^{\dagger}$, M_h are the rest masses of the particles α, h, as a generalised basic assumption of the IED model.

Assuming q' is not in excited mass state in $V_{\frac{1}{2}}$, and for the averaged $\mathcal{A}_{\mathrm{v}}$ used, (3a) gives also the actual total hamiltonian of q oscillation in V_{v} (the vacuum ground mass state), and (3b) that of the q oscillation - the excited mass state - in $V_{\frac{1}{2}}$. The respective particle rest masses are

$$M_\alpha^{\dagger} = M_\alpha + \mathcal{O}_\eta, \; M_{hN}^{\dagger} = NM_h + \mathcal{O}_\eta, \; M_h = M_\pi - M_{\nu_e} = \frac{\mathcal{E}_{h1}}{c^2} = \frac{2\langle V_h\rangle}{c^2} = \frac{9G\mathcal{A}_{h1}}{8\pi r_1^4 c^2} = K_h\mathcal{A}_{h1} \quad (4)$$

The second of Eqs (4c) is given by the definition for π^- rest mass (Sec 2) for $q = -e$ (cf Sec 4). Indicated by experiment [1] and in theory (Sec 4), the charge q generates a particle α either in a pure $n_{\mathrm{v}} = 1$ state in V_{v} (i.e. e^-), or a $N \geq 1$ excited state in V_{hN} (e.g. h^-, giving π^-), and not a mixed state of both. Suppose that q is in $N \geq 1$th excited mass state, and q' is in $N' = 0$ state. So q' is not in the $N \geq 1$th kinetic oscillation but shares the other half of the potential energy of (2), of a projected average $\frac{D_h}{D'}\langle V_{h'N}\rangle$, $\langle V_{h'N}\rangle = \langle V_{hN}\rangle = \frac{1}{2}\beta_h\langle\xi_q(t)^2\rangle_N = \frac{1}{2}\mathcal{E}_{hN}$. $D_h/D', = \frac{1}{2}$, reflects the fractional time that q', oscillating regularly in a $D' = 2$ plane, spends along the $D_h = 1$ ξ-axis. Including the shared potential in (3a) applied to q', the (rest) mass of α' is

$$M_{\alpha'}^{\dagger} = M_{\alpha'} + \frac{D_h\langle V_{hN}\rangle}{D'c^2} + \mathcal{O}_\eta' = M_{\alpha'} + \frac{1}{4}NM_h + \mathcal{O}_\eta', \quad \mathcal{O}_\eta' = \frac{\eta_{\mathrm{v}}D_{\alpha'}\langle V_{hN}\rangle}{D_h c^2}. \quad (5)$$

Eqs (4), (5) above, and (6) later, describe a common rest mass formation scheme due to quantised q oscillation in $\widetilde{V}_{\mathrm{v}}$, V_h or V_{hxy}, as contrasted to relativistic mass due to translation. Until Sec 4, we can not predict the M_h value, for in (4) $\mathcal{A}_{h1}$ is not known, also G, r_1 are not universal.

As illustrations we apply (4a,b-c),(5) to two composite particles for evaluating the η_{v} or mass. First, the neutron $n(e^-[\bar{\nu}_e], p)$[7] composed of $\alpha, \alpha' = e^-, p$ both in the $n_{\mathrm{v}}, n'_{\mathrm{v}} = 1, 1$ states; $D, D' = 2$. In their mutual presence the e^-, p masses are each given by (4a), $M_e^{\dagger} = M_e + \frac{\eta_{\mathrm{v}}DM_p}{2D'}$, $M_p^{\dagger} = M_p + \frac{\eta_{\mathrm{v}}D'M_e}{2D}$, where M_e, M_p are the free e^-, p rest masses; their P-O current has an apparent rest mass $M_{\nu_e}(\doteq 0)$. The n mass is the sum $M_n = M_e^{\dagger} + M_p^{\dagger} + M_{\nu_e} \doteq M_e + M_p + \mathcal{O}_\eta$, $\mathcal{O}_\eta = \frac{\eta_{\mathrm{v}}(M_e+M_p)}{2}$. Using in this the experimental M_n, M_e, M_p, hence $\mathcal{O}_\eta^{exp} = 0.782$ MeV, gives $\eta_{\mathrm{v}} \doteq \frac{2\mathcal{O}_\eta^{exp}}{M_e+M_p} = 1.67 \times 10^{-3}$. Second, the $\Lambda(h^-[\bar{\nu}_e], p)$ hyperon (Sec 2) composed of $\alpha = h^-$ in the $N = 1$th state in V_h, and $\alpha' = p$ in $n'_{\mathrm{v}} = 1$th state in $V_{\mathrm{v}'}$; $D_h = 1$, $D' = 2$. To the h^-, p masses (4b-c),(5) apply: $M_h^{\dagger} = M_h + \mathcal{O}_\eta$, $M_h = \frac{2\langle V_h\rangle}{c^2} = M_\pi - M_{\nu_e}$ (M_h is invariant), $M_p^{\dagger} = M_p + \frac{1}{4}M_h + \mathcal{O}_\eta'$, The Λ mass is thus $M_\Lambda = M_h^{\dagger} + M_p^{\dagger} + M_{\nu_e}^{\dagger} = \frac{5}{4}M_\pi + M_p + \sum\mathcal{O}_\eta \doteq 1113.36$ MeV, where $\sum\mathcal{O}_\eta = \eta_{\mathrm{v}}(\frac{D_h\langle V_{\mathrm{v},+e}\rangle}{D'c^2} + \frac{D'\langle V_h\rangle}{D_h c^2}) = \eta_{\mathrm{v}}(\frac{1}{4}M_p + M_\pi) \doteq 0.623$ MeV estimated using the η_{v} value of n; the experimental values for M_π, M_p, $M_{\nu_e}(\doteq 0)$ are used.

Finally we derive the μ^- mass from the projection of (4c). Consider a μ^- as produced from $\pi^- \to \mu^-\bar{\nu}_\mu$, of the original mass $M'_\pi = M_\pi - M_{\nu_\mu} = M_h + M_{\nu_e} + M_{\nu_\mu}$, generated by $-e$ along

$\nu_{y'z}$ (Sec 2.2). Its total EMR field has one component travelling on the continually reorienting $\nu_{y'z}$, which will not be readily re-absorbed by $-e$ and hence lost to the kinetic energies of h^-_{xy}, $\bar{\nu}_e$, ν_μ, and one (as two opposite but non-cancelling standing waves) on ν_{xy}, which can be more readily re-absorbed by $-e$. So the xy-projection of M'_π gives the μ^- mass, $M_\mu(\rho) = (M'_\pi)_{xy} = [M_h(1+\mathcal{O})]_{xy}$, where $\rho = r_{xy}$ is the radius of the elliptic ν_{xy}, $M_h = K_h \mathcal{A}_{h1}$ [Eq (4c)], and K_h is a constant. Provided taking $\mathcal{O}$ as a constant of ρ, the xy-projection of the mass M'_π reduces to a pure geometric xy-projection of $\mathcal{A}_{h1}$, $\mathcal{A}_{h1xy}(\rho)$, which is dependent on ρ. At the semi-major and semi-minor axes $\rho = a, b$, $\mathcal{A}_{h1_{xy}}(a) = \mathcal{A}_{h1}$ and $\mathcal{A}_{h1_{xy}}(b) = \mathcal{A}_{h1}\cos\theta_{\frac{1}{2}}$. Using their mean $\overline{\mathcal{A}}_{h1_{xy}} = (\mathcal{A}_{h1_{xy}}(a)\mathcal{A}_{h1_{xy}}(b))^{1/2} = \mathcal{A}_{h1}\sqrt{\cos\theta_{\frac{1}{2}}}$, $\cos\theta_{\frac{1}{2}} = \frac{1}{\sqrt{3}}$ from earlier, gives

$$M_\mu \doteq M'_\pi(\mathcal{A}_{h1_{xy}}(a)\mathcal{A}_{h1_{xy}}(b))^{1/2}/\mathcal{A}_{h1} = (M_\pi - M_{\nu_\mu})\sqrt{\cos\theta_{\frac{1}{2}}} = 105.860 \text{ MeV}. \tag{6}$$

4. *Ab initio* predictions of the muon and charged pion masses

Consider first two particles a, b of charges $-e, +e$ and masses $m_a = \gamma_a M_a$, $m_b = \gamma_b M_b$, moving at relative speed υ under their Coulomb force $F_c = -\nabla V_c = -\hbar\alpha c/r^2$ at a separation $r = r^0/\gamma$, where $V_c = -\frac{\hbar\alpha c}{r}$, $\alpha = \frac{e^2}{4\pi\epsilon_0\hbar c} = 1/137.036$. In the CM frame, the relativistic eom is $\mathscr{M} d_t^2\mathbf{r} = F_c\hat{r}$, where $\mathscr{M} = \gamma\mathscr{M}^0$, $\gamma = 1/\sqrt{1-\upsilon^2/c^2}$, $\mathscr{M}^0 = \frac{M_a M_b}{M_a+M_b}$, $\mathbf{r} = r\hat{r} = \mathbf{r}_a - \mathbf{r}_b$; and $\mathbf{r}_a = \frac{m_b\mathbf{r}}{m_a+m_b}$, $\mathbf{r}_b = -\frac{m_a\mathbf{r}}{m_a+m_b}$ relative to the CM at $\mathbf{R} = 0$ here. The total wavefunction ψ_{tot} of the fictitious particle of mass $\mathscr{M}$ obeys the Klein-Gordon equation (KGE). In the extreme case either $\gamma\doteq1$ or $\gamma >> 1$, the KGE reduces to SQR-KGE[7] $[\frac{\gamma((p_r^2)_{op}+(\mathcal{J}^2)_{op}/r^2)}{(\gamma+1)\mathscr{M}} + V_c]\psi = \mathscr{E}\psi$, where p_r and $\mathcal{J}$ are the radial and angular momentum operators of the kinetic motion of $\mathscr{M}$. For the central potential in spherical polar co-ordinates, $\psi = \mathcal{R}(r)\mathcal{Y}(\theta,\varphi)$; the SQR-KGE separates to $[-\frac{\gamma\hbar^2}{(\gamma+1)\mathscr{M}r^2}\frac{\partial}{\partial r}(r^2\frac{\partial}{\partial r}) + \frac{\gamma l(l+1)\hbar^2}{(\gamma+1)\mathscr{M}r^2} + V_c(r)]\mathcal{R}(r) = \mathscr{E}\mathcal{R}(r)$, $(\mathcal{J}^2)_{op}\mathcal{Y}(\theta,\varphi) = \mathcal{J}^2\mathcal{Y}(\theta,\varphi)$. The eigen functions are $\mathcal{R}(r) = \frac{(\kappa r)^\kappa}{r}e^{-\kappa r}\sum_{i=0,1\ldots}\frac{b_i}{(\kappa r)^i}$, $\kappa = \sqrt{-(\gamma+1)\mathscr{M}\mathscr{E}/\hbar^2}$, and the spherical harmonics $\mathcal{Y}_l^{m_l} = C_l^{m_l}P_l^{m_l}(\cos\theta)e^{im_l\varphi}$. The level n $(= 1, 2, \ldots)$ eigen energies are $\mathscr{E}_n = T_n + V_{cn} = -\frac{\mathscr{M}_n\upsilon_n^2}{(\gamma_n+1)} = \frac{V_{cn}}{\gamma_n+1} \equiv \frac{\hbar\alpha c}{(\gamma_n+1)r_n} = -\frac{T_n}{\gamma_n}$, $\upsilon_n = \frac{\alpha c}{n} = \frac{\upsilon_1}{n}$, $r_n = \frac{n^2\hbar}{\mathscr{M}_n\upsilon_1}$, $\mathscr{M}_n = \gamma_n\mathscr{M}^0$; the wavelength is $\lambda_n = \frac{h}{\mathscr{M}\upsilon_n} = \frac{2\pi r_n}{n}$. The relation $n^2 = \sum_{l=0}^{n-1}(2l+1)$ holds. Hence $\mathscr{E}_n = \frac{\mathscr{E}_n}{n^2}\sum_{l=0}^{n-1}(2l+1) = \sum_{l=0}^{n-1}\mathscr{E}_{nl}$; the l state projections of $\mathscr{E}_n$ and those of $\upsilon_n, r_n, \lambda_n, \mathcal{J}_n, \mathscr{M}_n$ are

$$\mathscr{E}_{nl} = \frac{\mathscr{E}_n(2l+1)}{n^2} = -\frac{\mathscr{M}_{nl}\upsilon_{nl}^2}{\gamma_{nl}+1} = -\frac{\mathscr{M}_{nl}\upsilon_n^2(2l+1)}{(\gamma_{nl}+1)n^2} = \frac{V_{cnl}}{\gamma_{nl}+1} = -\frac{\hbar\alpha c}{(\gamma_{nl}+1)r_{nl}} = -\frac{\hbar\alpha c(2l+1)}{(\gamma_{nl}+1)r_n n^2}, \tag{7}$$

$$\upsilon_{nl} = \frac{\upsilon_n\sqrt{2l+1}}{n} = \frac{\alpha c\sqrt{2l+1}}{n^2}, \quad r_{nl} = \frac{r_n n^2}{2l+1}, \quad \lambda_{nl} = \frac{h}{\mathscr{M}_{nl}\upsilon_{nl}} = \frac{2\pi r_{nl}\upsilon_{nl}}{\alpha c} = \frac{2\pi r_{nl}\sqrt{2l+1}}{n^2} = \frac{\lambda_n n}{\sqrt{2l+1}},$$

$$\mathcal{J}_{nl} = r_{nl}\mathscr{M}_{nl}\upsilon_{nl} = r_{nl}\mathscr{M}_{nl}\frac{\alpha c\sqrt{2l+1}}{n^2} = \sqrt{l(l+1)}\hbar, \quad \mathscr{M}_{nl} = \gamma_{nl}\mathscr{M}^0. \tag{8}$$

For the usual positronium, $M_b = M_a = M_e$, $\upsilon_n = \frac{\alpha c}{n}$, $\gamma_{nl}\doteq\gamma_n\doteq1$, $\mathscr{M}_{nl}\doteq\mathscr{M}^0 = \frac{1}{2}M_e$; $\lambda_{nl}^0 = \frac{2h}{M_e\upsilon_{nl}} = \frac{2hn^2}{M_e\alpha c\sqrt{2l+1}}$, $r_{nl}^0 = \frac{n^2}{\sqrt{2l+1}}\frac{\lambda_{nl}^0}{2\pi} = \frac{2\hbar c n^4}{M_e c^2\alpha(2l+1)} = \frac{1.058\times10^{-10}n^4}{2l+1}$ m, $\mathcal{J}_{nl}^0 = \frac{r_{nl}^0 M_e\upsilon_{nl}^0}{2} = \frac{r_{nl}^0 M_e\alpha c\sqrt{2l+1}}{2n^2}$.

We shall show that it is possible to create a positronium-like system, a double heavy positronium (DHP) composed of a relativistic electron and positron, $a, b = e^{-*}, e^{+*}$ moving at a relative speed $\upsilon_n^* = |\boldsymbol{\upsilon}_a - \boldsymbol{\upsilon}_b| = g_n\upsilon_n\doteq c$ in their Coulomb field V_c, such that e^{-*}, e^{+*} have each an apparent rest mass $M_{snl}^* = \mathscr{M}_{snl}^* + M_e$, that is given by the eigen state n, l relativistic mass energy $\mathscr{M}_{snl}^*c^2$ above the $n_v = 1$ vacuum level $M_e c^2$ (the e^- rest mass energy) for charge $-e$. This implies that, for level n, the apparent above-the-vacuum rest mass is $\mathscr{M}_{sn}^* = \gamma_{sn}^*\mathscr{M}^0 = \gamma_{sn}^*\frac{1}{2}M_e$ and it has an apparent rest wavelength $\Lambda_{sn}^* = \frac{h}{\mathscr{M}_{sn}^*\upsilon_n} = \frac{nh}{\gamma_{sn}^*\mathscr{M}^0\alpha c}$ equal to the wavelength $\Lambda_e = \frac{h}{M_e c}$ of the total EMR of e^-, $\Lambda_{sn}^* = \Lambda_e$. There thus requires $\frac{n}{\gamma_{sn}^*\frac{1}{2}\alpha} = 1$, or

$$\gamma_{sn}^* = \frac{2n}{\alpha} = 2\gamma_n^*, \quad \gamma_n^* = \frac{n}{\alpha}. \quad \text{So } \mathscr{M}_{sn}^* = \gamma_{sn}^*\mathscr{M}^0 = 2\gamma_n^*\mathscr{M}^0 = 2\mathscr{M}_n^* = \frac{nM_e}{\alpha}; \quad \text{and} \tag{9}$$

$$\mathscr{M}^*_{snl} = \mathscr{M}^*_{sn}\frac{2l+1}{n^2} = 2\gamma^*_{nl}\mathscr{M}^0 = 2\mathscr{M}^*_{nl}, \quad \gamma^*_{nl} = \frac{\gamma^*_n(2l+1)}{n^2} = \frac{(2l+1)}{n\alpha}, \tag{10}$$

given using (7); $g_n = \sqrt{\frac{\gamma^2-1}{\gamma^2}}\frac{c}{v_n} \doteq \frac{n}{\alpha}$. To achieve such dynamics for both the charges $-e, +e$ of e^{-*}, e^{+*}, there requires in the CM frame two fictitious particles, designating by $s = \mu, \bar{\mu}$, each having a mixed total wave $\psi_s = \psi_{\bar{\nu}}\psi_\nu e^{i\beta_s}$, with $\psi_{\bar{\nu}}, \psi_\nu$ satisfying the SQR-KGE. Accordingly $\mathscr{M}^*_{snl}$ is double the $\mathscr{M}^*_{nl}$ of a single relativistic positronium. The $\mathscr{M}^*_{snl}$'s are actually each moving at the speed $v^*_{nl} \doteq v^*_n \doteq c$; they have therefore each a relativistic wavelength

$$\lambda^*_{snl} = \frac{h}{\mathscr{M}^*_{snl}c} = \frac{\lambda^*_{nl}}{2}, \quad \lambda^*_{nl} = \frac{h}{\mathscr{M}^*_{nl}c} = \frac{(\alpha\sqrt{2l+1})\cdot h\cdot n^2}{n^2\cdot(\gamma^*_{nl}\frac{1}{2}M_e)\cdot(\alpha c\sqrt{2l+1})} = \frac{\alpha\sqrt{2l+1}\lambda^0_{nl}}{n^2\gamma^*_{nl}}. \tag{11}$$

The above system can be obtained by accelerating e^-, e^+ (from rest) to e^{-*}, e^{+*} of the CM frame linear momenta $\mathbf{p}_a = -M^*_{snl}v^*_{nl}\hat{z}$, $\mathbf{p}_b = -\mathbf{p}_a$ in $-z, z$ directions at time $t = 0$, and then subjecting them to a near head-on "quantum collision" at $\mathbf{r}_a$, $\mathbf{r}_b = \frac{1}{2}\mathbf{r}^*_{nl}, -\frac{1}{2}\mathbf{r}^*_{nl}$, at a separation

$$r^*_{snl} = \frac{1}{2}r^*_{nl} = r_a = r_b, \quad r^*_{nl} = |\mathbf{r}_a - \mathbf{r}_b| = \frac{n^2}{\sqrt{2l+1}}\frac{\lambda^*_{nl}}{2\pi} = \frac{r^0_{nl}\alpha\sqrt{2l+1}}{\gamma^*_{nl}n^2} = \frac{r^0_{nl}\alpha^2}{n\sqrt{2l+1}} \tag{12}$$

corresponding through (8c) to λ^*_{nl} of (11). In their central V_c field e^{-*}, e^{+*} are thus turned to rotations at the pre-defined eigen dynamical variables (9)-(11). The initial time condition $\mathbf{p}_{\bar{\mu}} = \mathbf{p}_b = -\mathbf{p}_\mu = -\mathbf{p}_a$ permits only states of the same n and $m, m' = -l,\ l$; $\psi_{\bar{\nu}}, \psi_\nu$ are thus travelling along two orbits $\bar{\nu}, \nu$ (two semiclassical effective circles of radii r^*_{nl}) in the $x'y'$, $x''y''$ planes of normals $\hat{z}', \hat{z}''$ at quantised angles $\pi - \theta_l, \theta_l$ to the z axis (Fig 2b); and $\beta_{\bar{\mu}} - \beta_\mu = \pi$. The angular momenta of the component motions of μ along $\bar{\nu}$, ν are $\mathbf{\mathfrak{J}}^*_{\mu\bar{\nu}nl} = |\mathbf{r}^*_{nl}\times\mathbf{p}^*_{nl}|\hat{z}' = \mathfrak{J}^*_{nl}(-\cos\theta_l\hat{z} + \sin\theta_l\hat{r}_{xy})$, $\mathbf{\mathfrak{J}}^*_{\mu\nu nl} = |\mathbf{r}^*_{nl}\times\mathbf{p}^*_{nl}|\hat{z}'' = J^*_{nl}(\cos\theta_l\hat{z} + \sin\theta_l\hat{r}_{xy})$,

$$\mathfrak{J}^*_{nl} = r^*_{nl}\mathscr{M}^*_{nl}c = \frac{(r^0_{nl}\alpha\sqrt{2l+1})(\gamma^*_{nl}\mathscr{M}^0)c}{\gamma^*_{nl}n^2} = \frac{r^0_{nl}M_e v_{nl}}{2} = \frac{r^0_{nl}M_e\alpha c\sqrt{2l+1}}{2n^2} = \sqrt{l(l+1)}\,\hbar = \mathfrak{J}^0_{nl} \tag{13}$$

for $v^*_{nl} \doteq c$; $\sin\theta_l = 1/\sqrt{(l+1)}$, $\cos\theta_l = \sqrt{l}/\sqrt{(l+1)}$; $\mathfrak{J}^*_{nl}$ is equal to $\mathfrak{J}^0_{nl}$ of the usual positronium. The orbits $\bar{\nu}, \nu$ undergo Thomas precession (TP) each about z in the lab frame (for a usual system the TP effect gives a small energy correction through g factor[8]); their normals z', z'' are thus turned to angles $\pi - \theta_j$, θ_j to the z axis (cf Fig 2a), so that $J^*_{\mu\eta nj} = |\mathbf{\mathfrak{J}}^*_{\mu\eta nl} \pm J^{TP}\hat{z}| = \sqrt{j(j+1)}\,\hbar$ for $\eta = \bar{\nu}, \nu$, $j = l - \frac{1}{2}$. For the superposed total motion of μ, $\mathbf{J}^*_{\mu nj} = \mathbf{J}^*_{\mu\bar{\nu}nj} + \mathbf{J}^*_{\mu\nu nj} = 2\sqrt{j(j+1)}$ $\sin\theta_j\ \hat{r}_{xy}$, $J_{\mu njz} = \mathfrak{J}^*_{\mu nlz} = 0$. The total motion of μ is along the elliptic $\nu_{y'z}$ in a $y'z$ or $y''z$ plane at a random angle φ to the x axis, and is at rest on the elliptic ν_{xy} in the xy plane. The xy-projected $\nu, \bar{\nu}$ motions cancel out; these also do not present for the incident e^{-*} (and e^{+*}) but (when actually decomposed) can be produced in an explosive collision. For $(n = 2)$ $l = 1$, the angular motion of μ ($\bar{\mu}$) resembles directly that of the muon μ^- (μ^+) in Sec 2.2.

e^{-*}, e^{+*}, or $\mu, \bar{\mu}$ in the CM frame, are energetically unstable, since at $t > 0$ their separation $(r^*_{nl} \rightarrow) r'$ along $\nu_{y'z}$ is variant. Under action $V'_c(< 0)$, e^{-*}, e^{+*} will continue to move closer, switching to magnetic or weak interaction $V_{\frac{1}{2}}$ dominant at r' comparable to the charge size $\bar{a}$ ($\approx 10^{-18} \sim 10^{-17}$ m [7]). Here the two quanta of the total apparent rest mass energy $M^*_{snl}c^2$ each are able to *in situ* (– at the same positions and velocities) convert to the more stable quantum oscillation energies of the charges $-e, +e$ in their $V_{\frac{1}{2}}$ field. Relative to the CM and $V_{\frac{1}{2}}$, the total mass energy and (instantaneous) four momentum (for the rotation) of $s = \mu$ (or $\bar{\mu}$) before the conversion are $\mathcal{E}' = T' + M_e c^2 = \gamma' M_e c^2 = \gamma_* M^*_{snl}c^2$, $p^{i'} = (\begin{smallmatrix}\mathcal{E}'/c \\ \gamma_* M^*_{snl}v_*\end{smallmatrix})$, where $\gamma_* = 1/\sqrt{1-(v_*/c)^2}$, v_* is the apparent velocity of $\mathscr{M}^*_{snl}$, $i = 0, 1, 2, 3$. Those after the conversion are $\mathcal{E}'' = \gamma'' M'' c^2$, $p^{i''} = (\frac{\mathcal{E}''/c}{\gamma'' M'' v''})$, where $M'' = M_{hN_{xy}} + \sum M_\nu$, $\gamma'' = 1/\sqrt{1-(v''/c)^2}$, and v'' is the velocity of M''. For the *in situ* conversion, $v_* = v''$, $\gamma_* = \gamma''$. Four momentum invariance $p_i'' p^{i''} = p_i' p^{i'}$ gives $M''^2 c^2 \gamma''^2 (1-(v''/c)^2) = M_{snl}^{*2} c^2 \gamma_*{}^2 (1-(v_*/c)^2)$, or $M'' = M_{snl}^* = \mathscr{M}_{snl}^* + M_e$.

In sum, $\mu, \bar{\mu}$ can be assigned with charges $-e, +e$, (the lab-frame) spins $S_{\mu z} = J_{\mu njz} + S_{\mp e} = \frac{1}{2}\hbar$ for $l = 1$, $j = \frac{1}{2}$ and have for $n = 2$ the masses $M_{s2,1}^*$ (at rest in the xy plane) given by (14), which are overall identifiable to those of the muons μ^-, μ^+. Thus when *in situ* converted to the $N = 1$th quantum oscillations of the charges $-e, +e$ in $V_{\frac{1}{2}}$, the converted resemble in all respects the muons μ^-, μ^+ (Sec 2.2). The un-projected $N = 1$ oscillation states of $-e, +e$ converted from the un-projected $n = 2$ states (of mass M_{sn}^* each) resemble then the pions π^-, π^+ (Sec 2.1) in theory. Using (10) for $\mathscr{M}_{snl}^*$, $n = 2, l = 1$, gives in the CM frame, and in the lab frame too where the $\nu, \bar{\nu}$ precessions cancel out, the rest mass of the muon μ^- (or μ^+), $M'' \equiv M_{\mu\mp}$,

$$M_{\mu^-} = M_{s2,1}^* = \mathscr{M}_{s2,1}^* + M_e = \frac{3}{4}\mathscr{M}_{s2}^* + M_e = \frac{3}{4}(\frac{2M_e}{\alpha}) + M_e = 105.549 \text{ MeV}. \quad (14)$$

$M_e = 0.511$ MeV, $\alpha = 1/137.036$ are used. Here $M_{\mu^-}^\dagger = M_{\mu^-} + \mathcal{O}_\eta = M_{\mu^-}$, apparently $\mathcal{O}_\eta = 0$, since the $M_{\mu^-}^\dagger, M_{\mu^+}^\dagger$ are defined in the xy plane and they have no relative motion in this plane. For the pion π^-, ($M'' \equiv$) $M_\pi^\dagger (= M_\pi + \mathcal{O}_\eta) = \mathscr{M}_{s2}^* + M_e = \frac{2M_e}{\alpha} + M_e = 140.525$ MeV in theory; $\mathcal{O}_\eta$ is a friction term from an actual π^-, π^+ production, not the $n = 2$ states here. In experiment, an e^-, e^+ collision can directly produce μ^-, μ^+, apparently owing to the symmetric partial $\bar{\nu}, \nu$ orbits for each charge, whilst for producing π^-, π^+, only one $\bar{\nu}$ or ν can be attributed to each charge. It instead takes an intermediate bound state $(e^- e^+ \rightarrow) \rho^0(770)$ to produce a $\pi^-\pi^+$ pair. ρ^0 can be represented as consisting primarily of a h_N^-, h_N^+ pair generated by $-e, +e$ in $N = 2$ oscillation states in $V_{\frac{1}{2}}$, converted from $n = 4, l = 1$ states of the DHP. Indicated by its still larger experimental mass, ρ^0 apparently also contains two e^-, e^+ pairs, which are present apparently to provide the $\bar{\nu}, \nu$ form of symmetry to each charge. Equations (9)-(13) are general. The $n > 2$ levels can be expected to give rise to all the higher masses of unstable leptons ($\tau^\mp$ are the only observed ones) and composite mesons, typically stabilised in presence of secondary e^-, e^+ pair(s).

The author expresses thanks to Chairman Professor C Burdik and the Organising Committee for the opportunity of presenting this work at the ISQS26, Tech Univ, Prague, 2019, to emeritus scientist P-I Johansson for continued moral support and private financial support to the author's unification research, to Professors B Johansson and I Lindgren for giving moral support to the author's unification research, and to Dr R Dahm and Professor S Catto for useful discussions.

References

[1] Perkins D H 2000 *Introduction to High Energy Physics* 4th ed (Cambridge: Cambridge University Press)
[2] Frauenfelder H and E M Henley 1991 *Subatomic Physics* 2nd ed (New Jersey: Prentice-Hall)
[3] Zheng-Johansson J X 2010 Internally electrodynamic particle model: its experimental basis and its predictions *Proc. 27th Int. Colloq. on Group Theoretical Methods in Physics* (Irevan 2008) ed G Pogosyan *Phys. Atom. Nucl.* **73** 571-81 (*Preprint* arxiv:0812.3951)
[4] Zheng-Johansson J X 2006 Dielectric theory of vacuum *Preprint* arXiv:physics/0612096
[5] Zheng-Johansson J X and P-I Johansson 2006 *Unification of Classical, Quantum and Relativistic Mechanics and of the Four Forces* (New York: Nova Science)
[6] Zheng-Johansson J X 2012 Vacuum potentials for the two only permanent free particles, proton and electron *Proc. Quantum Theory & Symmetries* (Prague 2011) ed C Burdik *J. Phys. Conf. Ser.* **343** 012135 (*Preprint* arxiv1111.3123)
[7] Zheng-Johansson J X 2016 A microscopic theory of the neutron *Proc. 23rd Int. Conf. on Integrable Systems and Quantum Symmetries ISQS23* (Prague 2015) ed C Burdik *J. Phys. Conf. Ser.* **670** 012056
[8] Jackson J D 1999 *Classical Electrodynamics* 3rd ed (New York: John Wiley and Sons)

Prime forms and higher genus deformed Eisenstein series

Alexander Zuevsky
Institute of Mathematics, Czech Academy of Sciences, Prague, Czech Republic
E-mail: zuevsky@math.cas.cz

Abstract. Using the theory of Szegő kernel on a genus g Riemann surfaces obtained as a result of the multiple ρ-parameter formalism of sewing of g handles to the complex sphere, we derive new formulas related prime forms, theta functions, and deformed Eisenstein series. We establish recurrent formulas for genus g prime forms and Szegő kernel as well as further identities. Using the above results, we introduce finally another definition of genus g counterpart of genus one deformed Eisenstein series. The results obtained are then useful in computation of vertex algebra related cohomologies.

1. Introduction: The Szegő Kernel on a Riemann Surface

Consider a compact Riemann surface $\Sigma^{(g)}$ of genus g with canonical homology cycle basis $a_1, \ldots, a_g, b_1, \ldots, b_g$. In general there exists g holomorphic one-forms $\nu_i^{(g)}$, $i = 1, \ldots, g$ which we may normalize by, e.g., [3]

$$\oint_{a_i} \nu_j^{(g)} = 2\pi i \delta_{ij}. \tag{1}$$

The genus g period matrix $\Omega^{(g)}$ is defined by

$$\Omega_{ij}^{(g)} = \frac{1}{2\pi i} \oint_{b_i} \nu_j^{(g)}, \tag{2}$$

for $i, j = 1, \ldots, g$. $\Omega^{(g)}$ is symmetric with positive imaginary part, i.e., $\Omega^{(g)} \in \mathbb{H}_g$, the Siegel upper half plane.

It is useful to introduce the normalized differential of the second kind defined by [5], [1]:

$$\omega^{(g)}(x, y) \sim \frac{dx\, dy}{(x-y)^2}, \tag{3}$$

for local coordinates $x \sim y$, with normalization $\int_{a_i} \omega^{(g)}(x, \cdot) = 0$ for $i = 1, \ldots, g$. Using the Riemann bilinear relations, one finds that $\nu_i^{(g)}(x) = \oint_{b_i} \omega^{(g)}(x, \cdot)$.

We also introduce the normalized differential of the third kind

$$\omega_{p_2 - p_1}^{(g)}(x) = \int_{p_1}^{p_2} \omega^{(g)}(x, \cdot), \tag{4}$$

for which $\oint_{a_i} \omega^{(g)}_{p_2-p_1} = 0$ and $\omega^{(g)}_{p_2-p_1}(x) \sim \frac{(-1)^a}{x-p_a}\, dx$ for $x \sim p_a$ and $a = 1, 2$.

We recall the definition of the theta function with real characteristics, e.g., [5], [1], [3]

$$\vartheta^{(g)} \begin{bmatrix} \alpha \\ \beta \end{bmatrix} \left(z | \Omega^{(g)} \right) = \sum_{m \in \mathbb{Z}^g} \exp \left(i\pi (m+\alpha).\Omega^{(g)}.(m+\alpha) + (m+\alpha).(z + 2\pi i \beta) \right), \tag{5}$$

for $\alpha = (\alpha_i), \beta = (\beta_i) \in \mathbb{R}^g$, $z = (z_i) \in \mathbb{C}^g$ and $i = 1, \ldots, g$.

There exists a (non-singular and odd) character $\left[\begin{smallmatrix} \gamma \\ \delta \end{smallmatrix}\right]$ such that [5],[1]

$$\vartheta^{(g)} \begin{bmatrix} \gamma \\ \delta \end{bmatrix} (0|\Omega) = 0, \qquad \partial_{z_i} \vartheta \begin{bmatrix} \gamma \\ \delta \end{bmatrix} (0|\Omega) \neq 0. \tag{6}$$

Let

$$\zeta(x) = \sum_{i=1}^{g} \partial_{z_i} \vartheta^{(g)} \begin{bmatrix} \gamma \\ \delta \end{bmatrix} (0|\Omega^{(g)}) \nu_i^{(g)}(x), \tag{7}$$

a holomorphic one-form, and let $\zeta(x)^{\frac{1}{2}}$ denote the form of weight $\frac{1}{2}$ on the double cover $\widetilde{\Sigma}$ of Σ. We also refer to $\zeta(x)^{\frac{1}{2}}$ as a (double-valued) $\frac{1}{2}$-form on Σ. We define the prime form $E^{(g)}(x, y)$ by

$$E^{(g)}(x,y) = \frac{\vartheta^{(g)} \left[\begin{smallmatrix} \gamma \\ \delta \end{smallmatrix}\right] \left(\int_y^x \nu | \Omega^{(g)} \right)}{\zeta(x)^{\frac{1}{2}} \zeta(y)^{\frac{1}{2}}} \sim (x-y) dx^{-\frac{1}{2}} dy^{-\frac{1}{2}} \quad \text{for } x \sim y, \tag{8}$$

where $\int_y^x \nu = (\int_y^x \nu_i) \in \mathbb{C}^g$. $E^{(g)}(x, y) = -E^{(g)}(y, x)$ is a holomorphic differential form of weight $(-\frac{1}{2}, -\frac{1}{2})$ on $\widetilde{\Sigma} \times \widetilde{\Sigma}$. Note that our definition differs from that of refs. [5], [1] by a factor of -1.

The normalized differentials of the second and third kind can be expressed in terms of the prime form [5]

$$\omega^{(g)}(x,y) = \partial_x \partial_y \log K^{(g)}(x,y)\, dx\, dy, \tag{9}$$

$$\omega^{(g)}_{p-q}(x) = \partial_x \log \frac{K^{(g)}(x,p)}{K^{(g)}(x,q)} dx. \tag{10}$$

Conversely, we can also express the prime form in terms of $\omega^{(g)}_{p-q}$ by [2]

$$E^{(g)}(x,y) = \lim_{p \to x,\ q \to y} \left[\sqrt{(x-p)(q-y)} \exp \left(-\frac{1}{2} \int_y^x \omega_{p-q} \right) \right] dx^{-\frac{1}{2}}\, dy^{-\frac{1}{2}}. \tag{11}$$

We define the Szegő kernel [10], [1] for $\vartheta^{(g)} \left[\begin{smallmatrix} \alpha \\ \beta \end{smallmatrix}\right] (0|\Omega^{(g)}) \neq 0$ as follows

$$S \begin{bmatrix} \theta \\ \phi \end{bmatrix} (x, y | \Omega^{(g)}) = \frac{\vartheta^{(g)} \left[\begin{smallmatrix} \alpha \\ \beta \end{smallmatrix}\right] \left(\int_y^x \nu | \Omega^{(g)} \right)}{\vartheta^{(g)} \left[\begin{smallmatrix} \alpha \\ \beta \end{smallmatrix}\right] (0|\Omega^{(g)}) E^{(g)}(x,y)} = \frac{\Theta^{(g)}(x,y)}{E^{(g)}(x,y)}, \tag{12}$$

where $\theta = (\theta_i)$, $\phi = (\phi_i) \in U(1)^n$ for $\theta_j = -e^{-2\pi i \beta_j}$, $\phi_j = -e^{2\pi i \alpha_j}$, $j = 1, \ldots, g$. The Szegő kernel has multipliers along the a_i and b_j cycles in x given by $-\phi_i$ and $-\theta_j$ respectively and is a meromorphic $(\frac{1}{2}, \frac{1}{2})$-form on $\widetilde{\Sigma} \times \widetilde{\Sigma}$ satisfying:

$$S^{(g)} \begin{bmatrix} \theta \\ \phi \end{bmatrix} (x,y) \sim \frac{1}{x-y} dx^{\frac{1}{2}}\, dy^{\frac{1}{2}} \quad \text{for } x \sim y, \tag{13}$$

$$S^{(g)} \begin{bmatrix} \theta \\ \phi \end{bmatrix} (x,y) = -S^{(g)} \begin{bmatrix} \theta^{-1} \\ \phi^{-1} \end{bmatrix} (y,x), \tag{14}$$

Journal of Physics: Conference Series **1416** (2019) 012044 doi:10.1088/1742-6596/1416/1/012044

where $\theta^{-1} = (\theta_i^{-1})$ and $\phi^{-1} = (\phi_i^{-1})$. Note that the skew-symmetry property (14) implies $S^{(g)}\begin{bmatrix}\theta\\ \phi\end{bmatrix}(x,y)$ has multipliers along the a_i and b_j cycles in y given by $-\phi_i^{-1}$ and $-\theta_j^{-1}$ respectively.

For a Riemann surface of genus one described by an oriented torus $\mathbb{C}/\Lambda$ for lattice $\Lambda = 2\pi i(\mathbb{Z}\tau \oplus \mathbb{Z})$ for $\tau \in \mathbb{H}_1$, the genus one prime form is

$$E^{(1)}(x,y) = K^{(1)}(x-y,\tau)dx^{-\frac{1}{2}}dy^{-\frac{1}{2}}, \quad K^{(1)}(z,\tau) = \frac{\vartheta_1(z,\tau)}{\partial_z \vartheta_1(0,\tau)}, \tag{15}$$

for $z \in \mathbb{C}$ and $\tau \in \mathbb{H}_1$ and where $\vartheta_1(z,\tau) = \vartheta\begin{bmatrix}\frac{1}{2}\\ \frac{1}{2}\end{bmatrix}(z,\tau)$.

For $(\theta,\phi) \neq (1,1)$ with $\theta = -e^{-2\pi i\beta}$ and $\phi = -e^{2\pi i\alpha}$ the genus one Szegő kernel is

$$S^{(1)}\begin{bmatrix}\theta\\ \phi\end{bmatrix}(x,y|\tau) = P_1\begin{bmatrix}\theta\\ \phi\end{bmatrix}(x-y,\tau)dx^{\frac{1}{2}}dy^{\frac{1}{2}}, \tag{16}$$

where

$$P_1\begin{bmatrix}\theta\\ \phi\end{bmatrix}(z,\tau) = \frac{\vartheta\begin{bmatrix}\alpha\\ \beta\end{bmatrix}(z,\tau)}{\vartheta\begin{bmatrix}\alpha\\ \beta\end{bmatrix}(0,\tau)}\frac{1}{K(z,\tau)} = -\sum_{k\in\mathbb{Z}}\frac{q_z^{k+\lambda}}{1-\theta^{-1}q^{k+\lambda}},$$

is a deformed Weierstrass function [9] for $q_z = e^z$ and with $\phi = \exp(2\pi i\lambda)$ for $0 \leq \lambda < 1$.

We also have a Laurant expansion [9]

$$P_1\begin{bmatrix}\theta\\ \phi\end{bmatrix}(z,\tau) = \frac{1}{z} - \sum_{n\geq 1} E_n\begin{bmatrix}\theta\\ \phi\end{bmatrix}(\tau)z^{n-1}, \tag{17}$$

for deformed Eisenstein series defined by

$$E_n\begin{bmatrix}\theta\\ \phi\end{bmatrix}(\tau) = -\frac{B_n(\lambda)}{n!} + \frac{1}{(n-1)!}\sum_{r\geq 0}\frac{(r+\lambda)^{n-1}\theta^{-1}q^{r+\lambda}}{1-\theta^{-1}q^{r+\lambda}} + \frac{(-1)^n}{(n-1)!}\sum_{r\geq 1}\frac{(r-\lambda)^{n-1}\theta q^{r-\lambda}}{1-\theta q^{r-\lambda}}, \tag{18}$$

for $n \geq 1$ and where $B_n(\lambda)$ is the Bernoulli polynomial defined by

$$\frac{q_z^\lambda}{q_z - 1} = \frac{1}{z} + \sum_{n\geq 1}\frac{B_n(\lambda)}{n!}z^{n-1}.$$

In particular, we have

$$\frac{\vartheta^{(1)}\begin{bmatrix}\alpha\\ \beta\end{bmatrix}(z,\tau)}{\vartheta^{(1)}\begin{bmatrix}\alpha\\ \beta\end{bmatrix}(0,\tau)} = -\sum_{k\in\mathbb{Z}}\frac{q_z^{k+\lambda}}{1-\theta^{-1}q^{k+\lambda}}\cdot\left[\frac{1}{z} - \sum_{n\geq 1}E_n\begin{bmatrix}\theta\\ \phi\end{bmatrix}(\tau)z^{n-1}\right],$$

For $(\theta,\phi) = (1,1)$ and $n \geq 2$ the deformed Eisenstein series reduce to the standard elliptic Eisenstein series with $E_n(\tau) = 0$ for n odd. Using the Laurant expansion (17) we find

$$P_1\begin{bmatrix}\theta\\ \phi\end{bmatrix}(x-y,\tau) = \frac{1}{x-y} + \sum_{k,l\geq 1} C\begin{bmatrix}\theta\\ \phi\end{bmatrix}(k,l)x^{k-1}y^{l-1}, \tag{19}$$

where for $k,l \geq 1$ we define

$$C\begin{bmatrix}\theta\\ \phi\end{bmatrix}(k,l,\tau) = (-1)^l\binom{k+l-2}{k-1}E_{k+l-1}\begin{bmatrix}\theta\\ \phi\end{bmatrix}(\tau), \tag{20}$$

for deformed Eisenstein series (18).

2. Prime form formula

In [8], in ρ-formalism [12], $\rho \in \mathbb{C}$ of self-sewing of Riemann surfaces, for $a, b = 1, 2$ and $k, l = 1, 2, \ldots$ they define weighted moments

$$Y^{(g)}_{\bar{a}b}(k,l) = \frac{\rho^{(k+l)/2}}{\sqrt{kl}} \frac{1}{(2\pi i)^2} \oint_{\mathcal{C}_a(u)} \oint_{\mathcal{C}_b(v)} u^{-k} v^{-l} \omega^{(g+1)}(u,v). \tag{21}$$

Note that $Y^{(g)}_{ab}(k,l) = Y^{(g)}_{\bar{b}\bar{a}}(l,k)$. They also define $Y^{(g)} = \left(Y^{(g)}_{ab}(k,l)\right)$ to be the infinite matrix indexed by the pairs a, k and b, l. They define a set of holomorphic one-forms on $\widehat{\Sigma}^{(g)}$

$$a_a(k,x) = \frac{\rho^{k/2}}{2\pi i \sqrt{k}} \oint_{\mathcal{C}_a(z_a)} z_a^{-k} \omega^{(g)}(x, z_a), \tag{22}$$

and define $a(x) = (a_a(k,x))$ and $\bar{a}(x) = (a_{\bar{a}}(k,x))$ to be the infinite row vectors indexed by a, k. In [8] they prove

$$\omega^{(g+1)}(x,y) = \omega^{(g)}(x,y) - a(x)\left(I - Y^{(g)}\right)\bar{a}(y)^T. \tag{23}$$

It follows from the Fay's representation (11) [2] of the prime form in terms of ω that

$$K^{(g+1)}\left(x, y, \Omega^{(g+1)}\right) = K^{(g)}\left(x, y, \Omega^{(g)}\right) e^{-\frac{1}{2} b_a \left(I - Y^{(g)}\right)_{\bar{a}\bar{a}}(\rho) b_a^T}. \tag{24}$$

Here $b_a(x, y; k) = \int_y^x a_a(\cdot, k)$, and

$$a_a(x)\left(I - Y^{(g)}_{\bar{a}\bar{a}}\right)(\rho) a_a^T(y) = \sum_{k,l \geq 1} a_a(x,k)\left(I - Y^{(g)}_{\bar{a}\bar{a}}\right)(k,l,\rho) a_a(y,l),$$

with $a_a(x,k)$ a certain one-form [7] on the initial Riemann surface $\widehat{\Sigma}^{(g_a)}$ of genus g. Here $\left(I - Y^{(g)}_{\bar{a}\bar{a}}\right)(k,l,\rho)$ is an infinite matrix determined from genus g data (see [6] for details). Thus from (24) for the multiple sewing of the sphere $\Sigma^{(0)}$ to form a genus g Riemann surface and $\rho_i \in \mathbb{C}$, we obtain

Proposition 2.1

$$K^{(g)}\left(x, y, \Omega^{(g)}\right) = K^{(1)}(x - y, \tau) \prod_{i=1}^{g-1} e^{-\frac{1}{2} b_a^{(i)} \left(I - Y^{(i)}\right)_{\bar{a}\bar{a}}(\rho_i) \left(b_a^{(i)}\right)^T} = K^{(1)}(x - y, \tau)\, \mathcal{Y}^{(g)}, \tag{25}$$

where $Y^{(i)}$ corresponds to genus i Riemann surface as above.

3. Recurrent formula for genus g prime forms and Szegő kernel

We have proved in [11] that the Szegő kernel $S^{(g+1)}$ of the genus $g+1$ Riemann surface obtained as a results of self-sewing a genus g Riemann surface is holomorphic in ρ for $|\rho| < r_1 r_2$ with

$$S^{(g+1)}(x,y) = S^{(g)}_{\kappa}(x,y) + \mathcal{D}^{(g)}, \tag{26}$$

$$\mathcal{D}^{(g)} = \xi h(x) D^{\theta}\left(I + \xi Y^{(g)} D^{\theta}\right) \bar{h}(y)^T,$$

for infinite diagonal matrix $D^{\theta}(k,l) = \begin{bmatrix} \theta_{g+1}^{-1} & 0 \\ 0 & -\theta_{g+1} \end{bmatrix} \delta(k,l)$. for $x,\, y \in \widehat{\Sigma}^{(g)}$ where $S^{(g)}_{\kappa}(x,y)$ is defined as follows: For $\kappa \neq -\frac{1}{2}$

$$S^{(g)}_{\kappa}(x,y) \;=\; \frac{\left(U^{(g)(x,y)}\right)^{\kappa} \vartheta^{(g)} \left[\begin{smallmatrix}\alpha^{(g)}\\ \beta^{(g)}\end{smallmatrix}\right] \left(\int_y^x \nu^{(g)} + \kappa z_{p_1,p_2}|\Omega^{(g)}\right)}{\vartheta^{(g)} \left[\begin{smallmatrix}\alpha^{(g)}\\ \beta^{(g)}\end{smallmatrix}\right] \left(\kappa z_{p_1,p_2}|\Omega^{(g)}\right) E^{(g)}(x,y)}, \tag{27}$$

where

$$U^{(g)}(x,y) = \frac{E^{(g)}(x,p_2)E^{(g)}(y,p_1)}{E^{(g)}(x,p_1)E^{(g)}(y,p_2)}, \tag{28}$$

for prime form $E^{(g)}$ and where $z_{p_1,p_2} = \int_{p_1}^{p_2} \nu^{(g)}$, for holomorphic 1-forms $\nu^{(g)}$. For $\kappa = -\frac{1}{2}$ then $S^{(g)}_{-\frac{1}{2}}(x,y)$ is given by

$$S^{(g)}_{-\frac{1}{2}}(x,y) = \Bigg(\frac{\left(U^{(g)}(x,y)\right)^{\frac{1}{2}}}{E^{(g)}(x,y)} \vartheta^{(g)} \left[\begin{smallmatrix}\alpha^{(g)}\\ \beta^{(g)}\end{smallmatrix}\right] \left(\int_y^x \nu^{(g)} + \tfrac{1}{2} z_{p_1,p_2}|\Omega^{(g)}\right)$$
$$-\theta_{g+1} \frac{\left(U^{(g)}(x,y)\right)^{-\frac{1}{2}}}{E^{(g)}(x,y)} \vartheta^{(g)} \left[\begin{smallmatrix}\alpha^{(g)}\\ \beta^{(g)}\end{smallmatrix}\right] \left(\int_y^x \nu^{(g)} - \tfrac{1}{2} z_{p_1,p_2}|\Omega^{(g)}\right)\Bigg).$$
$$\left(\vartheta^{(g)} \left[\begin{smallmatrix}\alpha^{(g)}\\ \beta^{(g)}\end{smallmatrix}\right] \left(\tfrac{1}{2} z_{p_1,p_2}|\Omega^{(g)}\right) - \theta_{g+1}\vartheta^{(g)} \left[\begin{smallmatrix}\alpha^{(g)}\\ \beta^{(g)}\end{smallmatrix}\right] \left(-\tfrac{1}{2} z_{p_1,p_2}|\Omega^{(g)}\right)\right)^{-1}. \tag{29}$$

We use also half-order differentials [11]

$$h_a(k,x) = h_a \begin{bmatrix}\theta^{(g)}\\ \phi^{(g)}\end{bmatrix} (\kappa;k,x) = \frac{\rho^{\frac{1}{2}(k_a-\frac{1}{2})}}{2\pi i} \oint_{\mathcal{C}_a(y_a)} y_a^{-k_a} S^{(g)}_{\kappa}(x,y_a) dy_a^{\frac{1}{2}}, \tag{30}$$

$$\bar{h}_a(k,y) = \bar{h}_a \begin{bmatrix}\theta^{(g)}\\ \phi^{(g)}\end{bmatrix} (\kappa;k,y) = \frac{\rho^{\frac{1}{2}(k_a-\frac{1}{2})}}{2\pi i} \oint_{\mathcal{C}_{\bar{a}}(x_{\bar{a}})} x_{\bar{a}}^{-k_a} S^{(g)}_{\kappa}(x_{\bar{a}},y) dx_{\bar{a}}^{\frac{1}{2}}, \tag{31}$$

and let $h(x) = (h_a(k,x))$ and $\bar{h}(y) = (\bar{h}_a(k,y))$ denote the infinite row vectors indexed by a, k. These are related by skew-symmetry with

$$h_a \begin{bmatrix}\theta^{(g)}\\ \phi^{(g)}\end{bmatrix} (\kappa;k,x) = -\bar{h}_{\bar{a}} \begin{bmatrix}(\theta^{(g)})^{-1}\\ (\phi^{(g)})^{-1}\end{bmatrix} (-\kappa;k,x). \tag{32}$$

These moments can be inverted to obtain

$$S^{(g)}_{\kappa}(x,y_a) \;=\; \sum_{k\geq 1} \rho^{-\frac{k_a}{2}+\frac{1}{4}} h_a(k,x) y_a^{k_a-1} dy_a^{\frac{1}{2}} \tag{33}$$

$$S^{(g)}_{\kappa}(x_{\bar{a}},y) \;=\; \sum_{k\geq 1} \rho^{-\frac{k_a}{2}+\frac{1}{4}} x_{\bar{a}}^{k_a-1} \bar{h}_a(k,y) dx_{\bar{a}}^{\frac{1}{2}}. \tag{34}$$

From the sewing relation [12] we have $dz_a^{\frac{1}{2}} = (-1)^{\bar{a}} \xi \rho^{\frac{1}{2}} \frac{dz_{\bar{a}}^{\frac{1}{2}}}{z_{\bar{a}}}$, for $\xi \in \{\pm\sqrt{-1}\}$.

On the other hand

$$\frac{\Theta^{(g+1)}(x,y)}{E^{(g+1)}(x,y)} = \frac{\Theta^{(1)}(x,y)}{E^{(}g)}(x,y) W^{(g)}(x,y) + \mathcal{D}^{(g)},$$

$$\frac{\Theta^{(g+1)}(x,y)}{Y^{(g)}E^{(g)}(x,y)} = \frac{\Theta^{(1)}(x,y)}{E^{(g)}(x,y)} W^{(g)}(x,y) + \mathcal{D}^{(g)}.$$

Thus (assuming $E^{(g+1)}(x,y)E^{(g)}(x,y) \neq 0$) we obtain

Proposition 3.1

$$E^{(g)}(x,y) = \left[\Theta^{(g+1)}(x,y)\left(Y^{(g)}\right)^{-1} - \Theta^{(1)}(x,y)W^{(g)}(x,y)\right]\left(\mathcal{D}^{(g)}\right)^{-1}.$$

Then it follows

$$E^{(1)}(x,y) = \left[\Theta^{(g+1)}(x,y)Y^{-1} - \Theta^{(1)}(x,y)W^{(g)}(x,y)\right]\mathcal{Y}^{(g)}\left(\mathcal{D}^{(g)}\right)^{-1}. \tag{35}$$

Depending how we express the prime form $E^{(1)}(x,y)$ in (35) we obtain identities for theta-functions (15), differentials (11) sum of Eisenstein series (17), or q-series (17).

Note that

$$\begin{aligned} S^{(g)}_{\kappa}(x,y) &= \frac{\left(U^{(g)}(x,y)\right)^{\kappa}\vartheta^{(g)}\begin{bmatrix}\alpha^{(g)}\\ \beta^{(g)}\end{bmatrix}\left(\int_y^x \nu^{(g)} + \kappa z_{p_1,p_2}|\Omega^{(g)}\right)}{E^{(g)}(x,y)\vartheta^{(g)}\begin{bmatrix}\alpha^{(g)}\\ \beta^{(g)}\end{bmatrix}\left(\kappa z_{p_1,p_2}|\Omega^{(g)}\right)} \\ &= S^{(g)}(x,y)\frac{\vartheta^{(g)}\begin{bmatrix}\alpha\\ \beta\end{bmatrix}(0|\Omega^{(g)})}{\vartheta^{(g)}\begin{bmatrix}\alpha\\ \beta\end{bmatrix}\left(\int_y^x \nu|\Omega^{(g)}\right)}\frac{\left(U^{(g)}(x,y)\right)^{\kappa}\vartheta^{(g)}\begin{bmatrix}\alpha^{(g)}\\ \beta^{(g)}\end{bmatrix}\left(\int_y^x \nu^{(g)} + \kappa z_{p_1,p_2}|\Omega^{(g)}\right)}{\vartheta\begin{bmatrix}\alpha^{(g)}\\ \beta^{(g)}\end{bmatrix}\left(\kappa z_{p_1,p_2}|\Omega^{(g)}\right)} \\ &= S^{(g)}(x,y)\, W^{(g)}(x,y). \end{aligned}$$

Thus

$$S^{(g+1)}(x,y) = S^{(g)}(x,y)\, W^{(g)}(x,y) + \mathcal{D}^{(g)}. \tag{36}$$

We then recurrently obtain

Proposition 3.2

$$S^{(g)}(x,y) = S^{(1)}(x,y)\prod_{i=1}^{g-1} W^{(i)}(x,y) + \sum_{j=1}^{g-1}\mathcal{D}^{(j)}\prod_{k=j+1}^{g-1} W^{(k)}(x,y). \tag{37}$$

Now let us express it in terms of prime forms $E^{(g)}$ and $E^{(1)}$. We obtain

$$\frac{\Theta^{(g)}(x,y)}{E^{(g)}(x,y)} = \frac{\Theta^{(1)}(x,y)}{E^{(1)}(x,y)}\prod_{i=1}^{g-1} W^{(i)}(x,y) + \sum_{j=1}^{g-1}\mathcal{D}^{(j)}\prod_{k=j+1}^{g-1} W^{(k)}(x,y),$$

from which we obtain the following identities.

For relation $E^{(g)}(x,y)$ and $E^{(1)}(x,y)$ again

$$\begin{aligned} \Theta^{(g)}(x,y)E^{(1)}(x,y) &= E^{(g)}(x,y)\Bigg(\Theta^{(1)}(x,y)\prod_{i=1}^{g-1} W^{(i)}(x,y) \\ &\quad + E^{(1)}(x,y)\sum_{j=1}^{g-1}\mathcal{D}^{(j)}\prod_{k=j+1}^{g-1} W^{(k)}(x,y)\Bigg), \end{aligned}$$

thus

$$E^{(g)}(x,y) = \frac{\Theta^{(g)}(x,y)\; E^{(1)}(x,y)}{\Theta^{(1)}(x,y)\prod_{i=1}^{g-1} W^{(i)}(x,y) + E^{(1)}(x,y)\sum_{j=1}^{g-1}\mathcal{D}^{(j)}\prod_{k=j+1}^{g-1} W^{(k)}(x,y)}.$$

Note that this is equivalent to

$$\Theta^{(g)}(x,y) = \mathcal{Y}\left[\Theta^{(1)}(x,y)\prod_{i=1}^{g-1} W^{(i)}(x,y) + E^{(1)}(x,y)\sum_{j=1}^{g-1}\mathcal{D}^{(j)}\prod_{k=j+1}^{g-1} W^{(k)}(x,y)\right],$$

or

$$E^{(1)}(x,y) = \frac{\Theta^{(g)}(x,y)\mathcal{Y}^{-1} - \Theta^{(1)}(x,y)\prod_{i=1}^{g-1} W^{(i)}(x,y)}{\sum_{j=1}^{g-1}\mathcal{D}^{(j)}\prod_{k=j+1}^{g-1} W^{(k)}(x,y)}.$$

Using (35) we obtain the identity

$$\left[\Theta^{(g+1)}(x,y)Y^{-1} - \Theta^{(1)}(x,y)W^{(g)}(x,y)\right]\mathcal{Y}^{(g)} = \frac{\Theta^{(g)}(x,y)\left(\mathcal{Y}^{(g)}\right)^{-1} - \Theta^{(1)}(x,y)\prod_{i=1}^{g-1} W^{(i)}(x,y)}{\sum_{j=1}^{g-1}\mathcal{D}^{(j)}\prod_{k=j+1}^{g-1} W^{(k)}(x,y)}\,\mathcal{D}.$$

4. Higher genus deformed Eisenstein series

In [4] they introduce a version of genus two counterpart of Eisenstein series. Here we give another possible definition of the deformed Eisenstein series in terms of the multiple-sewing construction. Note that

$$S(1)\begin{bmatrix}\theta\\ \phi\end{bmatrix}(x,y|\tau) = \frac{\vartheta^{(1)}\begin{bmatrix}\alpha\\ \beta\end{bmatrix}\left(\int_y^x \nu|\tau\right)}{\vartheta^{(1)}\begin{bmatrix}\alpha\\ \beta\end{bmatrix}(z,\tau)}\left[\frac{1}{z} - \sum_{n\geq 1} E_n\begin{bmatrix}\theta\\ \phi\end{bmatrix}(\tau)z^{n-1}\right]dx^{-\frac{1}{2}}dy^{-\frac{1}{2}}, \qquad (38)$$

i.e., the torus Szegő kernel can be expressed via deformed Eisenstein series. Using (37) we obtain by expanding the right hand side

$$\begin{aligned} S^{(g)}(x,y) &= \left[\frac{1}{x-y} + \sum_{k,l\geq 1} C\begin{bmatrix}\theta\\ \phi\end{bmatrix}(k,l)x^{k-1}y^{l-1}\right]\prod_{i=1}^{g-1} W^{(i)}(x,y)dx^{-\frac{1}{2}}dy^{-\frac{1}{2}} \\ &\quad + \sum_{j=1}^{g-1}\mathcal{D}(j)\prod_{k=j+1}^{g-1} W^{(k)}(x,y) \\ &= \frac{\vartheta^{(g)}\begin{bmatrix}\alpha\\ \beta\end{bmatrix}\left(\int_y^x \nu|\Omega^{(g)}\right)}{\vartheta^{(g)}\begin{bmatrix}\alpha\\ \beta\end{bmatrix}(z,\Omega)}\left[\frac{A(x,y)}{x-y} - \sum_{n,m\geq 1} E^{(g)}_{n,m}\begin{bmatrix}\theta\\ \phi\end{bmatrix}(\Omega^{(g)})\,x^{n-1}y^{m-1}\right], \end{aligned} \qquad (39)$$

where

$$A^{(g)}(x,y) = \frac{\vartheta^{(g)}\begin{bmatrix}\alpha\\ \beta\end{bmatrix}(z,\Omega^{(g)})}{\vartheta^{(g)}\begin{bmatrix}\alpha\\ \beta\end{bmatrix}\left(\int_y^x \nu|\Omega^{(g)}\right)}\prod_{i=1}^{g-1} W^{(i)}(x,y)\,dx^{-\frac{1}{2}}dy^{-\frac{1}{2}},$$

and by $E^{(g)}_{n,m}\begin{bmatrix}\theta\\ \phi\end{bmatrix}(\Omega^{(g)})$ we call the elements of the higher genus Eisenstein series

$$\sum_{n,m\geq 1} E^{(g)}_{n,m}\begin{bmatrix}\theta\\ \phi\end{bmatrix}(\Omega^{(g)})\, x^{n-1}y^{m-1} = -\frac{\vartheta^{(g)}\begin{bmatrix}\alpha\\ \beta\end{bmatrix}(z,\Omega^{(g)})}{\vartheta^{(g)}\begin{bmatrix}\alpha\\ \beta\end{bmatrix}\left(\int_y^x \nu|\Omega^{(g)}\right)}\left[\sum_{j=1}^{g-1}\mathcal{D}^{(j)}\prod_{k=j+1}^{g-1} W^{(k)}(x,y)\right.$$
$$\left. + \sum_{k,l\geq 1} C\begin{bmatrix}\theta\\ \phi\end{bmatrix}(k,l)x^{k-1}y^{l-1}\prod_{i=1}^{g-1} W^{(i)}(x,y)dx^{-\frac{1}{2}}dy^{-\frac{1}{2}}\right]. \tag{40}$$

5. Acknowledgments

Research of the author was supported by the GACR project 18-00496S and RVO: 67985840.

References

[1] Fay J D 1973 *Theta Functions on Riemann Surfaces*, Lecture Notes in Mathematics Vol 352 (Berlin-New York: Springer-Verlag)
[2] Fay J D 1992 *Kernel functions, analytic torsion, and moduli spaces* (Mem. Amer. Math. Soc. vol 96)
[3] Farkas H M and Kra I 1980 *Riemann Surfaces* (Springer-Verlag, New York)
[4] Gilroy T and Tuite M P 2015 Genus Two Zhu Theory for Vertex Operator Algebras *Preprint* math.QA/1511.07664
[5] Mumford D 1983 *Tata Lectures on Theta I and II* (Birkhäuser, Boston)
[6] Mason G and Tuite M P 2003 *Comm. Math. Phys.* **235** 47
[7] Mason G, Tuite M P 2010 *Comm. Math. Phys.* **300** 673
[8] Mason G and Tuite M P 2007 *Comm. Math. Phys.* **270** 587
[9] Mason G, Tuite, M P and Zuevsky A 2008 *Commun. Math. Phys.* **283** 305
[10] Schiffer M 1996 *SIAM J. Appl. Math.* **14** 922
[11] Tuite M P and Zuevsky A B 2011 *Commun. Math. Phys.* **306** 617
[12] Yamada A 1980 *Kodai Math. J.* **3** 114

AUTHOR INDEX